ADVANCES IN LIQUID CRYSTALS

A SPECIAL VOLUME OF ADVANCES IN CHEMICAL PHYSICS

VOLUME 113

ADVANCES IN LIQUID CRYSTALS

Edited by

JAGDISH K. VIJ

Trinity College, University of Dublin
Department of Electronic and Electrical Engineering
Dublin, Ireland

ADVANCES IN CHEMICAL PHYSICS
VOLUME 113

Series Editors

I. PRIGOGINE

Center for Studies in Statistical Mechanics
and Complex Systems
The University of Texas
Austin, Texas
and
International Solvay Institutes
Université Libre de Bruxelles
Brussels, Belgium

STUART A. RICE

Department of Chemistry
and
The James Franck Institute
The University of Chicago
Chicago, Illinois

AN INTERSCIENCE® PUBLICATION
JOHN WILEY & SONS, INC.
NEW YORK • CHICHESTER • WEINHEIM • BRISBANE • SINGAPORE • TORONTO

For ordering and customer service, call 1-800-CALL WILEY

Library of Congress Catalog Number: 58-9935

ISBN 0-471-18083-1

Printed in the United States of America.

10 9 8 7 6 5 4 3 2 1

CONTRIBUTORS TO VOLUME 113

L. M. BLINOV, Institute of Crystallography, Russian Academy of Sciences, Moscow, Russia

MARTIN BREHMER, Department of Chemistry and Institute of Materials Science, University of Wuppertal, Wuppertal, Germany

WILLIAM T. COFFEY, Department of Electronic and Electrical Engineering, Trinity College, University of Dublin, Dublin, Ireland

STEVE J. ELSTON, Department of Engineering Science, University of Oxford, Oxford, United Kingdom

H. FINKELMANN, Institut für Makromolekulare Chemie, Albert-Ludwigs-Universität Freiburg, Freiburg i.Br., Germany

ELISABETH GEBHARD, Department of Chemistry and Institute of Materials Science, University of Wuppertal, Wuppertal, Germany

DANIEL GUILLON, Institut de Physique et Chimie des Matériaux de Strasbourg, Groupe des Matériaux Organiques, Strasbourg Cedex, France

L. HARTMANN, Fakültat für Physik und Geowissenschaften, Universität Leipzig, Leipzig, Germany

YURI P. KALMYKOV, Institute of Radio Engineering and Electronics, Russian Academy of Sciences, Moscow Region, Russia

A. KOCOT, Institute of Physics, University of Silesia, Katowice, Poland

F. KREMER, Fakultät für Physik und Geowissenschaften, Universität Leipzig, Leipzig, Germany

W. LEHMANN, Fakültat für Physik und Geowissenschaften, Universität Leipzig, Leipzig, Germany

NIGEL J. MOTTRAM, Department of Engineering Science, University of Oxford, Oxford, United Kingdom

J. NACIRI, Naval Research Laboratory, Washington, D.C., U.S.A.

YU. P. PANARIN, Department of Electronic and Electrical Engineering, Trinity College, University of Dublin, Dublin, Ireland

T. S. PEROVA, Department of Electronic and Electrical Engineering, Trinity College, University of Dublin, Dublin, Ireland

B. R. RATNA, Naval Research Laboratory, Washington, D.C., U.S.A.

R. SHASHIDHAR, Naval Research Laboratory, Washington, D.C., U.S.A.

H. SKUPIN, Fakultät für Physik und Geowissenschaften, Universität Leipzig, Leipzig, Germany

P. STEIN, Institut für Makromolekulare Chemie, Albert-Ludwigs-Universität Freiburg, Freiburg i.Br., Germany

J. K. VIJ, Department of Electronic and Electrical Engineering, Trinity College, University of Dublin, Dublin, Ireland

RUDOLF ZENTEL, Department of Chemistry and Institute of Materials Science, University of Wuppertal, Wuppertal, Germany

PREFACE

This special volume of *Advances in Chemical Physics* is devoted to recent developments that have helped our understanding of the properties of liquid crystals. As now recognized, research on liquid crystals is of interest to chemists, physicists, materials scientists, biologists, and electrical and electronic engineers. Various aspects of research in this area, especially ferroelectric and antiferroelectric liquid crystals (FLCs and AFLCs), have grown rapidly. In particular, many basic research issues associated with FLCs and AFLCs achieving the goal of video rate full color displays have been addressed. This volume describes a number of important contributions that enhance our understanding of the properties of such liquid crystals. It also incorporates contributions to our understanding of the alignment and the self-order of discotics, and the relaxation phenomena in nematics. It is hoped the topics presented will stimulate new research in these aspects of liquid crystals.

The first chapter, by D. Guillon, addresses the topic of molecular engineering—specifically how to structurally engineer a molecule with ferroelectric properties. Guillon also demonstrates that molecules exhibiting such properties need not be chiral molecules, and that the molecules only need to induce chirality in the macroscopic structure. The second chapter, by R. Shashidhar, J. Naciri, and B. R. Ratna, describes recent developments in ferroelectric liquid crystalline materials with large electroclinic coefficients. These materials are highly promising for applications owing to their analog gray-scale capability and short response times. In the third chapter, L. M. Blinov describes the usefulness of the pyroelectric technique in its static and dynamic forms for investigating the polar mesophases and in searching for new liquid crystal materials, i.e., structures with ferroelectric/antiferroelectric properties. In the chapter by R. Zentel, E. Gebhard, and M. Brehmer, the authors address the synthesis and molecular structure of ferroelectric LC-elastomers (FLCE). FLCEs combine the liquid crystalline order of the ferroelectric phase and the rubber-like elasticity of a polymer network. In FLCEs, switching the electrical polarization leads to a mechanical deformation which, in turn, produces an internal stress in the polymer network. This field-induced stress results in a piezoelectric response; thus, the sample may vibrate as a result of the application of an AC electric field. The cross-linking needed to produce elastomeric behavior is carried out either within the siloxane sublayers (producing fast-switching elastomers) or between the separated siloxane sublayers (producing elastomers with favored ferroelectric switching).

In the chapter by F. Kremer, H. Skupin, W. Lehmann, L. Hartmann, P. Stein, and H. Finkelmann, the authors describe the experimental procedures and results of their measurements of the piezoelectric modulus of materials. The usefulness of time-resolved FTIR spectroscopy in providing a microscopic basis for understanding the macroscopic piezoelectricity is described. A. Kocot, J. K. Vij, and T. S. Perova describe the orientational effects in ferroelectric and antiferroelectric liquid crystals in the next chapter. They review the recent work on time-resolved FTIR spectroscopy of FLC polymers and show how polarized IR spectroscopy studies provide a microscopic basis for understanding of ferroelectricity and antiferroelectricity. The chapter by Yu. P. Panarin and J. K. Vij provides a review of the structure and properties, both static and dynamic, of antiferroelectric liquid crystals. The structure of ferrielectric phases is given in terms of existing models and a model being proposed, thereby suggesting that their experimental results should lead to new theories and/or models for the molecular structure of ferrielectric phases. In the contribution by S. J. Elston and N. J. Mottram, the authors discuss some of the fundamental issues associated with the order parameter variation in smectic liquid crystals caused by an externally applied mechanical stress. The chapter by T. S. Perova, J. K. Vij, and A. Kocot reviews issues concerning the alignment, order parameter, and structure of discotic liquid crystals studied by FTIR spectroscopy. Experiments on the switching mechanism in a ferroelectric liquid crystal dibenzopyrene using both polarized and time-resolved FTIR spectroscopy are reviewed. Finally, in the last chapter, W. T. Coffey and Y. P. Kalmykov discuss rotational diffusion and dielectric relaxation in nematics for a general case that the dipole moment is directed at an angle to the molecular axis of symmetry. Their approximate, but elegant formulas are presented in a form that can be compared with results from dielectric spectroscopy.

Trinity College   JAGDISH K. VIJ
Dublin, Ireland

INTRODUCTION

Few of us can any longer keep up with the flood of scientific literature, even in specialized subfields. Any attempt to do more and be broadly educated with respect to a large domain of science has the appearance of tilting at windmills. Yet the synthesis of ideas drawn from different subjects into new, powerful, general concepts is as valuable as ever, and the desire to remain educated persists in all scientists. This series, *Advances in Chemical Physics*, is devoted to helping the reader obtain general information about a wide variety of topics in chemical physics, a field that we interpret very broadly. Our intent is to have experts present comprehensive analyses of subjects of interest and to encourage the expression of individual points of view. We hope that this approach to the presentation of an overview of a subject will both stimulate new research and serve as a personalized learning text for beginners in a field.

I. Prigogine
Stuart A. Rice

ix

CONTENTS

MOLECULAR ENGINEERING FOR FERROELECTRICITY IN LIQUID CRYSTALS

DANIEL GUILLON

Institut de Physique et Chimie des Matériaux de Strasbourg,
Groupe des Matériaux Organiques, Strasbourg Cedex, France

CONTENTS

Advances in Liquid Crystals: A Special Volume of Advances in Chemical Physics, Volume 113,
edited by Jagdish K. Vij. Series Editors I. Prigogine and Stuart A. Rice.
ISBN 0-471-18083-1.   © 2000 John Wiley & Sons, Inc.

I. INTRODUCTION

Polar order in condensed matter represents an area of high fundamental and technological interest. Indeed, one of the most spectacular phenomena in crystals was revealed in 1921 in Rochelle salts (tartrate of potassium and sodium) [1]. In such materials a spontaneous polarization has been evidenced below a certain temperature called Curie temperature; the direction of this polarization could be changed by applying an external electric field. This property has been called ferroelectricity, from the analogy with ferromagnetism.

Many years later and with the development of optoelectronics, the transposition of such a ferroelectricity to anisotropic liquids is the subject matter of a large number of research studies. In particular, during the last two decades the possibility of taking advantage of the bistability of ferroelectric smectic C* liquid crystals, in fast electro-optical devices, has driven much interest in these systems [2]. The molecular conception of these compounds was directly deduced from the original work of R. B. Meyer [3], who predicts, using a symmetry argument, that tilted smectic phases (smectic C, I, or F) obtained with chiral molecules having a transverse dipole moment, should exhibit a spontaneous polarization. Meyer's predictions have been rapidly confirmed with the synthesis of the first smectic C* ferroelectric liquid crystal (DOBAMBC) [4], whose chemical structure and thermotropic behavior are shown in Fig. 1.

Then, the first generation of ferroelectric liquid crystals (FLC) materials was designed according to this concept in order to obtain a large spontaneous polarization, which generally leads to a short response time. In this context, a large number of smectic C* compounds with various polar groups and diverse chiral centers [5–9] that we will discuss in more detail below have been synthesized. More recently, other forms of chirality, such as axial chirality or planar chirality, have been explored in the design of FLC

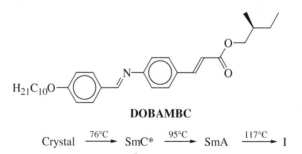

DOBAMBC

Crystal $\xrightarrow{76°C}$ SmC* $\xrightarrow{95°C}$ SmA $\xrightarrow{117°C}$ I

Figure 1. Chemical structure and mesomorphic behavior of the first smectic C* ferroelectric liquid crystal synthetized by P. Keller and co-workers [4].

materials. A new family of achiral molecules (banana shaped-molecules) that exhibit an electric polarization [10] when submitted to an electric field under the same conditions as do the more classical ferroelectric calamitic liquid crystals in the so-called SSFLC cells has also appeared in the last several years. This quite fascinating behavior has impelled several teams around the world to investigate this new type of materials, mainly from the experimental point of view [11]. Finally, other types of ferroelectricity in liquid crystals have been explored, such as the longitudinal ferroelectricity in orthogonal smectics resulting from a specific molecular design and the ferroelectricity in columnar liquid crystals.

Before going into these different aspects of ferroelectricity in liquid crystals further, let us first consider the few theoretical concepts which are currently used to describe the physical properties of FLC and to predict the type of molecular architecture needed for efficient electro-optical properties.

II. CONCEPTUAL AND THEORETICAL ASPECTS

A. Ferroelectricity in Tilted Smectic Phases

The introduction of chirality into calamitic molecules leads to a reduction in the symmetry of the classical smectic C phase and to the presence of a helical structure characteristic of the smectic C* phase. While the smectic C phase belongs to the C_{2h} symmetry group, i.e., possesses three symmetry elements [one mirror plane (m), one twofold axis (C_2), and one inversion center (i)], the smectic C* phase belongs to the symmetry group C_2 with only one symmetry element, the twofold rotation axis (Fig. 2).

Based on the Neumann's principle, it is known that any macroscopic physical property should adhere to the symmetry properties of the phase. If we consider a vector P, which can symbolize any physical property, and if we

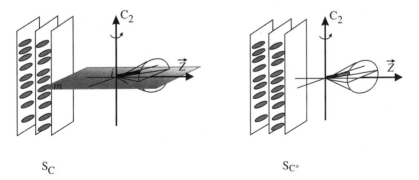

Figure 2. Symmetry elements of the smectic C and smectic C* phases.

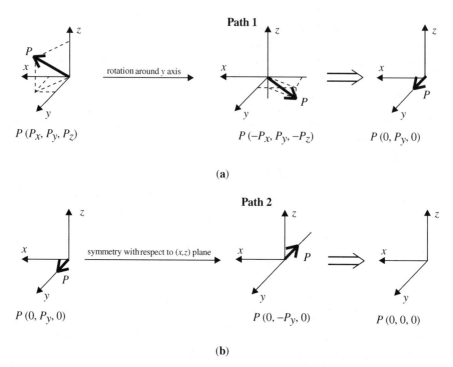

Figure 3. Two symmetry operations applied to the vector *P*. (a) Path 1: rotation around *y* axis; (b) path 2: reflection with respect to the (*x,z*) plane.

apply to it the symmetry operations (Paths 1 and 2 in Fig. 3) corresponding to the smectic A phase, it is clear that the resulting vector is cancelled. In the case of chiral smectic C* phases, the mirror plane does not exist any more, and the only symmetry element is the twofold axis, C_2, along *y*. After applying the corresponding single symmetry operation (path 1 in Fig. 3), one component of $\vec{P}$, along *y*, cannot be cancelled. In this case, if the vector $\vec{P}$ symbolizes a transverse dipole moment, there is a resulting spontaneous polarization in each smectic layer which acts along the C_2 axis, i.e., in a direction perpendicular to the tilt plane defined by the director $\vec{n}$ and the normal to the smectic layers $\vec{z}$. $\vec{P}$ verifies the relation:

$$\vec{P} = P_0(\vec{z} \wedge \vec{n}) \qquad (1)$$

The spontaneous polarization can only take two opposite directions. However, as the chirality of the molecules induces a helical variation of the director $\vec{n}$ from layer to layer, the direction of the spontaneous polarization $\vec{P}$ turns from one layer to the next one. As a consequence, the spontaneous polarization is cancelled as soon as the total thickness of the sample is larger than the helical pitch. This is the reason why some authors prefer to call this phase an "helielectric" phase [12] rather than a ferro-electric one. However, the spontaneous polarization can be observed when the helical structure of the smectic C* phase is broken. This can be achieved according to several different processes:

- When a shearing in the plane of the smectic layers is applied to the smectic C* phase, the helical structure is distorted, and a polarization appears perpendicularly to the shearing direction.

- The distortion, and even the cancellation of the helical structure, can be obtained by applying an electric field perpendicularly to the helical axis. The smectic C* phase then transforms into an unwound smectic phase corresponding to the classical biaxial smectic C phase. The threshold of the electric field needed to unwind the helical structure in the smectic C* phase is given by

$$E_c = \frac{\pi^4}{4} K_{33} \frac{\theta^2}{p^2 P} \tag{2}$$

where K_{33} is the elastic constant, p the helical pitch traced by $\vec{n}$, θ the tilt angle of the molecules in the smectic layers, and P the spontaneous polarization.

- The removal of the helical structure by the elastic interactions between the smectic C* phase and treated surfaces can be obtained when the sample is confined in a thin cell, for which the thickness is either less than or of the order of the helical pitch. This type of cell is called SSFLC (Surface Stabilized Ferroelectric Liquid Crystal) [13].

- In the case of mixtures of ferroelectric liquid crystals, the removal of the helical structure can be achieved by mixing at least two compounds having the opposite sense of helix, with the helical pitches being in the same order. The compensation of the right and left helices of these materials leads to an unwound smectic C* phase exhibiting a nonzero spontaneous polarization.

B. The Boulder Model

To analyze the smectic C phase from a molecular point of view, it is necessary to consider the molecules as objects that are more complex than a simple rigid rod. Indeed, through a comparison of the values of the tilt angle obtained by different techniques (X-ray diffraction, optical microscopy), it has been shown [14] that the molecules can be better described as objects having a zigzag shape in the smectic C phase. Thus the molecule within

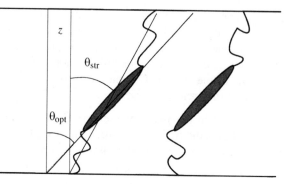

Figure 4. Schematic representation of a calamitic molecule in the smectic C phase [13]. θ_{opt} is the tilt angle which can measured by optical means, and θ_{str} corresponds to the tilt angle (steric angle) which can be deduced from X-ray investigations.

a smectic C layer is symbolized by a central part (corresponding to the polarized part) with a tilt angle θ_{opt} (as measured by optical method) and two terminal parts corresponding to the disorganized aliphatic chains, which are less tilted on average than the central one (Fig. 4).

According to the model developed by Walba and his collaborators [15], one molecule can be simulated by a surface with a constant molecular field. The interactions with the neighboring molecules are then averaged as a function of time. This surface, called "binding site," should hold the intrinsic characteristics to the symmetry of the phase. It would be a sphere in the case of the isotropic phase, a cylinder in the case of the nematic and smectic A phases. For the smectic C mesophase, the authors propose a bent cylinder. This model is in agreement with the "zigzag" model described. As shown in Figure 5, the optimal energetic conformation of the zigzag model fits quite well inside the bent cylinder. For such a molecular system, the polarization is then given by the following expression:

$$P = \sum_{n} D_i P_i \tag{3}$$

where D_i is the density of conformation i, and P_i is the total contribution of the conformation i to the polarization.

In this model, the authors consider the polarization as a manifestation of a form of molecular recognition [16–18]. The rotations of the chiral molecules are biased, resulting in nonzero dipole moment perpendicular to the tilt plane. Of course, the degree of rotation will depend upon the molecular architecture, and in particular upon the steric constraints induced by the chiral molecules.

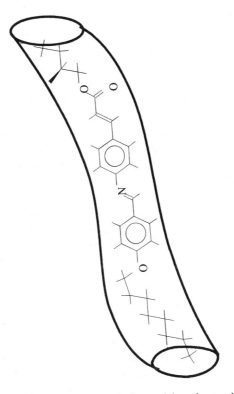

Figure 5. One "zigzag" shaped molecule inserted in a bent cylinder (adapted with permission from Ref. [15]).

C. Molecular–Statistical Theory of Ferroelectric Ordering

In a molecular–statistical theory of ferroelectric ordering in the smectic C* phases, it is claimed that the spontaneous polarization is caused by specific chiral and polar interactions between the chiral center of the molecule and the polarizable core of the nearest neighboring molecule [19]. When the molecules have large dipoles in the chiral center, the predominant inter- action of this nature is the induction interaction between the dipole and the polarizability of the neighboring molecule, modulated by the asymmetric shape of the molecule. The spontaneous polarization is thus given by the following expressions [20]:

$$\vec{P} = \mu(\vec{n} \cdot \vec{z})(\vec{n} \wedge \vec{z}) \tag{4}$$

$$\mu = -15/4(\rho^2/kT)(\vec{s} \cdot \vec{d})d^2\Gamma(D/L)^6(2\chi_\perp + 15\Delta\chi + \Delta\chi_\perp) \tag{5}$$

$$\Gamma = (\vec{d} \cdot \vec{a})[(\vec{m} \wedge \vec{d}) \cdot \vec{a}] \tag{6}$$

where $\vec{n}$ is the director and $\vec{z}$ is the smectic plane normal. The quantity μ is expressed in terms of molecular model parameters. The term $\vec{d}$ is the dipole moment of the chiral centre, $\vec{s}$ is the transverse dipole which reflects the polar asymmetry of the molecular shape in the transverse direction, $\chi_\perp = (\chi_{xx} + \chi_{yy})/2$ is the transverse molecular polarizability, $\Delta\chi = \chi_{zz} - \chi_\perp$ is the anisotropy of the molecular polarizability, and $\Delta\chi_\perp \cong (\chi_{xx} - \chi_{yy})$ is the anisotropy of the transverse polarizability. Finally, ρ is the number density of molecules, L is the molecular length and D is the average molecular diameter. The predominant contribution comes from the anisotropy of the polarizability, $\Delta\chi$, and as a consequence, the spontaneous polarization is approximately proportional to $\Delta\chi$. Γ is a parameter which measures the molecular chirality. It depends upon the orientation of the dipole in the chiral centre with respect to the long molecular axis $\vec{a}$ and the vector $\vec{m}$ pointing from the center of mass of the molecule to the chiral center. The vectors $\vec{a}$, $\vec{m}$, and, $\vec{d}$ are shown in Figure 6.

The most important factor in this theory is the value and the orientation of the dipole, $\vec{d}$, in the chiral center of the molecule. This polar bond directly attached to the chiral center takes part in the chiral and polar intermolecular interactions, and increases the spontaneous polarization substantially. In such a description, the spontaneous polarization is proportional to the cube of this dipole, $P_S \propto d^3$. According to the authors, this result can be used to explain the strong dependence of the spontaneous polarization as a function of the value of the dipole in the chiral center and of the location of other

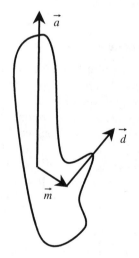

Figure 6. Schematic of a chiral molecule with a substitution group in the alkyl chain. The vectors are defined in the text (adapted with permission from Ref. [20]).

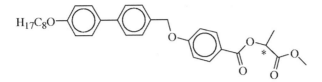

Figure 7. Molecular structure of 1MC1EOPBB.

dipole bonds around it [21]. In particular, when two polar bonds are attached on both sides of the chiral carbon, it has been shown that the spontaneous polarization can be considerably increased, as in the case of 1MC1EOPBB (Fig. 7).

D. Indigenous Polarization Theory

Starting from the symmetry arguments suggesting that chirality and transverse dipoles are sufficient to give rise to ferroelectricity but cannot predict the value of the polarization, Photinos and Samulski [22] have recently developed a model showing that polarity can be obtained even for the most symmetric flexible molecules that can form a nonchiral smectic C phase. According to this theory, which they called the Indigenous Polarity Theory (IPT), the intermolecular interactions giving rise to ferroelectricity are primarily excluded volume interactions which turn out to be the same as those producing the smectic phase itself.

In their model, the authors consider the mesogen molecule, as sketched in Figure 8. The molecule is constituted of three linear segments representing the mesogenic core (in the middle) and two terminal chains linked to the core. The latter are allowed to perform simultaneous 180° flips about the core. For clarity in Figure 8, one side of the planar molecule is black and the other white [23]. In order to respect the core–chain segregation constraints due to the amphipathic character of the constituent parts of the molecule [24], the cores would expose both of their sides with equal probability (no polarity) whereas the chain segments would expose their black side more than the white one, along the given direction of the C_2 axis. As a result, a transverse polar order occurs along the C_2 axis (indigenous polarity). If a dipole moment were attached perpendicular to the black side of the chain segments, then a spontaneous polarization would be obtained. The polarity does not result from the favoring of one conformation over the other one, but rather from the coupling of the conformations with orientations.

In summary, the IP theory states that the spontaneous polarization originates from a statistical biasing of mesogen conformations that derives from steric interactions in tilted smectic layers stacked one over the other. The temperature dependence of the polarization would simply be a direct consequence of molecular packing considerations. Recent computations of

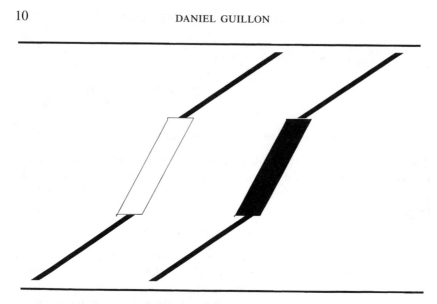

Figure 8. Two schematic dispositions of a mesogenic molecule (adapted with permission from Ref. [23]).

spontaneous polarization using this theory indicate a reasonable agreement with experimental values [25].

E. Ferroelectricity from Achiral Calamitic Molecules

A distinct approach to producing new polar materials consists in using a polyphilic effect for the formation of polar self-assemblies [26,27] exhibiting longitudinal ferroelectricty along the long molecular axis, in contrast to the previous structures explored where ferroelectricity is present in a direction perpendicular to the molecular axis. Polyphilic molecules with A–B–C–A structure contain three (or eventually more) different moieties with a distinct chemical nature; due mainly to the amphipathic character, these subunits have a tendency to segregate in microdomains in space [28] (intralamellar polyphilic effect; see Fig. 9). If the molecule is noncentrosymmetric with both ends being of the same chemical nature, then centrosymmetric or noncentrosymmetric stacking of individual layers leads to antiferroelectric or ferroelectric materials, respectively. As a consequence, longitudinal electric polarization could be obtained in smectic phases of polyphilic molecules where the centrosymmetric arrangement of the layers is disfavored. In such a phase, called also smectic A_p phase (p for polar) in the literature [29], the driving forces are chemical and steric, which lead to the appropriate sequence of layers. Moreover, if the molecules are tilted within each layer, the smectic C_p phase thus obtained is different from the classical smectic C^* widely discussed before. In this case, the polarization is

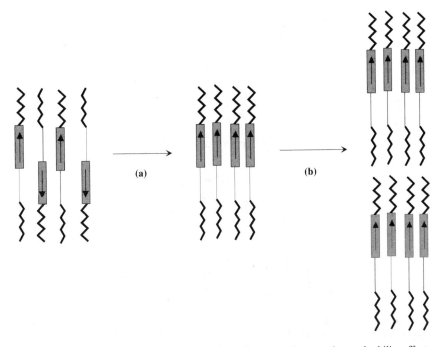

Figure 9. Schema of strategy to obtain polar mesophases using polyphilic effect. (a) Intralamellar polyphilic effect leading to the formation of polar smectic layers; (b) interlamellar polyphilic effect leading to the formation of polar condensed phase [with permission from *Nature* **359**, 621 (1992), Macmillan Magazines Ltd.].

expected to lie along a well-defined axis, which can be the layer normal or any other direction. As a consequence, it is possible to imagine tailoring mesomorphic materials with specific polarizations. However, it should be emphazised here that this longitudinal polarization cannot be easily rotated by an electric field.

Systems with more than three different types of layers can also exhibit polarization. Let us consider, for example, molecules containing two different rigid parts and two flexible chains of quite different lengths; they could self-assemble into layers formed with distinct sublayers (Fig. 10). Finally this concept can be extended to a number of nonchiral oligomers of mesogens, and of main chain polymers formed of mesogens and several nonchiral block copolymers. All these systems may form layered structures in which the molecular dipoles are oriented side by side and parallel. In order for the system to be stabilized, the incompatibility between the different sublayers should dominate with respect to the unfavorable dipolar lateral interactions.

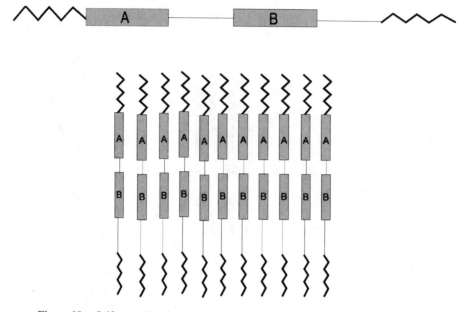

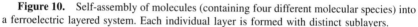

Figure 10. Self-assembly of molecules (containing four different molecular species) into a ferroelectric layered system. Each individual layer is formed with distinct sublayers.

More generally, from the point of view of the different forces acting in the case of molecules with A–B–C structure, the intralayer dipole–dipole interactions are expected to be much more important than the interlayer ones, since the interlayer spacing (about 30 to 50 Å, for example with calamitic mesogens) is much larger than the average lateral distance (about 5 Å between individual molecules within the smectic layers). The intralayer chemical interactions (amphiphilic interactions) have a tendency to promote polar order in a single layer. This kind of interaction should be more important than the intralayer dipole interactions which, in contrast, favor the antiferroelectric order inside one single layer. Finally, the interlayer dipole interactions (between fixed molecular dipoles) should be more important than the interlayer van der Waals interactions which promote antiferroelectric order between successive smectic layers, in order for the system to exhibit a global ferroelectric order. Theoretical calculations indicate that a ferroelectric arrangement of twin-molecules, formed by two rigid mesogenic moieties linked to a central flexible spacer and ended by two similar flexible aliphatic chains, is energetically allowed [30].

In a more recent paper [31], Prost and co-workers have shown that calamitic polyphilic molecules with A–B–C–A structure are better candidates for obtaining longitudinal ferroelectric smectic phases than calamitic molecules with A–B or A–B–C structure. They considered A–B–C–A and A–B'–C–A molecules (F^+ and F^- molecules) where the dipoles in the B moieties are respectively inverted and developed a Landau theory for a F^+/F^- mixture, where the competition between ferroelectricity and other forms of ordering such as antiferroelectricity are discussed. Depending upon the concentration of F^+ molecules in the mixture, a variety of different possible structures result, among them a "stripe" phase with an interlayer ferroelectric and an intralayer antiferroelectric arrangement, or a ferroelectric longitudinal phase (inter- and intralayer ferroelectric ordering).

F.   Induced Ferroelectricity from Achiral Banana-Shaped Molecules

In a structural study of main-chain polymers with two different spacers, Watanabe et al. [32] reported that a bilayer smectic phase could be obtained. This phase with a C_{2v} symmetry could thus be ferroelectric even if the system is nonchiral. This can be achieved if two different aliphatic spacers with odd numbers of carbon atoms are introduced into the polymer backbone in a regular, alternate way, so that they can segregate into specific microdomains. In such a system, the ferroelectricty is parallel to the layers.

Following the same direction but with low molecular weight compounds, a ferroelectric smectic phase has been obtained from banana-shaped achiral molecules [10] under electric field. In this case, the origin of the electric field-induced ferroelectricity was ascribed to the very efficient packing of the banana-shaped molecules within the smectic layers, resulting in a polar phase with C_{2v} symmetry. Because of their specific shape, the molecules can be closely packed and also aligned in the direction of bending (Fig. 11). Each layer is biaxial and the refractive indices are different in the bent direction (y axis) and in the direction normal to the y axis (x axis). There is a twofold axis along the y axis and two mirror planes perpendicular to the x and z (normal to the layers) axis, respectively, but there is no mirror plane perpendicular to the twofold axis. As a result, it is expected that spontaneous electric polarization will be obtained along the y axis, i.e., in the bent direction. This close packing of the molecules leads to a packing of the dipoles, producing a polar order parallel to the plane of the smectic layers; the same effect had been proposed earlier to explain the molecular origin of the spontaneous polarization in the *chiral* smectic C* phase of banana-shaped molecules [33].

In a very recent paper [11], D. R. Link et al. reported the results of optical microscopy studies of such compounds in freely suspended films and in bookshelf electro-optical geometries. They demonstrated that the bulk states are either antiferroelectric–racemic with a layer polar direction and handedness changing sign regularly from layer to layer, or antiferroelectric–chiral

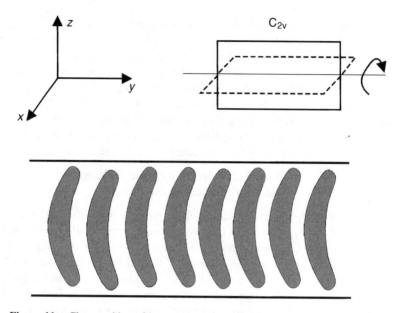

Figure 11. Close packing of banana-shaped molecules within a smectic layer, and C_{2v} symmetry elements.

with a uniform layer handedness. They conclude that the chirality of the smectic phase observed depends essentially on two factors: on the one hand, the intralayer collective polar ordering of the banana-shaped molecules, and on the other, the intralayer collective tilt of the molecular planes. The corresponding stabilization of adjacent layers can be easily changed by applying a weak electric field, resulting in a transition from an antiferro-electric state to a ferroelectric one.

III. MOLECULAR DESIGN FOR SMECTIC C* PHASES

One of the main objectives which has driven the synthesis of many smectic C* materials was the goal of obtaining high-spontaneous-polarization materials which at the same time exhibit good thermal stability. The absolute value of the spontaneous polarization is a critical parameter be-cause it determines the main parameters of the electro-optical properties. However, the synthesis of a large number of FLCs with different structures has shown that the values of the spontaneous polarization measured were fairly low with respect to the values which were expected from the magnitude of the molecular permanent dipole moment [34]. These low values were attributed to the free rotation of the molecules in the phase. It is clear that the degree of free rotation will depend upon the molecular architecture, and particularly upon the steric constraints induced by the molecules within the smectic layer. Other molecular physical properties may also play a role in the establishment of the spontaneous electric polarization in the smectic C* phases. These are the magnitude of the dipole moment, the positions inside the molecule of the chiral center and the dipole, the coupling between the dipole and the chiral center, and the value of the tilt angle of the molecules within the smectic layers.

A. Role of the Dipole Moment Strength

Figure 12 illustrates, through the investigation of three compounds with similar molecular structures, the effect of the lateral dipole moment on the value of the spontaneous polarization, P_S. In the case of DOBAMBC, the value of P_S at 76°C is 3 nC/cm^2. This value is very small when compared to those measured in ferroelectric crystals, in the order of the μC/cm^2. Indeed,

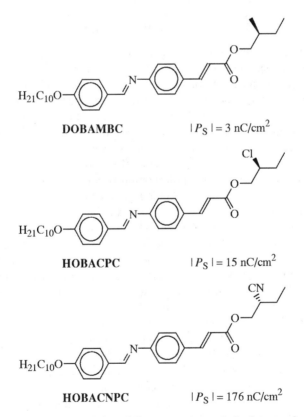

DOBAMBC $\quad | P_S | = 3$ nC/cm^2

HOBACPC $\quad | P_S | = 15$ nC/cm^2

HOBACNPC $\quad | P_S | = 176$ nC/cm^2

Figure 12. Example of variation of the spontaneous polarization as a function of the magnitude of the dipole moment.

to estimate the effective dipole moment per molecule, one has to use the expression $\mu_{eff} = P_S/\rho_0$, where ρ_0 is the number density of the molecules. Assuming that the volumic density of DOBAMBC is about 1 g/cm^3 (which is a classical value for calamitic liquid crystals [35]), the effective molecular dipole is quite low, close to 10^{-2} Debyes. Knowing the actual value of the transverse dipole of DOBAMBC, $\mu_\perp \approx 1.5$ D, the yield factor which measures the strength of the polar order leads to $f = \mu_{eff}/\mu_\perp \approx 10^{-2}$. This low value of the polar order parameter indicates clearly that the value of the spontaneous polarization in DOBAMBC is far from what would be expected were all dipoles oriented in the same direction. When the methyl group of the chiral center is replaced by a chlorine atom, the value of P_S at saturation, 15 nC/cm^2, is five times higher than that obtained for DOBAMBC. Moreover, with the substitution of the chlorine atom for a cyano group, well known to exhibit a strong dipole moment (about 4 Debyes), P_S reaches a saturated value of 176 nC/cm^2.

B.   Location of the Chiral Center

The magnitude of the dipole moment is not the single parameter for the variation of the spontaneous polarization. Until now, it had been shown and accepted that the positions of both the chiral center and of the transverse dipole moment inside the molecular architecture must be taken into account [36,37]. We have considered above (Sect. III.A) the case of molecules in which the transverse dipole moment is localized at the same position with respect to the rigid core (Fig. 12). Let us examine now the effect of the chiral center at different positions (Fig. 13). For example, in the case of DOBAMBC, the spontaneous polarization becomes 15 times higher when the chiral center is moved closer to the rigid core of the molecule (DOBA-1-MBC compound [38], $P_S = 42$ nC/cm^2). Another example is illustrated by the comparison of the MBRA8 and MORA8 compounds, which were the first ferroelectric liquid crystals exhibiting a stable smectic C* at room temperature [39]. In this case also, and despite the fact that the rigid cores are not exactly the same, the P_S values obtained are higher when the chiral center and the dipole moment are close to the rigid core [40].

These effects have to be analyzed in the framework of the general description of the smectic phases: the lateral arrangement of the molecules is mainly determined by the interactions between the rigid and polarizable parts, whereas the aliphatic chains are in a disorganized state. Thus, if the chiral center and the tranverse dipole moment are localized in the aliphatic chains far from the rigid core, the conformational freedom of the chains would induce a minimum coupling between these two fundamental molecular characteristics. Finally, the introduction of chirality very close to the rigid core will deeply break the symmetry of the smectic C phase

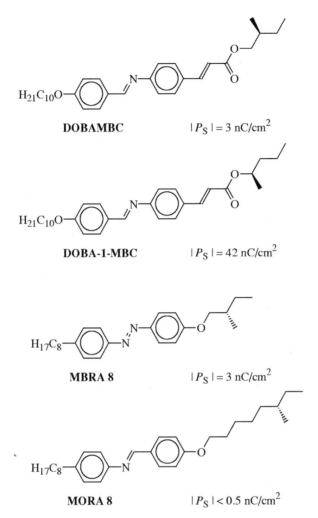

Figure 13. Variation of the spontaneous polarization as regards to the position of the chiral center and dipole moment with respect to the rigid core of the molecule.

towards the ideal case of the C_2 symmetry. In the same way, in order for the polar order to be maximized, the transverse dipole moment should be as close as possible to the rigid core, and in an even more ideal case localized at the same position as the chiral center.

C. Molecular Modeling

With a view to obtain stable FLCs exhibiting a large temperature domain of the ferroelectric phase and short electro-optical response times τ_r, the molecular design for the synthesis of such materials should be driven in a way to reach high values of the spontaneous polarization, P_S, as well as low values of the viscosity η. In other words, the ratio P_S/η should be as large as possible [41], τ_r being directly inversely proportional to it. It is difficult to predict the value of η for such a compound, since the viscosity depends not only upon the molecular architecture, but also upon intra- and inter-molecular interactions which are still too complex to model realistically. So, on the one hand, it is reasonable, in the synthesis of FLCs, to avoid functional groups able to induce intermolecular bonds (such as hydrogen bonding, for example), and to restrict the rigid core to two or three aromatic rings at the most. These molecular criteria are in favor of a low viscosity and of a smectic stability domain at relatively low temperatures, close to ambient. On the other hand, the value of the spontaneous polarization can be predicted to some extent. It is directly linked to the strength of the transverse dipole moment, and therefore strongly dependent on the nature of the main polar group in the molecule.

During the last few years, the efforts of chemists have led to a range of SmC* molecules bearing interesting chiral sources such as oxiranes, thiiranes [42], cyanohydrins [43], β-chlorohydrins [44], and some chiral and γ-lactone derivatives [45] (see Fig. 14). In most cases, a rigid coupling between the asymmetric center and a strong dipole (such as C=O, C–Hal, C–CN) was sought. This has also been achieved through the introduction of an heteroatom bearing a strong dipole and directly linked to the rigid part of the molecule into the molecular architecture, like in the sulfinate derivatives [46]. Indeed, the sulfur atom is chiral in its fourth oxidized state, such as in

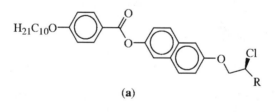

(a)

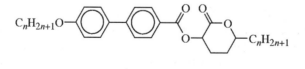

(b)

Figure 14. Examples of chiral sources used in the molecular design of ferroelectric liquid crystals : (a) β-chlrohydrin derivatives, (b) δ-valerolactone derivatives.

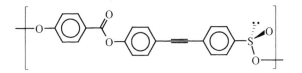

Figure 15. Chemical formula of the BTSO fragment, acting for the rigid part of the smectogenic molecule belonging to the *m*-BTS–O$_n$ homologous series.

sulfinyl groups, for which the dipole moment of the S=O bond reaches values close to 6 D. For example, molecular modeling has been performed on a sulfinate-based molecule (Fig. 15). Such a molecular design shows that the chiral center, which is also strongly polar, is localized as close as possible to the rigid aromatic part. Since the chiral sulfinate group is directly linked to the polarizable and rigid part of the molecule, it can be considered that all the local dipoles contributing to the total transverse dipole moment are localized within this chemical moiety, which in fact represents the rigid part of the more general smectogenic molecule presented in Figure 16 (the aliphatic end chains having no significant influence on the strength and on the orientation of the final overall molecular dipole).

The modelized molecule is represented in Figure 17 in its minimal conformational energy. It is interesting to notice that the geometry of the sulfinate group has been found the same as in the starting building block used for the synthesis of the *m*-(Benzene carbonyl) oxytolane Sulfinyl-O*n* compounds, for which the crystallographic atomic structure has been determined [47]. The overall value of the molecular dipole moment thus obtained is $\mu = 5.9$ D, whereas its transverse component is $\mu_\perp = 2.8$ D. Molecular modeling can thus be a powerful tool to predict FLCs with high P_S values prior to synthesize the molecules. In fact, this investigation could drive the synthesis towards molecules such as that shown in Figure 16, which led to P_S values up to 300 nC/cm [48].

A recent computational method for the prediction of the spontaneous polarization in ferroelectric liquid crystals has been developed from atomistic models of molecular structure [49]. Even if this method is not sensitive to the number of methylene groups in the nonchiral chain and to

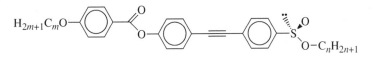

$$m\text{-BTS}{-}\text{O}_n$$

Figure 16. Molecular structure of the sulfinate series.

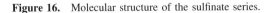

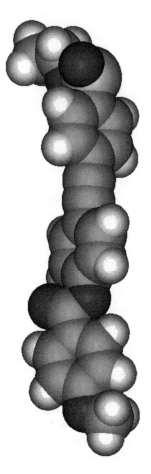

Figure 17. Molecular modelling of the rigid polarizable part of the m-BTS–O$_n$ molecules.

minor changes in the structure of the rigid part of the molecule, the first calculations performed yield semiquantitative predictions of the spontaneous polarization. It will be interesting to see how the model will be improved in the sense of realistic modelization of the structure and properties of ferroelectric mesophases, in order for the chemist to synthesize the appropriate molecule.

D. Role of the Enantiomeric Excess

If the interest of chemists was focused on the synthesis of stable, broad-range room-temperature smectic C* compounds, it must be noted, however, that the term "stability" referred mainly the *chemical* stability, whereas optical purity was rarely considered. Indeed, if the principal prerequisite for ferroelectricity is the introduction of optical activity into the SmC compound, this should be quantitatively associated with its ferroelectic parameters. In other words, for a given smectic C* compound, it is important to determine for which enantiomeric excess (ee) the ferroelectric quantities are measured. In fact, for the large majority of the published results, not much attention has been paid to the optical purity, the general assumption being that the measured ferroelectric properties correspond to the optically pure material even if the chiral smectic C compound may undergo partial chemical or thermal racemization during the initial synthesis or, for example, upon heating to the isotropic state without chemical transformation.

The relatively few published results in this domain show different behaviors and are ambiguous insofar as practically all of them consider the thermodynamically distinctive enantiomeric mixtures to have the same polar order. For example, a linear variation of the spontaneous polarization as a function of the enantiomeric excess is reported [50,51] even if the enantiomeric mixtures investigated are not thermodynamically the same. In fact, a continuous decrease of the SmC*–SmA transition temperature is observed on decreasing the enantiomeric excess. In another study [52], a correlation between the thermodynamic behavior and the variation of the spontaneous polarization was found, and it was stated that the P_S variation is at the origin of the transition temperature change between mesophases when varying the enantiomeric excess. However, such behavior should not be considered as due only to ferroelectricity, since it is frequently observed for nonferroelectric phases [53–54]. A nonlinear variation of P_S as a function of the enantiomeric excess has been observed in the case of investigations of pairs of *S*-enantiomers and racemic mixtures of FLC compounds based on a leucine chiral group [55]. In that case, the authors explained the P_S variation from a theoretical model based on different contributions originating from host–host, host–dopant and dopant–dopant interactions. Their model can lead to both linear and nonlinear $P_S = f(\text{ee})$.

More recently, it has been shown that enantiomeric excess can have large effects on both thermodynamic and electro-optical properties of ferroelectric liquid crystals [56,57]. The enantiomeric excess dependence of the saturated spontaneous polarization, P_S, for binary mixtures of (*R*)- and (*S*)-12-BTS–O_8 at different temperatures is shown in Figure 18. For the mixtures with ee values less than 30%, the spontaneous polarization is low. In fact, they do not exceed 3 to 4 nC/cm^2 in this ee range. For the pure enantiomer, the value of P_S is quite high (270 nC/cm^2) in the saturation regime. But a decrease of only 10% of ee induces a large decrease in P_S, by about 50%. This nonlinear decrease of P_S as a function of decreasing ee parallels an important decrease

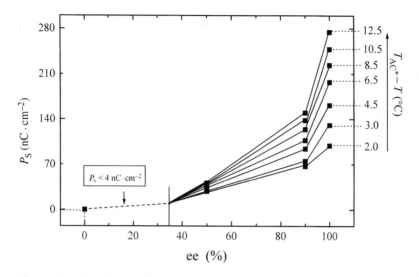

Figure 18. Variation of the spontaneous polarization, P_S, as a function of the enantiomeric excess, ee, for different temperatures (T_{AC^*} is the transition temperature between Smectic C* and A phases), for binary mixtures of (R)- and (S)-12-BTS−O$_8$ compounds (with permission of *Liq. Cryst.* **26**, 1315 (1999), Taylor and Francis, Ref. [57]).

of the melting transition temperature towards the smectic C* phase. This correlation can be explained by an increase of polar and steric effects, having a great influence on the polar order within the smectic C* mesophase, as does the dipole moment compensation between the two enantiomers. Thus, these two negative conjugated effects induce a steep decrease of the spontaneous polarization for ee contents slightly lower than 100%.

E. Role of the Terminal Aliphatic Chains

When varying the aliphatic chain lengths of the molecules, the macroscopic properties are in general likewise modified; for example, the rotational viscosity as well as the tilt angle of the SmC* molecules increase with increasing chain length. The case of the variation of the spontaneous polarization as a function of chain length is less obvious; indeed, increasing the length of the nonpolar molecular moiety contributes to the dilution of the transverse dipole moment in the material and thus to a decrease in the value of P_S, whereas at the same time, the increase of the tilt angle should have a tendency to increase P_S. As a consequence, it is interesting to investigate how these parameters act simultaneously on the magnitude of the spontaneous polarization.

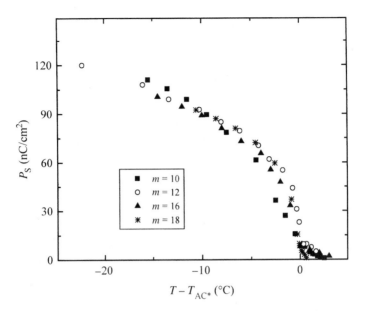

Figure 19. Spontaneous polarization, P_S, in the smectic C* phase as a function of temperature, $T - T_{AC^*}$, for different terms of the m-BTS–O$_{10}$ series (with permission from *J. Chem. Phys.* **106**, 7816 (1997), Ref. [58]).

The homologous series considered for such a study are those whose structure is represented in Figure 16. Only the terms with the same alkyl chain length attached to the sulfinyl group (n) have approximately the same enantiomeric excess (about 80%). In order to avoid any effect of this enantiomeric excess on the electrooptical properties, two homologous series of compounds, namely m-BTS–O$_{10}$ and m-BTS–O$_{12}$ have been studied by varying only the aliphatic chain length, m, between 10 and 18 [58], the results are similar for both series. The values of P_S for the m-BTS–O$_{10}$ homologous series are shown in Figure 19. It is worth noticing that the value of P_S does not seem to depend significantly upon the length of the aliphatic chain m, even far from the SmC*–SmA transition; however, a tendency to a small increase of P_S with m is visible in the neighborhood of the transition. For the same compounds, the corresponding optical tilt angles, θ, are shown in Figure 20. Now, it is obvious that the values of θ depend strongly upon the length of aliphatic chain; far from the SmC*–SmA transition, θ varies from 22 to 31° when m increases from 10 to 18 in the m-BTS–O$_{10}$ series. Such a change of θ as a function of m should have induced an increase of P_S, since P_S and θ are believed to be proportional in a first approximation [59]. However, as already pointed out, the variation of P_S with m is very small and needs, therefore to be analyzed in more detail.

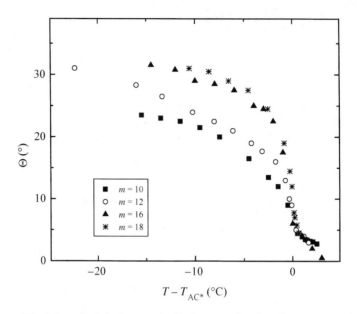

Figure 20. Tilt angle, θ, in the smectic C* phase as a function of temperature, $T - T_{AC^*}$, for different terms of the m-BTS–O_{10} series (with permission from *J. Chem. Phys.* **106**, 7816 (1997), Ref. [58]).

Let us recall the general expression of the spontaneous polarization as given by de Gennes and Prost [60]:

$$P_S = f\, n_b\, \mu_T \qquad (7)$$

where μ_T is the transverse molecular dipole, n_b the number of molecules per unit volume, and f a yield factor. In fact, this latter term represents the anisotropic distribution of the molecular dipole moments; this yield factor ranges from a few 10^{-3} to a few 10^{-1}, in general. This anisotropic distribution is directly connected to the anisotropy of the molecular field around each molecule, each of which should be the more pronounced the larger the tilt angle. On the one hand, when the chain length increases, the number n_b of molecules per unit volume decreases; on the other hand, the increase of the tilt angle with increasing chain length should contribute to improving the yield factor by changing the molecular rotation around the long axis. Concerning the electro-optical properties of the homologous series described just above, the two previous effects, acting in opposite ways, could roughly compensate each other so that P_S remains approximately constant for a given ΔT whatever the m value. Indeed, if we plot $n_b\theta$ as a function of the aliphatic chain length at different distances from the SmA–SmC* transition (see Fig. 21), it becomes clear that $n_b\theta$ is not constant

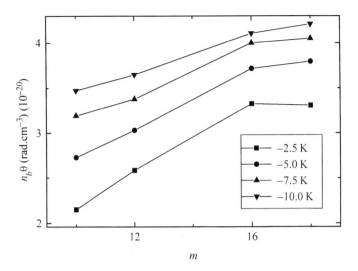

Figure 21. Values of $n_b\theta$ as a function of chain length, m, for the m-BTS–O$_{10}$ series (with permission from *J. Chem. Phys.* **106**, 7816 (1997), Ref. [58]).

with m. In other words, more subtle phenomena have to be taken into account to explain why P_S remains constant with m. In particular, it is interesting to study the variation of f with the tilt angle and with the temperature for all the terms of the two series.

The variations of the yield factor, f, as a function of the tilt angle of the molecules within the smectic layers are represented in Figure 22. Two main features can be identified. First, f increases regularly with θ, almost linearly in the domain which excludes the regions very close to the transition and very far from the transition where θ has a tendency to saturate. This linear increase may be explained by a better stabilization of the trihedron linked to the chiral group (and defined by the director of the molecules, the normal to the layers, and the dipolar vector) as the tilt angle increases. Second, for the same tilt angle θ, the decrease of f with the chain length of about 50% when m changes from 10 to 18 is quite important. This large effect can be attributed to larger orientational fluctuations of the chiral group for longer aliphatic chain lengths, which results from the larger volume given to it, then inducing a smaller anisotropy of the dipolar moments distribution.

Let us consider now the variations of the yield factor, f, as a function of temperature as represented in Figure 23. The yield factor, of about 0.05 to 0.15 a few degrees from the SmC*–SmA transition, slightly increases by 10 to 12% when m increases from 10 to 18. This small increase indicates a rough compensation between the effects discussed above: the direct decrease of f with m, and the increase of f with θ, itself increasing with m. The

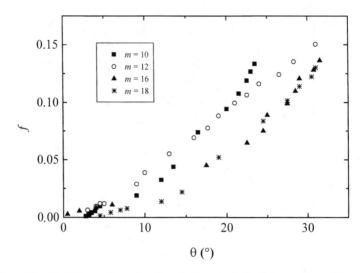

Figure 22. Yield factor, f, as a function of tilt angle, θ, for different terms of the m-BTS–O_{10} series (with permission from *J. Chem. Phys.* **106**, 7816 (1997), Ref. [58]).

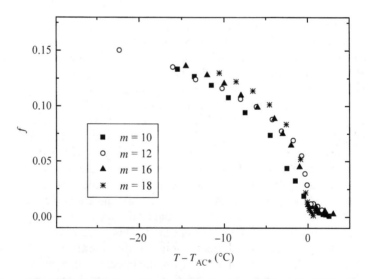

Figure 23. Yield factor, f, as a function of temperature, $T - T_{AC^*}$, for different terms of the m-BTS–O_{10} series (with permission from *J. Chem. Phys.* **106**, 7816 (1997), Ref. [58]).

overall slight increase of f indicates that a thicker aliphatic sublayer globally contributes to a better stabilization of the chiral group.

The above results show that different molecular contributions are involved in the distribution of the transverse dipole moments. First, increased orientational fluctuations and dilution effect, occuring when the aliphatic chain length m is increased, contribute to decrease in the spontaneous polarization. Second, the increase of tilt angle with increasing m contributes to increase P_S, due to a stabilization of the dipolar vectors. This latter stabilization can be related to a better defined building of the different smectic sublayers with a pronounced microsegregation in space of aliphatic and aromatic moieties of the molecules, when the aliphatic sublayer is thickened. In the case of the sulfinate series discussed in this section, all these effects act simultaneously and in contradictory ways ; in a whole, it seems that the effect of the tilt angle is slightly predominant over the two others (dilution and orientational fluctuations), so that the spontaneous polarization increases only slightly with aliphatic chain length for a given ΔT.

F.   Fluorination of Smectogenic Molecules

The introduction of fluorine atoms into the molecular architecture is also another way of tailoring materials to obtain the expected good physical properties. This approach has been explored by numerous chemists. First, by considering the chiral part of the molecule, it has been shown, in the case of the α-haloesters series, that the most stable smectic C* phases with the largest ferroelectric phase range were obtained when the halogen was a fluorine atom [61]. Second, the addition of fluorine in the achiral chain seems to raise the melting and clearing temperatures. Finally, certainly the most interesting point to be stressed in this section is the fact that lateral fluorination of the aromatic part of the molecule leads to a lowering of the viscosity of the materials [62–64], which is also a parameter of great importance in view of potential applications. This trend has been confirmed recently with the following series (see Figure 24), for which high values of the spontaneous polarization and low values of response times have been reported [65].

G.   Influence of Siloxane Groups

The different criteria mentioned above allowed the synthesis of materials with fast electro-optical switching, of the order of 10 to 100 μs. However, some problems still persist in using them in video devices, in the sense that switching times need to be faster (at least one order of magnitude) and a certain mechanical stability, implying an absence of buckling instabilities, has to be achieved. One can consider using side-chain liquid crystalline polymers with ferroelectric side groups, but if the mechanical properties are well improved, the switching time becomes of the order of the millisecond and even more, probably due to the higher viscosity of the polymer material as compared to low molar mass compounds [66,67]. An interesting approach

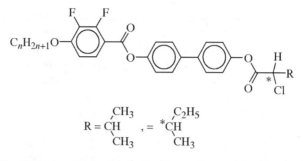

$$R = \underset{CH_3}{\overset{CH_3}{CH}} \ , = \ \underset{CH_3}{\overset{C_2H_5}{*CH}}$$

Figure 24. Example of ferroelectric liquid crystal material with two fluoro lateral substituents (after Ref. [65]).

to reaching a compromise between the good mechanical properties of polymers and the fast switching times of low molar mass liquid crystals has been proposed recently. It consists in introducing a few (generally 2 or 3) dimethylsiloxane segments into the molecular architecture [68,69], as sketched in Figure 25. In, fact, such materials have shown electro-optical properties similar to those of low molar mass materials [70], and have now become, the object of intense investigations [71,72]. The special feature of these compounds is that their molecules contain three distinct parts incompatible with each other. As a result, the siloxane moieties localize themselves in separate sublayers, reminiscent of the case of side-chain polymers for which the backbones are inserted between the smectic layers [73]. The elemental smectic layer is thus constituted by the superposition of three separate sublayers formed by the three molecular moieties [74], as shown in Figure 26 for the smectic C* phase.

Let us consider in more detail the case of the electro-optic characterization of two new ferroelectric organosiloxane liquid crystal materials (Si_nM) for which a short siloxane part is connected to the sulfinate mesogenic moiety 4-[[4-(decyloxy-(R)-($+$)-sulfinyl)phenyl]-ethynyl]-phenyl-4-hexyl-oxybenzoate (6-BTS-O_{10}). These compounds, noted as Si_2M ($n=1$), Si_3M ($n=2$) have the following chemical structure and polymorphic behavior (Fig. 27).

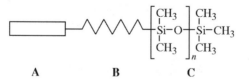

      A   B   C

Figure 25. Sketch of the three-blocks organosiloxane liquid crystal materials. A stands for the central rigid part, B for the aliphatic chain (chiral or nonchiral), and C the dimethylsiloxane part.

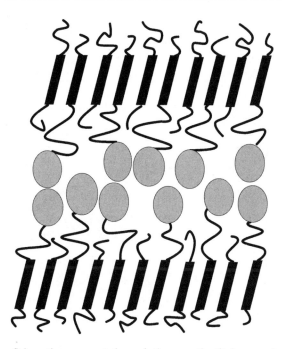

Figure 26. Schematic representation of the smectic C layers of organosiloxane compounds. The central part of the smectic layers is arbitrarily chosen to be formed by the siloxane sublayers, with the siloxane groups (ellipses) arranged in a partially bilayered structure. This central part is fringed with the disordered paraffin chains (wavy lines) arranged in single layers. The aromatic cores (rectangles) are confined in a distinct mono-arrangement sublayer, and tilted with respect to the normal layers.

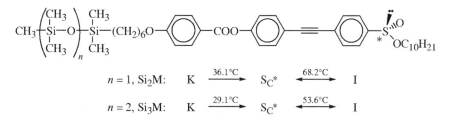

$$n = 1, Si_2M: \quad K \xrightarrow{36.1°C} S_{C^*} \xleftarrow{68.2°C} I$$

$$n = 2, Si_3M: \quad K \xrightarrow{29.1°C} S_{C^*} \xleftarrow{53.6°C} I$$

Figure 27. Chemical structure and polymorphic behavior of the two ferroelectric organosiloxane materials, Si_2M and Si_3M.

In particular, let us analyze the effect of the bulky siloxane group on the electro-optic properties when compared to the less bulky aliphatic chains of the m-BTS–O_{10} homologous series [75]. The comparison will be made mainly with the 10-BTS–O_{10} derivative (Fig. 28) which has a molecular length of the same order of the siloxane derivatives under consideration here.

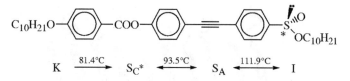

$$K \xrightarrow{\ 81.4°C\ } S_{C^*} \xleftrightarrow{\ 93.5°C\ } S_A \xleftrightarrow{\ 111.9°C\ } I$$

Figure 28. Chemical structure and polymorphic behavior of the 10-BTS–O_{10} derivative of the *m*-BTS–O_{10} series.

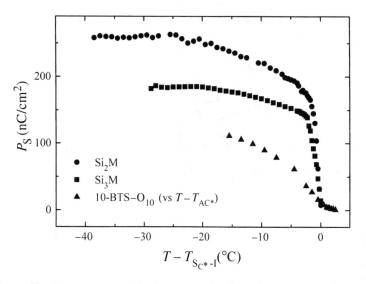

Figure 29. Spontaneous polarization as a function of temperature for the two organosiloxane compounds, Si_2M and Si_3M, and a comparison with the 10-BTS–O_{10} of structure in Fig. 28.

Figures 29 and 30 present the variation of the spontaneous polarization and optical tilt angle, respectively, as a function of temperature for the two organosiloxane compounds Si_2M and Si_3M. Far from the transition, they exhibit a spontaneous polarization higher (by a factor larger than 2) than for the similar compound without siloxane group. It can also be observed that the spontaneous polarization varies with the size of the siloxane moieties. It is important to remember that for the *m*-BTS–O_n series, P_S does not seem to depend on the size of the aliphatic chains. As regard to the variation of the optical tilt angle (Fig. 30), one can observe a quasi-independence of temperature (values between 38 and 44°), and that both materials (Si_2M and Si_3M) have similar tilt angle values, whatever the bulkiness of the siloxane moiety. These values are much higher than the ones (between 20 and 25°) shown by the nonsiloxane compound of the *m*-BTS–O_{10} series. They are also higher than those previously reported in other siloxane-based liquid

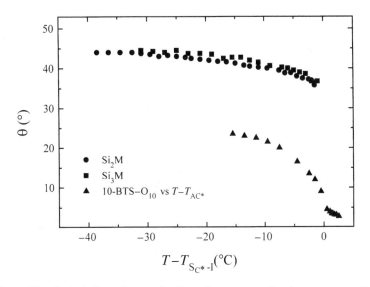

Figure 30. Optical tilt angle as a function of temperature for the two organosiloxane compounds Si_2M and Si_3M, and a comparison with the $10\text{-BTS}\text{-}O_{10}$ of structure in Fig. 28.

crystals [71]. However, this may not be surprising since our values of P_S are also larger and it is known that P_S and θ are strongly correlated [76]. On the other hand, since the siloxane units have a molecular area about twice as large as the aliphatic chains and the mesogenic cores, this forces the tilting of the mesogenic core with respect to the layer normal. In Figure 31 the temperature dependence of the switching time τ for the two siloxane compounds obtained for a maximum electric field of ~ 5 V/μm is shown. Indeed, the values are about of the same order of magnitude for Si_2M and $10\text{-BTS}\text{-}O_{10}$. The values for Si_3M are somewhat higher but still remain within reasonable limits from applications point of view. Moreover, the yield factor, as calculated in Section III.E, is of the order of 0.2 to 0.3, which represents a considerable improvement when compared to the mesogen deprived of siloxane moiety.

On the whole, it is observed that the presence of the siloxane group strongly contributes to the increase of the ferroelectric smectic C* phase temperature range (by greater than a factor of 2). Due to a larger transverse area of the siloxane groups as compared to the aliphatic chains, the optical tilt angle is higher in both siloxane compounds Si_2M and Si_3M than in the homologous series of $m\text{-BTS}\text{-}O_n$. The presence of the siloxane groups also contributes to a dilution of the molecular transverse electric dipole moments. However, due to the increase of the tilt angle and the decrease of the orientational fluctuations, the yield factor is higher by a factor of 2 for Si_2M

DANIEL GUILLON

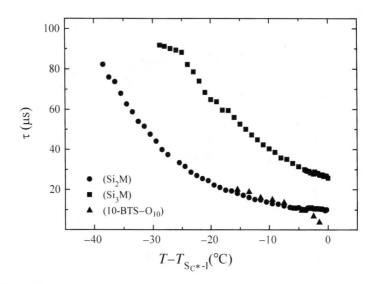

Figure 31. Switching time τ as a function of temperature for the two organosiloxane compounds Si$_2$M and Si$_3$M (for a maximum electric field of $\sim$5 V/μm), and a comparison with the 10-BTS–O$_{10}$ of structure in Fig. 28.

and Si$_3$M than for the nonsiloxane m-BTS–O$_n$ compounds. The overall conclusion points to a clear stabilization of the SmC* phases, with an increase of the spontaneous polarization and of the optical tilt angle.

H.   Other Forms of Chirality

All the above examples of smectic C* phases are obtained with molecules having a center of chirality. However, other forms of chirality, such as axial or planar chirality, can be considered in designing mesogens for smectic C* ferroelectric mesophases. Only a few examples of mesogenic compounds based on molecules with axial chirality have been reported in the literature [77–81], and the first chiral allene derivatives exhibiting smectic C* phases were described very recently [82]. The general chemical formula of the latter compounds is shown in (Fig. 32).

It was found that the broadest chiral smectic C phases were obtained for the disubstituted allene derivatives [83]. The investigation of their electro-optical properties showed that the spontaneous polarization can reach values close to 40 nC/cm^2. It is important here to stress out that these values are unexpectedly large, considering the weak value of the electric dipole moment associated with the stereogenic axis of the allenic moiety. As a matter of fact, the global dipole moment, μ, is nearly parallel to the rigid part of the molecule and has a relatively small value, $\sim$0.9 D (MOPAC calculations), resulting in a weak component on the axis perpendicular to the tilt plane.

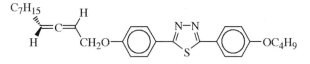

Figure 32.   Type of mesogen containing an allenic moiety.

Another form of chirality is the planar chirality. Some examples of compounds with such a feature have been shown to exhibit mesomorphic behavior in their racemic form [84,85]. However, separation of the corresponding two pure enantiomers was revealed to be quite difficult to achieve and remains an unsolved problem. Very recently, the synthesis of enantiomerically pure butadiene–tricarbonyliron complexes was achieved [86]. These compounds exhibit SmC* phases, and the measured spontaneous polarization for one of them was found to be 32 nC/cm^2, whereas the corresponding response time was relatively slow—about 9 ms. It would be interesting to investigate these new forms of chirality further to test their potentialities with regard to competitive electro-optical properties.

IV.   MOLECULAR DESIGN FOR OTHER POLAR PHASES

A.   Polyphilic Derivatives

Polyphilic compounds with a perfluoroalkyl chain in the central or in the terminal position have different behaviors. The presence of the perfluoro moiety close to the rigid polarizable part of the molecule is favorable to the occurrence of a smectic E phase (Fig. 33**a**). In this phase, the different constitutive parts of the molecule are localized in distinct segregated sublayers, as is expected of polyphilic compounds [87,88]. This structural organization is not observed any more, however, when the perfluoroalkyl chain is in a terminal position far from the rigid part of the molecule (Fig. 33**b**). In this case, a smectic A phase is observed with an interdigitation of aromatic and aliphatic parts within the smectic layers [89]. The difference is explained in terms of the steric constraints of the perfluoroalkyl moiety [90]. In the case of the smectic E phase, the biphenyl part determines the size of the lattice, and the perfuoroalkyl chain, due to its larger volume compared to that of the alkyl analogue, acts as a stop-block with respect to the other chemical parts. Thus, the formation of a polyphilic structure is highly favored if the bulkiest part (in the present case, the perfluoroalkyl chain) is located in the center of the molecule. The two other chemical moieties which are localized on both sides of the perfluoroalkyl chain cannot penetrate inside the perfluoro sublayer and themselves also form segregated sublayers.

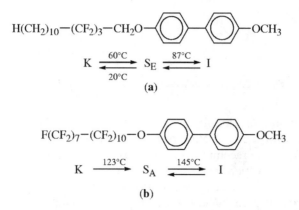

Figure 33. Two examples of polyphilic molecules with a perfluoroalkyl chain in the terminal (a) and central (b) position, respectively. From Ref. [87].

Some of these typical three-block polyphilic compounds have been investigated for their physical properties. Longitudinal ferroelectricity is demonstrated by the measurements of the acoustically induced piezoelectric response, pyroelectric effect, and repolarization currents [91]. However, the spontaneous polarization is low (only a few nC/cm^2), and such low values are not yet understood. Moreover, the structural studies performed on some of those materials do not reveal any clear relationships between molecular structure and macroscopic properties [92]. Finally, it is important to recall that it is necessary to use mixtures of these types of molecules in order to produce a stable longitudinal polar order; otherwise, spontaneous splay occurs, due the unfavorable lateral dipolar interactions between identical polar molecular moities, resulting in a bending of the smectic layers.

Another possible way to obtain polar systems with polyphilic molecules is to introduce associative groups into the molecular architecture. For example, if hydrogen-bond donor and acceptor groups are localized on both extremities of a polyphilic molecule, one can foresee the formation of polar rows of molecules in the liquid phases of such materials [93], as shown in Figure 34. Thus, the association of nonspecific interactions due to the polyphilic effect and of specific interactions due to the associative groups is able to overcome the natural tendency to form antiparallel arrangements. Some preliminary work seems to indicate that a mesomorphic supramolecular assembly obtained by using noncovalent bonding is a reasonable approach for polyphilic compounds. The compound shown in Figure 35 represents a first step toward this novel type of liquid crystalline material.

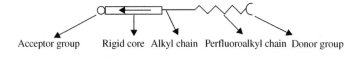

Acceptor group   Rigid core Alkyl chain Perfluoroalkyl chain Donor group

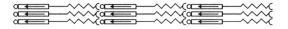

Figure 34. Polar rows of molecules expected from the self-assembly of polyphilic molecules containing specific associative donor and acceptor groups. (Adapted with permission of *Chem. Commun.* **441** (1997), Ref. [93]).

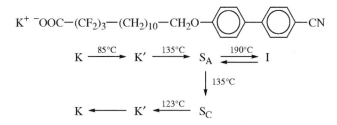

$$K^+ \; {}^-OOC-(CF_2)_3-(CH_2)_{10}-CH_2O-\!\!\left\langle\bigcirc\right\rangle\!\!-\!\!\left\langle\bigcirc\right\rangle\!\!-CN$$

$$K \xrightarrow{85°C} K' \xrightarrow{135°C} S_A \underset{\longleftarrow}{\overset{190°C}{\rightleftarrows}} I$$

$$\Big\downarrow 135°C$$

$$K \longleftarrow K' \xleftarrow{123°C} S_C$$

Figure 35. Example of a polyphilic molecule bearing specific associative groups at both ends and showing a mesomorphic behavior.

B. Perfluorinated Swallow-Tailed Compounds

Utilizing the experimental fact that perfluorinated chains are to a large extent immiscible with alkylated or aromatic parts, swallow-tailed molecules with inserted perfluorinated segments were used to tentitatively produce polar layers [94]. Two types of mesogens (A and B; see Fig. 36), different from their direction of the dipoles and from the position of the fluorinated moiety, were mixed, so that the formation of dimers AB were expected to be favored due to the combined effect of phase separation and steric repulsion.The antiparallel packing of the molecules, then, should lead to a polar layer (Fig. 37) assuming that the intralayer amphiphilic interactions should overcome the intralayer dipolar interactions and the entropic contributions. Unfortunately, no ferroelectricity could be detected on these A/B mixtures. This behavior can in turn be explained by several different behaviors: (1) an up-and-down orientation of the dimer pairs AB within each layer; (2) an up-and-down orientation of polarized smectic layers; or (3) a too-slow process of the switching time. Moreover, the X-ray investigations performed on these materials [94] seem to indicate that none of the interactions involved (steric, amphiphilic, and dipolar ones) are predominant. In conclusion, this approach to producing polar systems does not seem convincing, perhaps because two successive polarized layers prefer

A

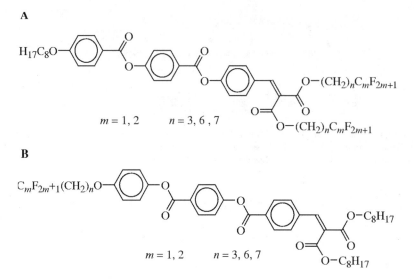

$$m = 1, 2 \qquad n = 3, 6, 7$$

B

$$m = 1, 2 \qquad n = 3, 6, 7$$

Figure 36. Example of two swallow-tailed molecules (A and B), used in binary mixtures to produce polar smectic layers.

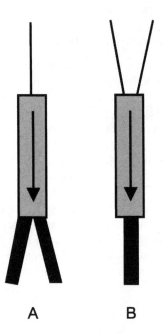

A   **B**

Figure 37. Expected possible dimer AB for the occurrence of polar smectic layers. (Adapted with permission from *Ferroelectrics* **180**, 341 (1996), Ref. [94]).

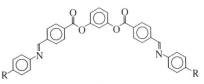

Figure 38. Molecular structure of the banana-shaped molecules.

to stack with their longitudinal dipoles in opposite directions in order to respect the natural tendency for the aliphatic and perfluorinated moieties along the normal direction to the smectic layers to face each other.

C. Achiral Banana-Shaped Molecules

The molecules of this type which have been reported up to now to produce polar smectic phases with an electric polarization parallel to the layers, are all based on the existence of a meta-disubstituted benzene ring in the center of their core. These molecules can adopt some banana-shaped conformations (Fig. 38). The first experimental results by T. Niori et al. [10] have been confirmed since by two other groups [95,96]. All the molecules investigated have a similar structure, with the same particular bent central part, except for two cases where the ester group is replaced by an azomethine group. However the distinct smectic phases identified by polarized microscopy and electro-optical measurements have not been fully characterized from the structural point of view, yet and these phases have been shown to be immiscible (or very weakly miscible) with classical smectic phases obtained from rod-like molecules. Moreover, two types of twisted smectic phases have been identified; in the first case, the helical axis is normal to the smectic layers, whereas in the second one, it is parallel to the layers [97]. In the first case, the structure is similar to that of the classical smectic C* phase, whereas in the second case, the structure is similar to that of the TGB phase [98]; however, it has to be mentioned that the latter appears at lower temperature than the normal helical phase. The origin of these helical structures was attributed first to conformational chirality, as suggested by ^{13}C NMR measurements; however, the difference in energy between two conformations of the two carbonyl groups is obviously too small compared to the thermal energy of the twisted phase. It seems more reasonable to assume that the twisted structure may be induced from the dipolar interactions [99], in a kind of a two-dimensional escape from a macroscopic polarization [100]. Finally, it is worth stressing out that quite high values of the electric polarization, several hundreds of nC/cm^2, have been reported for these materials [96,97,101], and that electric-field-induced ferroelectric as well as antiferroelectric phases have been observed.

Such fascinating compounds look very interesting from the fundamental point of view, but a lot of further experimental and theoretical studies are needed to clearly understand the exact role of the chemical architecture with respect to the structures of the different smectic phases encountered and to their electro-optical properties.

V. FERROELECTRICITY IN COLUMNAR MESOPHASES

A. Chiral Columnar Mesophases

Similarly to the molecular engineering of calamitic molecules to produce ferroelectric smectic C* phases, disk-like molecules with chiral peripheral chains tilted with respect to the columnar axis were predicted to lead to ferroelectric columnar mesophases [102]. Indeed, as is the case with all flat disk-shaped mesogenic molecules, the tilt is mainly associated with the flat rigid aromatic cores of the molecules, the side-chains being in a disordered state around the columnar core. Thus, the nearest part of the chains from the cores makes an angle with the plane of the tilted aromatic part of the molecules. If the chiral center and the dipole moment are located close to the core, then each column possesses a nonzero time-averaged dipole moment, and therefore a spontaneous polarization. For reasons of symmetry, this polarization must be, on average, perpendicular to both the columnar axis and to the tilt direction; in other words, the polarization is parallel to the axis about which the disk-shaped molecules rotate when they tilt (Fig. 39).

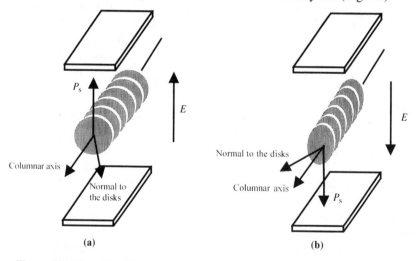

(a)   (b)

Figure 39. Ferroelectric switching in a columnar mesophase with electric field. (a) Electric field is up. (b) Electric field is down. (Adapted with permission from *J. Mat. Chem.* **5**, 417 (1995), Ref. [107]).

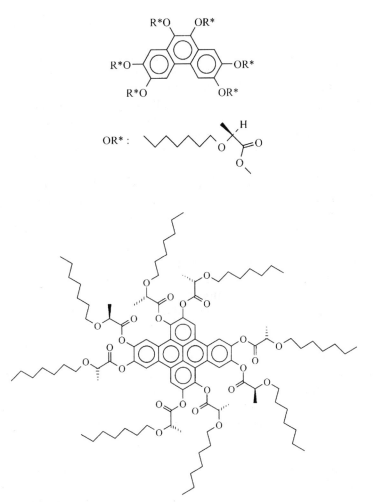

Figure 40. Examples of disk-shaped molecules with chiral peripheral chains leading to ferroelectric columnar mesophases (redrawn with permission from *Liq. Cryst.* **18**, 387 (1995), Ref. [105]).

The first report on ferroelectric switching in columnar mesophases was that of H. Bock and W. Helfrich [103]. Since then, a few other examples of compounds (Fig. 40) exhibiting switchable columnar mesophases have been found [104–107]. Their electro-optical properties have been investigated with a cell similar to that used for smectic C* phases. Because of the high viscosity of this type of columnar mesophase, a uniform alignment of the columns parallel to the glass plates of the cell is only obtained through shearing of the sample between the glass plates. If the measured spontaneous polarizations are of the same order as that of lamellar smectic C* phases, the switching times are, on the contrary, much slower (of the order of ms, compared to μs in the case of smectic C*). This is probably due to the high viscosity of the material related to the two-dimensional lattice of the columns, to the relatively large size of the molecules, and to the elliptic shape of the columns caused by the tilt of the aromatic cores with respect to the columnar axis. Thus, columns cannot rotate freely from one state to the other, or cannot rearrange themselves internally, depending on whether the switching mechanism involves rotation of the entire columns or a reversing of the tilt without molecular rotation.

Indeed, as pointed out by H. Bock and W. Helfrich [103], these two possible mechanisms of reversing the columnar polarization are similar to those observed in lamellar smectic C* ferroelectric phases. In the first mechanism, which consists of rotation of the columns around the columnar axis, there is an important friction between the columns. In the second mechanism, which implies an intermediate state where the columns are no longer elliptic but circular in shape (corresponding to a zero-tilt for the aromatic flat cores), there is an important elastic stress within one individual column and also between the columns due to the fact that the chains have to move from the up to the down (or vice versa) position along the direction of the columnar axis. However, it is not possible to determine, with the few experimental data known to date, which mechanism prevails.

It may appear rather hopeless from the applications point of view to continue research on this type of materials, since switching rates are slow and the orientation in the electro-optical cells, obtained by shearing, difficult to obtain in an industrial development. However, switchable ferroelectric columnar liquid crystals possess several advantages over the ferroelectric smectic C* liquid crystals. First, they are probably more resistant from the mechanical point of view, due to a certain stabilization induced by the two-dimensional arrangement of the columns, but as in the case of the classical smectic C* phases, one has to pay attention to avoiding the buckling of the columns. Second, there is no problem of chevrons since the columns are aligned parallel to the electrodes. Third, the tilt angle of the aromatic cores does not vary significantly with temperature (as far as we can state from the few known examples of such ferroelectric columnar mesophases). Finally, as the electro-optical properties seem to depend on the electric field strength [108], it is possible to imagine multistable switching devices by setting the tilt angle of the flat aromatic cores at different values.

B. Columnar Mesophases with Axial Polarity

Whereas in the previous examples, the spontaneous polarization is parallel to the C_2 axis (perpendicular to the columnar axis), ferro- and antiferro-

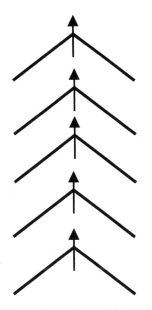

Figure 41. Schematic of a column with axial polarity obtained with pyramidic molecules. The vertical arrows stands for the molecular dipole.

electric arrangement of columns can also be envisaged with an electric polarization parallel to the columnar axis (Fig. 41). Several attempts involving bowl-shaped and pyramidic molecules [109,110] (Fig. 42) and metallomesogenic vanadyl complexes [111] have been reported to produce such a type of ferroelectricity. Even if these materials have the molecular requirements necessary to yield columnar mesophases with an axial polarity, unfortunately it was not possible to state clearly the existence of a spontaneous polarization. However, a very recent paper [112] seems to indicate that ferrolectric columnar mesophases with the polarization parallel to the optical axis are observed with pyramidic molecules subsituted with six chiral chains. In this case, the time switching would be in the range of seconds and the polarization about $10 \, nC/cm^2$.

It is worth pointing out in this section that polar columns located at the nodes of a two-dimensional lattice can all be oriented parallel or else alternately if the lattice is tetragonal [107]. In this case paraelectric, ferroelectric, or antiferrolectric mesophases can be obtained, depending on the specific interactions between columns. On the contrary, if the mesophase has the hexagonal symmetry, only a ferroelectric arrangement of the columns is possible. In this context, recent Monte Carlo simulations [113] of discotic mesogens with axial dipole have shown that nematic and hexagonal columnar mesophases can be generated. In the columnar mesophase, each

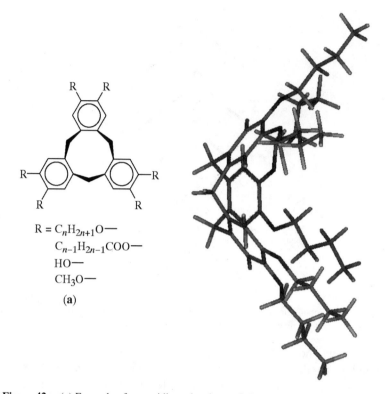

$R = C_nH_{2n+1}O-$
$\quad\quad C_{n-1}H_{2n-1}COO-$
$\quad\quad HO-$
$\quad\quad CH_3O-$

(a)

Figure 42. (a) Example of pyramidic molecule consisting of a rigid crown structure with a trigonal pyramidal symmetry; (b) molecular modeling of the same molecule showing its pyramidic shape (Biosym software).

column contains aligned dipolar domains, but any column is fully polarized, and the dipolar domains are paired with those of neighboring columns.

VI. ANTIFERROELECTRICITY IN LIQUID CRYSTALS

Among the numerous compounds giving rise to ferroelectric smectic C* mesophases exhibiting an optical bistability under electric field, one of them (now well known as MHPOBC [114]; see Fig. 43 [115]) was shown to present a tristable switching behavior [116]. The latter was attributed to the existence of an antiferroelectric phase (designated in the following as SmC_A^*) [117]. The structure of the SmC_A^* phase is shown in Figure 44. The molecules tilt in the same direction within one given layer, but in opposite directions in two adjacent layers. The spontaneous polarizations are parallel to the smectic layers, but in opposite directions in two succesive layers,

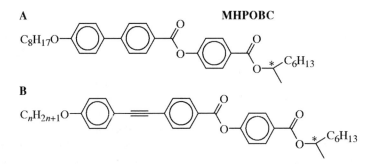

Figure 43. Two typical examples (A and B) of materials exhibiting antiferroelectric mesophases. (A) MHPOBC after [114]. (B) Tolane series after Ref. [115].

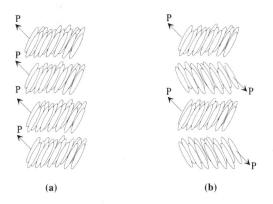

Figure 44. Schematic of the positions of molecules in (a) ferro- and (b) antiferroelectric phases.

canceling each other. As in the classical smectic C* phases described in Section III, there is a small precession of the tilt from layer to layer (a few degrees per layer caused by chirality) leading to the formation of a helical structure with the helical axis parallel to the layer normal. Such a structure had been described for another SO* antiferroelectric liquid crystal at the Second International Conference on Ferroelectric Liquid Crystals [118], and then proved to be general for the SmC$_A^*$ phase [119] of all the liquid crystal compounds synthesized so far. It is important to point out here the role of the external electric field which can reverse the direction of the dipoles and lead to the existence of ferroelectric phases with opposite polarizations, in addition to the SmC$_A^*$ phase with no polarization. The tristable switching thus established is very suitable for multiplex driving in fast-switching display devices; moreover, the viewing angle is much wider than that of bistable smectic C* phases.

It would be quite interesting to understand from the molecular engineering point of view why such antiferroelectric phases exist in liquid crystals and how they are stabilized. The main reason advanced today is the pairing of the transverse dipole moments localized on the end chains of the molecules [117]. Because of the thermal energy of the system, the holding time of a specific pair must be very short. Such a pairing is possible because antiferroelectric liquid crystal compounds have in general a strong polar group, and their corresponding spontaneous polarizations are fairly large. The pairing energy would stabilize the SmC_A^* mesophase, whereas a packing energy [120] resulting from temperature-dependent excluded volume effect would stabilize the SmC_A^* mesophase. However, and in spite of the great number of studies devoted to antiferroelectric mesophases, there is not yet a clear understanding of the molecular origin of these mesophases.

In order to avoid the helical macrostructure of the highly chiral materials used in the mixtures for the antiferroelectric display devices, using nonchiral antiferrolectric phases (SmC_A) as host materials [121] was proposed. Such a SmC_A structure has been obtained with a swallow-tailed compound [122]; the structure is the result of the formation of bent dimers due to the steric interactions between the swallow tails of one molecule and the single end chain of another molecule in adjacent layers. The packing of the bent dimers leads to a nonchiral antiferroelectric phase. The switching behavior under electric field is then observed by using a standard chiral antiferroelectric dopant in these host materials. Interestingly, antiferroelectric behavior has also been observed in mixtures of nonchiral side chain polymers with their monomers [123]. The antiferroelectric structure is a bilayer smectic C structure with an alternating tilt from one layer to the next, the role of the monomer being to induce the alternated structure. Field-induced macroscopic polarizations up to 400 nC/cm^2 have been measured.

VII. CONCLUSION

The different possibilities reported in this paper for tailoring materials chemically in order to obtain polar materials show unambiguously how rich the molecular engineering is in producing ferroelectricity in liquid crystals. Based on symmetry considerations, different kinds of polarizations may be envisaged, e.g., perpendicular or parallel to the long molecular axis of calamitic molecules, and perpendicular or parallel to the columnar axis in the case of columnar mesophases. Chiral as well as nonchiral molecules (or macromolecules) can be used as long as they are designed with the appropriate shape and chemical groups. However, if many interesting electro-optical properties have already been observed with a great variety of compounds, domains still remain where a large effort has to be undertaken in order to understand clearly the experimental observations (e.g., structure of the smectic phases of banana-shaped molecules, switching mechanism in columnar mesophases, etc.). Of course, simulations to predict the physical properties and to understand the structure-property relationships in these materials will become more and more important in the future, but it will take time before realistically modelizing ferroelectric liquid crystals is possible. Finally, it is important to point out that the way the materials are processed (oriented samples, thin films) can also greatly contribute to the enhancement of polar ordering (Langmuir–Blodgett films, shearing process, etc.).

Acknowledgments

The author would like to thank Prof. Y. Galerne, Prof. J. F. Nicoud and Dr. S. Méry for fruitful and helpful discussions.

References

1. J. Valasek, *Phys. Rev.* **17**, 475 (1991).

2. N. A. Clark and S. T. Lagerwall, in *Ferroelectric Liquid Crystals*, Gordon and Breach Science Publishers, New York, 1991, pp. 409–461.

3. R. B. Meyer, paper presented at the 5th International Liquid Crystal Conference, Stockholm, June 1974.

4. R. B. Meyer, L. Liébert, L. Strzelecki, and P. Keller, *J. Phys. (Paris) Lett.* **36**, 69 (1975).

5. (a) D. M. Walba, R. T. Vohra, N. A. Clark, M. A. Hanndschy, J. Xue, D. S. Parmar, S. T. Lagerwall, and K. Skarp, *J. Am. Chem. Soc.* **108**, 17424 (1986).
(b) M. A. Handschy, K. M. Johnson, G. Moddel, and L. A. Pagano-Stauffer, *Ferroelectrics* **85**, 279 (1988).
(c) J. W. Goodby, R. Blinc, N. A. Clark, S. T. Lagerwall, M. A. Osipov, S. A. Pikin, T. Sakurai, K. Yoshino, and B. Žekš, *Ferroelectric Liquid Crystals; Principles, Properties and Applications* Gordon and Breach Science Publishers, New York, 1991.

6. G. Scherowsky and J. Gay, *Liquid Crystals* **5**, 1253 (1989).

7. (a) D. M. Walba, K. F. Eidman, and R. C. Haltiwanger, *J. Org. Chem.* **54**, 4939 (1989).
(b) T. Kusumoto, T. Hanamoto, T. Hiyama, S. Takehara, T. Shoji, M. Osawa, T. Kuriyama, K. Nakamura, and T. Fujisawa, *Chem. Lett.* 311 (1991).

8. T. Sierra, M. B. Ros, A. Omenat, and J. L. Serrano, *Chem. Mater* **5**, 938 (1993).

9. (a) J. Nakauchi, M. Uematsu, K. Sakashita, Y. Kageyama, S. Hayashi, T. Ikemoto, and K. Mori, *Jpn. J. Appl. Phys.* **28**, L1258 (1989).
(b) M. Koden, K. Kuratate, F. Funada, K. Awane, K. Sakaguchi, Y. Siomo, and T. Kitamura, *Jpn. J. Appl. Phys.* **29**, L981 (1990).
(c) K. Sakashita, M. Shindo, J. Nakauchi, M. Uematsu, Y. Kageyama, S. Hayashi, T. Ikemoto, and K. Mori, *Mol. Cryst. Liq. Cryst.* **199**, 119 (1991).

10. T. Niori, T. Sekine, J. Watanabe, T. Furukawa, and H. Takezoe, *J. Mat. Chem.* **6**, 1231 (1996).

11. D. R. Link, G. Natale, R. Shao, J. E. Manlennan, N. A. Clark, E. Körblova, and D. M. Walba, *Science*, **278**, 1924 (1997).

12. J. W. Goodby, E. Chin, T. M. Leslie, J. M. Geary, and J. S. Patel, *J. Am. Chem. Soc.* **108**, 4729 (1986).

13. N. A. Clark and S. T. Lagerwall, *Appl. Phys. Lett.* **36**, 899 (1980).

14. R. Bartolino, J. Doucet, and G. Durand, *Ann. Phys.* **3**, 389 (1978).

15. D. M. Walba, S. C. Slater, W. N. Thurmes, N. A. Clark, M. A. Handschy, and F. J. Supon, *J. Am. Chem. Soc.* **108**, 5210 (1986).

16. D. M. Walba, R. T. Vohra, N. A. Clark, M. A. Handschy, J. Xue, D. S. Parmar, S. T. Lagerwall, and K. Skarp, *J. Am. Chem. Soc.* **108**, 7424 (1986).

17. D. M. Walba and N. A. Clark, *Ferroelectrics* **84**, 65 (1988).

18. For a review : D. M. Walba, *Adv. Synth. React. Solids* **1**, 173 (1991).

19. L. M. Blinov, L. A. Beresnev, M. A. Osipov, and S. A. Pikin, *Mol. Cryst. Liq. Cryst.* **158A**, 1 (1988).

20. M. A. Osipov, *Mol. Mat.* **1**, 217 (1992).

21. M. Taniguchi, M. Ozaki, K. Yoshino, K. Satoh, and M. Yamasaki, *Ferroelectrics* **77**, 137 (1988).

22. D. J. Photinos and E. T. Samulski, *Science* **270**, 783 (1995).

23. D. J. Photinos, A. F. Terzis, E. T. Samulski, T. J. Dingemans, A. Chen, and C.-D. Poon, *Mol. Cryst. Liq. Cryst.* **292**, 265 (1997).

24. A. Skoulios and D. Guillon, *Mol. Cryst. Liq. Cryst.* **165**, 317 (1988).

25. A. F. Terzis, D. J. Photinos, and E. T. Samulski, *J. Chem. Phys.* **107**, 4061 (1997).

26. F. Tournilhac, L. M. Blinov, J. Simon, and S. V. Yablonsky, *Nature* **359**, 621 (1992).

27. F. Tournilhac, L. Bosio, J. F. Nicoud, and J. Simon, *Chem. Phys. Lett.* **145**, 452 (1988).

28. D. Guillon, G. Poeti, A. Skoulios, and E. Fanelli, *J. Phys. Lett.* **44**, 491 (1983).

29. R. G. Petschke and K. M. Wiefling, *Phys. Rev. Lett.* **59**, 343 (1987).

30. D. R. Perchak and R. G. Petscek, *Phys. Rev. A* **43**, 6756 (1991).

31. J. Prost, R. Bruinsma, and F. Tournilhac, *J. Phys. II (France)* **4**, 169 (1994).

32. J. Watanabe, Y. Nakata, and K. Simizu, *J. Phys. II (France)* **4**, 581 (1994).

33. M. A. Osipov and S. A. Pikin, *Mol. Cryst. Liq. Cryst.* **103**, 57 (1983).

34. K. Skarp and M. Handschy, *Mol. Cryst. Liq. Cryst.* **165**, 439 (1988); D. M. Walba, M. B. Ros, T. Sierra, J. A. Rego, N. A. Clark, R. Shao, M. D. Wand, R. T. Vohra, K. E. Arnett, and S. P. Velsco, *Ferroelectrics* **121**, 247 (1991).

35. D. Guillon and A. Skoulios, *Mol. Cryst. Liq. Cryst.* **39**, 139 (1977).

36. J. W. Goodby and T. M. Leslie, *Mol. Cryst. Liq. Cryst.* **110**, 175 (1984).

37. J. W. Goodby, J. S. Patel, and E. Chin, *J. Phys. Chem.* **91**, 5151 (1987).

38. K. Yoshino, M. Osaki, T. Saturai, K. Sakamoto, and M. Honna, *Jpn. J. Appl. Phys.* **23**, 175 (1984).

39. A. Hallsby, M. Nilson, and B. Otterholm, *Mol. Cryst. Liq. Cryst.* **82**, 61 (1982).

40. D. M. Walba, S. C. Slater, W. N. Thurmes, N. A. Clark, M. A. Hanndschy, and F. Supon, *J. Am. Chem. Soc.* **108**, 5210 (1986).

41. S. T. Lagerwall and I. Dahl, *Mol. Cryst. Liq. Cryst.* **114**, 151 (1984).

42. G. Scherowsky and J. Gay, *Liq. Cryst.* **5**, 1253 (1989).

43. D. M. Walba, K. F. Eidman, and R. C. Haltiwanger, *J. Org. Chem.* **54**, 4939 (1989); T. Kusumoto, T. Hiyama, S. Takehara, T. Shoji, M. Osawa, T. Kuriyama, K. Nakamura, and T. Fujisawa, *Chem. Lett.* 311 (1991).

44. T. Sierra, M. B. Ros, A. Omenat, and J. L. Serrano, *Chem. Mater.* **5**, 938 (1993).

45. J. Nakauchi, M. Uematsu, K. Sakashita, Y. Kageyama, S. Hayashi, T. Ikemoto, and K. Mori, *Jpn. J. Appl. Phys.* **28**, 1258 (1989); M. Koden, K. Kuratate, F. Funada, K. Awane, K. Sakaguchi, Y. Siomo, and T. Kitamura, *Jpn. J. Appl. Phys.* **29**, 981 (1990); K. Sakashita, M. Shindo, J. Nakauchi, M. Uematsu, Y. Kageyama, S. Hayashi, T. Ikemoto, and K. Mori, *Mol. Cryst. Liq. Cryst.* **199**, 119 (1991).

46. M. Z. Cherkaoui, J. F. Nicoud, and D. Guillon, *Chem. Mater.* **6**, 2026 (1994).

47. M. Z. Cherkaoui and J. F. Nicoud, *New J. Chem.* **19**, 851 (1995).

48. M. Z. Cherkaoui, J. F. Nicoud, Y. Galerne, and D. Guillon, *J. de Phys. II Paris* **5**, 1263 (1995).

49. M. A. Glaser, V. V. Ginzburg, N. A. Clark, E. Garcia, D. M. Walba, and R. Malzbender, *Mol. Phys. Rep.* **10**, 26 (1995).

50. A. D. Chandani, Y. Ouchi, H. Takezoe, and A. Fukuda, *Jpn. J. Appl. Phys.* **27**, L276 (1988).

51. Ch. Bahr, G. Heppke and B. Sabaschus, *Ferroelectrics* **84**, 103 (1988).

52. Ch. Bahr, G. Heppke, and B. Sabaschus, *Liq. Cryst.* **9**, 31 (1991).

53. J. Jacques and A. Collet, *Enantiomers, Racemates and Resolution*, Krieger Publishing Company, Malabar, Florida, 1991.

54. M. Leclercq, J. Billard, and J. Jacques, *Mol. Cryst. Liq. Cryst.* **8**, 367 (1969).

55. H. R. Dübal, C. Escher, and D. Ohlendorf, *Ferroelectrics* **84**, 143 (1988).

56. M. Z. Cherkaoui, thesis, University of Strasbourg, France, 1993.

57. M. Z. Cherkaoui, J. F. Nicoud, Y. Galerne, and D. Guillon, *Liq. Cryst.* **26**, 1315–1324 (1999).

58. M. Z. Cherkaoui, J. F. Nicoud, Y. Galerne, and D. Guillon, *J. Chem. Phys.* **106**, 7816 (1997).

59. B. Otterholm, C. Alstermark, K. Latishler, A. Dahlgren, S. T. Lagerwall, and K. Skarp, *Mol. Cryst. Liq. Cryst.* **146**, 189 (1987).

60. P. G. de Gennes and J. Prost, *The Physics of Liquid Crystals*, 2nd ed., Clarendon Press Oxford, U.K., 1993.

61. R. Twieg, K. Betterton, R. DiPietro, C. Nguyen, H. T. Nguyen, A. Babeau, and C. Destrade, *SPIE* **1445**, 86 (1991).

62. J. Nakauchi, M. Uematsu, K. Sakashita, Y. Kagemaya, S. Hayashi, and K. Mori, *Liq. Cryst.* **7**, 41 (1990).

63. H. T. Nguyen, R. J. Twieg, M. F. Nabor, N. Isaert, and C. Destrade, *Ferroelectrics* **121**, 87 (1991).

64. H. T. Nguyen, A. Bouchta, L. Navailles, P. Barois, N. Isaert, R. Twieg, A. Maaroufi, and C. Destrade, *J. Phys.* **2**, 1889 (1992).

65. I. Zaréba, H. Allouchi, M. Cotrait, M. F. Nabor, C. Destrade, and H. T. Nguyen, *Liq. Cryst.* **21**, 565 (1996).

66. H. J. Coles, H. F. Gleeson, G. Scherowsky, and A. Schliwa, *Mol. Cryst. Liq. Cryst. Lett.* **7**, 117 (1990).

67. S. Pfeiffer, R. Shashidar, J. Naciri, and S. Mery, *SPIE Proc.* **1665**, 166 (1992).

68. H. J. Coles, H. Owen, J. Newton, and P. Hodge, *Liq. Cryst.* **15**, 739 (1993).

69. E. Wischerhoff and R. Zentel, *Liq. Cryst.* **18**, 745 (1995).

70. H. J. Coles, I. Buttler, K. Raina, J. Newton, J. Hannington, and D. Thomas, *SPIE Proc.* **2408**, 22 (1995).

71. J. Newton, H. J. Coles, P. Hodge, and J. Hannington, *J. Mater. Chem.* **4**, 869 (1994); P. Kloess, J. McComb, and H. J. Coles, *Ferroelectrics* **180**, 233 (1996).

72. J. Naciri, J. Ruth, G. Crawford, R. Shashidar, and B. R. Ratna, *Chem. Mater.* **7**, 1397 (1995).

73. L. Noirez, P. Keller, and J. P. Cotton, *Liq. Cryst.* **18**, 129 (1995).

74. M. Ibn-Elhaj, A. Skoulios, D. Guillon, J. Newton, P. Hodge, and H. J. Coles, *J. Phys. II France* **6**, 1807 (1996); M. Ibn-Elhaj, H. J. Coles, D. Guillon, and A. Skoulios, *J. Phys. II France* **3**, 1807 (1993).

75. P. Sebastiao, S. Mery, M. Sieffert, J. F. Nicoud, Y. Galerne, and D. Guillon, *Ferroelectrics* **212**, 133–141 (1998).

76. R. Blinc and B. Zeks, *Phys. Rev. A* **18**, 740 (1978); B. Zeks, *Mol. Cryst. Liq. Cryst.* **114**, 259 (1984).

77. G. Solladié and R. Zimmermann, *Angew. Chem.* **97**, 70 (1985).

78. H. Poths, R. Zentel, S. U. Vallerien, and F. Kremer, *Mol. Cryst. Liq. Cryst.* **203**, 101 (1991).

79. F. Yang and R. F. Lemieux, *Mol. Cryst. Liq. Cryst.* **260**, 247 (1995).

80. G. Heppke, D. Lötzsch, and F. Oestereicher, *Z. Naturforsch.* **41a**, 1214 (1986).

81. G. Solladié, P. Hugelé, R. Bartsch, and A. Skoulios, *Angew. Chem. Int. Ed. Engl.* **260**, 247 (1996).

82. K. Zab, H. Kruth, and C. Tschierske, *J. Chem. Soc. Chem. Comm.* 977 (1996); R. Lunkwitz, C. Tschierske, A. Langhoff, F. Giesselmann, and P. Zugenmaier, *J. Mater. Chem.* **7**, 1713 (1997).

83. R. Lunkwitz and C. Tschierske, *6th International Conference on Ferroelectric Liquid Crystals*, Brest, France, July 20–24, 1997.

84. L. Ziminsky and J. Malthête, *J. Chem. Soc., Chem. Commun.* 1495 (1990).

85. R. Deschenaux and J. Santiago, *Tetrahedron Lett.* **35**, 2169 (1994).

86. P. Jacq and J. Malthête, *Liq. Cryst.* **21**, 291 (1996).

87. S. Pensec, F. G. T. Tournilhac, and P. Bassoul, *J. Phys. II (France)* **6**, 1597 (1996).

88. S. Diele, D. lose, H. Kruth, G. Pelzl, F. Guittard, and A. Cambon, *Liq. Cryst.* **21**, 603 (1996).

89. S. Pensec, F. G. Tournilhac, P. Bassoul, and C. Durliat, *J. Phys. Chem. B* **102**, 52–60 (1998).

90. S. Pensec, thesis, University of Paris 6, 1997.

91. F. Tournilhac, L. M. Blinov, J. Simon, D. B. Subachius, and S. V. Yablonsky, *Synthetic Metals* **54**, 253 (1993).

92. L. M. Blinov, T. A. Lobko, S. N. Ostroskii, S. N. Sulianov, and F. G. Tournilhac, *J. Phys. II (France)* **3**, 1121 (1993); Y. Shi, F. G. Tournilhac, and S. Kumar, *Phys. Rev. E* **55**, 4382 (1997).

93. S. Pensec and F. G. Tournilhac, *Chem. Commun.* 441 (1997).

94. E. Dietzmann, W. Weissflog, S. Marscheffel, A. Jakli, D. Lose, and S. Diele, *Ferroelectrics* **180**, 341 (1996).

95. G. Heppke, A. Jakli, D. Krüerke, C. Löhning, D. Lötzsch, S. Paus, S. Rauch, and N. K. Sharma, ECLC97, Zakopane, Poland.

96. W. Weissflog, Ch. Lischka, I. Benne, T. Scharf, G. Pelzl, S. Diele, and H. Kruth, ECLC97 Abstract Book and Proceedings of ECLC97, SPIE Vol. 3319, 14–19 (1998).

97. T. Sekine, T. Niori, J. Watanabe, T. Furukawa, S. W. Choi, and H. Takezoe, *J. Mater. Chem.* **7**, 1307 (1997).

98. J. W. Goodby, M. A. Wauch, S. M. Stein, E. Chin, R. Pindak, and J. S. Patel, *Nature* **337**, 449 (1989).

99. T. Sekine, T. Niori, M. Sone, J. Watanabe, Suk-Won Choi, Y. Tahanishi, and H. Takezoe, *Jpn. J. Appl. Phys.* **36** (1997).

100. A. G. Khachaturyan, *J. Phys. Chem. Solids* **36**, 1055 (1975).

101. T. Sekine, Y. Takanishi, T. Niori, J. Watanabe, and H. Takezoe, *Jpn. J. Appl. Phys.* **36**, L 1201 (1997).

102. J. Prost, in *Symmetries and Broken Symmetries*, N. Boccara ed., 1981, p. 159.

103. H. Bock and W. Helfrich, *Liq. Cryst.*, **12**, 697 (1992).

104. G. Scherowsky and X. H. Chen, *Liq. Cryst.* **17**, 803 (1994).

105. H. Bock and W. Helfrich, *Liq. Cryst.* **18**, 387 (1995); T. S. Perova, J. K. Vij, and H. Bock, *Mol. Cryst. Liq. Cryst.* **263**, 293 (1995).

106. G. Heppke, D. Lötzch, M. Müller, and H. Sawade, *6th International Conference on Ferroelectric Liquid Crystals*, Brest, France, July 20–24, 1997.

107. G. Scherowsky and X. H. Chen, *J. Mat. Chem.* **5**, 417 (1995).

108. H. Bock and W. Helfrich, *Liq. Cryst.* **18**, 707 (1995).

109. H. Zimmermann, R. Poupko, and Z. Luz, and J. Billard, Z. *Naturforsch.* **40a**, 149 (1985).

110. J. Malthête and A. Collet, *Nouv. J. de Chimie* **9**, 151 (1985); L. Lei, *Mol. Cryst. Liq. Cryst.* **146**, 41 (1987); R. Poupko, Z. Luz, N. Spielberg, and H. Zimmermann, *J. Am. Chem. Soc.* **111**, 5094 (1989).

111. B. Xu and T. M. Swager, *J. Am. Chem. Soc.* **115**, 8879 (1993).

112. A. Jakli, A. Saupe, G. Scherowsky, and X. H. Chen, *Liq. Cryst.* **22**, 309 (1997).

113. R. Berardi, S. Orlandi, and C. Zannoni, *J. Chem. Soc., Faraday Trans.* **93**, 1493 (1997).

114. D. L. Chandani, Y. Ouchi, H. Takezoe, A. Fukuda, K. Terashima, K. Furukawa, and A. Kishi, *Jap. J. Appl. Phys.* **28**, L-1261 (1989).

115. P. Cluzeau, H. T. Nguyen, C. Destrade, N. Isaert, P. Barois, and A. Babeau, *Mol. Cryst. Liq. Cryst.* **260**, 69 (1995).

116. N. Hiji, D. L. Chandani, S. Nishiyama, Y. Ouchi, H. Takezoe, and A. Fukuda, *Ferroelectrics* **85**, 99 (1988); H. Takezoe, J. Lee, D. L. Chandani, E. Gorecka, Y. Ouchi, A. Fukuda, K. Terashima, and K. Furukawa, *Ferroelectrics* **114**, 187 (1991).

117. A. Fukuda, Y. Takanishi, T. Isozaki, K. Ishikawa, and H. Takezoe, *J. Mater. Chem.* **4**, 997 (1994).

118. Y. Galerne and L. Liebert, *Abs. 2nd Int. Conf. Ferroelectric Liq. Cryst.*, Göteborg, 1989, p. O27.

119. Y. Galerne and L. Liebert, *Phys. Rev. Lett.* **66**, 2891 (1991); Ch. Bahr and D. Fliegner, *Phys. Rev. Lett.* **70**, 1842 (1993).

120. T. Isozaki, T. Fujikawa, H. Takezoe, A. Fukuda, T. Hagiwara, Y. Suzuki, and I. Kawamura, *Phys. Rev. B* **48**, 13439 (1993).

121. D. D. Parghi, S. M. Kelly, J. W. Goodby, *6th International Conference on Ferroelectric Liquid Crystals*, Brest, France, July 20–24, 1997.

122. I. Nishiyama and J.W. Goodby, *J. Mat. Chem.* **2**, 1015 (1992).

123. E. A. Soto Bustamante, S. V. Yablonskii, B. I. Ostrovskii, L. A. Beresnev, L. M. Blinov, and W. Haase, *Liq. Cryst.* **21**, 829 (1996).

LARGE ELECTROCLINIC EFFECT AND ASSOCIATED PROPERTIES OF CHIRAL SMECTIC A LIQUID CRYSTALS

R. SHASHIDHAR, J. NACIRI, AND B. R. RATNA

Naval Research Laboratory, Washington, D.C., U.S.A.

CONTENTS

I. INTRODUCTION

Garoff and Meyer [1,2] first demonstrated that when an electric field is applied to a Smectic A liquid crystal composed of chiral molecules along the layer plane, the transverse dipole of the molecules couple to the electric field and tilt the molecules in a plane perpendicular to the electric field direction (Fig. 1). This field-induced tilting of molecules is known as the electroclinic effect (ECE). It is also referred to as the "soft mode" in analogy with the softening of a vibration mode near the paraelectric–ferroelectric transition in solid ferroelectrics like barium titanate. The ECE was demonstrated as a pretransition effect in the smectic A phase near the smectic A–smectic C*

Advances in Liquid Crystals: A Special Volume of Advances in Chemical Physics, Volume 113, edited by Jagdish K. Vij. Series Editors I. Prigogine and Stuart A. Rice.
ISBN 0-471-18083-1. © 2000 John Wiley & Sons, Inc.

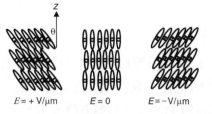

Figure 1. Schematic diagram of the electroclinic effect in a smectic A liquid crystal. The molecules are oriented normal to the layer plane in the absence of electric field. The molecules tilt with the application of an electric field (in a direction perpendicular to the plane of the paper). The tilt direction is reversed on changing the E-field direction.

transition. It can be described, in its simplest form, in terms of a Landau phenomenological description with a free energy (F) that is expressed in terms of the tilt (θ) as the order parameter and including a term to account for the coupling of the tilt to the electric field.

$$F = 1/2\, a\theta^2 + 1/4\, b\theta^4 - sE\theta, \qquad (1)$$

where a, b, and s are the Landau coefficients. For small values of θ, the θ^4 term can be neglected. Then the equilibrium value of θ, obtained by minimizing Eq. (1), is

$$\theta = sE/\alpha(T - T_c) \qquad (2)$$

Here s is the chiral coupling coefficient and $a = \alpha(T - T_c); T_c$ is the smectic A–smectic C* phase transition temperature and α is a parameter that determines the nature of this transition.

The amount of tilt induced for a given field can be defined in terms of the electroclinic coefficient $(e_c) = \theta/E = s/\alpha(T - T_c)$.

Excellent reviews by Clark and Lagerwall [3] and by Lagerwall et al. [4] have covered all aspects of the physics of ECE. Since those reviews, molecular engineering of electroclinic materials has made considerable progress, leading to the development of materials with large electroclinic coefficients. The aim of the present article is to review these recent developments.

II.   DEVELOPMENT OF MATERIALS WITH LARGE ELECTROCLINIC COEFFICIENTS (ECE)

The combination of the fast electro-optic response time and inherent analog capability makes the electroclinic liquid crystal materials very attractive for a wide variety of applications. However, in order that these materials be useful for any device, it is necessary to have large field-induced tilt angles (at least 22.5°) at sufficiently low voltages. It is also important to have sufficiently fast switching times at ambient temperatures. Although the ECE has been known for a long time, materials exhibiting this combination of properties did not exist until recently. This was due to the fact that the electroclinic effect is essentially a pretransition effect. The tilt fluctuations in the smectic A phase gain strength as the transition to the smectic C* phase is approached [1,2]. Consequently, induced tilt angles of a few degrees were observed only very close to the smectic A–smectic C* phase transition temperature. The first report of a large electroclinic effect appears to be due to Nishiyama et al. [5] who reported induced tilt angles of about 16° for applied voltages of 20 V/micron in the smectic A phase of MHPOBC close to the transition to the smectic C* phase. However, the smectic C* – smectic A transition temperature in this material was too high at 122°C. The next important step in material development was due to Walba et al. [6–8] and Andersson et al. [9]. They reported induced tilt angles greater than 20°C in the smectic A phase of some mixtures around ambient temperature. However, the electric field strengths required to achieve such large tilt angles were still very high (20–40 V/µm). The observation of Williams et al. [10] was unusual in that they observed a pronounced electroclinic effect in W317, a material that exhibits only a smectic A phase, with no smectic C* phase at a lower temperature. Further, the large tilt angles were observed over a wide range of temperatures (20–60°C) and they were only weakly temperature dependent (Fig. 2). This remarkable property made the material W317 most attractive from the application point of view. However, the large electric fields required to achieve sufficient tilt angles made the material unsuitable for most applications, especially for silicon-based active matrix devices.

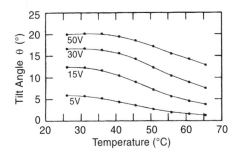

Figure 2.   Electroclinic tilt vs. temperature for various applied voltages in the smectic A Phase of W317. From Ref. [10].

Recognizing the importance of the results for W317, research groups at the University of Colorado at Boulder and at the Naval Research Laboratory (NRL), Washington, DC pursued the synthesis of more materials that are variants of the structure of W317. Walba et. al., [11] observed that simple homologues of W317 did not exhibit a large electroclinic effect or the desired smectic A phase at room temperatures. The next study was therefore aimed at developing new achiral tails for lowering melting points and for broadening the liquid crystal phase ranges. The idea was to incorporate fatty acid-derived tails with cis double bonds to lower the melting points, an approach used successfully in other classes of materials by Kelley [12]. Indeed all these materials (Fig. 3) did exhibit wide smectic A phase range and large induced tilt angles, with electroclinic coefficients >1° Vμm^{-1}. Around the same time, the NRL group also reported electroclinic coefficients of 1.2–1.4°/V/μm^{-1} in materials, abbreviated as KNnm [13–16], possessing a chemical structure similar to W317. The NRL group also studied the effect of varying the chain length (n) at the nonchiral end of the

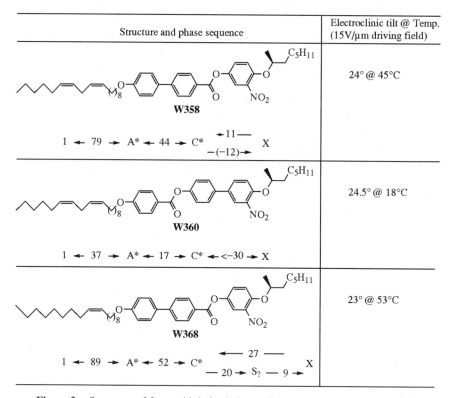

Structure and phase sequence	Electroclinic tilt @ Temp. (15V/μm driving field)
W358 1 ← 79 → A* ← 44 → C* ←11— —(−12)→ X	24° @ 45°C
W360 1 ← 37 → A* ← 17 → C* ← <−30 → X	24.5° @ 18°C
W368 1 ← 89 → A* ← 52 → C* — 27 — — 20 → S? — 9 → X	23° @ 53°C

Figure 3.   Structures of fatty acid-derived electroclinic materials. From Ref. [11].

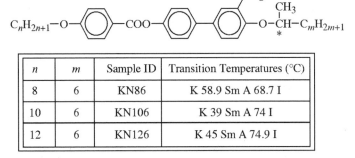

n	m	Sample ID	Transition Temperatures (°C)
8	6	KN86	K 58.9 Sm A 68.7 I
10	6	KN106	K 39 Sm A 74 I
12	6	KN126	K 45 Sm A 74.9 I

Figure 4. Summary of transition temperatures for the even members of the KNnm series. From Ref. [13].

molecule on the electroclinic coefficient, while keeping the alkyl chain length ($m = 6$) near the chiral part unchanged (Fig. 4). The induced tilt angle data (Fig. 5) shows a strong dependence on the length of n; KN126 showing the largest ($1.4°/V\mu m^{-1}$) electroclinic coefficient at about 30°C [13]. Essentially the same result was observed [16] for $m = 5$ also. Crawford et al. [15] also studied the effect of varying m, by keeping n fixed and found that the larger value of m also leads to a larger induced tilt angle. The value of the induced tilt for KN125 is about 15° (for applied field of 10 V/μm) compared to 4° for KN123 [15]. Hence a longer chain length on the chiral or achiral part of the molecule, always results in a large electroclinic tilt angle.

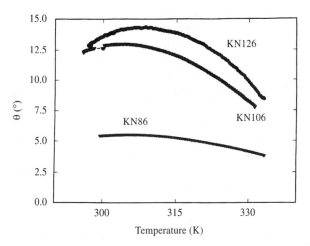

Figure 5. The electroclinic tilt angle θ as a function of temperature for the KNn6 series recorded for an applied electric field strength of 10 V/μm. From Ref. [13].

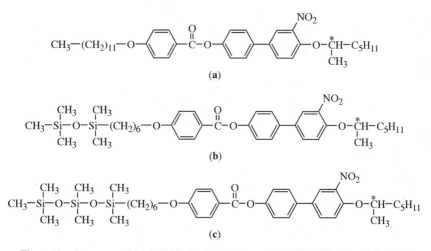

Figure 6. Structures of (a) KN125, (b) DSiKN65, and (c) TSiKN65. From Ref. [19].

A significant improvement in the electroclinic coefficient has been achieved in a new class of organosilane materials developed by Naciri et al. [17]. The feature of the molecular structure in these materials, which are structurally similar to those reported by Coles et al. [18], is that one or more siloxy groups are attached to the hydrocarbon chain at the nonchiral end of the molecule (Fig. 6). The substitution of the siloxy group leads to a very low melting point ($<5°C$). It also leads to the highest electroclinic coefficients (EC) reported so far for any materials [17,19]. TSiKN65, with three siloxy groups, exhibits an electroclinic coefficient of $5°/V\mu m^{-1}$ while DSiKN65 (with two siloxy groups) shows a value of $4°/V\mu m^{-1}$ (Fig. 7). The data for KN125 is also shown in Figure 7 for comparison. Thus, addition of the siloxy groups leads to a 4- to 5-fold increase in the electro-clinic coefficient. Coupled with the wide temperature range of the smectic A phase, these materials exhibit properties that are very attractive for applications.

Figure 8 shows the response time (τ) vs. electric field (E) at different temperatures in the smectic A phase of KN125. At higher temperatures, τ is independent of E as expected from Landau theory [20]. However, at low temperatures, τ shows a sharp increase with decreasing field. It may be recalled that such a behavior of τ is associated with the critical slowing down on approaching the smectic A–smectic C* transition [20]. We can therefore infer that KN125 has a "virtual" smectic C* phase (at lower temperatures) whose occurrence is precluded by crystallization or a higher ordered phase. A similar deduction has been made for the behavior of W317 on the basis of dielectric spectroscopic studies [21].

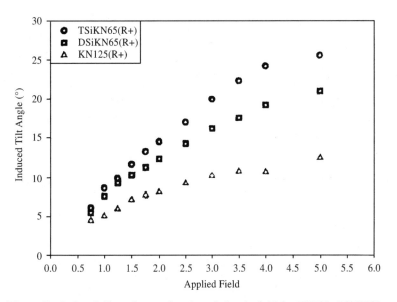

Figure 7.   Induced tilt angle as a function of electric field for KN125, DSiKN65, and TSiKN65 From Ref. [19].

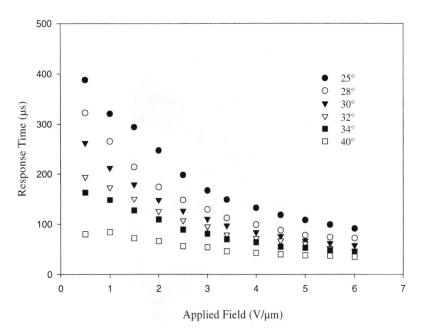

Figure 8.   Field dependence of soft mode response time at different temperatures in the smectic A phase of KN 125.

An important question that arises from these studies is "What is the reason for such a large electroclinic tilt in the KNnm and SiKNnm series of materials?" It must be pointed out that a common feature of the chemical structure of the molecules of all the compounds mentioned above exhibiting the large ECE, is that the chiral center is in close proximity to the large transverse (NO_2) dipole group. It is conceivable that this increases the steric hindrance to the rotation of the NO_2 group which in turn makes the coupling of the dipole moment to the applied field more efficient resulting in a large induced tilt angle.

III. LAYER BUCKLING AND ASSOCIATED EFFECTS

It has been observed that electroclinic materials with large induced tilt exhibit a "stripe texture" (Fig. 9) when observed in a planar configuration under crossed polarizers [10,22]. It has been suggested that these stripes arise from the reduction of the layer spacing due to tilting of the molecules and the consequent buckling of the smectic layers. Detailed X-ray studies conducted under in situ electric fields have shown that this indeed is the case [23,24]. These results as well as optical and light scattering studies on stripes are discussed in the following sections.

Figure 9. Optical microscopic picture of the stripe texture seen in the smectic A phase of KN125.

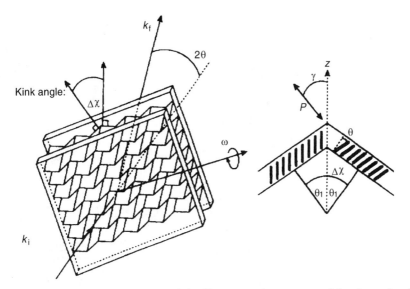

Figure 10. Scattering geometry of the X-ray scattering set-up used for the study of buckled layers. The θ, ω and χ-scan directions shown enabled probing of the layer thickness, chevron texture, and kink angle of the buckle layers, respectively. From Ref. [23].

A. X-Ray Studies of the Striped Texture

X-ray studies have been carried out on two materials—KN125 [23] and W317 [24]. The main conclusion of the two studies is the same (i.e. "rigid-rod-like" tilting of the molecules under an electric field). We will discuss the results here on KN125. The sample was aligned in a bookshelf geometry between transparent conducting glass electrodes. The cell was mounted on a 4-circle goniometer and the experiments performed at the National Synchrotron Light Source (Brookhaven). The scattering geometry is shown in Figure 10. The following experimental scans were conducted for each electric field to determine the orientation of the layer in the striped domains. The smectic mass density wave vector $q(= 2\pi/d)$, where d is the layer thickness, was probed to determine the change in the layer spacing at different electric fields. A rocking scan in the ω-direction enabled probing of the Chevron distortions while the χ-angle scans were used to probe the direction of the smectic mass density wave in the plane of the cell.

The stripe deformation as probed by the χ-scan is shown in Figure [11(a)] for two values of the electric field. For zero field a broad peak is seen while for large field ($E = 10$ V/μm), the χ-scan splits into two peaks. The value of $\Delta\chi$ which is the difference in the position of the two χ-scan peaks, gives the "kink angle" of the stripe deformation. ω-scans taken for the same E-field

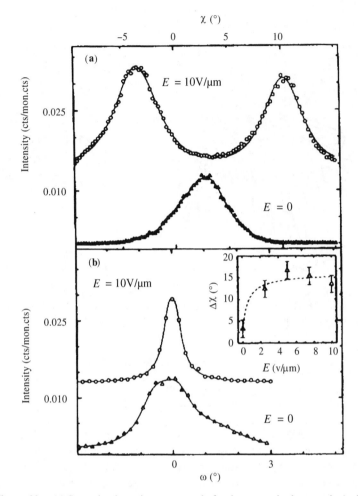

Figure 11.   (a) Scattering intensity versus angle for the χ-scan in the smectic A phase of KN125. Top figure: Applied field of 10 V/μm, Bottom figure: No electric field. (b) Scattering intensity versus angle for the ω-scan in the smectic A phase of KN125. Top Figure: 10 Vs/μm, Bottom figure: No applied field. From Ref. [23] The inset shows the variation of kink angle as a function of E-field.

values showed a single peak in both cases [Fig. 11(b)] indicating thereby that the chevron layer distortion is absent here. The presence of the stripes can have a significant effect on the intensity of light transmitted through the sample and hence on the values of the optical tilt angle.

When the bookshelf structure of smectic A is uniform, the optical tilt angle can be derived from the well-known Lee–Patel formalism [25]. However, when the layers are distorted (buckled), evaluation of the true

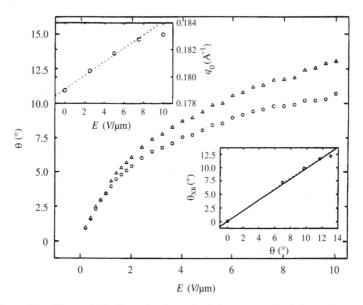

Figure 12. The optical tilt angle determined from transmitted intensity assuming bookshelf (circles) geometry and stripe (triangles) domains. For low values of $[E]$, the true tilt angle, from the striped domain, corresponds to the value obtained from the bookshelf model, while at higher fields there is a significant deviation. The variation of $|q_0|$ as a function of electric field (top inset). The optical tilt angle vs. the X-ray tilt angle with a line of slope one (bottom inset). From Ref. [23].

optical tilt angle is more complicated. For this purpose the layer profile modulation, as obtained by the X-ray scattering data, was used to obtain the true tilt angles from optical transmission measurements, averaging the intensity of adjacent domains. These data are shown in Figure 12. For low fields the optical tilt angle corresponds to that derived from the bookshelf model [23] but at high fields, there is a significant deviation.

Another important result emerges from these studies. The plot of optical tilt angle (evaluated as described above) versus the X-ray tilt angle shows a line of slope unity (see bottom inset of Fig. 12). This implies that the molecules tilt as rigid rods. Exactly similar results have been obtained by Rappaport et al. [24] whose work on W317 was reported at the same time as Crawford et al. [23]. This rigid rod-like tilting makes the layer buckling extremely pronounced when the induced tilt angles are large.

A detailed quantitative analysis of the profile of the buckled layers has been carried out by Geer et al. [26]. They followed the position and shape of the χ-scans of the X-ray scattering for several values of the electric field ranging from 1 to 14 V/μm (Fig. 13). With increasing field, the peak separation increases with a concomitant decrease in the intensity of

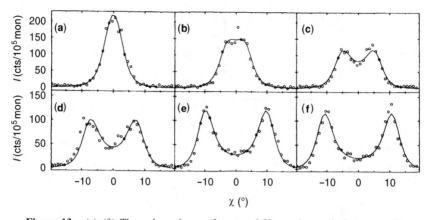

Figure 13. (a)–(f) The x-dependence of scattered X-rays (open circles) at the Bragg reflection at $E = 1, 2, 4, 7, 10$, and 14 V/μm, respectively. The evolution of the peak separation is clearly evident. Solid lines represent best fits to the model described in the text. Agreement is very good for all values of E. From Ref. [26].

scattering at the $\chi = 0$ position. Except at the highest E, the scattered intensity for the $\chi = 0°$ position exceeds the background scattering expected from a simple superposition of the two separated peaks. This implies that a simple soliton picture [22] is not adequate to describe the layer profile at different fields. For a better comparison with theory, a continuum free energy for larger distortion has been developed. The essence of this approach is to express the free energy in terms of $U(r)$, the displacement of the layers from their equilibrium position. Since the layer modulation is observed only along the x-direction, the free energy is minimized over displacement functions of a simple variable $U(x)$. This leads to an Euler–Lagrange equation in terms of $w = \partial u/\partial x$. One solution of this was used by Pavel and Glagorova [22]. However, this model was not able to account for the scattering near $\chi = 0$ for most of the E-values. To analyze the data for all of the fields, a general minimization of the free energy was performed in a manner analogous to that used by Singer [27] in another context. The approach was essentially to perform nonlinear least squares fits to the data for different values of E simultaneously, with three field-independent parameters, namely, the length $\lambda (= (K/B)^{1/2}$ where K and B are the splay and compressive elastic constants), coefficient C (that relates the dilative strain to the field E), and the layer mosaicity. The fitting results (solid lines in Fig. 13) show that the agreement with experiment is excellent. The angular profile $\omega(x)$ and the layer displacement profiles $u(x)$ derived from the fits described above are shown in Figure 14. An important result that emerges from this figure is that the layer profile evolves continuously from a low-amplitude sinusoidal profile at low fields to a high amplitude soliton (triangular) profile at high fields.

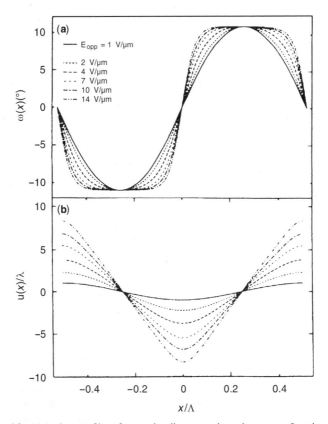

Figure 14. Angular profile of smectic director orientation as a function of E, as determined from analysis of the data in Figure 13. The narrowing of the defect or kink region with increasing E is clearly evident. (b) Corresponding layer-buckling profile. An evolution from a low-amplitude sinusoid at low fields to a high-amplitude soliton profile at high fields can be seen. From Ref. [26].

B. Optical Studies on Striped Domain Textures

It is to be expected that the stripes can affect the contrast ratio. To understand these effects a spatially resolved optical characterization of the stripes has been carried out [28,29] using a focused laser beam of size less than the width of a single stripe and the data compared with those obtained by using a much larger beam covering several stripes. The extinction (I_{min}/ I_{max}) ratio, measured by rotating the cross-polarizers in tandem with respect to the optic axis of the sample sandwiched between them, is shown in Figure 15 as a function of applied field. Contrary to what is expected the extinction is nearly the same regardless of the beam size relative to the stripe width.

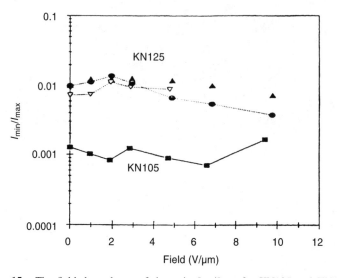

Figure 15. The field dependence of the ratio I_{min}/I_{max}, for KN125 and KN105. The squares represent large beam measurements on KN105. The triangles represent large beam data on KN125. The inverted triangles and the circles represent small beam data on KN125. From Ref. [28].

These results suggest that the optical contrast is affected by an angular distribution of the molecular director on a scale much smaller than the stripe width. This is consistent with the results of X-ray scattering experiments described in the previous section. The contrast ratio, which is the inverse of the ratio shown in Figure 15, is an order of magnitude larger for KN105 compared to KN125. KN105 being a lower homologue, has a much lower tilt angle and hence a much weaker stripe deformation. The large contrast ratio measured in KN105 (in spite of its low electroclinic coefficient) is a clear indication of the detrimental effect the layer buckling can have on the electro-optics of a device.

The effect of sample thickness and electric field on the spatial periodicity of the stripes has also been investigated (Fig. 16). The stripe periodicity is shown to scale as twice the thickness of the cell, while it is invariant with respect to the electric field [30]. The former result clearly establishes that the stripe formation in KN125 is indeed due to the geometric factors associated with the tilting of the molecules and the shrinkage in layer spacing.

More recently, a second set of stripes with a spatial frequency that is higher than that of the stripes described earlier has been observed [30,31] in cells thicker than 2 μm for KN125. Although a model based on surface-induced effects has been proposed [30] to explain the stripes, alternate explanations cannot be ruled out.

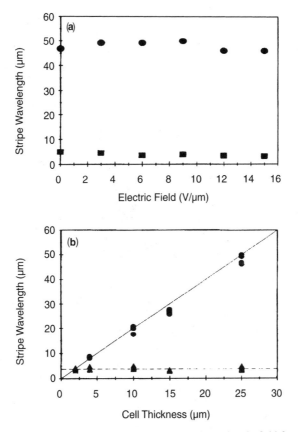

Figure 16. (a) The measurement of stripe wavelength vs electric field for a 25 μm cell, (b) the measured stripe wavelength Λ vs cell thickness D. Data for all electric fields that were studied are shown together, because the wavelength is found to be independent of electric field. The solid line indicates the theoretical predictions $\Lambda = 2D$ for the longer-wavelength stripes; the dashed line indicates the constant value of $\Lambda = 3.8\,\mu m$ for the shorter-wavelength stripes. From Ref. [30]

C. Dynamic Light-Scattering Study of Stripes

The optical stripes of KN125 break the in-plane symmetry of the usual smectic A. The resulting structure has a symmetry of the rectangular columnar phase [32]. Sprunt et al. [33] reported the first light-scattering study of a chiral smectic A with a pronounced modulation of both the layer orientation and director. The modulated structure is shown to have a profound influence on the fluctuation modes. Data for the normalized autocorrelation function vs. time (Fig. 17) shows two decays associated with overdamped fluctuations that are separated in time scales by 3 orders of

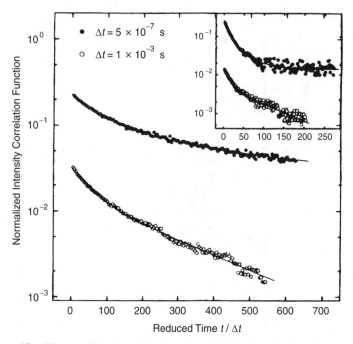

Figure 17. The normalized intensity–intensity correlation function for 15 μm samples of KN125 (main figure) and the lower homolog KN105 (inset) at 25°C and $\theta_s = 24°$. The solid and open circles represent separate measurements taken at different sample times. The solid lines are doubled-exponential fits (see text). From Ref. [33].

magnitude (1 to $10\,S^{-1}$ and 10^3–$10^4\,S^{-1}$). Each decay can be fitted to a double exponential. Hence the layer-modulated chiral smectic A phase exhibited four overdamped fluctuation modes in a thin cell—a pair at low frequency and another pair at high frequency. In contrast, the uniform, nonchiral smectic A in a bulk geometry exhibits only one mode at time scales $> 10^5\,S^{-1}$ [34].

To explain the observed dynamics of the modulated structure, a phenomenological model has been developed. The layer modulation may be described (Fig. 18) by two parameters—a periodicity d_x and a kink angle α. The vertical dashed lines indicate "domain walls" of the modulation. The fluctuating degrees of freedom of the system can hence be taken as the director **n**, the smectic layer displacement (U_z) about an equilibrium layer spacing and the stripe wall displacement U_x (about d_x). Considering the rectangular symmetry in the x–z plane, the bulk elastic free-energy density has been developed taking into account the anisotropy in the elastic constant D and a difference in structural characteristics of the surface and bulk

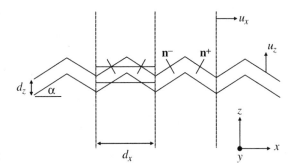

Figure 18. Schematic of the layer modulation in the chiral smectic A phase. The xz-plane projection of a unit cell for the structure is outlined as a rectangle (dimensions d_x and d_y). Along the x direction, domain walls of the layer modulation, indicated by dashed lines bound the unit cell. u_z and u_x are fluctuating components of the smectic layer and stripe domain wall displacement, respectively. From Ref. [33].

regions. Data on frequencies of the overdamped modes plotted against the scattering angles are shown in Figure 19. The solid lines are the fits to the equations for the fast (U_2) and slow (U_1) modes as given by the continuum model [33] which takes into account both bulk and surface components.

In summary, the dynamic light-scattering study of the modulated smectic-A shows novel, low frequency hydrodynamic mode which restores the symmetry of the flat layers. Also observed is an anomalously slow non-hydrodynamic director fluctuation mode. Both these modes are split into two components due to the bulk and surface contributions.

IV. OTHER SYMMETRY-BREAKING EFFECTS

The field-induced tilt breaks the rotational symmetry in a smectic A liquid crystal and is expected to be accompanied by other (field-induced) symmetry-breaking effects that arise from the interaction of the applied field with the transverse dipole of the molecule. Several such symmetry-breaking effects have been observed. These are discussed in this section.

A. Dielectric Biaxiality

The first observation of field-induced biaxiality in a smectic A was by Kimura, Akahave, and Kobayashi [35] in 764E, near the smectic A–smectic C* transition. For quantitative evaluation of the biaxiality, the principal dielectric constant for a EC liquid crystal can be defined [36,37] by ε_3 parallel to the director **n**, ε_2 parallel to the applied field and ε_1, normal to these two axes (Fig. 20). In the absence of an electric field, the dielectric

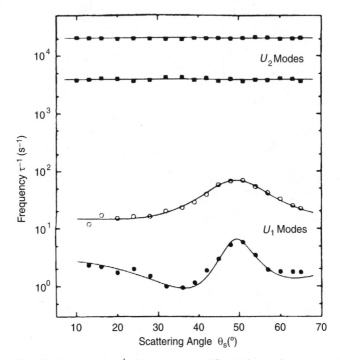

Figure 19. The frequencies τ^{-1} of the overdamped fluctuation modes vs. scattering angle θ_s for the 15 μm sample of KN125 at 25°C. The solid lines are fits to the model for the fast (U_2) modes (filled squares), and for the slow (U_1) modes, where filled and open circles correspond to bulk and surface components, respectively. From Ref. [33].

constants measured in the planar geometry is $\varepsilon_\perp = (\varepsilon_1 + \varepsilon_2)$. In the presence of an electric field, the degeneracy is removed and the measured dielectric constant will correspond to ε_2. These measurements enabled the determination of the field-induced biaxiality $\partial\varepsilon = \varepsilon_2 - \varepsilon_1$. The temperature dependence of $\partial\varepsilon$ is shown in Figure 21. The same figure also shows the temperature-dependence of the field-induced tilt angle (θ). The similarity in the behavior of θ and $\partial\varepsilon$ are explained quantitatively in terms of a Landau-type free energy including a term which represents the effect of induced dielectric biaxiality.

B. Optical Biaxiality

More recently field-induced optical biaxiality in chiral smectic A has been reported [38,39]. The material studied was KN125 [16]. Optical transmission between crossed polarizers was measured (for the sample in a bookshelf geometry) as a function of angle between the optic axis and the polarization

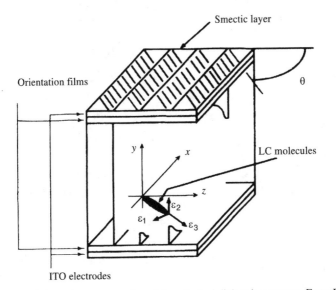

Figure 20. Schematic representation of the principal dielectric constants From Ref. [35].

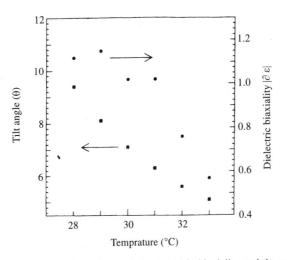

Figure 21. Temperature dependence of the dielectric biaxiality and the molecular tilt in the smectic A Phase of 764 E. From Ref. [35].

direction for different values of the applied electric field (Fig. 22). Two important features of these data are: (1) The shift in the angular position of the minima and maxima with electric field as a measure of the induced tilt angle (θ) and (2) The change in the amplitude of the transmission maximum.

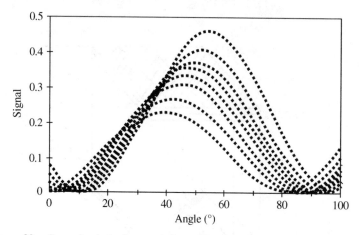

Figure 22. Crossed-polarized transmission as function of polarization angle for KN125, subjected to electric fields of 0.0V (lowest data set) 1.0 V/μm, 5.2 V/μm, 7.3 V/μm and 10.5 V/μm (highest data set) From Ref. [38].

The increase in the amplitude with increasing field gives $\delta(\Delta n)$ which in turn associated with induced optical biaxiality. As in the case of dielectric biaxiality [35], the tilt angle (θ) and $\partial(\Delta n)$ show similar field dependence (Fig. 23). It should also be mentioned that field-induced birefringence (as opposed to biaxiality) could possibly be playing a role in these studies and should be taken into account.

C.   Second Harmonic Generation in the Smectic A Phase

The Ferroelectric smectic C^* phase has C_2 symmetry and has been shown to exhibit second harmonic generation (SHG) [40–44]. The electroclinic smectic A in the presence of an electric field is also expected to show induced SHG. Kobayashi et al. [45] did observe SHG in the smectic A phase of a material exhibiting a large electroclinic effect. The angular dependence of the SHG signal has been measured for different applied fields (Fig. 24). A strong SHG signal is seen at 5 kV/cm at an angle of 20–25°, which is interpreted as being the phase-matching angle. The phase-matching angle itself shifts to higher angles and the amplitude of the SHG signal decreases with decreasing field. In fact, the strength of the SHG signal exhibits an E^2-dependence. Another important result is that in the same material, the SHG signal in the smectic C^* phase shows a saturation at 0.04 V/μm which corresponds to the unwinding of the helical structure. On the other hand SHG in the smectic A phase shows a monotonic dependence on E (Fig. 25). It should also be mentioned that in the angular dependence of SHG measurement, extraordinary waves have been used for the fundamental and

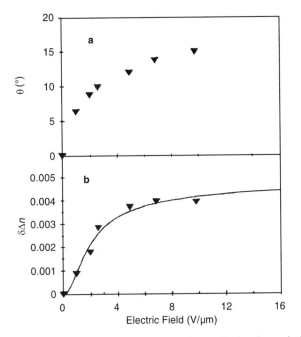

Figure 23. Variation of (a) the optical tilt angle θ and (b) the change in birefringence $\partial(\Delta n)$ as a functional of applied electric field in the smectic A phase of a $10\,\mu$m-thick KN125 liquid crystalline sample. The curve in (b) is a fit of the theoretical curve based on a simple molecular model involving reorientation about the long molecular axis to the experimental data. From Ref. [38].

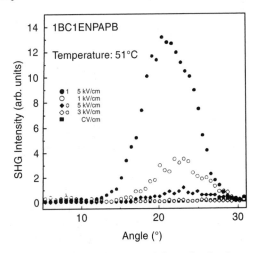

Figure 24. Angular dependence of the SHG intensity in the smectic A phase of 1BC1ENPABB. From Ref. [45].

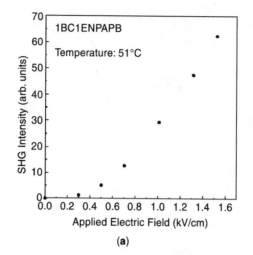

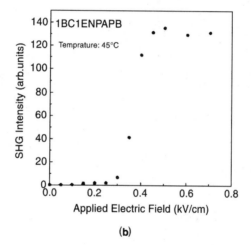

Figure 25. Applied field dependence of SHG peak intensities in the smectic A phase (a) and the smectic C* phase (b). (c) E^2 dependence of SHG peak intensity in the smectic A phase From Ref. [45].

ordinary waves for the SHG, confirming thereby type I phase-matching. The value of the effective nonlinear coefficient d_{eff} in the smectic A phase, estimated under the condition of phase matching comes out to be about 0.01 pm/V. This value is a factor of 5 smaller than that in the smectic C* phase of the same material.

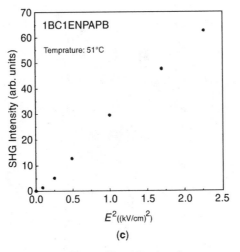

Figure 25. *(Continued)*

V.   APPLICATION OF ELECTROCLINIC LIQUID CRYSTALS

The soft mode electro-optic effects associated with electroclinic materials has attracted considerable attention due to their analog gray-scale capability and short response times. The different aspects of ELCs that are relevant to the use in electrically addressed spatial light modulator (SLM) have been addressed in detail by Davey and Crossland [46]. The use of ECE in optically addressed SLM was demonstrated by Collings et al. [47]. Andersson et al. [48] have discussed a variety of electro-optic components and devices that can be constructed utilizing the electroclinic material in a bookshelf geometry. In particular, multiple electroclinic cells with potential to give achromatic full-modulation light valves as well as electrically controlled high-speed color filters, reflective single-cell devices suitable for integration with semiconductors to give high speed SLMs, and general optical processing components have been discussed [48]. Tunable filter capable of spatial discrimination, has been demonstrated by Sharp et al. [49]. Such filters have application in optical signal processing, wavelength division multiplexing/demultiplexing, and colorimetry and full color displays. Using an electrically addressed SLM the video rate operation as well as the optical transmission of full color NTSC video signal has been shown by Okada et al. [50]. Johnson and co-workers have demonstrated a variety of other devices—a high-speed refractive index modulator using a homeotropically aligned EC material [51], high-speed analog complex-

Figure 26. A line element array patterned in ITO showing continuously varying gray shade in successive lines. These lines have been addressed with a continuously varying square wave. The cell gap is $2\,\mu m$. The optic axis is oriented with respect to the polarization direction such that the line appears bright (the top lines in the two repeated patterns). The dark state at the bottom of the patterns corresponds to a 4V square wave.

amplitude SLM using a planar geometry [52] and a high-speed continuously tunable filter for wavelength division multi-access optical network [53]. All these device-related work has been carried out using materials which do not have a large electroclinic coefficient. Recently the feasibility of obtaining 256 gray levels in a material with large electroclinic coefficent was shown and the gray scale capability was demonstrated using a 1×64 array device (Fig. 26) driven by a dc balanced circuit [54]. Use of such materials will lead to a simplification of the devices in the case of stacked filters and to a wider dynamic range that is important for all SLM applications. Efforts in this direction have just begun.

References

1. S. Garoff and R. B. Meyer, *Physical Review A* **18**, 2739–40 (1978).

2. S. Garoff and R. B. Meyer, *Physical Review A* **19**, 338–47 (1979).

3. N. A. Clark and S. T. Lagerwall, in *Ferroelectric Liquid Crystals: Principles, Properties and Applications*, Vol. 7, J. W. Goodby, et al. (eds.), Gordon and Breach, Amsterdam, 1991, pp. 1–94.

4. S. T. Lagerwall, M. Matuszczyk, P. Rodhe, and L. Odman, in *The Optics of Thermotropic Liquid Crystals*, S. Elston and R. Sambles (eds.), Taylor and Francis, London 1998, pp. 155–194.

5. S. Nishiyama, Y. Ouchi, H. Takezoe, and A. Fukuda, *Jap. J. Appl. Phys., Part 1*, L1787–11789 (1987).

6. D. M. Walba, M. B. Ros, N. A. Clark, R. Shao, K. M. Johnson, M. G. Robinson, J. Y. Liu, and D. Doroski, in *Materials for Nonlinear Optics: Chemical Perspectives*, 1991, p. 484. G. D. Stucky (ed.), American Chemical Society, Washington DC, 1991, p. 484.

7. D. M. Walba, M. B. Ros, N. A. Clark, R. Shao, M. G. Robinson, J. Y. Liu, K. M. Johnson, and D. Doroski, *J. Am. Chem. Soc.* **113**, 5471 (1991).

8. D. M. Walba, M. B. Ros, T. Sierra, J. A. Rego, N. A. Clark, R. Shao, M. D. Wand, R. T. Vohra, K. E. Arnett, and S. P. Velsco, *Ferroelectrics* **121**, 247 (1991).

9. G. Andersson, I. Dahl, L. Komitov, M. Matuszczyk, S. T. Lagerwall, K. Skarp, B. Stebler, *Ferroelectrics* **114**, 137–150 (1991).

10. P. A. Williams and N. A. Clark, *Ferroelectrics* **121**, 143 (1991).

11. D. M. Walba, D. J. Dyer, R. Shao, N. A. Clark, R. T. Vohra, K. More, W. N. Thurmes, and M. D. Wand, *Ferroelectrics* **148**, 435 (1993).

12. S. M. Kelly, *Liquid Crystals* **14**, 675 (1993).

13. B. R. Ratna, G. P. Crawford, S. K. Prasad, J. Naciri, P. Keller, and R. Shashidhar, *Ferroelectrics* **148**, 425–434 (1993).

14. B. R. Ratna, G. P. Crawford, J. Naciri, and R. Shashidhar, *Proc. SPIE Int. Soc. Opt. Eng.* **2175**, 79–86 (1994).

15. G. P. Crawford, J. Naciri, R. Shashidhar, P. Keller, and B. R. Ratna, *Mol. Cryst. Liq. Cryst.* **263**, 2155–2164 (1995).

16. G. P. Crawford, J. Naciri, R. Shashidhar, and B. R. Ratna, *Jpn. J Appl. Phys. Pt 1* **35**, 2176–2179 (1996).

17. J. Naciri, J. Ruth, G. Crawford, R. Shashidhar, and B. R. Ratna, *Chem. Mater.* **7**, 1397–1402 (1995).

18. H. J. Coles, H. Owen, J. Newton, and P. Hodge, *Liquid Crystals* **15**, 739 (1993).

19. K. Nelson, J. Naciri, P. Bey, D. Cuttino, and B. R. Ratna, *SPIE 3143*, 51 (1997).

20. I. Abdulhalim and G. Moddel, *Liq. Cryst.* **9**, 493–518 (1991).

21. P. A. Williams and N. A. Clark, *J. Appl. Phys.* **78(1)**, 413 (1995).

22. J. Pavel and M. Glogarova, *Liq. Cryst.* **9**, 87–93 (1991).

23. G. P. Crawford, R. E. Geer, J. Naciri, R. Shashidhar, and B. R. Ratna, *Appl. Phys. Lett.* **65**, 2937–2939 (1994).

24. A. G. Rappaport, P. A. Williams, B. N. Thomas, N. A. Clark, M. B. Ros, D. M. Walba, *Appl. Phys. Lett.* **67**, 362–364 (1995).

25. S. Lee and J. S. Patel, *Appl. Phys. Lett.* **54**, 1653–1655 (1989).

26. R. E. Geer, S. J. Singer, J. V. Selinger, B. R. Ratna, and R. Shashidhar, *Physical Review E* **57** 3059 (1998).

27. S. J. Singer, *Physical Review E* **48**, 2796 (1993).

28. J. R. Lindle, A. T. Harter, S. R. Flom, F. J. Bartoli, R. Shashidhar, and B. R. Ratna, *Proc. SPIE Int. Soc. Opt. Eng.* **2651**, 178–185 (1996).

29. J. R. Lindle, F. J. Bartoli, S. R. Flom, A. T. Harter, B. R. Ratna, and R. Shashidhar, *Appl. Phys. Lett.* **70**, 1536 (1997).

30. F. J. Bartoli, J. R. Lindle, S. R. Flom, B. R. Ratna, G. Rubin, J. V. Selinger, and R. Shashidhar, *Physical Review E* **58**, 5990 (1998).

31. A. Tang and S. Sprunt, *Physical Review E* **57**, 3050 (1998).

32. J. Prost and N. A. Clark, *Proc. Int. Conf. Liq. Cryst.*, Bangalore, 1979; S. Chandrasekhar, Ed., Heyden, Philadelphia, 1980. p. 53.

33. S. Sprunt, J. V. Selinger, G. P. Crawford, B. R. Ratna, and R. Shashidhar, *Phys. Rev. Lett.* **74**, 4671 (1995).

34. J. D. Litster, *Light Scattering Near Phase Transitions*, H. Cummins and A. P. Levanyuk (eds.) North Holland. New York 1983, p. 112.

35. M. Kimura, T. Akahane, and S. Kobayashi, *Jpn. J. Appl. Phys. Pt 1* **32**, 3530–3533 (1993).

36. J. Hoffmann, W. Kuczynski, J. Malecki, and J. Pavel, *Ferroelectrics* **76**, 61–67 (1987).

37. J. C. Jones, E. P. Raynes, and M. J. Towler, *Mol. Cry. Liq. Cryst.* **199**, 277 (1991).

38. F. J. Bartoli, J. R. Lindle, S. R. Flom, J. V. Selinger, B. R. Ratna, and R. Shashidhar, *Physical Review E* **55**, 1271 (1997).

39. J. R. Lindle, F. J. Bartoli, S. R. Flom, B. R. Ratna, and R. Shashidhar, *SID*, **28**, 719 (1997).

40. N. M. Shtykov, M. I. Barmik, L. M. Blinov, and L. A. Beresnev, *Mol. Cry. Liq Cryst.* **124**, 379 (1985).

41. A. Taguchi, H. Ouchi, H. Takezoe, and A. Fukuda, *Jpn. J. Appl. Phys.* **28**, L997 (1989).

42. M. Ozaki, M. Utsumi, Y. Uchiyama, and K. Yoshino, *Ferroelectrics* **148**, 337 (1993).

43. J. Y. Liu, M. G. Robinson, K. M. Johnson, D. M. Walba, M. B. Ross, N. A. Clark, R. Shao, and D. Doroski, *J. Appl. Phys.* **70**, 3426 (1991).

44. K. Schmitt, R. P. Herr, M. Schadt, J. Fünfschilling, R. Buchecker, X. H. Chen, and C. Benecke, *Liquid Crystals* **14**, 1735 (1993).

45. K. Kobayashi, T. Watanabe, S. Uto, M. Ozaki, K. Yoshino, M. Svensson, B. Helgee and K. Skarp, *Jpn. J. Appl. Phys., Part 2* **35**, L104–L107 (1996).

46. A. B. Davey and W. A. Crossland, *Ferroelectrics* **114**, 1–4 (1991).

47. N. Collings, W. A. Crossland, R. C. Chittick, and M. F. Bone, *Proc. SPIE* **46**, 963 (1988).

48. G. Andersson, I. Dahl, L. Komitov, S. T. Lagerwall, K. Skarp, and B. Stebler, *J. Appl. Phys.* **66**, 4983–4995 (1989).

49. G. D. Sharp, K. M. Johnson, H. J. Masterson, D. Doroski, *Ferroelectrics* **114**, 1–4 (1991).

50. H. Okada, K. Kurabayashi, Y. Kidoh, H. Onnagawa, K. Miyashita, *Jpn. J. Appl. Phys. Pt 1* **33**, 4666–4672 (1994).

51. A. Sneh, J. Y. Liu, and K. M. Johnson, *Optics Letters* **19**, 305–307 (1994).

52. G. D. Sharp and K. M. Johnson, *Optics Letters* **19**, 1228–1230 (1994).

53. A. Sneh and K. M. Johnson, *J. Lightwave Technol.* **14**, 1067–1080 (1996).

54. B. R. Ratna, H. Li, K. S. Nelson, J. Naciri, P. P. Bey, *SID Digest* **Vol. XXVIII**, 207 (1997).

PYROELECTRIC STUDIES OF POLAR AND FERROELECTRIC MESOPHASES

L. M. BLINOV

*Institute of Crystallography, Russian Academy of Sciences,
Moscow, Russia*

CONTENTS

Advances in Liquid Crystals: A Special Volume of Advances in Chemical Physics, Volume 113,
edited by Jagdish K. Vij. Series Editors I. Prigogine and Stuart A. Rice.
ISBN 0-471-18083-1. © 2000 John Wiley & Sons, Inc.

I. INTRODUCTION

There are certain single crystals that upon being dropped into a Dewar vessel with liquid nitrogen show effects of electrical breakdown. The crystal appears to be punctured by tiny lighting. Such crystals are pyroelectric, and the breakdown is a result of a thermally induced accumulation of charges on the opposite faces of the crystal perpendicular to its polar axis **h**. The crystal is characterized by the spontaneous polarization $P_s(T)$, which is a net dipole moment of a unit volume of the crystal (in C/m^2 units) that is indentically equal to the surface density of charges $\pm q$ on each of the opposite faces.

$$P_s \equiv q \qquad (1)$$

In the steady-state regime the intrinsic charges on the surface are compensated by foreign charges coming from the air or through an external circuit. However, when both temperature and the density of the intrinsic charges undergo rapid change but the compensation takes a long time (open-circuit regime), the net charge density at the opposite electrodes becomes high enough to create an electric field for crystal breakdown (for a typical pyrocoefficient $\partial P/\partial T$ value of about 5 nC/cm^2 K and fast cooling by $\Delta T = 200$ K, $E = q/\varepsilon\varepsilon_0 = 4 \cdot 10^6$ V/cm.)

It should be noted that for pyroelectrics in general and for ionic crystals in particular, the appearance of spontaneous polarization is not evident. Pyroelectricity can be treated without introducing a finite P_s similarly to the appearance of electric polarization upon the heating or cooling of a sample [1]. In molecular crystals, however, the electric dipoles "belong" to molecules and are well separated from each other due to relatively weak inter-molecular interactions. The crystal as a whole resembles an "oriented gas" of dipoles, and the calculation of P_s is straightforward. We will use this concept throughout this article.

The pyroelectric technique, which is based on the measurements of the pyroelectric current—that is, electric current passing through an external circuit due to a temperature variation of the surface charge density—has proved to be a powerful tool for investigation of crystalline ferroelectrics [2]. Because of its nondestructive nature, it is especially useful when one wants to study the virgin domain structure of a ferroelectric without switching the polarization. The other principal advantage is a possibility for obtaining polarization measurements in the field off-regime which is not, in general influenced by any current flow. The local structure of domains can also be investigated by a probe light beam using the pyroelectric technique [3]. Moreover the pyroelectric technique is a sensitive method for the study of a spatial distribution of the spontaneous polarization along the polar axis [4]. The main disadvantage of the pyroelectric technique is the difficulty of making absolute measurements of the spontaneous polarization; however, it can be overcome in most cases of any importance.

The present article is devoted to pyroelectric investigations of liquid crystals. Unfortunately, after the first observation of the pyroelectric effect in a ferroelectric liquid crystals [5], only few attempts were made at adopting the method worldwide. The technique, however, took deep root in our group in Moscow and proved to be very powerful and flexible for studying various systems. Its benefit became more and more recognized after the discovery of ferroelectric liquid crystals polymers [6], which as a rule are too viscous to be studied by the conventional repolarization technique. Thus the aims of this article are to describe the basic principles and specific features of the pyroelectric measurements on liquid crystals and to summarize the most important results obtained using the technique.

In general, the pyroelectric technique is sensitive to the appearance of any polar state characterized by macroscopic polarization, spontaneous or any other. We deal with the macroscopic polarization P in many systems, the most familiar examples being crystals belonging to one of 10 polar symmetry classes [2] and ferroelectric liquid crystals with C_2 point group symmetry [7]. Both manifest spontaneously polarized states. X- and Z-type (but not Y-type) Langmuir–Blodgett films are another example of the system with spontaneous polarization P_s [8], although in this case the polar symmetry is artificially imposed by the preparation process. We can prepare an artificial polar structure from a nonpolar nematic liquid crystal using special hybrid boundary conditions, homeotropic on one limiting plate and planar on the opposite one. Because of the flexoelectric effect [9], such a deformed structure is also macroscopically polarized, and the flexoelectric polarization can be measured by the pyroelectric technique [10].

Macroscopic polarization can be induced by an external dc electric field. The polarization then is proportional to the dielectric susceptibility of the substance. Such polarization can be frozen on cooling the substance, and it remains in this state even after the external field is switched off. Poled polymers, or *electrets*, manifest this kind of macroscopic (but not spontaneous) polarization. A similar situation is encountered in antiferroelectrics where spontaneous polarization vanishes. This is due to the compensation effect of the two oppositely directed polarizations of the two sublattices. When the dc electric field is applied, the macroscopic polarization becomes finite and, in certain substances, very high [2].

In the examples mentioned above we dealt with bulk polarization that is homogeneous in space. In the more general approach, one finds the net dipole moment in many surface layers, for example, at the interface between two different substances. The contact between two metals, the *p–n* semiconductor junction, or even the double electric layer at the interface between a solid substrate and a liquid electrolyte can be considered examples of structures with imposed spontaneous polarization that are strongly inhomogeneous in space. With changing temperature the surface density of charges changes, and the effect can be regarded as a special case of the apparent, or "false," pyroelecric effect [11]. The "surface" pyroelectric effect in centrosymmetric crystals originates solely from their polar boundaries (for details on this, see Tagantsev [1]).

This article is organized as follows: After we introduce the basic concepts of spontaneous and field-induced polarization and the corresponding pyroelectric coefficients, we describe the experimental techniques for pyroelectric measurements, both the steady state and pulsed. Next we discuss a study of the polar surface layers at the isotropic liquid-solid interface in general terms relevant to all liquid crystals. Flexo-electric polarization in a nematic cell with a nonuniform director field is considered in our investigation of the nonferroelectric polar system. For ferroelectric liquid crystals we discuss the temperature behavior of spontaneous polarization, including the special case of polarization sign inversion, soft-mode dynamics, susceptibility related to the different phase transitions and nonuniform structures, among other phenomena. Finally, we look at a few "unusual" structures that have been studied for their pyroelectric effect. Among them are achiral ferroelectrics (strongly disordered phases formed by polyphilic compounds, and other meso-genic substances) and achiral antiferroelectrics (monomer-polymer mixtures).

II. POLARIZATION AND PYROELECTRIC EFFECT: BASIC CONCEPTS

A. Cells and Equivalent Electric Circuits

In the simplest case, a single crystal pyroelectric or single-domain ferro-electric material has no own characteristic time (the polarization follows temperature without any delay). Consider, for example, a pair of electrodes at opposite interfaces perpendicular to the polar axis and without an applied dc bias field (the field off-regime). By convention (see [1]), the pyroelectric coefficient is the first derivative of the spontaneous polarization with respect to temperature:

$$\gamma_s(T) = \frac{dP_s(T)}{dT} \tag{2}$$

A small change in temperature ΔT that is uniform over the whole sample causes a change in the spontaneous polarization ΔP_s and, consequently, a change in the surface density of charge. However, the case is not that simple, since the temperature change causes thermal expansion in the sample. If the sample is "unclamped," the extension results in a piezoelectric effect, that is, in a piezoelectric voltage across the same electrodes. This is known as the secondary pyroelectric effect [2]. For certain crystals it can be of the same order of magnitude as the primary pyroelectric effect defined in Eq. (2) (but in the most cases it is of the order of 10% of the primary effect).

Pyroelectric studies of liquid crystals are usually performed in capillary cells, as shown in Fig. 1(a). The gap between the two glasses with ITO transparent electrodes is fixed by teflon strips of various thicknesses, and the two opposite ends of the capillary are kept open. In open capillary cells, the liquid crystal appears to be clamped, though it is still under atmospheric pressure, and the secondary pyroelectric effect does not arise. In sealed cells, the effect of stress (or pressure) has to be analyzed in each case.

In pyroeffect measurements the cell is connected to an external electric circuit, as shown in Fig. 1(a) for both the external field on- and off-regimes. When the external source is switched off, the current is generated by the temperature change in the surface densities of the charges. As the current charges the system to full capacitance ($C = C_s + C_L$), it flows through both the sample R_s and the load R_L resistance, Fig. 1(b). Thus the equation expressing pyroelectric current is

$$A\frac{\partial q}{\partial t} = C\frac{\partial U_p}{\partial t} + \frac{U_p}{R_L} + \frac{U_p}{R_s} \tag{3}$$

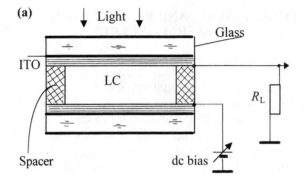

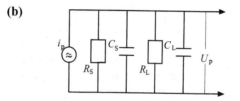

Figure 1. Typical liquid crystal cell for pyroeffect measurements (a) and the equivalent electric circuit (b).

where A is electrode area and U_p is the pyroelectric voltage across the electrodes. Combining Eqs. (1) to (3), we have the main equation for the pyroelectric effect [12]:

$$A\gamma_s \frac{\partial T}{\partial t} = C \frac{\partial U_p}{\partial t} + \frac{U_p}{R} \tag{4}$$

where $R = R_s R_L / (R_s + R_L)$.

Often it is desirable to have a uniform single domain state. Then liquid crystal can be poled with an external voltage source. If a dc voltage $U_0 \gg U_p$ is applied to a cell, we have the more general equation

$$A\gamma_s \frac{\partial T}{\partial t} = C \frac{\partial U_p}{\partial t} + \frac{U_p}{R} + U_0 \frac{\partial C_s}{\partial t} \tag{5}$$

where the last term gives the thermally induced change in a sample capacity. This term exists even in centrosymmetric systems where it describes the so-called capacitive bolometer effect [13].

Alternatively, we can take the total polarization

$$P = P_s + P_i \tag{6}$$

where the field-induced polarization

$$P_i = \varepsilon_0 \chi E \tag{7}$$

is added to the spontaneous polarization. The induced polarization is temperature dependent due to temperature dependence of susceptibility $\chi(T)$. Usually this contribution is small, and it can be taken into account in the framework of an equivalent circuit, and Eq. (4), by introducing a total field-dependent pyrocoefficient:

$$\gamma = \frac{\partial P}{\partial T} = \gamma_s + \gamma_i = \gamma_s + \varepsilon_0 E \frac{\partial \chi}{\partial T} \tag{8}$$

The actual spontaneous pyrocoefficient can be found from experiment by extrapolating the total coefficient to the zero field condition.

B. Time Dependencies

Equation (4) has the following general solution [1]:

$$U_p = \frac{A\gamma}{C} \exp\left(-\frac{t}{RC}\right) \int_0^t \exp\left(\frac{t'}{RC}\right)\left(\frac{dT}{dt'}\right) dt' \tag{9}$$

When the rate of temperature change is very slow compared with electric time constant $\tau_E = RC$, then

$$U_p = AR\gamma \frac{dT}{dt}\left[1 - \exp\left(-\frac{t}{RC}\right)\right] \tag{10}$$

It follows immediately from Eq.(10) that if $T = \beta t$ and if resistance R is very small (short-circuit regime), the pyroelectric voltage is constant, and the absolute magnitude of the pyrocoefficient may be calculated simply as

$$\gamma = \frac{U_p}{AR\beta} \tag{11}$$

The simplest linear variation of temperature (called the *static regime*) was often used in the past for studying solid pyroelectrics. A problem arose, however, when it was necessary to keep a single domain structure of a ferroelectric material with an external dc bias voltage, since there is an extra ohmic current through the sample that usually exceeds the weak pyroelectric current caused by slow temperature variation. The way around the problem proved to be to use a faster variation of temperature; as a result this is referred to as the *dynamic regime* of measurements.

To obtain a very fast (in comparison with RC) jump of temperature by ΔT, we can neglect the exponential term under the integral in Eq. (9), so the time variation of the pyroelectric voltage again becomes very simple:

$$U_p = \frac{A\Delta T\gamma}{C} \exp\left(-\frac{t}{RC}\right) \tag{12}$$

A voltage pulse has zero-time magnitude $A\Delta T\gamma/C$; it relaxes to zero with the RC time constant. The *current* response of a pyroelectric to an instantaneous change of temperature is just a δ-function.

In practice, however, it is not possible to provide an instantaneous change of temperature, since every sample has its own thermal time constant which is dependent on its heat capacity and on conditions of heat transfer to the surrounding medium, in particular, to substrates confining the liquid crystal layer. In every case the problem of calculation of the pyroelectric response has to be divided into two subproblems: First, one must find the dependence of the sample temperature on time and, in some cases, on coordinates, and second, one must calculate the pyroelectric response with a known $T(t, x, y, z)$ function. Fortunately, in many important cases we have either the one-dimensional problem, $T = T(t, z)$, or a uniform temperature distribution, $T = T(t)$.

Let us consider the case where the temperature is changed by a uniform absorption (absorption coefficient η) of incident light with power density $W(t)$ in W/cm^2. It is convenient to write the energy balance equation (for the uniform distribution of temperature) together with the equation for the continuity of an electric current. This is just another form of Eq. (4):

$$\frac{d\Delta T}{dt} + \frac{\Delta T}{\tau_T} = \frac{A\eta}{C_T} W(t) \tag{13}$$

$$\frac{dU_p}{dt} + \frac{U_p}{\tau_E} = \frac{A\gamma_s}{C} \frac{d\Delta T}{dt} \tag{14}$$

The two equations contain the source terms (on their right-hand sides), and the reactive and active losses.

The electric time constant is

$$\tau_E = RC \tag{15}$$

The thermal time constant in Eq. (13) is an analogous product of the thermal resistance (inverse thermal conductivity of the sample to the background medium) and heat capacity:

$$\tau_T = R_T C_T = \frac{c_T \rho d^2}{\lambda} = \frac{C_T d}{\lambda A} \tag{16}$$

where c_T, ρ, and λ are the specific heat, the density, and the specific thermal conductivity of the pyroelectric. Value τ_T ranges from 100 µs to 10 ms for a liquid crystal layer thickness d varied from 10 to 100 µm. Strictly speaking, the influence of the glass substrates confining the liquid crystal layer has also to be taken into account.

As a rule, in dynamic experiments the temperature is controlled optically by pulses of laser light of various duration, t_p. Let us look at two important cases: One deals with short single-laser pulses and the other deals with an infinite periodic sequence of long pulses. The first technique is better suited to absolute measurements of γ_s and studies of P_s kinetics; the second technique where the use of periodic temperature variation is made is more convenient for automatic measurements controlled by a computer. We will discuss first the response of a pyroelectric to a single light intensity step.

1. Pyroelectric Response to Light Intensity Step

The time dependence of an incident light intensity ($W(t) = 0$ for $t < 0$ and $W(t) = W_0$ for $t > 0$) is shown at the top of Fig. 2. The energy-balance equation (13) has the simple solution

$$\Delta T(t) = \frac{A\tau_T\eta W_0}{C_T}\left[1 - \exp\left(-\frac{t}{\tau_T}\right)\right] \qquad (17)$$

The exponential growth of temperature is also shown in Fig. 2. We want to examine the shape of the current response. In the short-circuit regime where $\tau_T \gg \tau_E$, the second (active) term on the left-hand side of Eq. (14) is much

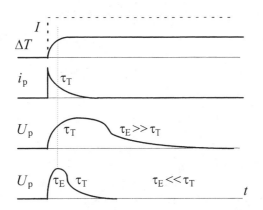

Figure 2. Qualitative oscillograms of exciting light intensity I, temperature increment ΔT, pyroelectric current i, and pyrovoltage U_p for the case of a single light intensity step.

greater than the first (reactive) term. For the pyroelectric current flowing through resistance R, we have

$$i_p = A\gamma_s \frac{d\Delta T}{dt} \qquad (18)$$

Substituting $d\Delta T/dt$ from Eq. (17) into Eq. (18), we get the form of the current response:

$$i_p = \frac{A^2\gamma_s\eta W_0}{C_T} \exp\left(-\frac{t}{\tau_T}\right) \qquad (19)$$

which is shown in the second oscillogram of Fig. 2. For the short-circuit regime, $U_p = i_p R$, and the shape of the voltage response repeats that of the current.

In the general case the voltage response depends on the two time constants. The problems has been analyzed by Ref. [14], and for $\tau_T \neq \tau_E$, the following solution was obtained:

$$U_p(T) = \xi \left(\frac{1}{\tau_E} - \frac{1}{\tau_T}\right)^{-1} \left[\exp\left(-\frac{t}{\tau_T}\right) - \exp\left(-\frac{t}{\tau_E}\right)\right] \qquad (20)$$

where $\xi = \gamma_s\eta W_0 A^2/CC_T$. The two voltage oscillograms predicted by Eq.(20) for $\tau_T \gg \tau_E$ and $\tau_T \ll \tau_E$ are also shown in Fig. 2. For the special case of $\tau_T = \tau_E = \tau$, we have $U_p = t \cdot \exp(-t/\tau)$ [14].

2.   Response to a Light Pulse

The rectangular pulse of duration t_p is a sum of the two equal steps, positive and negative, shifted by t_p with respect to each other. Correspondingly the pyroelectric response is a superposition of the positive and negative responses shown in Fig. 2.

Quasi-static Regime. One important case corresponds to very long optical pulses, $t_p \gg \tau_T, \tau_E$. The oscillograms of temperature, pyroelectric current, and pyroelectric voltage (only for a typical condition $\tau_E \geq \tau_T$) are shown in Fig. 3. We call this regime *quasi-static*. It does not matter that we have a single long pulse or a sequence of such pulses, since, upon switching the light on and off, all relaxation processes in the sample are terminated and the amplitudes of both the pyrocurrent and pyrovoltage reach their maximum steady-state values. In the literature this regime is called dynamic to distinguish it from the static regime mentioned earlier, Eq. (11).

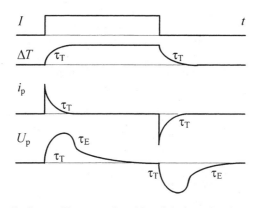

Figure 3. Qualitative oscillograms of exciting light intensity I, temperature increment ΔT, pyroelectric current i, and pyrovoltage U_p in the case of a rectangular light pulse of long duration, $t_p \gg \tau_E, \tau_T$ (a quasi-static regime).

Dynamic Regime. The other important case corresponds to very short optical pulses, $t_p \ll \tau_T, \tau_E$. The temperature increases linearly during the action of light and then slowly decreases with the thermal time constant, Fig. 4. The maximum temperature increment is t_p/τ_T times less than the steady-state value. Since the pyroelectric current is the time derivative of ΔT, it repeats the shape of the light pulse (with accuracy of the above inequality), then changes sign and, being very small, follows the decreasing temperature. Note that the positive and negative areas of the current oscillogram shown in Fig. 4 are always equal (the same charge flows

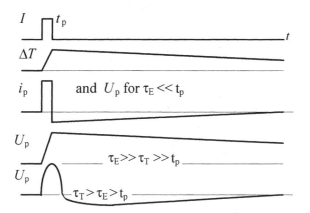

Figure 4. Qualitative oscillograms of exciting light intensity I, temperature increment ΔT, pyroelectric current i, and pyrovoltage U_p in the case of a rectangular short pulse, $t_p \ll \tau_E, \tau_T$ (dynamic regime).

through the external circuit on heating and on cooling) and that the ratio of positive to negative current amplitudes is equal to τ_T/t_p. In this figure the pyroelectric voltage is shown for two cases, $\tau_E > \tau_T$ and $\tau_T > \tau_E$. In the first case U_p changes sign around $t = \tau_T$ (not shown in the figure), and in the second case, around $t = \tau_E$.

3.   Response to Harmonically Modulated Light

This case was originally discussed by Cooper [12] with a slightly different equivalent scheme for the pyroelectric. We will follow the arguments presented in papers [15]. Periodic optical excitation was used originally by Chynoweth [16] and for the first time applied to liquid crystals by Glass et al. [17].

Let the incident light power density be modulated harmonically by a chopper with angular frequency ω,

$$W(\omega) = W_0(1 - \cos \omega t) \tag{21}$$

The solution of the energy-balance equation (13) is

$$\Delta T(t) = \frac{A\eta W_0 \tau_T}{2C_T} \left(1 - \frac{\cos \omega t}{(1 + \omega^2 \tau_T^2)^{1/2}}\right) \tag{22}$$

Only the second time-dependent term is responsible for the pyrocurrent. Combining Eqs. (18) and (22), we get the current amplitude:

$$i_p = \frac{\gamma_s \eta W_0 A^2 \omega \tau_T}{2C_T(1 + \omega^2 \tau_T^2)^{1/2}} \tag{23}$$

The frequency dependence of the pyroelectric current is shown qualitatively in Fig. 5. In the low frequency limit where $\omega \ll 1/\tau_T$ the current amplitude linearly grows with frequency until the saturation is reached where $\omega \gg 1/\tau_T$ and

$$i_p(\omega \to \infty) = \frac{\gamma_s \eta W_0 A^2}{2C_T} \tag{24}$$

The pyroelectric voltage can be obtained as a product of i_p and the real part of the electric circuit impedance:

$$U_p = \frac{\gamma_s \eta W_0 A^2 \omega \tau_T}{2C_T(1 + \omega^2 \tau_T^2)^{1/2}} \cdot \frac{R}{(1 + \omega^2 \tau_E^2)^{1/2}} \tag{25}$$

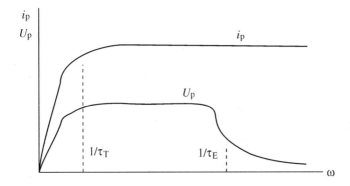

Figure 5. Frequency dependencies of the pyroelectric current and voltage for harmonic light excitation.

At high frequencies, $\omega \gg 1/\tau_T, 1/\tau_E$, the voltage decays with electric time constant (Fig. 5), and obeys the asymptotic expression

$$U_p(\omega \to \infty) = \frac{\gamma_s \eta W_0 A^2}{2 C_T C \omega} \tag{26}$$

In the plateau region, where $1/\tau_E \gg \omega \gg 1/\tau_T$,

$$U_p(\text{plateau}) = \frac{\gamma_s \eta W_0 A^2 R}{2 C_T} \tag{27}$$

In conclusion of this section, it is worth recalling that all expressions derived here are assumed to have (1) a uniform temperature distribution (ΔT) over the sample, (2) a uniform distribution of the pyrocoefficient over the sample with the polar axis perpendicular to the electrodes, (3) an absence of the bias dc voltage [but see Eqs. (6–8)], and (4) an absence of any relaxation processes in the material (only the thermal time constant of the sample and the electric time constant of the circuit are taken into account). For nonuniform pyroelectirc structures (e.g., at interfaces) and materials with additional relaxation processes, the pyroelectric response will briefly be discussed later on. For a nonuniform temperature distribution, at least one more thermal time constant must be taken into account, one that is related to the heat transfer equalizing the temperature gradient. For example, when light is absorbed in an electrode, instead of the plateau shown in Fig. 5, an increasing or decreasing curve will be observed depending on the interplay between the layer thickness and the length of the temperature wave.

III. PYROELECTRIC TECHNIQUES

A. Pulse Technique

1. Dynamic Regime

A short pulse technique allows us to study kinetic effects. It may be based, for example, on a Q-switched Nd^{3+} YAG laser (wavelength $\lambda = 1.06\,\mu m$). The laser provides powerful (about $W_m \approx 0.1\,MW/cm^2$) bell-shaped pulses of $t_p = 30\,ns$ duration with a repetition rate of 1 to 20 Hz. The light is absorbed either in the ITO electrodes or in the bulk of the liquid crystal if a special light-absorbing dye is introduced as a dopant into the liquid crystal. The dye will provide complete absorption of light in the cell. Then, since the pulse is short compared to the thermal constant of the sample, the second term in energy-balance equation (13) is small (no dissipation of energy into background medium, e.g., into ITO covered glasses), the temperature increment can be easily calculated for any shape of the pulse

$$\Delta T = \frac{E_p}{C_T} = \frac{A\eta}{C_T} \int_0^{t_p} W(t)dt \qquad (28)$$

if absorbed power $\eta A W(t)$ (or energy E_p) of the pulse and heat capacity C_T of a sample are known. This case corresponds to Fig. 4. As a rule, the laser beam should be focused on a sample in such a way that the whole area A is uniformly illuminated, but this is not too critical far from the phase transitions where the pyroelectric response is a linear function of irradiation intensity and the response is the same independently of the place where the heat is absorbed. The crucial condition is the complete absorption of light energy by the dye molecules. As for specific heat, its temperature dependence is generally not known, so for many practical purposes it can be assumed to be a constant even in the vicinity of ferroelectric phase transitions where it varies typically within a few percent [18] (often $c_T \approx 2.5\,J/gK$ is taken as a good approximation).

In any case, it is not trivial to find a dye that absorbs IR light and is soluble enough in a liquid crystal. In our early experiments [19] a dithiyne Ni-complex was used in concentration 0.5 wt% which provided full absorption of $1.06\,\mu m$ laser light in a $200\,\mu m$ thick cell. For simplicity, in most practical cases, the absorption is provided by the ITO electrodes (they absorb few percents of laser irradiation), and the absolute measurements can only be made by comparison with a reference substance placed between the glasses with the same type of absorbing electrodes.

A block diagram of the experimental setup is shown in Fig. 6(a). A laser pulse is formed by an active or passive Q-switch and monitored by a

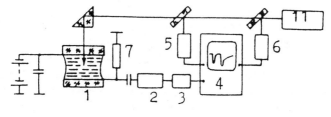

(a)

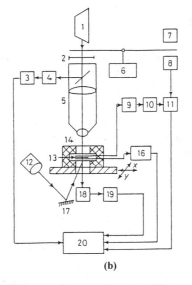

(b)

Figure 6. Block diagrams of setups for pulse (a) and steady-state (b) pyroelectric measurements. (a) 1: Liquid crystal cell; 2: amplifier; 3: delay line; 4: oscilloscope; 5: photomultiplier for pulse control; 6: photomultiplier for triggering; and 7: load resistor 0.1–1 kOhm. (b) 1: He–Ne laser, 2: shutter; 3, 10, 19: selective amplifiers, 4, 8, 18: photodiodes; 5: microscope; 6: light chopper; 7: light-emitting diode; 9: preamplifier; 11: lock-in detector; 12: lamp; 13: liquid crystal sample; 14: thermostate; 15: $X - Y$ movable table; 16: digital voltmeter for a thermocouple; 17: mirror; 20: computer.

photomultiplier. A liquid crystal cell is connected to a dc bias source and a load resistance typically as low as $R_L = 100$ Ohm. In this case the shape of the pyroelectric response pulse repeats the shape of the laser pulse, Fig. 4, if no relaxation process longer than $t_p = 30$ ns is involved in the liquid crystal. The pyroelectric pulse is amplified by a wideband amplifier and observed at a double-beam storage (digital) oscilloscope together with a pulse from the photomultiplier. To avoid an interference of the observed signal with an electromagnetic noise coming from the laser pump pulse front, a delay line has to be used to delay the pyroelectric response for about 100 ns (with passive Q-switching the noise also decreases). This version of the experiment was used, for example for the first investigation of the dynamics of the soft mode in classical compound DOBAMBC (see below). Its main drawback is the complexity of the experimental setup and some difficulty in controling the short-pulse response with a computer.

A dc bias source is necessary to orient all domains of a multidomain ferroelectric sample in one direction (monodomenization process). We will also find it necessary when we study field-induced polarization (e.g., in pyroelectric soft-mode susceptibility measurements). If necessary, the field dependence of the pyroelectric response can be extrapolated to a zero field condition (spontaneous polarization measurements).

2. A Quasi-static Regime

In another version of the pulse technique, the Q-switch is removed and a cw laser pulse (consisting of many spikes) of total duration about 100 μs is used. In this case the load resistance may be enhanced up to 100 kOhm, and the bandwidth of the amplifier may be reduced in order to improve the signal-to-noise ratio. The spikes are integrated by an enhanced electric time constant, and a smooth pulse is observed on the oscilloscope screen. Now it is difficult to accumulate the whole energy absorbed during the pulse action without its scattering into the background medium, and the absolute measurements are only possible by comparison with a known substance [20]. However, this technique is more convenient for the measurement of the temperature dependencies of the pyro-response and spontaneous polarization, since (1) the amplifier ouput may be connected to an interface of a computer to store and process oscillograms and (2) the much higher signal-to-noise ratio allows for a search of new pyroelectrics with a rather low pyrocoefficient. In the Institute of Crystallography (Moscow) we have been using both pulse techniques since 1970. Both techniques were also recently used in Osaka University [20,21].

B. Steady-State Technique

In the steady-state (actually quasi-static) version [Fig. 6(b)], which is similar to Chynoweth's technique [16,17], a He–Ne laser with power $P = 10–15$ mW and wavelength $\lambda = 0.63$ μm (or a laser diode with power $P = 30$ mW and $\lambda = 0.69$ μm [22]) is used to provide a small temperature change ΔT in a sample due to a weak uncontrollable absorption of light in both ITO layers and the liquid crystal. As in the pulse version, a sample is placed in a thermal jacket and connected to a bias voltage source. The laser beam is modulated by a chopper (or from a function generator in the case of a laser diode),

$$\Delta T = T_0 + T_\text{m} \sin \omega t \tag{29}$$

The circular frequency $\omega = 2\pi f$ may be varied (typically $f = 70\,\text{Hz}$ or $7\,\text{kHz}$). The pyroelectric response is measured at the same frequency f by a lock-in amplifier as a two-component (X and Y) sine voltage across the load resistor $R_\text{L} = 10\,\text{MOhm}$. Both outputs are connected to a PC computer, and the modulus and phase of the signal are calculated.

The operating frequency has to be properly selected. For example, at a frequency of $70\,\text{Hz}$ for $R_\text{L} = 10\,\text{MOhm}$ the short-circuit condition is fulfilled $(R_\text{L} \ll R', (\omega C)^{-1}$, where R' and C are the parallel resistance and capacity of the sample and input circuit), and the pyrovoltage amplitude is proportional to the load resistance; see Eq. (25). At lower frequencies the condition $\omega \gg 1/\tau_\text{T}$ is violated, the pyrovoltage decreases, and the signal-to-noise ratio becomes lower. Thus a frequency of around $100\,\text{Hz}$ sems to be the optimum for typical liquid crystal cells.

However, in some cases, it is beneficial to use higher frequencies, such as when it is necessary to filter out some parasitic contribution from the temperature modulation of the conductance current. We encountered this problem when we studied polarization sign inversion phenomena in ferroelectric liquid crystals. To fulfill the short-circuit condition at frequencies, say, $7\,\text{kHz}$ and higher, a much lower load resistance should be used; this results in a decrease in the signal-to-noise ratio. Thus, to quit the short-circuit regime and work with a pure capacitance load, it is better to have the same $R_\text{L} = 10\,\text{Mohm}$ when the condition $R' \gg R_\text{L} \gg (\omega C)^{-1}$ is valid. In the latter case, according to Eq. (26), the pyrovoltage decreases with increasing frequency. This decrease coming from a decrease in the input impedance is compensated by a proportional decrease in noise, so the signal-to-noise ratio appears to be constant. Thus a proper selection of modulation frequency depends on the particular task.

The steady-state technique is convenient for cumbersome measurements of the temperature dependencies of polarization at various fields. However, it does not allow us to study kinetic effects, and it needs a reference sample for the absolute measurements of spontaneous polarization. This technique is very simple, and it can be combined with texture and electrooptical studies of liquid crystals under a microscope (this is the technique used, e.g., at the Technische universität Darmstadt). An even simpler technique is based on an incandescent lamp (instead of a laser) with a beam mechanically chopped at a certain frequency. Due to the strong absorption of IR light in glass, the slightly focused radiation of a 25 W incandescent lamp provides a 10 times higher temperature modulation amplitude (at 70 Hz) than the laser diode mentioned earlier. In this case the reference signal for the lock-in amplifier is provided by a photodiode which detects part of the chopped beam reflected by the glass plate [64,65].

IV. "PYROEFFECTS" IN NONPOLAR SUBSTANCES

A. Apparent Pyroeffect at the Nonpolar Dielectric–Metal Interface

Irrespective of the crystalline structure of a dielectric material, the temperature dependence of the surface density of charges (apparent or "false" pyroelectric effect [11]) can be observed at the interface between the two substances. For example, when an insulator is in contact with a metal, the surface density of charges arises in order to equalize Fermi levels of the two materials. In thermodynamic equilibrium the energy levels of an insulator are curved near the interface and the space charge density (in the bulk) have a certain distribution $\rho(z)$ along the normal to the surface.

For a thin layer of a dielectric material, we can introduce the macroscopic polarization and pyroelectric coefficient averaged over a sample:

$$\langle P \rangle \equiv \frac{\sigma_1 - \sigma_2}{2} = \frac{1}{d} \int_{-d/2}^{+d/2} z\rho(z)dz \qquad (30)$$

$$\langle \gamma \rangle = \frac{d\langle P \rangle}{dT} \qquad (31)$$

Here σ_1 and σ_2 are surface density of charges at opposite surfaces of the dielectric with coordinates $\pm d/2$.

The sufficient condition for the apparent pyroelectric effect is a temperature-dependent asymmetric distribution of the space charge density. In the ideal case of two symmetric interfaces, $\sigma_1 = \sigma_2$ [a particular case of the asymmetric distribution shown in Fig. 7(a)], or for an infinitely thick layer, $d \Rightarrow \infty$, the "pyroeffect" vanishes. In the case of two different electrodes (e.g., ohmic and injection contacts), the surface charge density has opposite signs at opposite electrodes [Fig. 7(b)], and for a very thin film, $d \approx L_1 + L_2$, such an antisymmetric density distribution is very close to that of a genuine crystalline pyroelectric whose surface phenomena are disregarded [Fig. 7(c)].

Therefore, in thin layers, the steady-state temperature dependence of the sum of two thermoelectromotive forces (thermo-emf) at opposite contacts, which originated from Fermi statistics, may be indistinguishable from the pyroelectric effect. Of course, the kinetics of the two effects is different: For the genuine pyroelectric effect, it is determined by molecular or lattice

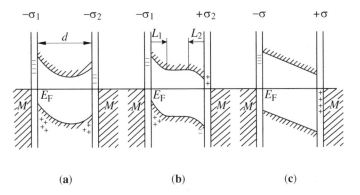

Figure 7. Potential diagrams for thin film metal–insulator–metal structures (L_i: Debye lengths). (*a*) Asymmetric distribution of the space charge (centrosymmetric for $L_1 = L_2$; (*b*) antisymmetric distribution; and (*c*) genuine (volume) pyroelectric with spatially uniform P_s.

motions (sublattices vibrations, dipolar reorientation, conformational changes, intramolecular charge transfer, etc.) with characteristic times of the order of 10^{-8}–10^{-10} s; for thermo-emf, it is determined by the rate of redistribution (diffusion) of the charge carriers between an electrode and surface states, and in thin interfacial layers (about $1000 \,\overset{\circ}{\text{A}}$) this process can take 10^{-9}–10^{-12} s. Hence, in a typical experimental situation, where the kinetics of the pyro-response is controlled by much longer thermal τ_T and electric τ_E times, the two effects are indistinguishable.

This phenomenon is quite general for thin inhomogeneous layers such as those of organic semiconductors and Langmuir-Blodgett films, weak electrolytes in contact with metals, and surface layers of liquid crystals [23]. Estimates show that magnitude of the apparent pyroelectric effect is very sensitive to ambient conditions and can be as high as $30 \,\text{nC/cm}^2 \,\text{K}$ [11]. On the one hand, this phenomenon must be taken into account when discussing electric properties of genuine polar mesophases. On the other hand, it may be used for probing potential barriers at various interfaces.

The idea of the technique has been demonstrated by an example of thin film metal–tetracene–metal sandwich structures [4]. Tetracene (T) is a centrosymmetric organic crystal. Its polycrystalline films are known to form electron injection contacts with aluminium and hole injection contacts with gold [24]. Thus Al–T–Al and Al–T–Au sandwich structures are good candidates to use in modeling the symmetric and antisymmetric diagrams shown in Fig. 7.

Now, imagine that a short pulse of light is absorbed in the top Al electrode at $z = 0$ whose temperature increases by ΔT for time t_p. Then the temperature front begins to spread into the dielectric film. In this case the

temperature is not constant over the sample and the velocity of the front is
determined by the thermal conductivity equation

$$\frac{\partial T(z, t)}{\partial t} = \frac{\lambda}{c_T \rho} \cdot \frac{\partial^2 T(z, t)}{\partial z^2} \tag{32}$$

with initial and boundary conditions

$$T(z, 0) = 0; \quad \frac{\partial T(0, t)}{\partial z} = -\frac{\eta}{\lambda} W(t)$$

$$W(t) = 0 \quad \text{for} \quad t < 0, \, t > t_p$$

$$W(t) = W_0 \quad \text{for} \quad 0 < t < t_p \tag{33}$$

and all parameters defined earlier; see Eqs. (13) to (16).

First, let us qualitatively discuss the process for the nearly centrosymmetric structure [Fig. 7(a)]. At time t_1 the temperature front reaches distance
$z = -d/2 + L_1$ and the Al–T barrier is, on average, heated by ΔT_1. Layer
L_1 behaves as a thin pyroelectric film with, say, a positive pyroelectric
coefficient. As this layer is heated and cooled, it generates a single positive
pyroelectric pulse.

Then the thermal front travels over the middle, "dead," zone of the film
between the two barriers $(-d/2 + L_1 < z < d/2 - L_2)$ where no response is
anticipated. Eventually the front reaches the bottom Al electrode, and the
temperature of the T–Al barrier increases by $\Delta T_2 (\Delta T_2 < \Delta T_1)$. The T–Al
barrier has the opposite sign of the potential slope and generates a negative
pyroelectric pulse. Thus we anticipate a pyroresponse oscillogram in the
form of two peaks, a positive proportional to ΔT_1 and the negative
proportional to ΔT_2 [4]. When the bottom Al electrode is irriadiated, the
negative pulse comes ahead of the positive one. If $\Delta T_2 = \Delta T_1$, the two
peaks have the same amplitude for the centrosymmetric structure, and with
the poor time resolution (or in a steady-state regime), no pyroelectric
response is observed.

Similarly, in the consideration of an antisymmetric structure, which
results in an anticipated oscillogram with two peaks of the same sign, it does
not matter from what side a sandwich is irradiated (to change the sign of the
oscillogram, the signs of both barriers have to be exchanged. With poor time
resolutions (thin films), the two pulses merge into one, and such a response
is indistinguishable from the response of a genuine pyroelectric sample.

The qualitative picture discussed has been confirmed experimentally [4]
for Al–T–Al and Al–T–Au sandwiches of various thickness. A short (30 ns)
pulse of a Q-switched Nd^{3+} YAG laser was used in those experiments. The

thickness of potential barriers was estimated to be about 100 nm. In addition it has been shown how an external dc electric field modifies the potential diagrams in Fig. 7. The electric field results in a systematic slope of the diagram and, consequently, to the induced pyroelectric effect. The interplay between the field-induced and "surface" pyroeffects can be seen on the screen of an oscilloscope.

A theory for the pyroelectric response of a metal-dielectric interface to a triangular pulse was proposed in Ref. [4] using two examples, the Schottky barrier and the double electric layer, and the time dependencies of the response were discussed. The pyroelectric voltage (in the short-circuit regime) was calculated by integrating over the film normal z,

$$U(t) = \int_{d/2}^{+d/2} \gamma(z) \frac{\partial}{\partial t} T(z, t) dz \qquad (34)$$

with function $T(z, t)$ found from the thermal conductivity equation (32). The expression above assumes that for the short time under discussion, 10^{-7}–10^{-6} s, the thermal energy is conserved within the system.

The "surface" pyroelectric effect is a general phenomenon. It was also studied [10,22] on a nematic liquid crystal in contact with a transparent SnO_2 electrode. The oscillogram of the pyroelectric response, and hence the slope of the potential barrier, were strongly dependent on the dc electric field applied to the liquid crystal cell. A field-dependent pulse pyroelectric response has also been observed on a NaCl water solution in contact with an ITO electrode [25].

B. Flexoelectric Polarization in Nematics

The macroscopic dipole moment of a unit volume of a nematic liquid crystal [9],

$$\mathbf{P}_f = e_1 \mathbf{n} \operatorname{div} \mathbf{n} + e_3 (\mathbf{n} \times \operatorname{curl} \mathbf{n}) \qquad (35)$$

called *flexoelectric polarization*, depends on the curvature of the director field (div $\mathbf{n}$ and curl $\mathbf{n}$) at a constant modulus of the order parameter S. The coefficients e_1 and e_3 are flexoelectric coefficients for the splay and bend distortions, respectively. For a uniform director orientation (director $\mathbf{n}$=const) $\mathbf{P}_f = 0$. A simplified molecular picture of the phenomenon is discussed in Ref. [26].

Thus, if a nematic liquid crystal is mechanically distored, the sample acquires macroscopic polarization. In fact, what is usually measured in experiments is not the flexoelectric polarization but a converse flexoelectric effect, that is, the appearance of distortion caused by the application of an electric field to a sample. The direct flexoelectric effect has been measured only dynamicaly as an electric response of an acoustically distorted thermotropic nematic layer [27,28] or a bilayered lyotropic membrane [29]. In both cases the absolute magnitude of the flexoelectric polarization has not been found.

There exists, however, a direct way to measure the temperature behavior of the flexoelectric polarization of a distorted nematic structure by integrating its pyroelectric coefficient. The method is based on the fact that $\mathbf{P}_f$ vanishes in the isotropic phase, so the transition temperature T_{NI} can be taken as the initial point for the integration [10,22,30].

$$P_f(T) = \int_{T_{NI}}^{T} \gamma(T)dT \tag{36}$$

1. Measurements of Flexoelectric Polarization and Flexoelectric Coefficients

Consider, for instance, a hybrid nematic cell (plane or wedge shaped) with asymmetric boundary conditions, angles θ_0 and θ_d, at two opposite interfaces with coordinates 0 and d (Fig. 8). Then the flexoelectric polarization is due to a combined splay and bend distortion and its z-component follows from Eq. (35):

$$P_f^z = (e_1 + e_3)\sin\theta\cos\theta\frac{\partial\theta}{\partial z} \tag{37}$$

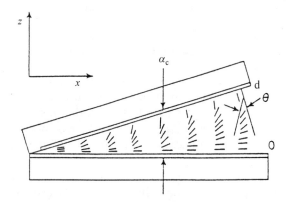

Figure 8. Wedge-form, hybrid nematic cell with asymmetric boundary conditions, and angles $\theta_0 = \pi/2$ and $\theta_d = 0$ at two opposite interfaces with $z = 0$ and d.

and

$$\langle P_f^z \rangle = \frac{1}{d} \int_0^d P_f^z dz = \frac{e_1 + e_3}{4d} (\cos 2\theta_d - \cos 2\theta_0) \tag{38}$$

In a hybrid cell with planar ($\theta_0 = \pi/2$) and homeotropic ($\theta_d = 0$) director orientations at the opposite interfaces, Eq. (38) transforms into

$$\langle P_f^z \rangle = \frac{e_1 + e_3}{2d} \tag{39}$$

and the magnitude of the sum of the two flexo-coefficients follows from the z-component of $\langle \mathbf{P}_f \rangle$ found from the experimentally measured pyroelectric coefficient (36).

Such an experiment has been carried out on homeo-planar cells filled with classical nematic liquid crystals 5CB and MBBA The pulse pyroelectric technique with a YAG laser operating in the continuous wave regime ($t_p = 100\,\mu s$) was used. The IR light was absorbed by ITO electrodes. The relative magnitude of the z-component of $\langle \mathbf{P}_f \rangle$ was measured by integrating the pyroelectric response, and its absolute magnitude was found by comparison with the response of classical ferroelectric liquid crystal DOBAMBC (the latter was placed in a planar cell made of the same IR absorbing ITO glasses). The absolute magnitudes of the pyroelectric coefficient and the flexoelectric polarization are shown in Fig. 9 for 5CB.

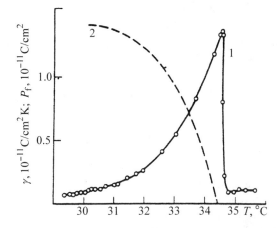

Figure 9. Absolute magnitudes of the pyroelectric coefficient γ (curve 1) and the z-component of the flexoelectric polarization (curve 2) in terms of temperature for a hybrid cell with 5CB.

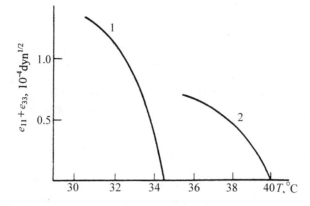

Figure 10. Tempterature dependencies of the sum of two flexoelectric coefficients for 5CB (curve 1) and MBBA (curve 2) found by the pyroelectric technique.

The pyroelectric signal in the isotropic phase has been attributed to the apparent pyroelectric effect due to interfacial polarization at the ITO-5CB contacts, discussed earlier in this section.

The final result of paper [10], namely the determination of the temperature dependencies of the sum of two flexoelectric coefficients for 5CB and MBBA, is shown in Fig. 10. See also Ref. [20,22].

2. *Measurements of the Anchoring Energy*

Another version of the pyroelectric technique, the so-called local technique [3,30,31], was used to measure the thickness dependence of flexoelectric polarization in order to extract the corresponding dependence of angle θ_d at the homeotropic boundary and, as a consequence the anchoring energy for the director at that boundary. To this effect a wedge-shaped cell, shown in Fig. 8, was used. It is well known that the director anchoring at the planar boundary is much stronger than that at the homeotropic boundary, and with decreasing thickness, the whole layer acquires a planar orientation. In other words, angle θ_d becomes equal to $\theta_0 = \pi/2$ at the critical thickness d_c. Then, according to Eq. (38), the flexoelectric polarization (and pyroelectric response) vanishes.

Thus it is convenient to study the spatial distribution of the pyroelectric response along the x-coordinate of the wedge cell, using a small diameter optical probe to produce the temperature increment [30]. A probe beam of a low-power He–Ne laser was focused by a microscope and modulated by a chopper; see Eq. (21) for the form of excitation. The beam provided a spatial resolution of about 15 μm. To enhance the sensitivity of the technique, a small amount (less than 3%) of a dye absorbing at $\lambda = 633$ nm was dissolved in nematic 5CB. The power density ηW_0 absorbed by the dye was measured along the wedge independently. The average pyroelectric coefficient $\langle \gamma \rangle$ as a function of thickness d was found from the amplitude of the pyroelectric response [30],

$$U_p = \frac{\langle \gamma \rangle A^2 \eta W_0 d(\omega/D_T)^{1/2} R}{\sqrt{2} C_T \sqrt{1 + \omega^2 \tau_E^2}} \tag{40}$$

This equation differs from the high-frequency limit of Eq. (26), since it was derived under the assumption [32] that the diffusion of heat into thick glass plates is the dominant thermal scattering process, and therefore $T \neq$ constant over a sample. At frequency 200 Hz the characteristic thermal distance in glass, $L_T = (D_T/\omega)^{1/2} \approx 30\,\mu\mathrm{m}$, considerably exceeds $d/2$ even at the thick end of the wedge ($\omega = 2\pi \cdot 200$ 1/s, $D_T = 2\lambda/c\rho \approx 10^{-7}$ m²/s, and the thermodynamic parameters defined above; see Eq. (16)). Thus the thermal

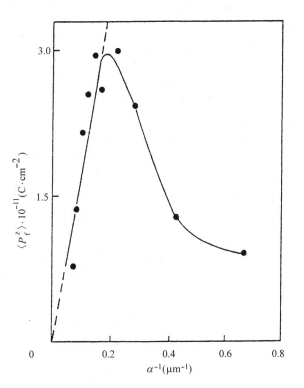

Figure 11. Flexoelectric polarization of the hybrid-oriented 5CB as a function of inverse thickness ($T = 27°\mathrm{C}$).

conductivity equation (32) might be solved with initial and boundary conditions in the form

$$\Delta T(z, t = 0) = 0, \quad \frac{\partial \Delta T(z = d/2, t)}{\partial z} = -\eta\lambda^{-1}W_0(1 - \cos\omega t) \quad (41)$$

This was applied earlier to a thin light-absorbing film on a heat conductive substrate [32], but here two glass plates are taken into account by the extra factor 1/2 in Eq. (40).

The absolute magnitude of $\langle P_f^z \rangle(x)$ has been calculated by integrating $\langle \gamma \rangle$ over temperature, Eq. (36). Fig. 11 shows flexoelectric polarization at room temperature as a function of inverse thickness. The critical thickness corresponds to the polarization maximum. For increasing thickness (the left slope) the polarization decreases proportionally to $1/d$, as expected by Eq. (38), with θ-angles independent of d. In thinner part of the wedge the polarization decreases dramatically because of the transition to the uniform (planar) state. The anchoring energy as a function of the director deviation angle at the homeotropic boundary is calculated from the polarization curve. In fact the shape of the potential well for the director deviation is found by the pyroelectric technique [30,33].

V. CHIRAL FERROELECTRICS AND ANTIFERROELECTRICS

The first measurements of the steady-state pyroelectric response of liquid crystal, DOBAMBC, were carried out [5] right after the discovery of its

ferroelectric properties [7]. However, quantitative pyroelectric measurements of the spontaneous polarization of ferroelectric liquid crystals became possible only after the development of the pulse technique [19,34,35].

A. Spontaneous Polarization in the Smectic C^* Phase

1. Quasi-static Measurements of P_s on Classical Materials

These measurements are performed either with a laser pulse longer than the soft-mode relaxation times of a ferroelectric or with the low-frequency sine wave modulation of a light beam. Curve *a* in Fig. 12 shows the first plot [35]

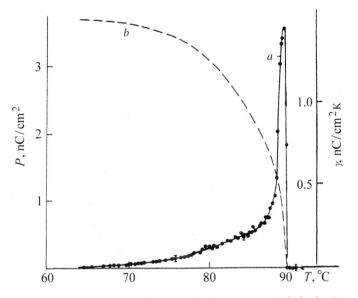

Figure 12. Pyroelectric response in (curve *a*) and spontaneous polarization (curve *b*) of the classical ferroelectric liquid DOBAMBC.

of the pyroelectric response to a $t_p = 100\,\mu s$ bell-shaped laser pulse of a rather thick ($d = 200\,\mu m$) DOBAMBC layer doped with a dye. The dye provides absorption $\eta \approx 0.7$ of light with wavelength $1.06\,\mu m$, and the growth of temperature has been calculated assuming that all the absorbed power $\eta W(t)$ of the pulse is stored in the liquid crystal layer during the pulse action. By analogy with Eq. (17) we have

$$\frac{\partial \Delta T}{\partial t} \approx \frac{\eta W(t) A}{C_T} \tag{42}$$

where C_T is the heat capacity of the DOBAMBC layer. Taking Eq. (42) with Eq. (18), from the pyroelectric voltage in the short-circuit regime,

$$U_p(t) = R_L A \gamma \frac{\partial \Delta T}{\partial t} = \frac{\gamma \eta W(t) A^2 R_L}{C_T} \tag{43}$$

it is easy to calculate the absolute magnitude of the pyrocoefficient γ and P_s (see curve *b* in Fig. 12) from the voltage oscillograms. When paper [35] was published, the P_s value of DOBAMBC (and ferroelectric liquid crystals (FLCs), in general) was measured only twice, by the polarization switching technique [36, 37]. The results of all three measurements for the smectic C* phase were very close to each other. The pyroelectric technique allowed P_s to be found in the low temperature, very viscous H-phase of the same compound. This value was not measured earlier at all. Steady-state measurements of the P_s value for another classial FLC HOBACPC (hexyloxy-homolog of DOBAMBC with a Cl substitutent in the chiral center) were reported a little later [19].

In experiments with surface stabilized (SSFLC) cells where the helical structure is unwound, in the field off-regime there still exist two equally probable states for the P_s orientation: up and down domains separated by domain walls. When areas occupied by domains of opposite sign are equal, macroscopic polarization vanishes. With the increasing bias field, the domains of one sign grow at the cost of the other, and the macroscopic polarization and pyroelectric effect appear. Then the field dependence of the macroscopic polarization becomes extremely nonlinear, and the P_s value has to be found by extrapolation of the $P(E)$ curve to the zero field.

In thin cells the $P(E)$ curve is actually a hysteresis loop which makes it similar to that usually measured, for example, by the Sawyer-Tower technique [2]. However, the fundamental reason for the hysteresis in ferroelectric liquid crystals is quite different. FLCs do not possess the volume bistability that exists in solid ferroelectrics where, under field action, one sublattice "moves through" the another sublattice passing a zero polarization state. In FLCs, the field just reorients the vector of spontaneous polarization whose magnitude remains the same.

Because of the tilt-polarization coupling, the director must be reoriented together with P_s along the conical surface with 2θ angle at the apex (θ is the director tilt, the order parameter of the smectic C phase). The field-induced torque exerted on the director is opposed by the surface anchoring forces. There exists a threshold for the polarization (and the director) field-induced reorientation that is dependent on the anchoring energy. The stronger the anchoring, the higher is the threshold and the wider is the hysteresis loop for P_s. A simple model based solely on the consideration of dispersion (nonpolar) anchoring describes the hysteresis in the pyroelectric response of a cell with a so-called bookshelf orientation [38]. The "inverse hysteresis" observed (sometimes) in thin cells, especially with chevron structure of smectic C layers, requires some additional assumptions, for example, on the role of the depolarization field caused by an ionic charge [38].

A set of voltage-temperature dependencies of the pyroelectric response and polarization will also be discussed in Section V. B. for multicomponent FLC mixtures with high P_s.

Both the spontaneous polarization and pyroelectric coefficient may have different signs in different substances. A compensated nonferroelectric mixture may be prepared by mixing two substances with opposite polarization signs. At a slightly different composition of the same components, we have a ferroelectric mixture wth compensated chirality (nonhelical ferroelectric). For the first time such mixtures have been prepared from left-handed DOBAMBC and right-handed HOBACPC, and their ferroelectric properties were studied by the long-pulse steady-state pyroelectric technique [39]. The spontaneous polarization and helical wave vector for these mixtures are shown in Fig. 13. The two "magic" points, zero P_s and zero q, are indicated in the figure.

2. Polymer FLCs

Another example of application of the pulse steady-state technique is the first observation of ferroelectricity in polymer liquid crystals [6]. Figure 14 shows the temperature dependencies of the pyroelectric response and spontaneous polarization in the smectic C^* phase of a chiral side-chain metacrylate polymer (P5*M):

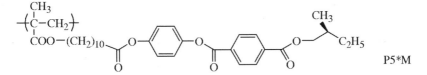

In later observations other ferroelectric polymethacrylates were synthesized with enhanced P_s, and they were also characterized by the pyroelectric technique [40]. Much higher spontaneous polarization (up to $200 \, C/cm^2$) has been found in the polysiloxane liquid crystal polymer [41]. A characteristic feature of polymer FLCs is blurred ferroelectric transition.

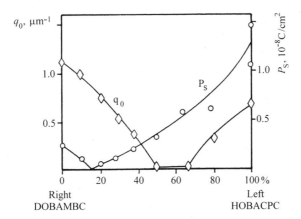

Figure 13. Spontaneous polarization P_s and the helical wave vector q for mixtures of left-handed DOBAMBC with right-handed HOBACPC.

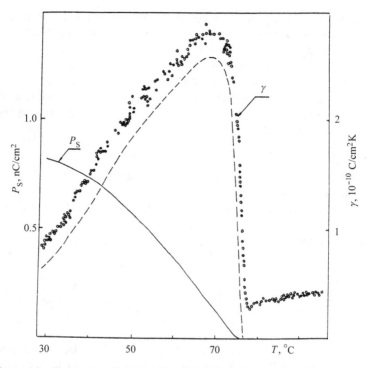

Figure 14. Temperature dependencies of the pyroelectric coefficient and spontaneous polarization in the smectic C^* phase of a chiral side-chain metacrylate polymer P5*M (1 mm cell thickness, 300 V bias voltage). The dashed line shows the extrapolation to zero voltage.

This is the reason why the peak of the pyroelectric response of a polymer is not as sharp as that of any low molecular weight counterpart. A detailed comparison has been made of the pyroelectric behavior of a low molecular weight mesogen capable of polymerization (monomer) with its polymethacrylate counterpart [40]. The mesogenic moiety contained a chiral center with the dipolar C^*–Cl group which provided an enhanced P_s. In this case the polymer FLC was viscous, and with the pulse technique it was difficult to fulfill conditions for the quasi-static regime. That explains the somewhat lower P_s values measured for a polymer as compared with its monomer [40,42].

Cross-linking of a ferroelectric polymer considerably reduces its spontaneous polarization. This result is probably due to an incomplete orientation of the cross-linked sample by the electric field. The cross-linking effect was shown with the help of the pulse pyroelectric technique [43].

3. FLC with Temperature Inversion of Polarization Sign

In conventional ferroelectric liquid crystals, at the transition from the smectic A to the chiral smectic C^* phase, spontaneous polarization appears as a second-order parameter simultaneously with the tilt angle θ of the longitudinal molecular axes with respect to smectic layers. In most cases P_s is proportional to θ over a wide temperature range of the C^* phase. However, in some substances a strong deviation from the linear $P_s (\theta)$ dependence is observed. Moreover a sign inversion of spontaneous polarization (P_s) at a certain temperature in the smectic C^* phase was reported some years ago [44]. It happened that the tilt angle was increasing monotonically with T. This phenomenon is intriguing. Up until now it has only been observed in two systems: (1) in single-component substances whose molecules have very flexible tails with at least one CH_2 chain between a chiral center and a rigid molecular core [44,45] and (2) in two-component mixtures with rather rigid molecules [46]. Strictly speaking, what is measured by the switching technique is not the spontaneous but the total (spontaneous and field-induced) polarization P. The pyroelectric technique, in principle, allows the measurements of macroscopic polarization to be performed in both the field on-regime (P) and field off-regime (P_s). Therefore a comparison of the temperature dependencies of P and P_s in the smectic C^* phase can yield interesting results.

Such comparative measurements have been carried out [47] by the quasi-static pyroelectric technique on a Chisso phenylpyrimidine derivative (8PPyO6). The conventional switching technique showed polarization sign inversion at $T_i(P_s) = 37.7°C$ [48]:

$$C_8H_{17} \overline{} \langle \text{ring} \rangle \overline{} \langle \text{pyrimidine ring} \rangle \overline{} OCH_2 - \overset{\overset{F}{|}}{C^*}H - C_6H_{13} \qquad 8PPyO6$$

The substance manifests the following phase sequence:

$$Cr-46.2C-(S_X^*-25C-S_c^*-45C)-S_A-80C-Iso$$

with $P_s = 27\,nC/cm^2$ at $T = 30°C$.

Figure 15 displays the temperature behavior of the pyroelectric coefficient (above) and total polarization (below) measured on the $8\,\mu m$ thick cell both at zero bias and at an applied dc voltage of $+10\,V$. Curve $\gamma_s(T)$ for the zero field has one zero point, and after integrating, spontaneous polarization P_s (below) shows a sign inversion at $39.8°C$. For the bias voltage of $10\,V$ the pyroelectric curve shows two zero points. The second point corresponds to the sign inversion point because here a strong field

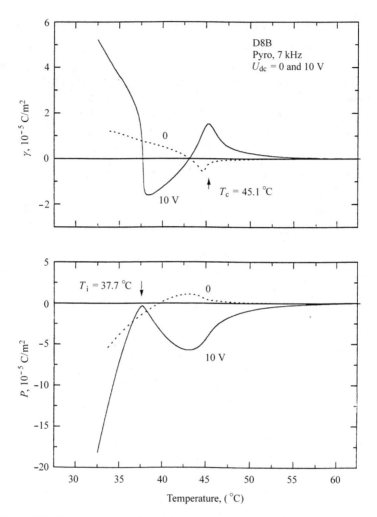

Figure 15.   Temperature behavior of the total pyroelectric coefficient γ (*above*) and total polarization P (*below*) for 8PPyO6 measured at frequency 7 kHz either without bias or with a dc voltage of + 10 V applied to the cell electrodes.

reorients the polarization and changes the sign of the pyro-response. The new location of T_i coincides with that determined from electrooptical measurements. The curve for P at 10 V does not cross the abscissa for the reason mentioned: both below and above the inversion point polarization is directed along the external field irrespectively of its sign. In the smectic A phase, a typical "tail" of the field-induced polarization attributed to the ferroelectric soft-mode (Section V.B.3) is observed.

Now we can comment on the zero-field curve for polarization measured by the pyroelectric effect. When the pyroeffect is measured at the high frequency of 7 kHz, we observe a sign inversion of P_s. With measurements at 70 Hz it is difficult to separate the true pyroelectric effect from the thermal modulation of the space charge double layers adjacent to electrodes. These layers may manifest themselves also in the smectic A or isotropic phase, and after application of a voltage to a cell, they provide an external long-living zero-field current which originates from the polarization of the electrodes. Certainly such layers can also influence the polarization measurements by a switching technique. This effect is likely responsible for the difference in T_i points at zero and finite bias voltage (Fig. 15). The pyroelectric effect is a powerful tool for studying such phenomena.

4. Study of Spatially Inhomogeneous Structures

In thick cells, ferroelectric liquid crystals form a helical structure with zero macroscopic polarization and pyroelectric coefficient. With an electric field applied, the helical structure becomes distorted, and at a certain field it unwinds [26]. The process of helix unwinding depends on many factors (field strength, the ratio of helical pitch h_0 to cell thickness d, surface anchoring, structural defects, etc.). With surface anchoring disregarded and in the weak field limit, $E \ll E_c$, the polarization and the quasi-static pyroelectric response (in the short-circuit regime) are calculated [38] as

$$P(E) = \frac{P_s^2}{K_\phi q_0^2 \sin^2\theta} E \qquad (44)$$

$$U_p = k \frac{\gamma_s P_s}{K_\phi q_0^2 \sin^2\theta} E \qquad (45)$$

where $q_0 = 2\pi/h_0$ is the equilibrium wave vector of the helix, K_ϕ and θ are the twist elastic modulus and the tilt angle of the smectic C phase, k is a coefficient that depends on the field frequency and the thermal and electric time constants. Thus in the weak field approximation we have a linear dependence between the pyroelectric voltage and a bias field. With increasing field, it is substituted by a superlinear growth of U_p with subsequent saturation.

Another example of a inhomogeneous structure is an FLC cell with the previously formed linear domain structure [49]. A very stable spatially periodic optical pattern arises in layers of FLCs with molecular orientation parallel to substrates after a dc field treatment of the cell. The stripes are oriented perpendicular to the director orientation, and their period is inversely proportional to the polarization magnitude squared and independent of cell thickness. The optical pattern may be observed for several months after switching the dc field off. The theoretical model is based on consideration of a stabilizing role of structural defects (dislocations) interacting with free charges in a FLC layer [50]. On measuring the pyroelectric response, one applies a dc bias field, and the domains disappear one after another. Hence the macroscopic polarization changes by steps. To some extent the process recalls that of Barkhausen pulses in multidomain solid ferroelectric samples [2]. The steps (or sharp changes in the slope of the pyrovoltage-bias curves) were recently observed in experiment [51].

B. Soft-Mode Susceptibility and Landau Coefficients at the Ferroelectric Phase Transition

1. General Consideration

The phase transition from the smectic A to smectic C phase is characterized by the appearance of a tilt in the director θ with respect to the smectic layer normal which is considered to be the order parameter for both the achiral C and chiral C^* phases [52,53]. The latter is ferroelectric [9,54] with spontaneous polarization P_s coupled to the tilt angle. Dynamically the SmA–SmC (C^*) transition is characterized by two relaxation modes for the director fluctuations [55, ch. 5], the fluctuations of the amplitude of the tilt θ (soft mode) and fluctuations of the azimuthal angle of the director projection onto the smectic plane (Goldstone mode). Due to the P_s–θ coupling the dielectric strength of both modes in the ferroelectric smectic C^* phase is much higher than that in the C-phase.

Theoretically the temperature and field behavior of ferroelectric liquid crystals is described in terms of the classical [52] or revised [55] Landau mean field model. Experimentally the Landau coefficients were measured for rather narrow spectrum of ferroelectric liquid crystals [36,56–63]. It is evident that the parameters of the SmA–SmC* phase transition have to be determined on both sides of the ferroelectric transition. To this effect, one can use the tilt-polarization measurements complemented by a study of the soft-mode susceptibility. However, it is very difficult to measure small polarization values in the vicinity of a transition (inasmuch as a high field repolarization technique is used) and even more difficult to measure the soft-mode susceptibility on the smectic C^* side of the transition (since it is overlapped by the Goldstone-mode susceptibility).

Both of these difficulties can be overcome by the pyroelectric technique [56,64,65]. First, both spontaneous and field-induced polarization can be measured at very low fields (no switching necessary), and second, the Goldstone-mode contribution can be filtered out during the soft-mode susceptibility measurements. The reason for the latter is quite fundamental:

In dielectric spectroscopy the perturbation of a structure is provided by an electric field that interacts strongly with both the soft and Goldstone modes. In the pyroelectric technique the perturbation is provided by a pulse or a periodic modulation of temperature, and the response is proportional to the temperature derivative of the polarization. Thus this technique is especially sensitive to the soft-mode contribution to the polarization because the latter manifests a very narrow peak near the SmA–SmC* transition.

In framework of the simplest Landau model, the free energy of a ferroelectric liquid crystal taken in the SI unit system is

$$g = \frac{1}{2}a_0\theta^2 + \frac{1}{4}b\theta^4 + \frac{P^2}{2\varepsilon_0\chi_\perp} - CP\theta - PE + \text{high-order terms} \qquad (46)$$

where θ [in rad] is the tilt of the director, $a_0 = \alpha(T - T_0)$ [in J/m^3 K] describes the elasticity for θ changes, P [in C/m^2] is the polarization, C [in V/m] is the P–θ coupling constant, E is an external electric field, $\chi_\perp = \varepsilon_\perp - 1$ is the background relative dielectric susceptibility that may be taken from the smectic A phase well above T_0, ε_0 [in F/m] is the dielectric constant of vacuum. During the tilt, polarization, and soft-mode susceptibility measurements the helical structure is assumed to be unwound, and all terms related to chirality are disregarded in Eq. (46).

In the case of an achiral smectic C phase, in absence of an external field, after minimization of (46) with respect to the tilt angle, we have

$$\theta_s = \left(\frac{\alpha}{b}\right)^{1/2}(T_0 - T)^{1/2} \qquad (47)$$

where $T_0 = T_{CA}$ is the phase transition point for a racemic compound. Equation (47) allows us to determine coefficient b if α is known.

For the smectic A phase of a chiral compound, with an exception of very strong fields we can assume small θ angles and a linear relationship between the additional polarization (over the background value $\chi_\perp E$) and the tilt:

$$P = \chi_\perp\varepsilon_0 C\theta \qquad (48)$$

Then terms 3 and 4 in the expansion (46) merge with the first term and renormalize the transition temperature. After minimization with respect to the tilt angle, we arrive at the Curie-Weiss type of expression for the linear electroclinic tilt angle with respect to the electric field:

$$\theta = \frac{\chi_\perp\varepsilon_0 CE}{a_0 - \chi_\perp\varepsilon_0 C^2} = \frac{\chi_\perp\varepsilon_0 CE}{\alpha(T - T_c)} \qquad (49)$$

Since the dipole-dipole interactions stabilize the smectic C* phase (which, in this particular case, is polar but not helical), the phase transition temperature increases and the field-induced tilt angle diverges at temperature

$$T_c = T_0 + \frac{\chi_\perp \varepsilon_0 C^2}{\alpha} \tag{50}$$

In reality, a growth of the induced tilt at the phase transition is limited by two factors. In compensated (nonhelical) ferroelectrics or in thin cells, only $b\theta^4/4$ term in the expansion (46) is limiting [65,67]. In other cases, where the helix cannot be compensated over the whole range of the smectic C* phase and a finite wave vector q_0 remains, the chiral terms in the expansion (46) become important. These renormalize the transition temperature and may limit the divergence of the induced tilt in Eq. (49) [52]:

$$\theta = e_c E = \frac{\chi_\perp \varepsilon_0 C E}{\alpha(T - T_{ch}) + K q_0^2} \tag{51}$$

Here e_c is the so-called electroclinic coefficient, K is renormalized bend elastic modulus, and the transition temperature for helical (and polar) ferroelectric is

$$T_{ch} = T_0 + \frac{\chi_\perp \varepsilon_0 C^2}{\alpha} + \frac{K q_0^2}{\alpha} \tag{52}$$

Combining Eqs. (48) and (51), we have an expression for the low field soft-mode susceptibility in the smectic A phase:

$$\chi_{sm} = \frac{P}{\varepsilon_0 E} = \frac{\chi_\perp^2 \varepsilon_0 C^2}{\alpha(T - T_{ch}) + K q_0^2} \tag{53}$$

For very strong fields the helical structure does not exist, the $b\theta^4/4$ term in expansion (46) dominates over $a_0\theta^2$ term, and the induced tilt

$$\theta = \left(\frac{\chi_\perp \varepsilon_0 C E}{b}\right)^{1/3} \tag{54}$$

and susceptibility

$$\chi = \frac{1}{\varepsilon_0}\frac{\partial P}{\partial E} = \frac{1}{3}\left(\frac{\chi_\perp^4 \varepsilon_0 C^4}{b}\right)^{1/3} E^{-2/3} \tag{55}$$

become independent of temperature (condition (48) is still assumed to be valid).

In the chiral smectic C^* phase, for $E = 0$ and $\theta = \text{const} = \theta_s$, minimization of Eq. (46) with respect to polarization provides the relationship between the spontaneous polarization and equilibrium tilt angle:

$$P_s = \chi_\perp \varepsilon_0 C \theta_s \qquad (56)$$

and the low field susceptibility consists of two parts (the soft and the Goldstone mode):

$$\chi = \chi_{SM} + \chi_G = \frac{(1/2)\chi_\perp^2 \varepsilon_0 C^2}{2\alpha(T_{ch} - T) + Kq_0^2} + \frac{(1/2)\chi_\perp^2 \varepsilon_0 C^2}{Kq_0^2} \qquad (57)$$

Thus, if we combine Eqs. (53) and (57) for the two phases, the low field susceptibility manifests a quasi–Curie type behavior with a cusp of height

$$\chi_{SM}^m = \frac{\chi_\perp^2 \varepsilon_0 C^2}{Kq_0^2} \qquad (58)$$

at temperature T_{ch}. With independently measured values of P and θ, from the temperature dependence of the inverse susceptibility,

$$\frac{\partial(\varepsilon_0 \chi_{SM})^{-1}}{\partial T} = \frac{k\alpha}{(\chi_\perp \varepsilon_0 C)^2} \approx \frac{k\alpha}{(P/\theta)^2} \qquad (59)$$

in the smectic A $(k = 1)$ and smectic $C^*(k = 4)$ phases, we can calculate Landau coefficients C, α (or a), and b (using Eq. (48))

When the classical Landau approach (46) disagrees with experimental data, additional contributions have been included in the Landau free energy, for example, terms that reflect a quadrupolar order of the short axes of lathlike molecules; for example, see [55, ch. 5] and [66] for theory and [47,59] for experimental data.

To clarify the principle of the pyroelectric measurements of the soft-mode susceptibility, let us consider the field derivative of the total pyroelectric coefficient which was defined earlier; see Eq. (8):

$$\frac{\partial\gamma(T, E)}{\partial E} = \frac{\partial}{\partial E}\left[\frac{\partial P(T, E)}{\partial T}\right] = \frac{\partial}{\partial T}\left[\frac{\partial P(T, E)}{\partial E}\right] = \frac{\partial\chi(T, E)}{\partial T} \times \varepsilon_0 \qquad (60)$$

Thus, if at each temperature the field derivative of pyrocoefficient is measured, the temperature derivative of the total susceptibility might be calculated as

$$\chi_{SM}(T) = \frac{1}{\varepsilon_0} \times \int_{\infty}^{T} \frac{\partial \gamma(T, E)}{\partial E} dT \qquad (61)$$

If the polarization is known from Eq. (36) (with some temperature $T > T_{C^{\cdot}A}$ as an initial point for integrating pyrocoefficient), function $\chi_{SM}(T)$ is known from Eq. (61), and $\theta(T)$ is measured independently, then the temperature dependence of Landau coefficient $\alpha(T)$ may be found from Eq. (59) over a wide temperature range on both sides of the A–C^* transition. As was mentioned earlier, this technique is especially useful for the study of temperature and field dependence of soft-mode susceptibility near the smectic A–C^* phase transition for unwound helical structures (it does not matter by what means, surfaces, electric fields, or compensation of helical pitch with specially chosen additives). In this case the Goldstone mode is suppressed, and the temperature derivative of the susceptibility is very high. We will show several examples of application of this technique below.

2.   Soft-Mode Susceptibility of DOBAMBC

We believe that work [56] was the first successful attempt to measure the whole temperature dependence of the soft-mode susceptibility in a ferroelectric liquid crystal, on both sides of the SmA–SmC* transition. The experiment was made on a rather thick $(d = 110\,\mu m)$ cell with a dye (0.2%) dissolved as was explained at the beginning of Section III. The complete absorption of IR light of a Nd-glass laser, operating in the cw regime (with total pulse duration of $200\,\mu s$), was provided by the dye, and this made it possible to carry out the absolute measurements of the total polarization.

Figure 16 shows some field dependencies of a pyroelectric coefficient measured at various temperatures. As can be seen, the sign of the slopes on opposite sides of the A–C^* phase transition is different. The magnitude of the slope has been converted in the temperature derivative of the soft-mode susceptibility according to Eq. (60). This is displayed in Fig. 17 (CGS units).

Upon integrating this curve over the temperature, we obtain the temperature dependence of the soft-mode susceptibility as is also shown in the figure. The shape of the curve is similar to the cusp predicted in Ref. [55, ch. 5]. It is also worth noting that the maximum value of $\chi_{SM} = 0.012$ is 20 times less than the subtracted background value ($\chi_{\perp} = 0.28$), and this smallness characterizes the sensitivity of the pyroelectric measurements. If

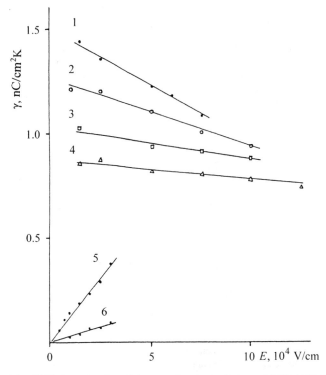

Figure 16.   Field dependencies of the pyroelectric coefficient of DOBAMBC at various temperatures and $E \gg E_c$. Temperatures: 90.2 (1), 90.05 (2), 89.8 (3), 89.6°C (4) in the smectic C^* phase and 91.0 (5), 91.4°C (6) in the smectic A phase.

the inverse soft-mode susceptibility is multiplied by $(P/\theta)^2$, we get an effective elastic modulus for tilt K_θ. For an assumed temperature-independent ratio (P/θ), the temperature dependene $K_\theta(T)$ obeys the Landau theory (straight lines in Fig. 17).

3.   Landau Coefficients for a Multicomponent Mixture with High P_s

Recently the pyroelectric technique has been applied to the measurements of Landau expansion coefficients α, b, and C in Eq. (46) for several multicomponent mixtures [64,65,68]. Some of the mixtures studied showed an anomalous electroclinic effect (as FLC 441 and 442; see [68]), while others (as FLC 273 and 363) showed very high P_s, which are results of great interest for comparison with the former. Here we will present only the results of the investigation of FLC 273 in order to illustrate the technique; for details, see [64].

L. M. BLINOV

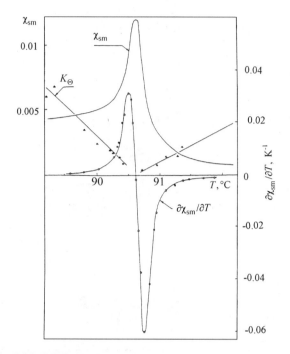

Figure 17. DOBAMBC: Temperature dependence of the soft-mode susceptibility χ_{SM} (in CGS units), its temperature derivative, and the elastic modulus K_θ proportional to the inverse susceptibility (in arbitrary units).

The mixture is based on achiral compounds possessing a wide range smectic C phase and two chiral nonmesogenic additives. The phase transition temperatures are $T_{C^*A} = 53.8°C$ and $T_{AI} = 67°C$. The uniform planar orientation of the substance in a flat capillary cell (area $0.15\,cm^2$, a gap of about $10\,\mu m$) was achieved by means of an ac field treatment at the smectic A–C* transition. The ITO electrodes were covered with polyvinyl-alcohol layers and rubbed unidirectionally.

The director tilt was measured by the technique based on observation of an oscillogram of the electrooptical response of a cell placed on a rotating table under a polarizing microscope [69].

The polarization was measured with a 30 Watt incandescent lamp as a heat source. The focused light was mechanically chopped at frequency 70 Hz (see Section III.B). The sample temperature was measured with a Pt resistor and a digital multimeter (with accuracy of about 0.03°C). The temperature dependence of the spontaneous polarization (on an arbitrary scale) was calculated by integrating the pyroelectric voltage according to

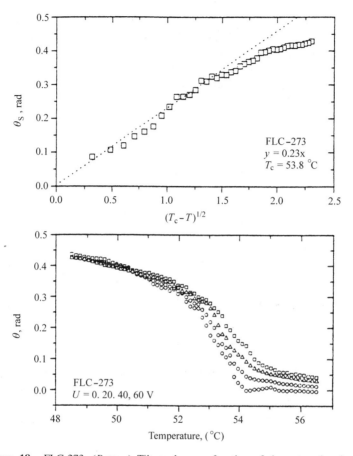

Figure 18. FLC-273. (*Bottom*) Tilt angle as a function of the external voltage and temperature. Voltages (*from bottom to top*): 0 (extrapolated values), 20, 40, and 60 V. (*Top*). The spontaneous tilt angle (at zero voltage) as a function of the square root of the "temperature distance" from the A–C* transition.

Eq. (36) (with T_{NI} replaced by T_{CA}). Then the correct scale for P_s was introduced by fitting P_s at severel temperatures to the values measured independently with the standard repolarization current technique. At 20°C the P_s value is about 100 nC/cm².

Temperature dependencies of the field-induced tilt angle of the mixture FLC-273 at three different voltages are displayed in Fig. 18 (bottom). The fourth (lowest) curve was obtained by extrapolation to the zero field. It thus represents the equilibrium (spontaneous) tilt in the smectic C* phase. In the upper part of Fig. 18 the equilibrium tilt angle is plotted as a function of the square root of the temperature deviation from the SmA–SmC* transition.

Despite some point scattering, linear dependence can be seen, at least within 2°C from the transition. According to Eq. (47), the measured slope 0.23 rad/$K^{1/2}$ allows for the calculation of the ratio of the first and second Landau coefficients: $\alpha/b = 0.053\,K^{-1}$. The deviation of the curve from the linear dependence at enhanced tilt angles will be discussed later.

Tilt-polarization Coupling Constant. To find P/θ, we compare first the spontaneous tilt with the spontaneous polarization. Fig. 19 illustrates our

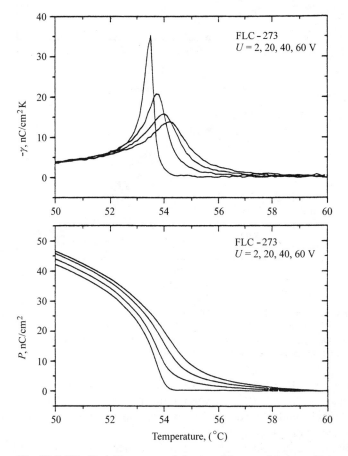

Figure 19. FLC-273. (*Top*). Temperature behavior of the pyroelectric coefficient $-\gamma$ (the negative value is taken for convenience of plotting) on cooling in the presence of an electric voltage. The voltages correspond to each subsequent curve. With the maxima shifted from the left to the right, they are 2, 20, 40, and 60 V. (*Bottom*). Temperature dependencies of the polarization of FLC-273 calculated by integrating γ. Voltages (*from bottom to the top*): 0 (spontaneous polarization), 20, 40, and 60 V.

approach. First, the pyroelectric coefficient γ was measured as a function of temperature and applied voltage, Fig. 19 (above). Then, after integrating the pyrocoefficient over temperature, starting from $T = 60°C$ (where γ is negligible), we calculate the polarization, Fig. 19 (bottom). The lowest of the four curves is the spontaneous polarization found by extrapolation of the other curves to a zero field.

To verify Eq. (56), the spontaneous polarization of the smectic C^* phase is plotted as a function of the equilibrium tilt angle taken from Fig. 18. The results are displayed in Fig. 20. (below). The linear relationship between the spontaneous polarization and tilt holds within a range of approximately

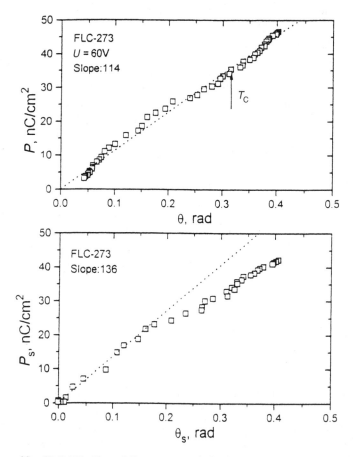

Figure 20. FLC-273. (*Bottom*) Spontaneous polarization as a function of the equilibrium tilt angle in the smectic C^* phase (zero voltage). (*Top*) Total polarization as a function of the total tilt angle, $U = 60$ V.

10°C below the phase transition. The slope is $136\,\text{nC/cm}^2\text{rad}$. At lower temperatures the coupling constant considerably decreases.

In the smectic A phase, only the field-induced polarization–tilt ratio can be measured. For a field of $6\,\text{V/}\mu\text{m}$ the curve is shown in the upper part of Fig. 20. Again, almost linear dependence is observed near the phase transition. Now the slope is $114\,\text{nC/cm}^2\text{rad}$, which is 20% smaller than for spontaneous polarization–tilt ratio. In general, the tilt-polarization measurements in the vicinity of the SmA–SmC* transition are very difficult, and accuracy of about 20% is acceptable. We can conclude that for FLC-273 we have approximately the same value of the coupling constant in both the smectic A and C* phases. In other substances the slopes in the SmA and SmC* phases can be quite different (e.g., the ratio of SmC* to SmA slopes in FLC-363 exceeds 2 [64]). A strong temperature dependence of P/θ ratio has also been found for the polysiloxane ferroelectric liquid crystal [41].

Soft-Mode Susceptibility. The first step is to take field derivatives of the $\gamma(T, E)$ curves, Fig. 19 (top), to obtain the temperature derivative of the soft-mode susceptibility according to Eq. (60). The next step is to integrate these derivatives over temperature to obtain the soft-mode susceptibility as a function of temperature and field, Eq. (61). The new plots are displayed in Fig. 21 (top). For fields higher than $0.4\,\text{V/}\mu\text{m}$, the Goldstone mode is suppressed almost completely, and the peaks of the soft-mode susceptibility are well pronounced. Also seen in the figure is how higher electric fields suppress the maximum of the soft-mode susceptibility approaching the high field limit, Eq. (55). This phenomenon resulting from the same pyroelectric technique was studied in detail in papers [65,67]. It was shown that the position T_m and the height χ_m of the susceptibility maximum change with an external field according to the laws

$$T_\text{m} = T_\text{c} + 3\left(\frac{bk^2}{16\alpha^3}\right)^{1/3} E^{2/3} \tag{62}$$

and

$$\chi_\text{m}^{-1} = 3\varepsilon_0\left(\frac{b}{2k^4}\right)^{1/3} E^{2/3} \tag{63}$$

and are mostly controlled by the second Landau coefficient b.

To calculate the first Landau coefficient a, we have to plot the inverse susceptibility $\Delta\chi^{-1}(T, E)$ and estimate the temperature slope. Figure 21 (bottom) illustrates the technique for several selected voltages. The inverse susceptibility manifests a field-dependent minimum that might be considered an "apparent" phase transition point. The linear extrapolation lines of the inverse susceptibility curves from the smectic A side down to the zero

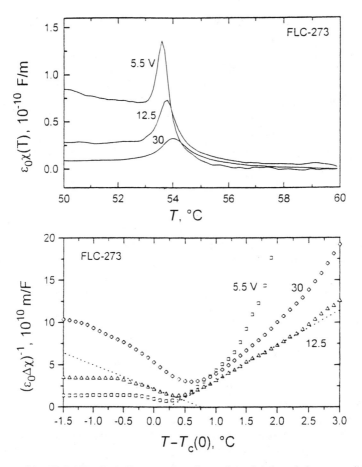

Figure 21. FLC-273. (*Top*) Temperature-voltage dependencies of the susceptibilities calculated from the pyroelectric data. (*Bottom*) Temperature-voltage dependencies of the inverse susceptibilities near the A–C* transition. $T_c(0)$ is the transition temperature in the zero field. voltages are shown at the curves.

ordinate cross the abcissa close to the points of the susceptibility minima (e.g., see the curve taken at $U = 12.5\,\text{V}$). The same does not hold for the smectic C phase where intersections with abscissa take place at temperatures higher (sometimes few degrees) than the "apparent" transition point. Such a phenomenon has been observed earlier [59] and attributed to the influence of the biquadratic tilt–polarization coupling term in the Landau expansion.

The slope at the smectic A side is dependent on voltage. In particular, this dependence is pronounced at small voltages. For intermediate fields about $10\,\text{V}/\mu\text{m}$, the magnitude of the slope within 1.5–2°C up from the "apparent" transition is nearly constant and, together with the tilt–polarization coupling constant for the A phase, can be used in calculating the first Landau coefficient; see Eq. (59), $\alpha = 5 \times 10^4 \, \text{J/m}^3 \, \text{K}$. From the α-coefficient measured in the A phase and the α/b ratio measured in the C^* phase, the Landau b-coefficient can be found, $b = 93 \times 10^4 \, \text{J/m}^3$.

With decreasing temperature the experimental points for the equilibrium tilt angle deviate down from the curve predicted by Eq. (47). It is a common opinion that the third Landau coefficient c in the $c\theta^6/6$ term is positive; thus, indeed, the curve should deviate down. Therefore with b known we can calculate the coefficient c from the inflection point as observed in Fig. 18 (top). The c coefficients found is $9 \times 10^6 \, \text{J/m}^3$ for FLC-273.

Thus the temperature behavior of the tilt, polarization, and susceptibility of the multicomponent liquid crystal mixture FLC-273 showing the second-order smectic A–ferroelectric smectic C^* phase transition and the high spontaneous polarization in the chiral smectic C^* phase were shown to follow predictions of the simple Landau theory. Since the purpose of this section is to show the methodological aspects of the Landau coefficient measurements, we will not discuss the magnitudes of the obtained parameters. The order of magnitude of all the parameters is quite reasonable (e.g., it is not the case for FLC-441), and a comparison with other materials has been done [64].

4. Landau Coefficients for an FLC with Polarization Sign Reversal

Because of the strong deviation from linear $P - \theta$ dependence, Eq. (56), in FLCs with polarization sign reversal at a certain temperature, strictly speaking, the Landau expansion in the form of Eq. (46) is no more valid. However, from the experimental point of view, it is of great interest to measure the effective "temperature-dependent Landau expansion coefficients" over a wide temperature range in order to understand better the reason for the discrepancy with the simplest theory. Such measurements were carried out recently on 8PPyO6, the substance already discussed in Section V.A. 3. The experiment was similar to that described in the previous section, but the setup was controlled by a computer [47].

The samples (area $= 11.5 \, \text{mm}^2$) were prepared from ITO covered glass plates separated with 8 µm thick teflon spacers and filled with a liquid crystal in the isotropic phase. One of the ITO electrodes was covered with a polyvinylalcohol layer and rubbed unidirectionally. The absence of any diffraction on the helical pitch indicates that the helix is unwound. The temperature was measured with accuracy of 0.01°C. The thermal jacket was installed on a polarizing microscope which allowed the observation of the texture and the installation of a Si photodiode or a laser diode instead of a microscope ocular. Thus electrooptical, pyroelectric, and capacitance measurements could be carried out exactly in the same thermal regime (on sample cooling) by the same lock-in amplifier.

The tilt angle was measured indirectly from the field-induced optical transmission of a cell placed between crossed polarizers and oriented at a certain angle (about 10°) with respect to a polarizer. The absolute calibration was made by the conventional technique [69]. For polarization measurements, a beam of a 30 mW laser diode ($\lambda = 690$ nm) was used to provide a small local temperature change in a sample. The light was chopped at frequency 7 kHz by square pulses from the TTL output of the function generator and directed onto a FLC cell through the microscope. The soft-mode dielectric susceptibility was calculated from the temperature dependence of the capacitance component of the ac current through the cell, measured as a voltage across a load resistor of 2 kOhm (at frequency 300 Hz).

Figure 22 shows dielectric permittivity of an 8 μm cell measured with very low ac electric voltage either without a dc bias or with $U_{dc} = 1$ V. The peak of the soft-mode susceptibility is clearly seen, and it corresponds to the zero-field A–C* phase transition point, $T_C = 45.1$°C. In addition a deep minimum near 37–38°C points to the polarization sign inversion temperature T_i. Application of the bias voltage, as expected, suppresses the soft-mode peak.

In Fig. 15 shown earlier, $P_s(T)$ dependence was plotted for 8PPyO6 at zero and finite (10 V) bias voltages. Figure 23 shows the measured (electrooptical) tilt angles under similar conditions (0.1 V and 10 V of the 3 Hz square wave pulses applied). The inversion point is clearly seen because, at this point, $P_s = 0$ and the director is not switched. In fact the real tilt angle C the order parameter for the smectic C* phase) changes monotonically with the temperature [48].

As we can see, the qualitative behavior of susceptibility, the field-induced polarization, and the electroclinic effect in the smectic A phase are consistent with what can be expected by the Landau description. Let us try to estimate the most important Landau coefficients C and α. To avoid the integration procedure which may reduce accuracy, we can present the pyroelectric coefficient from Eq. (53) (without the chiral term) as follows:

$$\gamma = \frac{dP}{dT} = -\frac{\chi_\perp^2 \varepsilon_0^2 C^2 E}{\alpha (T - T_c)^2} \tag{64}$$

Using Eq. (49), we should have a temperature-independent apparent coefficient α_{app}:

$$\alpha_{app} = \frac{\gamma(T)E}{\theta^2(T)} \tag{65}$$

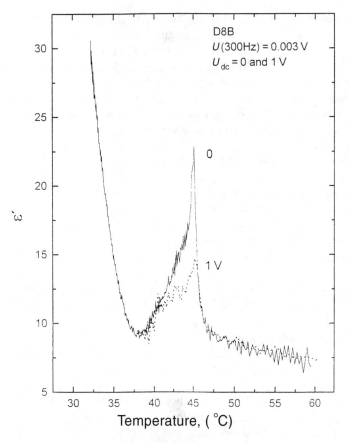

Figure 22. Dielectric permittivity of PPy806 measured at ac voltage $U = 0.003$ V (300 Hz) without bias and with a dc voltage of 1 V.

The result of such a calculation is shown in Fig. 24. Clearly, the apparent α-coefficient strongly depends on temperature. Even in the smectic A phase, it changes within one order of magnitude. Thus, strictly speaking, the Landau description (46) is not valid. We can, however, proceed further and verify the validity of the Curie-Weiss law separately for the tilt angle and induced polarization, Eqs. (49) and (53). Figure 25 shows the reciprocals of both the induced tilt angle and polarization as functions of temperature. The behavior of the two curves is quite different; the inverse tilt obeys the Curie-Weiss law within a range of 2–6°C from the A–C* transition, but the inverse polarization does not. Thus, a deviation from the Landau-type behavior can mostly be attributed to the temperature dependence of the

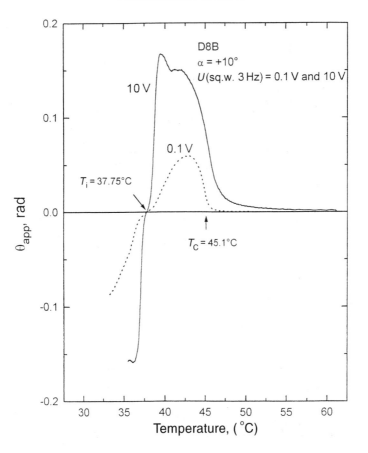

Figure 23. Apparent field-induced tilt angle of PPy806 measured with a square wave voltage of ±0.1 V and ±10 V at frequency 3 Hz.

induced polarization (or field-dependent susceptibility). Indeed, a field dependence of the soft-mode susceptibility is seen in Fig. 22.

It is of interest to calculate from Eq. (48) the apparent polarization-to-tilt ratio. As expected, the product $\varepsilon_0 \chi_\perp C$ depends on temperature, Fig. 26. It has a maximum well above the SmA–SmC* transition, at $T-T_c \approx 5°C$. The polarization-to-tilt ratio also shows a certain anomaly (an inflection point) right at the phase transition. Such an anomaly has also been observed in an experiment on the ferroelectric mixture FLC-363 which showed no inversion phenomenon [64]. Both the maximum of C in the A phase and the inflection point at the transition can be understood in framework of the theory [66], since it takes tilt-induced biaxiality into account (it does not matter whether the spontaneous or field-induced tilt is considered).

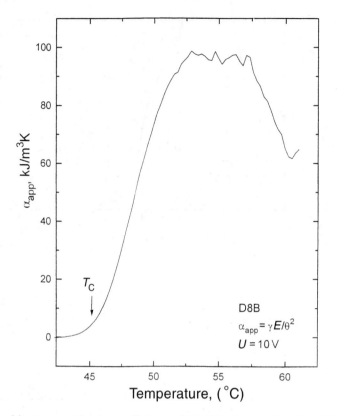

Figure 24. Apparent Landau coefficient α of PPy806 calculated from Eq. (65). γ and θ measurements were carried out at $U = 10$ V.

5.   Field-Induced Polarizaztion in a Nematic at the N–C* Transition

The anomalous electric properties of the chiral nematic phase in the vicinity of a N*–C* phase transition have been the subject of several investigations. Under an external electric field, the nuclei of the smectic C* phase were shown to form in the nematic phase [70,71] a linear electrooptical effect typical of polar mesophases observed in the chiral nematic phase [72,73]. The aim of our work is to make direct measurements of the field-induced polarization in the chiral nematic phase and to compare the values of the susceptibility and the electroclinic coefficient (close to the N*–C* transition) with those typical of the smectic A phase (close to the SmA–SmC* transition) [74].

We want to find to what extent the simplest Landau approach developed for the SmA–SmC* transition (see above) might be useful in this particular

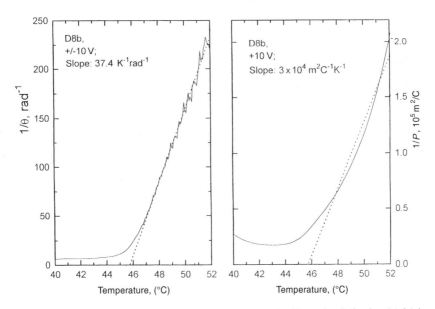

Figure 25. Fitting of the reciprocal tilt angle (*left*) and reciprocal polarization (*right*) in the smectic A phase of PPy806 to the Curie-Weiss law.

case. So far in framework of phenomenological theory [71] of the frist-order N^*–smectic C^* transition, the relevant electrical parameters of the chiral nematic phase have not been considered.

Ferroelectric properties have been studied for a phenyl-pyridine derivative FFP (E. Merck, Darmstadt)

$$C_8H_{17}-O-\!\!\!\!\!\bigcirc\!\!\!\!\!\!\!\bigcirc\!\!\!\!\!-OCH_2-\underset{\underset{F}{|}}{C^*}H-C_6H_{13} \qquad (FFP)$$

having the first-order nematic–smectic C^* phase transition at $T_c = 47°C$ [70]. The substance was placed in a flat capillary cell of $7\,\mu m$ thickness. The ITO electrodes were covered with polyvinylalcohol layers and rubbed unidirectionally to achieve uniform planar orientation in the field on-regime. In the field off-regime the sample had a helical structure. The pyroelectric coefficient was measured dynamically with a chopped (70 Hz) beam of a 30 W incandescent lamp. The pyro-response was measured over a wide temperature range with a lock-in amplifier and a computer.

The temperature dependencies of the pyroelectric response for various bias voltages applied to the cell are show in Fig. 27. Because of the helical

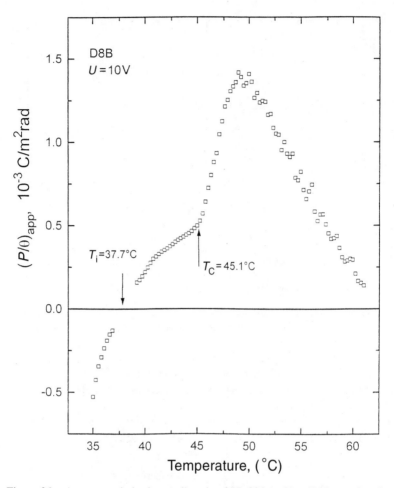

Figure 26. Apparent polarization-to-tilt ratio of PPy806 (at $U = 10$ V) as a function of temperature.

structure of the material, the zero-field pyroelectric signal is negligible. It increases considerably with the increasing bias voltage. For $U = 15$ V the helical structure is completely unwound. After integrating over the temperature, the macroscopic (a sum of the spontaneous and field-induced terms) polarization is found (Fig. 28).

Since, a relevant theory is lacking, we used the simplest Landau approach to smectic A–C^* transition, Eqs. (51) and (53), to calculate the static value of the soft-mode dielectric susceptibility and the electroclinic coefficient [75] for the chiral nematic phase. We can compare them with the same

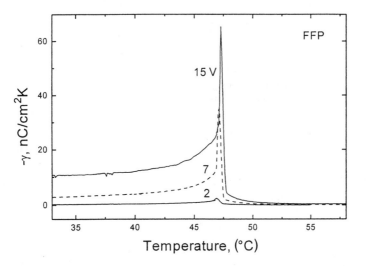

Figure 27. Temperature dependencies of the pyroelectric coefficient of FFP for various bias voltages applied to the cell of 7 μm thickness (the absolute y-axis scale was found after fitting the magnitude of the polarization in Fig. 28).

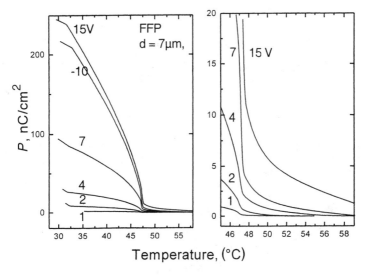

Figure 28. Temperature dependencies of the macroscopic polarization of FFP at various bias voltages. (*Left*) Polarization in a wide temperature range necessary for the absolute calibration. (*Right*) Pretransitional behavior of the macroscopic polarization in the nematic phase of FFP. Voltages are shown at the curves.

parameters typical of the smectic A phase. The C constant is taken from the tilt-polarization ratio in the smectic C^* phase. Figure 29 (left) shows the electroclinic coefficient in the chiral nematic phase calculated from data of Fig. 28. The magnitude of the coefficient is field dependent; it grows with the unwinding helix in the smectic C^* phase. At 15 V the helix is unwound, so we can compare the e_c values for the chiral nematic and smectic A phases. From Fig. 28 (left) we have $e_c = 1.4 \cdot 10^{-8}$ rad.m/V for $T - T_{C^*N} = 1$ K. This is comparable to the figure found in [72] for another chiral nematic material and one order of magnitude less than e_c values in the best smectic A electroclinic materials [76].

For further comparison of pretransitional properties of the N^* and A phases, the temperature slope of the inverse soft-mode susceptibility

$$\frac{d\chi^{-1}}{dT} = \frac{\alpha}{\chi_\perp^2 \varepsilon_0 C^2} \tag{66}$$

is plotted in Fig. 29 (right). As in the case of the electroclinic effect at the first-order smectic $A-C^*$ transition [77, 78], the inverse susceptibility for the completely unwound structure linearly depends on temperature, and the divergence of the susceptibility takes place at a temperature T^* by 2 K lower than the first-order transition point. With the *field-independent C-constant*,

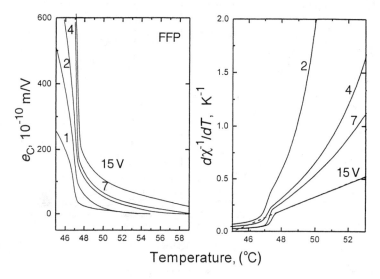

Figure 29. Temperature dependencies of electroclinic coefficient (*left*) and the slope of the inverse susceptibility (*right*) in the chiral nematic phase of FFP for various bias voltages.

$C \approx 2.5 \cdot 10^{-3}\,\mathrm{C/m^2}$, we have from Fig. 29 that both the slope and the apparent elastic modulus α are field dependent. The field dependence of modulus α is discussed in [74]. For sufficiently high fields of the order of 10^7 V/m the α-value approaches the level typical of the smectic A–C* phase transition, $\alpha \approx 5 \cdot 10^4\,\mathrm{J/m^3 K}$ (for smaller fields the apparent modulus increases, e.g., see [64]).

C. Dynamics of Polarization

1. Soft Mode

In general, a ferroelectric phase transition is characterized by a peak in dielectric susceptibility χ_{sm} due to softening of a certain fluctuation mode. In the particular case of the smectic A–C (or C*) transition the dynamics of the tilt fluctuations is described in terms of the Landau-Khalatnikov equation for the balance of the elastic Γ^{θ} and viscous torques Γ^{v} [55]:

$$\Gamma = \Gamma^{\theta} + \Gamma^{v} = \frac{\partial F}{\partial \theta} - \gamma_{\theta}\frac{\partial \theta}{\partial t} = 0 \qquad (67)$$

where $F(\theta, T, P, E)$ is the free energy expressed by Eq. (46), and γ_{θ} is the "tilt viscosity" which describes energy dissipation when the θ-angle (the order parameter of the transition) changes.

In the linear approximation, when the deviation of angle θ from its equilibrium value is small ($b = 0$ in Eq. (46)), we have

$$\frac{\partial \theta}{\partial t} = \frac{\alpha(T - T_0)}{\gamma_{\theta}}\theta = \frac{\theta}{\tau_{SM}} \qquad (68)$$

where τ_{SM} is the soft-mode relaxation time (inverse of the soft mode relaxation frequency f_{SM})

$$\tau_{SM} = \frac{1}{2\pi f_{SM}} = \frac{\gamma_{\theta}}{\alpha(T - T_0)} \qquad (69)$$

Equation (69) can be used to find the viscosity coefficient γ_{θ} after the Landau coefficient α and the soft-mode relaxation time are determined.

According to Eq. (69), fluctuations of the tilt relax with time resulting in a characteristic divergence at the SmA–SmC* transition T_0 (the Curie-Weiss law). However, in experiments, a peak is observed in τ_{SM}, and its amplitude is limited by higher terms in Landau expansion. Among them the most important is term $b\theta^4$, but chiral (or derivative) terms might also be essential. The peak has to be suppressed in the strong field limit.

Verification of the Landau theory and determination of the γ_θ coefficient by measurements of temperature dependencies of χ_{SM} and τ_{SM} is one of the key tasks in FLC dielectric spectroscopy. The problem, however, is very difficult for an FLC with small spontaneous polarization, since χ_{SM} is proportional to the square of the tilt-polarization constant C, which in turn is proportional to P_s. For example, to our knowledge, up to now all attempts to study soft-mode dynamics in DOBAMBC by dielectric spectroscopy were unsuccessful, and the data on χ_{SM} and τ_{SM} were obtained from the electro-clinic effect only on the smectic A side of the SmA–SmC* transition [75].

On the other hand, pyroelectric measurements can be performed with the electric field applied as the azimuthal fluctuations of the director (the Goldstone mode) are suppressed at the field much lower than those suppressing the soft-mode fluctuations. The change in temperature monitors the shape of the χ_{SM} peak with high sensitivity (thermal modulation instead of the field modulation used in dielectric spectroscopy). That is why the dynamics of the soft mode in the smectic C* phase of DOBAMBC has been studied by the pyroelectric effect as far back as 20 years ago [79,80].

Because of the coupling of the tilt and polarization, the soft-mode dynamics in FLCs are studied by observating the polarization kinetics. The simplest kinetic equation [79] for the polarization of an FLC unwound by a bias dc field is

$$\frac{dP}{dt} = -\frac{P - P_0}{\tau_{SM}} \tag{70}$$

For a short rectangular pulse, $t_p \ll \tau_T$ (thermal time constant) with power density W_0, $\partial T / \partial t \approx 0$ for $t > t_p$, and it is defined by Eq. (42) for $t < t_p$. Assuming a small temperature increment and an expanding equilibrium polarization P_0 in the Taylor series, we obtain

$$U(t) = \frac{\gamma R_L \eta W_0 A^2}{C_T} \left(1 - \exp\left(-\frac{t}{\tau_{SM}} \right) \right), \qquad t < t_p \tag{71}$$

and

$$U(t) = \frac{\gamma R_L \eta W_0 A^2}{C_T} \left(1 - \exp\left(-\frac{t_p}{\tau_{SM}} \right) \right) \exp\left(-\frac{t - t_p}{\tau_{SM}} \right), \qquad t > t_p \tag{72}$$

For a long pulse, $\tau_{SM} < t_p$, we have the steady-state value

$$U(t_p) = \frac{\gamma R_L \eta W_0 A^2}{C_T} \tag{73}$$

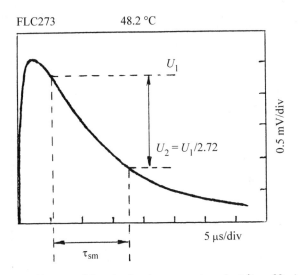

FLC273   48.2 °C

Figure 30. Oscillogram of the pyroelectric response to a short ($t_p = 30\,\mathrm{ns}$) pulse of a Q-switched Nd^{3+} YAG laser and determination of τ_{SM} (FLC-273, cell thickness $= 10\,\mu\mathrm{m}$, bias voltage $= 8\,\mathrm{V}$).

which coincides with Eq. (43). The front edge has the time constant τ_{SM}. For a short pulse, $\tau_{SM} > t_p$,

$$U(t_p) = \frac{\gamma R_L \eta W_0 A^2}{C_T} \cdot \frac{t_p}{\tau_{SM}} \tag{74}$$

In the both cases, for $t > t_p$, the pyroelectric response decays with time constant τ_{SM}. A typical oscillogram is shown in Fig. 30 for a mixture with high spontaneous polarization (FLC-273). Both the pulse amplitude and relaxation time grow from both sides on approach to the ferroelectric phase transition. Next we consider the results on soft-mode relaxation time and "tilt viscosity" measurements for several FLCs.

Classical Compounds. Figure 31 reproduces the temperature behavior of the soft-mode relaxation time in the SmA and SmC* phases of DOBAMBC which is measured as a pyroelectric response to the short pulse ($t_p = 30\,\mathrm{ns}$) of a Q-switched Nd^{3+} YAG laser [56]. The relaxation was observed directly on the screen of a storage oscilloscope. The critical exponent found earlier for the smectic C* phase, $\nu = 0.67$ [79], is rather far from the $\nu = 1$ predicted by the simple Landau technique. For this reason the "tilt viscosity" coefficient γ_θ calculated from the data on τ_{SM} (Fig. 31) strongly depends on temperature (it must be constant in the Landau approach).

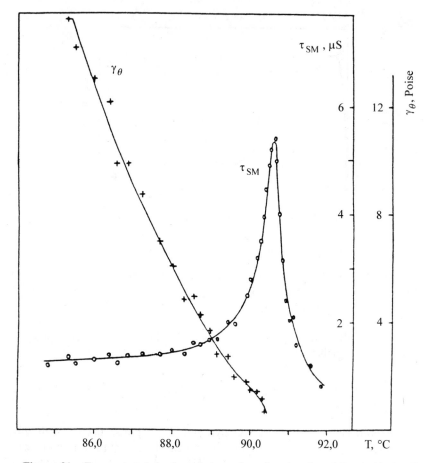

Figure 31. Temperature behavior of the soft-mode relaxation time τ_{SM} and "tilt viscosity" γ_θ in the C^* phase of DOBAMBC found from the pyroelectric response decay.

Another example of soft-mode relaxation time measured by the pyroelectric response is shown in Fig. 32, where relaxation times in the vicinity of two transitions in HOBACPC are displayed [81]. The smectic A–SmC* transition is known to be second order, and the C^*–G^* transition was previously assigned on the basis of the DSC data as a first-order transition. From the critical behavior of the relaxation time near the T_{CG} point, the transition can be treated as a second-order or a weak first-order transition. The critical exponents were found to be equal to 1 (with 10% accuracy) on both sides of the SmC*–SmG* transition.

Determination of the "tilt viscosity" (from pyroelectric data) γ_θ and the "azimuthal viscosity" γ_φ (from electrooptical data) in the smectic C^* phase

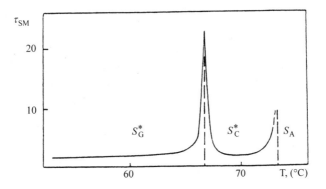

Figure 32. Temperature behavior of the inverse soft-mode relaxation time for the smectic A–C* and C*–G* transitions in HOBACPC obtained from the pyroelectric response relaxation.

of another FLC (HOPEOOBA) leads to the conclusion that it is γ_θ (and not γ_φ) that is comparable with nematic twist viscosity, and γ_φ is about 1 or 2 orders of magnitude less (Fig. 33) [82]. In addition to SmA and SmC*, HOPEOOBA has also a nematic phase.

Mixture with High P_s (FLC-273). The data on the static Landau expansion coefficients for this mixture were discussed in Section V.B.3. Recently we carried out measurements of soft-mode relaxation times, and we will briefly demonstrate them here (for the details, see [21]).

We used a load resistor of an intermediate value, $R_L = 1.64 \, \text{k}\Omega$. With the capacitance of the SmC* phase, it provided an optimum electric time constant of about 0.5 µs. The result is shown in Fig. 34. The temperature was changed very slowly (upon cooling, a temperature range of 55–45°C was maintained for 6 hours). In this way a well-defined peak of τ_{SM} could be measured. After accumulating a series of 32 laser pulses, at each temperature as oscillogram was taken from the screen.

The τ_{SM} time was calculated as the time when the response decreases $e = 2.72$ times from the point where exponential decay starts; see Fig. 30. The pyrovoltage is also shown in Fig. 34, but it is not steady state and cannot be used in calculating P_s. The inverse soft-mode time is plotted as a function of the temperature deviation from the SmA–SmC* phase transition shown in Fig. 35. It appears to follow a linear dependence on both sides of the transition, with the slope ratio SmC*/SmA $= -2.7$, which is not too far off from the ratio 2 expected in terms of the simplest Landau theory. The "tilt viscosity" coefficients estimated from Fig. 35 using Eq. (69), ($\gamma_\theta = \alpha/\text{slope}$ for the smectic A and $\gamma_\theta = -2\alpha/\text{slope}$ for smectic C* phase) are 3.6 P for A and C* phases, respectively ($\alpha = 5 \times 10^4 \, \text{J/m}^3 \, \text{K}$, as in Section V.B.3). For

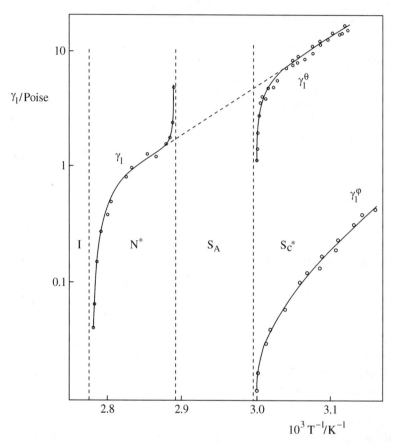

Figure 33. Comparison of temperature dependencies of tilt γ_θ and azimuthal γ_φ viscosities in the smectic C* phase with twist viscosity of the nematic phase in HOPE OOBA.

comparison, in HOPE OOBA (Fig. 3.33), the same coefficient varied between 1 and 10.

2. Fast Mode

During out soft-mode investigations of DOBAMBC and certain ferroelectric mixtures, we found another dynamic mode which we observed over a whole range of the SmC* phase [83]. The inset to Fig. 36 shows an oscillogram of the pyroelectric response of DOBAMBC to a 30 ns laser pulse. The oscillogram consists of two parts: The fast response (shorter than 10 ns) follows the shape of the pulse, and the slow one (in the μs range) is the soft-mode relaxation shown above (see Fig. 30).

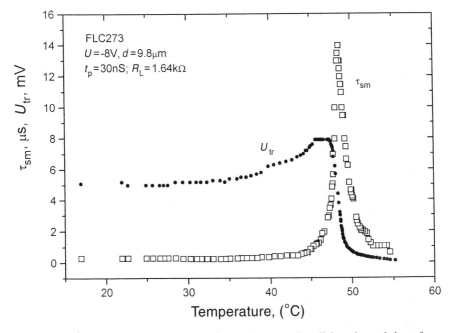

Figure 34. Amplitude of the pyroelectric response to a short-light pulse and the soft-mode relaxation time as functions of temperature for FLC-273.

Conventionally the spontaneous polarization (the secondary order parameter of an FLC) is assumed to be proportional to the tilt angle θ (the primary order parameter). Let us call this part of P_s the slow component, P_θ. With a short exciting pulse, $t_p \ll \tau_{SM}$, the tilt angle cannot reach its new steady-state value to the end of the pulse, and hence the steady-state P_s value cannot be measured in this regime (the long pulse is necessary, as discussed in Section V.1). On the other hand, the steady-state value of the fast polarization component P_t with a characteristic time $\tau_t \ll t_p$ can be measured over a wide temperature range. The temperature dependencies of the fast components of the pyroelectric coefficient and polarization are shown in Fig. 36 for DOBAMBC [81]. The slow P_θ component was measured independently (by the pulse steady-state pyroelectric technique), and the total polarization is

$$P_s = P_\theta + P_t \qquad (75)$$

In Fig. 37 both components are presented in the form of the tilt angle dependencies, $P_\theta(\theta)$ and $P_t(\theta^3)$. The experiment shows that the slow

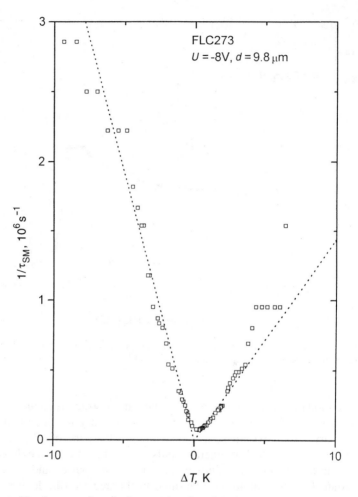

Figure 35. Inverse soft-mode time as a function of temperature deviation from the A–C*
phase transition for FLC-273. Slopes are $1.4 \cdot 10^5$ s^{-1}K^{-1} in A-phase and $-3.8 \cdot 10^5$ s^{-1}K^{-1} in
C-phase).

component is proportional to θ (as anticipated) and the fast component
is proportional to θ^3. The latter dependence has been predicted theoretically
for the transverse component of the spontaneous polarization caused by the
ordering of the short molecular axes. It is this polarization component that
originates from the biaxiality of the smectic C* phase that has been widely
discussed in recent literature [66,84].

It should be noted that in DOBAMBC the fast polarization component is
not very strongly pronounced, $P_\theta \approx 20P_t$ at 70°C, whereas in some other

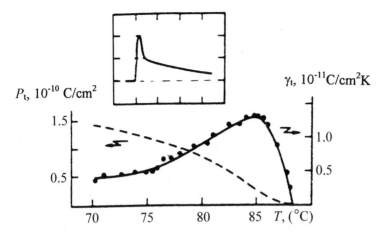

Figure 36. Temperature dependencies of the fast components of the pyroelectric coefficient γ_t and the fast component of the spontaneous polarization P_t for DOBAMBC. (*Inset*) Oscillogram of the pyroelectric response. The time scale is 100 ns/div; the vertical scale is 0.1 mV/div.

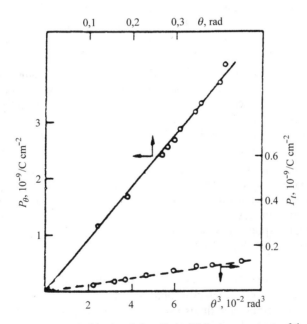

Figure 37. Fast P_t (*dashed line*) and slow P_θ (*solid line*) components of the spontaneous polarization of DOBAMBC as functions of the molecular tilt angle.

mixtures the fast component may be only two times smaller than the slow one [81].

D. Pyroelectricity in Chiral Antiferroelectrics

Recently, largely due to an enormous contribution of the Tokyo group (see the excellent review article by Fukuda [85]), chiral antiferroelectric mesophases have attracted a lot of attention as possible materials for display technologies. The remarkable electrooptical properties of these materials were discovered in 1988 [86]. Earlier there were recorded some observations of antiferroelectric properties, and this history is discussed in [85]. In fact the very first paper on this subject was published as early as 1982 [87], but it was overlooked. In that completely forgotten paper, the pyroelectric response was measured and explicitly referred to as an indication of antiferroelectricity. Below we briefly reproduce the most important results of the paper [87].

The experiment was conducted using two component mixtures of achiral 4-nonyloxy-benzylidene-4'-amino-pentylcinnamate (NOBAPC) with the chiral additives HOBACNPC and HOBAPCPC (hexyloxy-homologues of DOBAMBC with dipolar substituents -CN and Cl in the chiral moiety, respectively). The pyroelectric measurements were carried out with a Nd-glass laser in the cw (quasi-static) regime, and they showed that the mixture manifested much richer polymorphism than the pure matrix (NOBAPC). The macroscopic polarization calculated from the pyroelectric response is shown in Fig. 38 for mixtures of NOBAPC with 20 wt% of HOBACNCP and HOBACPC. In the pure matrix only smectic I, A, C, and B phases were observed on cooling before crystallization. In the mixtures, the new smectic phases S_3^*–S_7^* appear with unusual pyroelectric properties. For example, even in the 1 mm thick cells, the pyroelectric response in S_7^* phase appears at a certain threshold voltage, while in the $S5^*$ phase (which is probably a conventional smectic C^* phase) there is no threshold. With the increasing field, the response saturates, even if the field is very weak (Fig. 39). With a change in the bias field direction, the polarization changes its sign. The $(+)$ and $(-)$ states are characterized by a certain memory dependent on a particular phase. Although the electrooptical response has not been observed in [86], the antiferroelectric structure shown in Fig. 40 has been suggested for explanation of the observed results.

Recently the pyroelectric properties of compound AS573 possessing ferro-, ferri-, and antiferroelectric phases were studied over a wide temperature range at various bais fields [88]. The field-induced transition between the smectic C^* and a ferrielectric phase was more precisely detected by the pyroelectric technique than by repolarization current method. A discrepancy

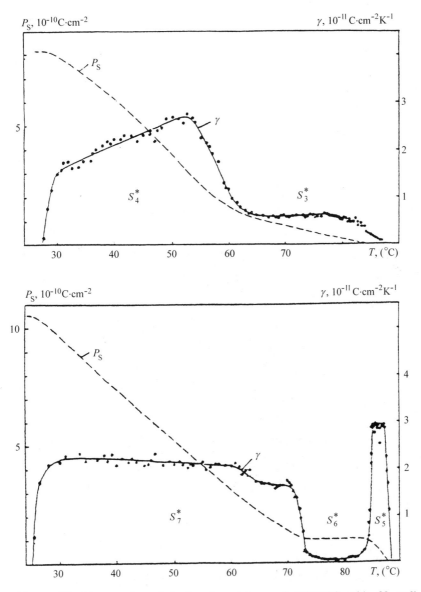

Figure 38. Macroscopic polarization in mixtures of NOBAPC with 20 wt% of HOBACNCP (*top*) and HOBACPC (*bottom*). The cell tickness 1 mm; the bias voltage is 100 V.

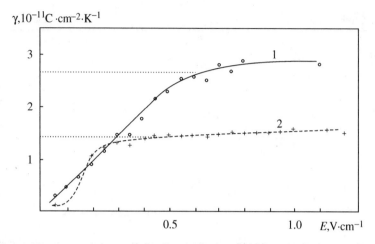

Figure 39. Pyroelectric response versus bias voltage in the ferroelectric (86°C, curve 1) and antiferroelectric (67°C, curve 2) phases of NOBAPC-HOBACPC (80 : 20) mixture.

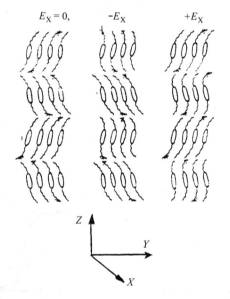

Figure 40. Model for an antiferroelectric and its field behavior [86].

was observed between the P_s values obtained by the two methods. This was explained as caused by too slow response times of the antiferroelectric and ferrielectric phases, which result in a violation of the steady-state regime at the modulation frequency (125 Hz) used for pyroelectric measurements.

VI.　ACHIRAL FERRO- AND ANTIFERROELECTRIC LIQUID CRYSTALS

As a rule, achiral (with mirror symmetry) crystalline ferro- and antiferro-electrics [2] manifest so-called *proper* ferroelectricity when the spontaneous polarization arises as a primary order parameter. Proper ferroelectrics are due to dipole-dipole interactions, and there are some examples of *improper* crystalline ferroelectrics. Among the proper ferroelectrics there have been found crystalline polymers such as polyvinylidene fluoride (PVF_2) and its copolymers.

Almost all liquid crystalline ferroelectrics [7] and antiferroelectrics [85], including the polymer ones [6], are chiral (tilted mesophases SmC^*, SmF^*, SmI^*, SmG^*, etc.). The mechanism of ferroelectricity in these systems is quite specific: A tilt of the elongated chiral molecules, that is, the order parameter of a tilted phase, results in a polar ordering of their short axes (and transverse dipole moments) perpendicular to the tilt plane. The P_s vector stays in the plane of the smectic layer perpendicularly to the tilt plane. Such materials also belong to a class of improper ferro- and anti-ferroelectrics.

The search for *achiral* analogues of *mesomorphic* ferro- and antiferro-electrics is still a challenge to researchers, both the theoreticians and the experimentalists. Prost and Barois [89] predicted that where the spontaneous polarization arises due to the dipole-dipole interaction, a "longitudinal" polarization is directed along the normal to the smectic layers; from the symmetry point of view, this is allowed even in the smectic A phase. Improper ferroelectric and antiferroelectric ordering can arise from certain intermolecular interactions that favor parallel alignment of electric dipoles, as was discussed in [90]. In the tilted smectic phase of an achiral substance, the P_s must lie in the tilt plane, as was discussed theoretically by Brand, Cladis, and Pleiner [91, 92]. A ferroelectric order has also been predicted for achiral discotic mesophases formed by bowl-like molecules [93]. A brief review of other theoretical works on the subject can be found in Ref. [94].

Few recent experiments seem to be in accordance with theoretical predictions (see also [94]). An achiral mesophase composed of so-called polyphilic compounds [95] was the first achiral mesogenic system that manifested (a small) spontaneous polarization, of the order of few nC/cm^2. The polarization was found by peizoelectric and pyroelectric measurements. We will briefly discuss these measurements below. Recently there were some publications on the ferroelectric properties of an achiral smectic mesophase composed of banana-type molecules measured by the repolar-ization current technique [96]. The pyroelectric measurements revealed a polar crystalline phase in a mesogenic substance 4OCB [97]. A polar nematic phase appears to have been found in the main-chain polymer by the optical SHG technique [98]. Finally, the antiferroelectric behavior of achiral mixtures of the side-chain polymers with their low molecular mass counterparts has again been observed with the pyroelectric technique [99, 100]. In the latter case very high macroscopic polarization, of the order of $300 \, nC/cm^2$ was measured. Of particular interest is the fact that none of the two counterparts alone manifests this behavior. Below we will also demonstrate the main results of paper [100].

A. Polyphilic Compounds

A chemical approach in the construction of new polar materials was suggested in [90]. The basic idea is to use the polyphilic effect to form the "building elements" of a polar phase. By this concept, chemically different moieties of a molecule tend to segregate to form polar aggregates or lamellas that result in the formation of polar phases. The direction of the spontaneous polarization is allowed to be in the plane of the lamellas or layers (transverse ferro- or antiferroelectrics) or perpendicular to the layers (longitudinal ferro-or antiferroelectrics).

The structure of the polyphilic compound PC1 is shown at the base of Fig. 41. It consists of three distinct parts: perfluoroalkyl and alkyl chains and

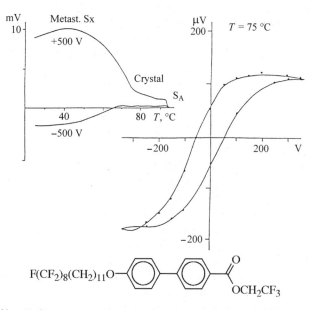

Figure 41. (*Left*) temperaure dependencies of the piezoelectic response for PC1 poled with either $+500$ or -500 V on cooling down from the isotropic phase. (*Right*) The hysteresis of the piezovoltage taken at 75°C. (*Bottom*) Polyphilic compound PC1.

a biphenyl rigid core. The X-ray structure investigations [101] showed the following phase sequences: on heating—crystal $\rightarrow$ 98°C$\rightarrow$ smectic A $\rightarrow$ 115°C$\rightarrow$ isotropic; on cooling—isotropic $\rightarrow$115°C$\rightarrow$ smectic A $\rightarrow$82°C $\rightarrow$ smectic X′. SmX′ is stable down to room temperature but crystallizes in the next heating cycle.

Two fundamental properties of ferroelectric substances were studied, namely the field-induced switching of the piezoelectric response [95] and the pyroelectric effect [102]. The polarization was calculated by integrating $\gamma(T)$ over temperature, starting from the phase transition to the smectic A phase.

In experiments with compound PC1, the polar state can be induced using a poling voltage. If the sample is poled during cooling from the isotropic phase (e.g., for 1 hour), the piezoelectric response observed in the metastable smectic X′ phase is stable for days. Its amplitude is comparable to the signal observed for conventional (chiral) ferroelectrics, which is three orders of magnitude higher than the typical background level of nonpolar compounds. Upon heating, the piezovoltage of PC1 decreases (Fig. 41). It is impossible to repolarize the sample at room temperature, but repolarization is easily attained with increasing temperature. The higher the temperature, the faster is the process of repolarization. The corresponding hysteresis loop (in coordinates of applied dc voltage–piezovoltage) is shown in the same figure. The characteristic time of the repoling process at the temperature 75°C is on the order of a few minutes. The temperature behavior of the pulse pyroelectric response of a layer of PC1 poled with a voltage of $+30\,\mathrm{V}$ is shown in Fig. 42. The calculated value of the field-induced macroscopic polarization is given in the same figure. The magnitude of P_s at room temperature is rather small, about $4\,\mathrm{nC/cm^2}$.

Thus the X′-phase of the polyphilic compound PC1 manifests some ferroelectric properties. The direction of the macroscopic polarization can be switched by an external field with the hysteresis loops typically seen in ferroelectrics. No electrooptical effect is observed, and this means that the polarization and the director are decoupled. X-ray studies have shown that the X′-phase is spatially modulated and greatly disordered [101]. The disorder may be related to the domain structure at the microscopic level [102a].

B. Pyroelectric Properties of 4OCB

Mesogenic compounds very rarely form polar crystals. One of the few examples is the crystal of 4-butyloxy-4′-cyanobiphenyl (4OCB) whose polar structure is well established by the X-ray technology [103]. This substance has a monotropic nematic phase, and in the slow cooling from that phase, it crystallizes into another solid ("opaque") modification that manifests

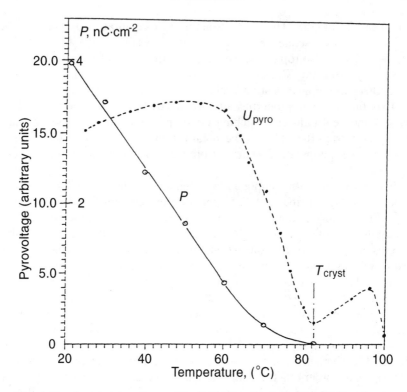

Figure 42. Pyroelectric voltage and the field-induced polarization of PC1 as functions of temperature (poling dc voltage is +30 V; cell thickness is 11 μm).

ferroelectric switching [97]. The pyroelectric coefficient of a previously poled "opaque" phase can be measured upon cell heating without the bias field. Above 53°C the sign of the pyroelectric response is changed reversibly by the external bias field, with a hysteresis effect typical of ferroelectrics (see Fig. 43). The structure of the "opaque" phase is not yet known, and the observed phenomenon requires further investigation.

C. Antiferroelectric Polymer-Monomer Mixtures

1. Materials and Measurements

Extremely unusual electric behavior has been observed in a mixture of an achiral side-chain methacrylate polymer PM6R8 with its monomer M6R8 (Fig. 44), and also in mixtures of the shorter alkyl chain homologues of the original pair and in the acrylate analogues of the same pair. All three pairs mentioned show antiferroelectric switching. In comparison, a polymer-monomer pair having no hydroxy group is "electrically inactive." The

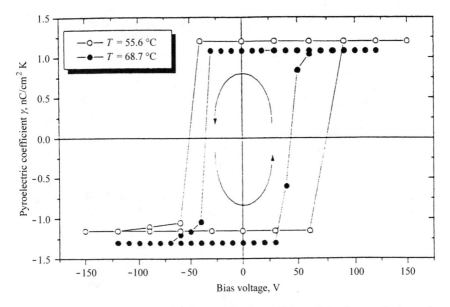

Figure 43. Hysteresis loops for the pyroelectric coefficient of the "opaque" phase of 4OCB measured at two temperatures [97]. The cell thickness is 10 μm.

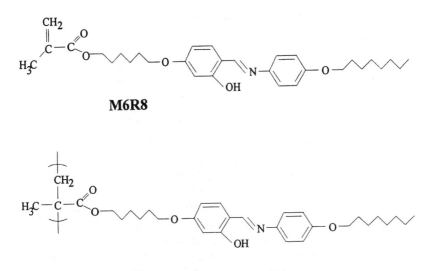

Figure 44. Chemical formulas of the investigated monomer and polymer with their abbreviations.

materials were characterized by DSC, chromatography, and optical rotation measurements. In the more important cases (pure PM6R8, pure M6R8, and their mixture in the 74:26 ratio), X rays were used to determine the structures. Here we present only data on the pyroelectric behavior of mixtures composed of the compounds shown in Fig. 44.

In the study of the electric behavior of the materials, two techniques have been used; both are sensitive to the appearance of any (spontaneously or field-induced) polarized state, namely, the measurements of the pyroelectric and piezoelectric [68, 104] response. The cells consisted of two ITO covered, untreated glass plates with thicknesses of 8 and 110 μm for pyroelectric and piezoelectric measurements, respectively. For pyroelectric measurements the pulse technique described in Section III.A.2 was used along with software newly developed by Dr. S. P. Palto. The data on the pulse response were recorded and processed with an IBM/PC computer. The temperature dependence of the macroscopic polarization was calculated by integrating the pyroelectric voltage, starting from transition temperature T_i to the isotropic phase,

$$P(E, T) = \int_{T_i}^{T} \gamma(E, T)dT \tag{76}$$

Then the correct scale for γ and P_s was introduced by a comparison of the pyroelectric response at a certain temperature with the value measured for a well-known ferroelectric substance. The advantage of the pyroelectric technique over the conventional transient current or Sawyer-Tower hysteresis measurements, especially in this case, is that it enables one to measure the quasi-static polarization at virutally zero frequency. Thus it can be applied to very viscous, slowly switched materials.

2. Pyroelectric Data for PM6R8–M6R8 Mixtures

Figure 45 shows the pyroelectric coefficient as a function of temperature for mixtures of PM6R8 and M6R8 at various concentration (in wt%) of the monomer. All the measurements have been carried out on cooling 8 to 10 μm thick cells under the dc bias field $E = 12$ V/μm. At zero field no pyroelectric response is observed on cooling the isotropic phase down to room temperature. Even after the bias field is applied, the field-induced pyroresponse (and polarization) in the isotropic phase ($T > 170°C$) is still negligible for all of the mixtures. On transition to the bilayer smectic C phase, a sharp increase in the signal is easily seen, in some cases the signal increases by more than an order of magnitude. The polarization permanently increases on cooling, though its growth is less pronounced in the glassy state.

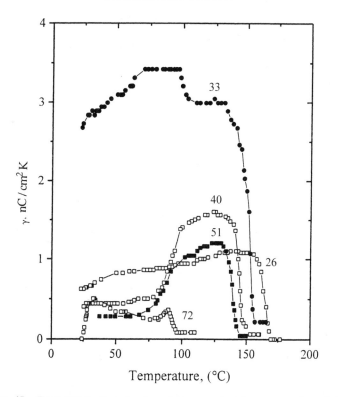

Figure 45. Temperature dependencies of the pyroelectric response for mixtures of PM6R8 and M6R8 at various concentration (in wt%) of the monomer. The temperature scans were made upon cooling under the dc bias field $E = 12\,V/\mu m$.

The value of the macroscopic polarization both in the mesophase and in the glassy state has a well-pronounced maximum as a function of monomer concentration displayed in Fig. 46. Neither the monomer nor the pure polymer shows a pyroelectric response, and the optimum response is achieved for a mixture with approximately 33% of the monomer.

The voltage dependence of the macroscopic polarization in the smectic C phase is extremely nonlinear. At low voltages the field-induced polarization increases linearly with the field and rapidly relaxes after the field is switched off. However, above a certain threshold the polarization grows in a superlinear way, and the field-induced state has a certain memory (about half a minute). At these voltages, a double hysteresis loop in coordinates showing the dc bias–pyroresponse (instead of the polarization in conventional techniques) is easily measured (Fig. 47). Each subsequent point in the loops shown in the figure was taken half a minute after the application of the corresponding voltage. Such loops are typical of antiferroelectrics. They

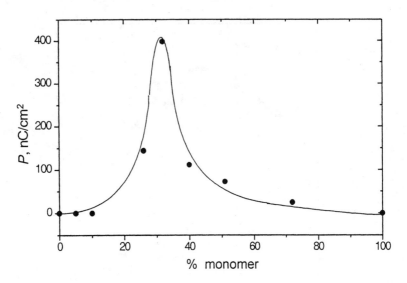

Figure 46. Maximum values of the macroscopic polarization (taken at room temperature after cooling to a glassy state) for mixtures of PM6R8 and M6R8 versus concentration of the monomer. The line is a guide for the eye.

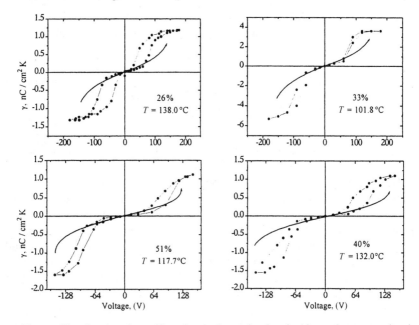

Figure 47. Quasi-static antiferroelectric loops in the dc bias voltage–pyroelectric coefficient coordinates obtained for mixtures PM6R8 and M6R8 at different concentrations of the monomer (cell thickness is about 10 μm).

correspond to three stable states: one with zero polarization (the field off-state) and two states with the electric polarization oriented along two possible directions of the external field. The essential difference of this system from the solid and liquid crystalline (chiral) antiferroelectrics is the rather slow response. It should be noted that no electrooptical response is observed under ac or dc voltage. Strong field nonlinearity was confirmed by measurements of the piezoelectric response of the PM6R8–M6R8 (74:26) mixture.

In the glassy state the pyroelectric response remains stable for a long time. Some of the samples stored at room temperature keep their polarization for two years. The best magnitude of the pyrocoefficient in the glassy state, $\gamma = 2.4\,\text{nC/cm}^2\text{K}$ measured for a PM6R8–M6R8 mixture with 33% of the monomer, is comparable with that observed for strongly poled PVF_2 films and at least one order of magnitude higher than the value typical of chiral polymer SmC^* ferroelectrics in their glassy state. On heating, with no field applied, the material losses its memory. This can be seen in Fig. 48 in the heating scans, where the temperature dependencies of the remnant pyrocoefficient (which was induced by the field during the

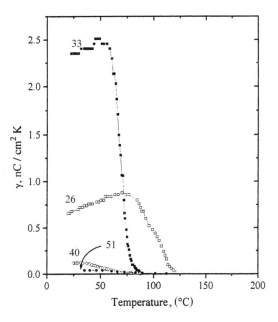

Figure 48. Temperature dependencies of the remnant pyroelectric coefficient for PM6R8-M6R8 mixtures at various concentrations of the monomer. The measurements were made upon heating without a dc bias field after the poling process has been terminated in the glassy state.

cooling scans; Fig. 45) are shown. Long term memory is lost in the mesophase, before the transition into the isotropic liquid. The latter is consistent with the existence of three stable states in the antiferroelectric mesophase one of which (the field off-state) corresponds to the zero polarization.

3. Structure

According to X-ray data [100], the mixture PM6R8–M6R8 (74:26) which manifests the antiferroelectric behavior shows a phase transition from the isotropic into the bilayered smectic C phase. The X-ray studies were carried out separately on the pure polymer PM6R8 and its monomer M6R8. The only mesophase found in the monomer has been referred to as a smectic A. The pure polymer was shown to have the same bilayered smectic C phase as the (74:26) mixture. However, high-order reflections are more pronounced for the mixture (e.g., the ratio of the second to first harmonic intensities for the mixture is twice as large as that for the pure polymer). The in-plane structure of the smectic layers was shown to be liquid-like in both cases, for the polymer and the mixture.

It was concluded that both the pure polymer and their mixture with the monomer have a bilayer C-structure in which the mesogenic side groups are tilted with respect to the layer normal of the smectic C phase. However, there must be a great difference between the polymer, which is electrically inactive, and the mixture, which shows the antiferroelectric properties. Four possible structures consistent with the X-ray data are sketched in Fig. 49: Case (a) is the conventional smectic C; cases (b) and (c) correspond to the smectic C with a regularly alternating direction of the tilt and fractures in the region of the tail (structure b) or the head (structure c) of the mesogenic groups. Such alternating structures have earlier been reported for chiral smectics C^* [85]. In case (d) a change of the molecular tilt occurs in the both tail and head regions.

Note the remarkable difference between structures (b) and (c) and structure (d). Imagine that the dipole moment of a mesogenic group is directed from its head, which is attached by the main chain, to its free end and that the mesogen groups are free to rotate around their longitudinal axes (liquid-like smectic layers). Then, because of the projections of the dipoles on the smectic layer plane in cases (b) and (c), respectively, the in-plane polarization occurs only in the regions of the tails or the heads, P_t or P_h. The possible directions of polarization are shown by arrows. In structure (d) both P_t and P_h are allowed. Both (b) and (c) bilayered structures are always antiferroelectric. Structure (d) is also *antiferroelectric* as long as $P_t = P_h$. If, however, a free rotation is not assumed and the transverse components of dipole moments play a dominant role, polarization could vanish in the

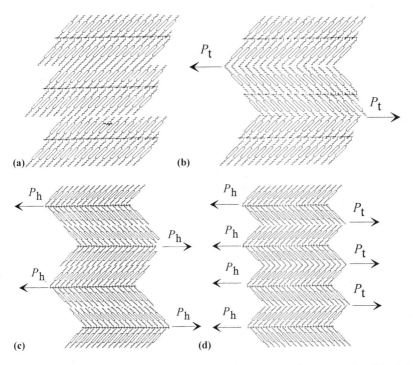

Figure 49. Four possible packing structures of the mesogenic groups of a side-chain polymer in the bilayer smectic C phase: (a) uniform tilt; (b, c) alternating tilt with in-plane antiferroelectric order; and (d) alternating tilt with possible in-plane antiferro-, ferro-, or ferrielectric orders. P_t and P_h are virtual polarizations in the regions of tails and heads of mesogenic units.

regions of, say, heads ($P_h = 0$) and a net macroscopic polarization P_t would appear in structure (d) which becomes *ferroelectric*, as is discussed in [91, 92]. In the more general case of finite but not equal magnitudes of P_t and P_h, structure (d) is *ferrielectric*.

In this example structure (d) is rejected, since no ferroelectric properties are observed. Conventionally structure (a) is attributed to the pure polymer and one of the two structures, (b) or (c); both are compatible with antiferroelectricity and are taken as the most probable structures of polymer monomer mixtures. The role of the monomer additives in provoking new structures was reported earlier [105]. The maximum magnitude of the macroscopic polarization measured ($400\,nC/cm^2$) can be accounted for if one assumes that all mesogenic units of both the polymer and the monomer (with molecular weight about 500) have their dipole moment projections on the field direction of about 1 Debye, and this is quite reasonable. On transition to the glassy state a field-induced macroscopic polarization becomes frozen, and the material manifests a pyroelectric response that is comparable to the typical response of the proper polymer ferroelectric.

VII. CONCLUSIONS

This article has presented a review of the basic concepts of spontaneous and field-induced polarization and the pyroelectric coefficients. The corresponding equations for the pyroelectric current were given for several simple but important cases, and experimental techniques for pyroelectric measurements were described. The problem of polar surface layers at the isotropic (nonpolar) dielectric–metal interface was discussed and also the problem of nonpolar nematic liquid crystal under asymmetric boundary conditions. For chiral ferroelectric liquid crystals, the temperature behavior of the spontaneous polarization was considered including soft-mode susceptibility measurements in both static and dynamic regimes. Finally, a few "exotic" achiral structures discovered and studied by pyroelectric effect were demonstrated. In fact, to the author's knowledge, nearly all the literature relevant to the pyroelectric studies of liquid crystals has been included in the reference list, and the author hopes that the article will be useful for readers interested in starting pyroelectric measurements in order to investigate polar mesophases in more detail or in searching for novel polar liquid crystalline systems.

Acknowledgments

I would like to thank many colleagues who assisted my work on the pyroelectric properties of liquid crystals, especially Dr. L. A. Beresnev, Dr. E. P. Pozhidayev, Dr. E. B. Sokolova, and Dr. S. V. Yablonsky for numerous useful discussions; Prof. W. Haase (Technische Universität Darmstadt), Prof. J. Simon (ESPCI, Paris), and Prof. K. Yoshino (Osaka University) for the generous use of their laboratories where much of the work presented in this article was carried out. The help of Dr. S. P. Palto (Moscow) and Dr. Th. Weyrauch (Darmstadt) with software for the measurement of the pyroelectric response is also greatly acknowledged. I am grateful to Prof. S. A. Pikin for many stimulating discussions about ferroelectricity in general. The work was partially supported through INTAS Grant no.95-IN-RN-128.

References

1. A. K. Tagantsev. *Phase Transitions* **35**, 119 (1991).
2. M. E. Lines and A. M. Glass, *Principles and Application of Ferroelectric and Related Materials* Clarendon Press, Oxford, 1977.
3. A. Hadni, *Ferroelectrics* **140** 25 (1993).
4. L. A. Beresnev, L. M. Blinov, and E. B. Sokolova, *Mikroelektronika* **4**, 422 (1975).

5. L. J. Yu, H. Lee, S. C. Bak, and M. M. Labes, *Phys. Rev. Lett.* **36**, 388 (1976).

6. V. P. Shibayev, M. V. Kozlovsky, L. A. Beresnev, L. M. Blinov, and N. A. Plate, *Polym. Bull.* **12**, 299 (1984).

7. R. B. Meyer, L. Liebert, L. Strzelecki, and P. Keller, *J. Phys. (Paris) Lett.* **36**, L-69 (1975).

8. L. M. Blinov, N. N. Davydova, V. V. Lazareva, and S. G. Yudin, *Fiz. Tverd. Tela (Leningrad)* **24**, 2686 (1982).

9. R. B. Meyer, *Phys. Rev. Lett.* **22**, 918 (1969).

10. L. A. Beresnev, L. M. Blinov, S. A. Davidyan, S. G. Kononov, and S. V. Yablonsky, *Pis'ma Zh. Eksp. Teor. Fiz.* **45**, 592 (1987).

11. L. M. Blinov, S. V. Yermakov, and L. M. Korolyev, *Fiz. Tverd Tela, (Leningrad)* **14**, 3671 (1972).

12. J. Cooper, *Rev. Sci. Instr.* **33**, 92 (1962).

13. V. P. Singh and A. Van der Ziel, *Ferroelectrics* **15**, 135 (1977).

14. M. Simphony and S. Shaulov, *J. Appl. Phys.* **42**, 3741 (1971).

15. J. C. Joshi and A. L. Dewar, *Phys. Stat. Sol.* **A70**, 353 (1982).

16. A. G. Chynoweth, *J. Appl. Phys.* **27**, 78 (1956).

17. A. M. Glass, J. S. Patel, J. W. Goodby, D. H. Olson, and J. M. Geary, *J. Appl. Phys.* **60**, 2778 (1986).

18. B. I. Ostrovski, S. A. Taraskin, B. A. Strukov, A. Z. Rabinovich, A. S. Sonin, and N. I. Chernova, *Fiz. Tverd. Tela, (Leningrad)* **19**, 3686 (1976).

19. L. M. Blinov, L. A. Beresnev, N. M. Shtykov, and Z. M. Elashvili, *J. Physique (France)* **40**, Colloque C3, C3-269 (1976).

20. L. M. Blinov, M. Ozaki, and K. Yoshino, *Pis'ma Zh. Eksp. Teor. Fiz.* **69**, 220 (1999).

21. L. M. Blinov, M. Ozaki, S. Okazaki, and K. Yoshino, *Ferroelectrics* **212**, 37, (1998); *Mol. Mater.* 9, 163 (1997).

22. L. Blinov, M. Barnik, M. Ozaki, N. Shtykov, and K. Yoshino, *Phys. Rev. E.*

23. P. Guyot-Sionnest, H. Hsiung, and Y. R. Shen, *Phys. Rev. Lett.* **57**, 2963 (1986).

24. F. Gutmann and L. Lyons, *Organic Semiconductors.* Wiley, New York, 1967.

25. L. A. Beresnev, PhD thesis, Inst. of Solid State Physics, Chernogolovka, 1979.

26. L. M. Blinov, and V. G. Chigrinov, *Electrooptic Effects in Liquid Crystal Materials* Springer Verlag, New York, 1993.

27. S. V. Yablonsky, L. M. Blinov, and S. A. Pikin, *Pis'ma Zh. Eksp. Teor. Fiz.* **40**, 226 (1984).

28. S. V. Yablonsky, L. M. Blinov, and S. A. Pikin, *Mol. Cryst. Liq. Cryst.* **127**, 381 (1985).

29. A. G. Petrov, A. Derzhanski, and Y. V. Pavloff, *J. Phys. (Paris) Lett.* **42**, L-119 (1981).

30. L. M. Blinov, D. Z. Radzhabov, S. V. Yablonsky, and S. S. Yakovenko. *Nuovo Cimento* **12D**, 1353 (1990).

31. L. M. Blinov, L. A. Beresnev, D. Z. Radzhabov, and S. S. Yakovenko, *Mol. Cryst. Liq. Cryst.* **191**, 363 (1990).

32. L. M. Blinov, L. V. Mikhnev, E. B. Sokolova, and S. G. Yudin, *Pis'ma Zh. Tekh. Fiz.* **9**, 1494 (1983).

33. L. M. Blinov, D. Z. Radzhabov, D. B. Subachius and S. V. Yablonsky, *Pis'ma Zh. Eksp. Teor. Fiz.* **53**, 223 (1991).

34. L. A. Beresnev, L. M. Blinov, and Z. M. Elashvili, *Pis'ma Zh. Tekh. Fiz.* **4**, 225 (1978).

35. L. A. Beresnev and L. M. Blinov, *Pis'ma Zh. Tekh. Fiz.* **4**, 931 (1978).

36. K. Yoshino, T. Uemoto, K. G. Balakrishnan, S. Yanagida, and Y. Inuishi, *Tech. Rep. Osaka Univ.* **27**, 427 (1977).

37. B. I. Ostrovsky, A. Z. Rabinovich, A. S. Sonin, and B.A. Strukov, *Zh. Eksp. Teor. Fiz.* **74**, 1748 (1978).

38. J. W. O'Sullivan, Yu. P. Panarin, and J. K. Vij, *J. Appl. Phys.* **77**, 1201 (1995).

39. L. A. Beresnev, V. A. Baikalov, L. M. Blinov, E. P. Pozhidayev, and G. V. Purvanetskas, *Pis'ma Zh. Eksp. Teor. Fiz.* **33**, 553 (1981).

40. M. V. Kozlovsky, L. A. Beresnev, L. M. Blinov, S. G. Kononov, and V. P. Shibayev, *Fiz. Tverd. Tela (Leningrad)* **29**, 98 (1987).

41. A. Kocot, R. Wrzalik, J. K. Vij, and R. Zentel, *J. Appl. Phys.* **75**, 728 (1994).

42. V. P. Shibayev, M. V. Kozlovsky, N. A. Plate, L. A. Beresnev, and L. M. Blinov, *Liq. Cryst.* **8**, 545 (1990).

43. M. Mauzac, H. T. Nguyen, F. Tournilhac, and S. V. Yablonsky, *Chem. Phys. Lett.* **240**, 461 (1995).

44. N. Mikami, R. Higuchi, T. Sakurai, M. Ozaki, and K. Yoshino. *Jpn. J. Appl. Phys.* **25**, L-833 (1986); J. S. Patel and J. W. Goodby, *Phil. Mag. Lett.* **55**, 283 (1987).

45. R. Eidenschink, T. Geelhaar, G. Andersson, A. Dahlgren, K. Flatischler, F. Gouda, S. T. Lagerwall, and K. Skarp, *Ferroelectrics* **84**, 167 (1988).

46. H. Stegemeyer, A. Sprick, M. A. Osipov, V. Vill, and H.-W. Tunger, *Phys. Rev. E* **51**, 5721 (1995).

47. L. M. Blinov, L. A. Beresnev, D. Demus, S. V. Yablonsky, and S. A. Pikin, *Mol. Cryst. Liq. Cryst.* **292**, 277 (1997).

48. S. Saito, K. Murashiro, M. Kikuchi, T. Inukai, D. Demus, M. Neundorf, and S. Diele, *Ferroelectrics* **147**, 367 (1993).

49. L. Beresnev, M. Pfeiffer, S. Pikin, W. Haase, and L. Blinov, *Ferroelectrics* **132**, 99 (1992).

50. S. A. Pikin, L. A. Beresnev, S. Hiller, M. Pfeiffer, and W. Haase, *Mol. Mat.* **3**, 1 (1993).

51. S. A. Pikin, L. M. Blinov, L. A. Beresnev, S. Hiller, E. Schumacher, B. I. Ostrovsky, and W. Haase, *Ferroelectrics* **181**, 111 (1996).

52. S. A. Pikin and V. L. Indenbom, *Usp. Fiz. Nauk* **125**, 251 (1978).

53. S. A. Pikin, *Structural Transformations in Liquid Crystals.* Gordon and Breach, New York, 1993.

54. L. A. Beresnev, L. M. Blinov, M. A. Osipov, and S. A. Pikin, *Mol. Cryst. Liq. Cryst.* **158A**, 1– 150 (1988).

55. J. W. Goodby, R. Blinc, N. A. Clark, S. T. Lagerwall, M. A. Osipov, S. A. Pikin, T. Sakurai, K. Yoshino, and B. Zeks, *Ferroelectric Liquid Crystals: Principles, Properties and Applications.* Gordon and Breach, Philadelphia, 1991.

56. E. P. Pozhidayev, L. M. Blinov, L. A. Beresnev, and V. V. Belyayev, *Mol. Cryst. Liq. Cryst.* **124**, 359 (1985).

57. S. Dumrongrattana, C. C. Huang, G. Nounesis, S. C. Lien, and J. M. Viner, *Phys. Rev. A* **34**, 5010 (1986).

58. F. Gouda, K. Skarp, and S. T. Lagerwall, *Ferroelectrics* **113**, 165 (1991).

59. F. Gouda, T. Carlsson, G. Andersson, S. T. Lagerwall, and B. Stebler, *Liq. Cryst.* **16**, 315 (1994).

60. C. Bahr, G. Heppke, and B. Subaschus, *Ferroelectrics* **84**, 103 (1988).

61. T. Chan, C. Bahr, G. Heppke, an C. W. Garland, *Liq. Cryst.* **13**, 667 (1993).

62. C. Bahr, and G. Heppke, *Phys. Rev. A* **44**, 3669 (1991).

63. F. Gießelmann and P. Zugenmaier, *Phys. Rev. E* **52**, 1762 (1995).

64. L. M. Blinov, L. A. Beresnev, and W. Haase, *Ferroelectrics* **174**, 221 (1995).

65. L. M. Blinov, L. A. Beresnev, and W. Haase, *Ferroelectrics* **181**, 187 (1996).

66. M. A. Osipov and S. A. Pikin, *J. Phys. II France* **5**, 1223 (1995).

67. M. Glogarova and J. Pavel, *Liq. Cryst.* **6**, 325 (1989).

68. K. Saxena, L. M. Blinov, L. A. Beresnev, and W. Haase, *Ferroelectrics* **213**, 73 (1998).

69. V. A. Baikalov, L. A. Beresnev, and L. M. Blinov, *Mol. Cryst. Liq. Cryst.* **127**, 397 (1985).

70. W. Haase, S. Pikin, L. A. Beresnev, and S. Hiller, *Liq. Cryst.*, **15**, 779 (1993).

71. S. Pikin, L. A. Beresnev, S. Hiller, and W. Haase, *Ferroelectrics* **147**, 263 (1993).

72. L. Komitov, S. T. Lagerwall, B. Stebler, G. Anderson, and K. Flatischer, *Ferroelectrics* **114**, 167 (1991).

73. C. Legrand, N. Isaert, J. Hmine, J. M. Busine, J. P. Parneix, H. T. Nguyen, and C. Destrade, *Ferroelectrics* **121**, 21 (1991).

74. L. M. Blinov, L. A. Beresnev, and W. Haase, *Ferroelectrics* **181**, 211 (1996).

75. S. Garoff and R. B. Meyer, *Phys. Rev. Lett.* **38**, 848 (1977).

76. A. B. Davey and W. A. Crossland, *Ferroelectrics* **144**, 167 (1991).

77. Ch. Bahr, G. Heppke, and U. Klemke, *Ber. Bunsenges. Phys. Chem.* **95**, 761 (1991).

78. Ch. Bahr and G. Heppke, *Phys. Rev. Lett.* **65**, 3297 (1990).

79. L. A. Beresnev, L. M. Blinov, and E. B. Sokolova, *Pis'ma Zh. Eksp. Teor. Fiz.* **28**, 340 (1978).

80. L. A. Beresnev and L. M. Blinov, *Ferroelectrics* **33**, 129 (1981).

81. L. A. Beresnev, L. M. Blinov and G. V. Purvanetskas, *Pis'ma Zh. Eksp. Teor. Fiz.* **31**, 37 (1980).

82. L. M. Blinov, V. A. Baikalov, M. I. Barnik, L. A. Beresnev, E. P. Pozhidayev, and S. V. Yablonsky, *Liq. Cryst.* **2**, 121 (1987).

83. E. P. Pozhidayev, L. A. Beresnev, L. M. Blinov, and S. A. Pikin, *Pis'ma Zh. Eksp. Teor. Fiz.* **37**, 73 (1983)

84. M. A. Osipov, R. Meister, and H. Stegemeyer, *Liq. Cryst.* **16**, 173 (1994).

85. A. Fukuda. *J. Mater. Chem.* **4**, 997 (1994).

86. N. Hiji, A. D. L. Chandani, S. Nishiyama, Y. Ouchi, H. Takezoe, and A. Fukuda, *Ferroelectrics* **85**, 99 (1988).

87. L. A. Beresnev, L. M. Blinov, V. A. Baikalov, E. P. Pozhidayev, G. V. Purvanetskas, and A. I. Pavlyuchenko, *Mol. Cryst. Liq. Cryst.* **89**, 327 (1982).

88. J. W. O'Sullivan, Yu. P. Panarin, J. K. Vij, A. J. Seed, M. Hird, and J. W. Goodby, *J. Phys: Condensed Matter.* **8**, L551 (1996); J. W. O'Sullivan, Yu. P. Panarin, J. K. Vij, A. J. Seed, M. Hird, and J. W. Goodby, *Mol. Cryst. Liq. Cryst.* **301**, 111 (1997); N. M. Shtykov, J. K. Vij, V. P. Panov, R. A. Lewis, M. Hird, and J. W. Goodby, *J. Mater. Chem.* **9**, 1383 (1999).

89. J. Prost and Ph. Barois, *J. Chim. Physique* **80**, 65 (1983).

90. F. Tournilhac, L. Bosio, J.-F. Nicoud, and J. Simon, *Chem. Phys. Lett.* **145**, 452 (1988).

91. H. R. Brand, P. Cladis, and H. Pleiner, *Macromolecules* **25**, 7223 (1992).

92. P. Cladis and H. R. Brand, *Liq. Cryst.* **14**, 1327 (1993).

93. L. Lin, *Mol. Cryst. Liq. Cryst.* **146**, 41 (1987).

94. L. M. Blinov, *Liq. Cryst.* **24**, 143 (1997).

95. F. Tournilhac, L. M. Blinov, J. Simon, and S. V. Yablonsky, *Nature* **359**, 61 (1992).

96. T. Niori, T. Sekine, J. Watanabe, T. Furukawa, and H. Takezoe, *J. Mater. Chem.* **6**, 1231 (1996).

97. S. V. Yablonskii, T. Weyrauch, W. Haase, S. Ponti, A. Strigazzi, C. A. Verachini, and C. Gandolfo, *Ferroelectrics* **188**, 175 (1996).

98. T. Watanabe, S. Miyata, T. Furukawa, H. Takezoe, T. Nishi, M. Sone, A. Migita, and J. Watanabe, *Jpn. J. Appl. Phys.* **35**, L505 (1996).

99. E. A. Soto Bustamante, S. V. Yablonski, B. I. Ostrovskii, L. A. Beresnev, L. M. Blinov, and W. Haase, *Chem. Phys. Lett.* **260** 447 (1996).

100. E. A. Soto Bustamante, S. V. Yablonskii, B. I. Ostrovskii, L. A. Beresnev, L. M. Blinov, and W. Haase, *Liq. Cryst.* **21**, 829 (1996).

101. L. M. Blinov, T. A. Lobko, B. I. Ostrovskii, S. N. Sulianov, and F. Tournilhac, *J. Phys. II, (France)* **3**, 1121 (1993).

102. F. Tournilhac, L. M. Blinov, J. Simon, D. Subachius and S. V. Yablonsky, *Synth. Metals* **54**, 253 (1993).

102a. J. Prost, R. Bruinsma, and F. Tournilhac, *J. Phys. II, France* **4**, 169 (1994).

103. L. Walz, H. Paulus, and W. Haase, *Z. für Kristallographie* **180**, 97 (1987).

104. S. V. Yablonskii, E. I. Katz, M. V. Kozlovskii, T. Weyrauch, E. A. Soto Bustamante, D. B. Subachius, and W. Haase, *Mol. Mat.* **3**, 311 (1994).

105. H. Leube and H. Finkelman, *Makromol. Chem.* **192**, 1317 (1991).

FERROELECTRIC LC-ELASTOMERS

RUDOLF ZENTEL, ELISABETH GEBHARD, AND
MARTIN BREHMER

*Department of Chemistry and Institute of Materials Science,
University of Wuppertal, Wuppertal, Germany*

CONTENTS

I. INTRODUCTION

Ferroelectric materials are a subclass of pyro- and piezoelectric materials (Fig. 1). They are very rarely found in crystalline organic or polymeric materials, because the observation of a ferroelectric hysteresis requires enough molecular mobility for a reorientation of molecular dipoles in space. So semicrystalline polyvinylidenefluoride (PVDF) is the only definitely known compound [1]. However, ferroelectric behavior is very often observed in chiral liquid crystalline materials, both low molar mass and polymeric. Here we will discuss materials (LC-elastomers) that combine a liquid crystalline phase with ferroelectric properties (preferably the chiral smectic C^* phase) with a polymer network structure (Fig. 2). The coupling

Advances in Liquid Crystals: A Special Volume of Advances in Chemical Physics, Volume 113,
edited by Jagdish K. Vij. Series Editors I. Prigogine and Stuart A. Rice.
ISBN 0-471-18083-1. © 2000 John Wiley & Sons, Inc.

159

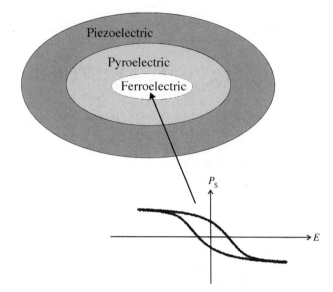

Figure 1. Ferroelectric hysteresis showing the spontaneous polarization P_S of a ferroelectric material reversed by an applied electric field E.

(a)

Figure 2. (*Continued*)

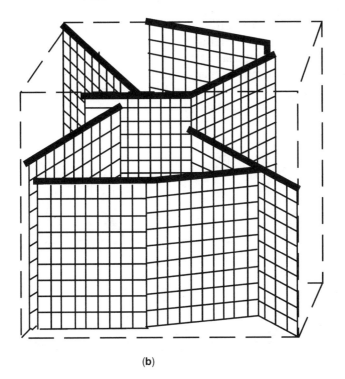

(b)

Figure 2. Network properties in a polymer network structure are that it is (1) soft, i.e., it can be transformed like rubber band, but retains its shape, and (2) that it couples to director orientation, director (a) preferably being parallel (or perpendicular) to polymer chains (LC-elastomer) [3], or else the (b) director aligns (parallel) to chains in the oriented phase-separated polymer network structure (low molar mass LC in LC-thermoset) [6,28].

of the liquid crystalline director to the network or the softness of the network will thereby be chosen, so that a reorientation of the polar axis is still possible. Thus, densely cross-linked systems, which possess a polar axis but cannot be switched [2], will be excluded. It will be the role of the network (1) to form a rubbery matrix for the liquid crystalline phase and (2) to stabilize a director configuration. LC-materials with these properties can be made (see Fig. 2) either by covalently linking the mesogenic groups to a slightly cross-linked rubbery polymer network structure [Fig. 2(a)] [3–5] or by dispersing a phase-separated polymer network structure within a low molar mass liquid crystal [see Fig. 2(b)] [6]. Both systems possess locally very different structures. They may show, however, macroscopically very similar properties.

LC-elastomers [Fig. 2(a)] have been investigated in detail [3–5]. Although the liquid crystalline phase transitions within the network are nearly unaffected by the network, the network retains the memory of the phase and director pattern during cross-linking [7]. In addition, it freezes fluctuations of the smectic layers and leads to a real long-range order in one dimension [8]. An attempt to change the director pattern by applying electric or magnetic fields leads, in the case of LC-elastomers, both to a deformation of the network and to an elastic response (Fig. 3). As a consequence of this,

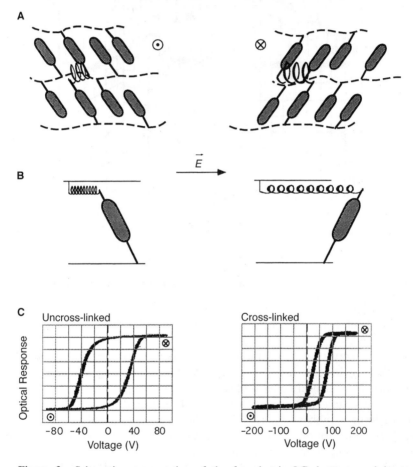

Figure 3. Schematic representation of the ferroelectric LC-elastomer and its two switching states. [11] (A) polymer chain acts as cross-linking point by connecting different mesogenic groups, attached to the main polymer chains. A ferroelectric switching in this elastomer extends polymer chains. (B) The entropy elasticity arising from this acts like a spring, which stabilizes one state. (C) For the uncross-linked system (*left*) the hysteresis is symmetric to zero voltage and both states are equal. After cross-linking in one polar state (*right*) only that state is stable with no electric field, and the hysteresis is no longer symmetric to zero voltage.

nematic LC-elastomers could never be switched in electric fields, if the shape of the elastomer were kept fixed. For freely suspended pieces of nematic LC-elastomers, shape variations in electric fields have been observed sometimes [9,10]. For ferroelectric liquid crystals the interaction with the electric field is, however, much larger. It has thus been possible to prepare real ferroelectric LC-elastomers (see Fig. 3) [11,12]. In these systems the polymer network stabilizes one switching state like a soft spring. It is, however, soft enough to allow ferroelectric switching. The ferroelectric hysteresis can therefore be measured in these systems; it is, however, shifted away from zero voltage (see Fig. 3).

II. SYNTHESIS OF FERROELECTRIC LC-ELASTOMERS

The ferroelectric LC-elastomers described so far [11–13] are prepared from cross-linkable ferroelectric polysiloxanes (Fig. 4), which are prepared by hydrosilylation of precursor polysiloxanes [14]. The cross-linking is finally initiated by irradiating a photoradical that is generated, leading to a oligomerization of acrylamide or acrylate groups (see Fig. 4). The functionality of the netpoints is thus high (equal to the degree of polymerization) and varies with the cross-linking conditions.

The advantage of this photochemically initiated cross-linking is that the cross-linking can be started—at will—after the liquid crystalline polymer is oriented and sufficiently characterized in the uncross-linked state (Fig. 5). The advantage of using polymerizable groups (acrylates) for cross-linking is that small amounts of these groups are sufficient to transform a soluble polymer into a polymer gel and that the chemical reactions happen far away from the mesogen. Cinnamoyl moieties, on the other hand [15] require a high concentration of these groups for cross-linking. The dimers thus formed are, in addition, nonmesogenic. Figure 6 summarizes the ferroelectric LC-elastomers discussed in this paper. Two different positions of cross-linkable groups are used. In polymer *P1* the cross-linking group is close to the siloxane chains, which are known to microphase separately from the mesogenic groups [14,16]. The cross-linking should, therefore, proceed mostly within the siloxane sublayers. In polymers *P2* and *P3* the cross-linking group is located at the end of mesogens. The cross-linking should, therefore, proceed mostly between different siloxane layers (see Fig. 6). A comparison of these elastomers allows the evaluation of structure property relations.

III. FERROELECTRIC CHARACTERIZATION
(UNCROSS-LINKED SYSTEMS)

Before cross-linking, polarization, tilt angle, and switching times can be determined in the usual way [14,17]. Figure 7 shows the temperature-

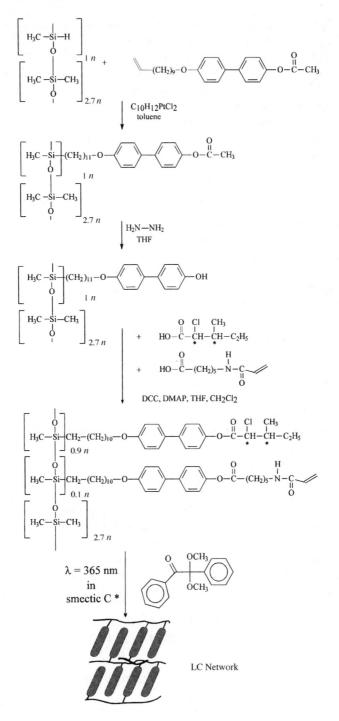

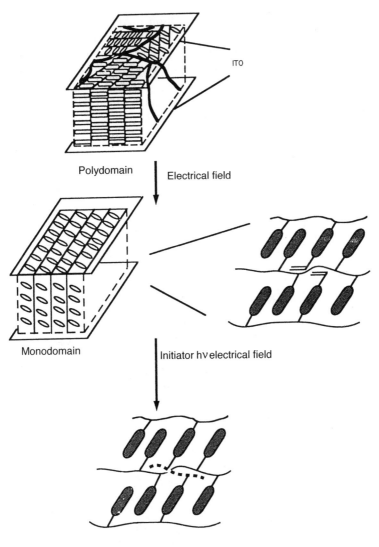

Polydomain | Electrical field

Monodomain | Initiator hν electrical field

Oriented s_c*-network

Figure 5. Preparation of polar smectic C^* monodomains [11,12]. (ITO: indium tin oxide)

Figure 4. Synthetic route to the cross-linkable polysiloxane *P2* and the following preparation of the oriented smectic C^* network using UV-light in the presence of a photoinitiator (1,1-dimethoxy-1-phenyl-acetophenone) [11].

dependence of the spontaneous polarization for polymers *P1–P3*. For the homopolymer related to polymer *P2* all relevant parameters were determined in a careful study by Kocot et al. [18]. It seems that the electroclinic effect is especially strong in these polysiloxanes (see Fig. 8) [12]. This has implications for the freezing of a memory of the tilt angle present during cross-linking and is discussed later in relation to the observation of a piezo-effect in the smectic A phase, in Section VII.

Cross-linking Within Siloxane Sublayer (Intra-layer)

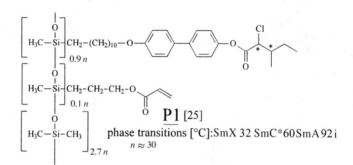

P1 [25]

phase transitions [°C]:SmX 32 SmC*60 SmA 92 i

$n \approx 30$

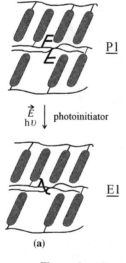

Figure 6.   (*Continued*)

Cross-linking Between Siloxane Sublayers (Interlayer)

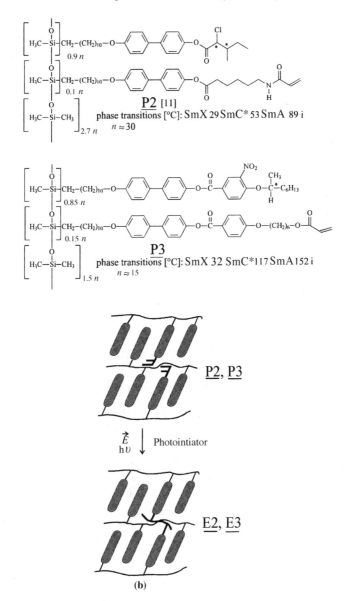

P2 [11]

phase transitions [°C]: SmX 29 SmC* 53 SmA 89 i

$n \approx 30$

P3

phase transitions [°C]: SmX 32 SmC* 117 SmA 152 i

$n \approx 15$

P2, P3

$\vec{E}$
$h\upsilon$ Photointiator

E2, E3

(b)

Figure 6. Chemical structure and phase transition temperatures of polymers *P1-3*. (a) *P1* designed to favor an intra layer crosslinking; (b) *P2-3* forming an inter layer network.

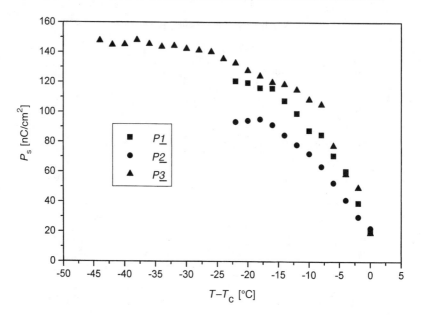

Figure 7. Temperature dependence of the spontaneous polarization P_S for the polymers *P1-3* measured by the triangular wave method.

IV. PROPERTIES OF FERROELECTRIC LC-ELASTOMERS

A. Elastomer Properties

The cross-linking reaction was studied by FTIR-spectroscopy for a series of copolymers analogues to polymer *P2*, each differing in the amount of cross-linkable groups [13]. These measurements show a decrease of the acrylamide double bond (Fig. 9) on irradiation. Conversions between 60 to 84% were observed. The uncertainity of the conversion is, however, high, because only very small number of double bonds are present in polymer *P2*, and they are visible in an infrared spectrum with very low absoption. Mechanical measurements, which show how this cross-linking (conversion of double bonds) leads to an elastic response, are, however, still missing. This is because the photocross-linking can only be performed in thin layers of some μm, in thickness, and is best performed between two glass slides to keep oxygen away.

Mechanical measurements have, however, been made for chemically cross-linked LC-elastomers [3,4,19,20]. For these systems it can be shown that stretching allows the orientation of the liquid crystalline phase to take place. In ideal situations it is thus possible to prepare a ferroelectric

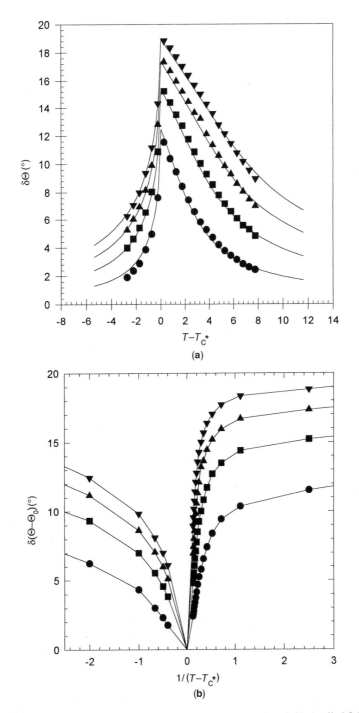

Figure 8. (a) Temperature dependence of the electroclinic effect [12]. Applied field: (●) 5 MVm⁻¹, (■) 10 MVm⁻¹, (▲) 15 MVm⁻¹,(▼) 20 MVm⁻¹. (b) Inverse electroclinic effect. [12] Applied field: (●) 5 MVm⁻¹, (■) 10 MVm⁻¹, (▲) 15 MVm⁻¹, (▼) 20 MVm⁻¹.

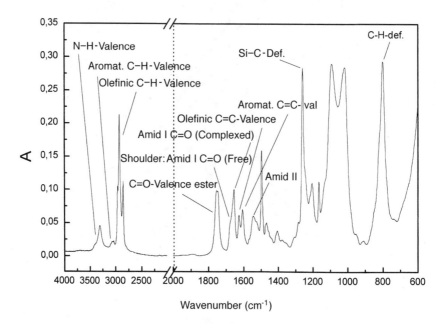

Figure 9. FTIR-spectrum of a polymer related to *P2* showing the valence vibration of the C=C-bond at a wavenumber of 1626 cm^{-1} [13].

monodomain by stretching (Fig. 10) [21,22]. This result can be rationalized as a two-stage deformation process (Fig. 11) [23]. This possibility of orienting or reorienting the polar axis mechanically is the basis for the piezoelectric properties to be discussed later. However, a ferroelectric switching could be observed for none of the chemically cross-linked systems. This may be because the chemically cross-linked films are too thick (several hundred μm compared to about 10 μm for the photochemically cross-linked systems) and the electric field applied therefore too small. In addition, the cross-linking density in the chemically cross-linked systems is presumably higher.

B. Ferroelectric Properties

The ferroelectric properties of the photochemically cross-linked elastomers *E1* to *E3* differ strongly and depend on the topology of the network formed. For the systems with interlayer cross-linking (see Fig. 6, *E2* and *E3*) the switching time is strongly increased. The spontaneous polarization can, therefore, no longer be determined with the triangular wave method. A slow switching is, however, still possible and therefore the ferroelectric hysteresis can be measured optically (Fig. 12) [11,12]. After photochemical cross-linking in a ferroelectric monodomain, the ferroelectric hysteresis shows a

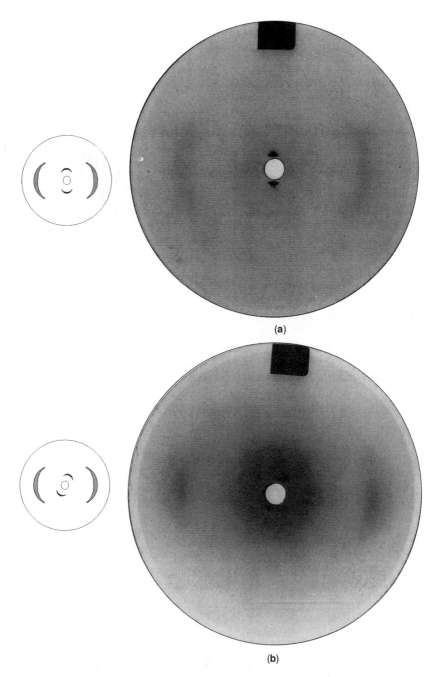

Figure 10. X-Ray pattern of mechanically oriented smectic C^* elastomers [22]. (a) Smectic C^* structure with helical superstructure, smectic layers oriented perpendicular to the vertical fibre axis strain $\approx 300\%$. (b) Smectic C^* structure with untwisted helical superstructure, smectic layer structure tilted with respect to the fibre axis, macroscopic polarization strain $> 400\%$.

First deformation   Second deformation

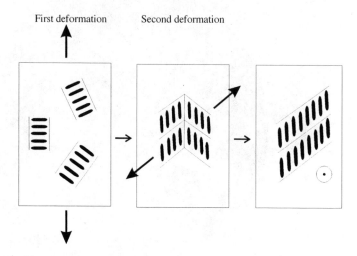

Figure 11.   Two-step deformation process of a chiral smectic C* elastomer displaying a macroscopic polarization in the end [27].

stabilization of the orientation present during cross-linking. At zero external voltage only this state is stable. The second switching state can, however, be reached. The network acts, therefore, like a spring that has stabilized one state, because a switching to the other state leads to a deviation from the most probable conformation of the polymer chain [24] (see Fig. 3). The shift of the center of the hysteresis loop away from zero voltage gives then the magnitude of the electric field necessary to balance the mechanical field of the network. Increasing the temperature of this ferroelectric elastomer leads to a narrowing of the hysteresis loop, which is lost at the transition to the smectic A phase (Fig. 13).

The elastomer with preferable intralayer cross-linking (*E1*, see Fig. 6) shows a completely different behavior (Fig. 14) [25]. In this case the switching time increases by less than a factor of two, the polarization can still be determined, and a measurement of the ferroelectric hysteresis shows no stabilization of the switching state present during cross-linking. The coupling between the orientation of the mesogens and the network conformation in this case is obviously very week. This is the result of the network topology (Fig. 6), in which an interlayer cross-linking is rare.

V.   IMAGING OF FERROELECTRIC LC-ELASTOMERS

Free-standing films can be prepared from the uncross-linked polymers. They can be transferred to a solid substrate and photocross-linked (Fig. 15).

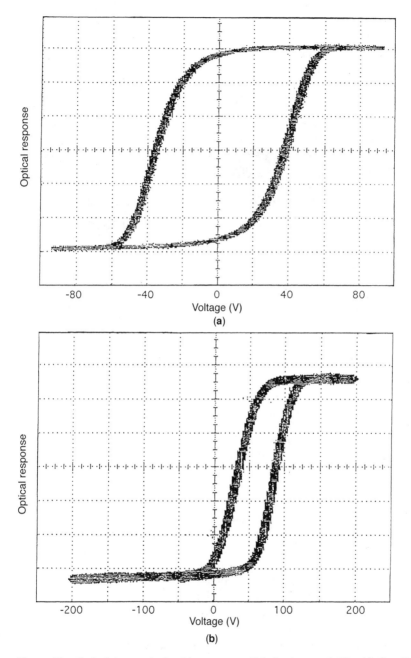

Figure 12. Optical hysteresis for (a) an uncross-linked mixture of *P2* with 2 wt% photoinitiator at 40°C, 175 V_{pp}, 1 Hz [11]; (b) elastomer *E2* at 40°C, 400 V_{pp}, 0.1 Hz [11,12].

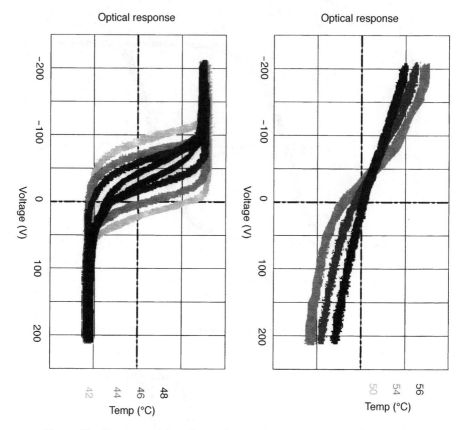

Figure 13. Temperature dependence of the optical hysteresis of elastomer *E2* (SmC*
49°C S$_A$) [12]. (a) Ferroelectric behavior of the S$_C^*$ phase (42, 44, 46, and 48°C, respectively).
(b) electroclinic behavior of the SmA phase(50, 54, and 56°C, respectively).

Thereafter the topology of the films can be imaged by AFM, which gives a
direct visualization of the smectic layer structure at low temperatures. The
uncross-linked polymers can only be imaged deep inside the smectic phase
and in the tapping mode, which does not induce strong lateral forces. At
higher temperatures the sample is too soft and mobile to allow an imaging.
The cross-linked elastomer *E1*, on the contrary, is mechanically stable, and
films sustain the tapping mode as well as the contact mode of the atomic
force microscope. The cross-linked films are stable up to temperatures corresponding to the isotropic phase of the uncross-linked material [26].

Because the measurements can be done in all phases, it is also possible to
determine the change of the smectic layer thickness at the phase transitions
in a direct way. For elastomer *E1*, for example, the smectic layer thickness is

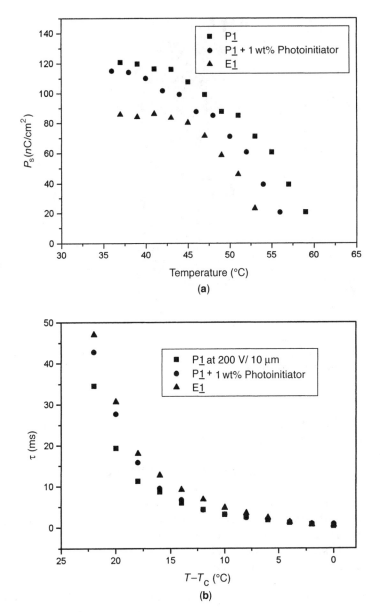

Figure 14. (a) Temperature dependence of the spontaneous polarization P_S for *P1* and *E1*. (b) Temperature dependence of the switching time τ (defined as 0–100% change in transmission) for *P1* and *E1*.

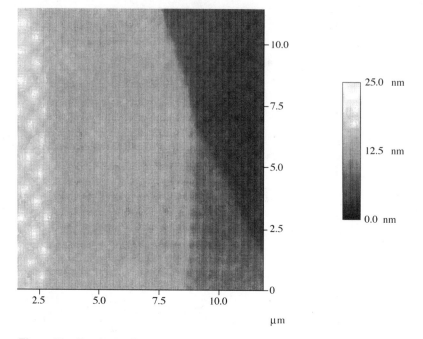

Figure 15. Tapping mode AFM image of transferred freely suspended film of *E1*. The surface is made up of plateaus corresponding to domains of definite thickness separated by steps [26].

4.2 nm in the smectic C^* phase (36°C, tilt angle about 30°). It increases to 4.4 nm at 50°C in the smectic A phase [26].

VI.   PIEZOELECTRIC PROPERTIES

Because a ferroelectric material has to be piezoelectric, the observation of a piezoresponse is natural. It has been observed for the elastomers described here and also for more densely cross-linked systems [27–30] for which no ferroelectric switching could be observed. For the elastomers described here [11] it is, however, possible to change the piezoresponse by a reorientation of the polar axis in an external field (see *E2* in Fig. 16). For this experiment the polar axis was kept in one orientation during cross-linking. This resulted in a positive piezoresponse (see Fig. 16). Thereafter the direction of the polar axis was inverted by applying an external field of opposite direction. Then the external field was removed and the piezocoefficient was measured. At first a piezoresponse of opposite sign (negative), but identical value is determined (see Fig. 16). In the field free state, this piezoresponse conti-

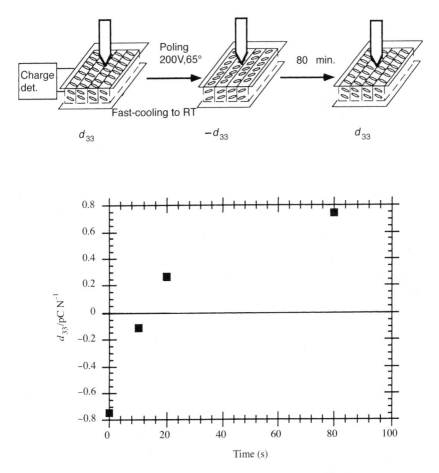

Figure 16. Relaxation of the piezocoefficient d_{33} of elastomer *E2* at room temperature after reversal poling at 65°C (SmA phase) [11].

nuously decreases, then it goes through zero and increases again, finally reaching the original positive value. This experiment is comparable to the hysteresis measurements of Figures 12 and 13 insofar as it shows that two polar states are accessible, but the one present during cross-linking is stabilized.

The magnitude of the piezocoefficients reported so far (up to 7 pC/N) are in the range of those observed for PVDF, but a bit smaller. They depend strongly on the director configuration inside the cell. In fact, for a cell with a perfect bookshelf geometry only a very small piezocoefficient is expected, because a compression along the polar axis (d_{33}) should not lead to a change of the director. For other geometries much higher values are expected. For an estimate of an upper limit, the change of the polarization associated with a complete reorientation of the polar axis (ferroelectric switching), which are about 90–150 nC/cm^2 (*P1* to *P3*), should be borne in mind.

Another example, which shows the coupling of liquid crystalline and elastomer properties nicely, is the observation of a piezoeffect in the smectic A phase of elastomer *E2*. After cross-linking of a ferroelectric monodomaine in the smectic C* phase, a piezoeffect d_{33} is observed even in the smectic A phase (see Fig. 17). From symmetry consideration such a piezoeffect is, however, not allowed for an untilted (orthogonal) smectic phase. One possible explanation of this would be that the sample is still in the smectic C* phase, which has been stabilized by network formation. This possibility is however excluded by the hysteresis measurements (Figs. 12 and 13), which show clearly the disappearance of the ferroelectric hysteresis about the Curie temperature (49°C). We believe that the basis for this is a mechanoclinic effect in analogy to the well-known electroclinic effect in the smectic A phase [12]. This idea is best represented by plotting the liquid crystalline potential, the elastic potential of the network, and their superposition in one graph [12] (see Fig. 18).

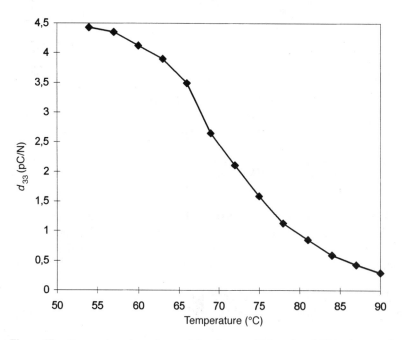

Figure 17.   Temperature dependence of the piezo-coefficient d_{33} of *E2* in the smectic A phase [11,12].

As the network is formed in the smectic C* phase, an internal elastic field is created, which has its minimum value for the tilt angle and tilt direction present during cross-linking. Other tilt directions are destabilized [Fig. 18(a)]. This explains the shift of the hysteresis loop of Figure 12. As the temperature changes, the elastic field tends also to preserve the tilt angle present during cross-linking. In the smectic A phase [see Fig. 18(b)] the situation is similar to the electroclinic effect. The liquid crystalline potential is centered at zero tilt angle; the elastic field tries to induce a nonzero tilt. If this scenario is right, then the temperature dependence of the piezo effect in the smectic A phase should be similar to the temperature dependence of the electroclinic effect, which increases strongly on approaching the Curie temperature. Figure 17 shows that this is indeed the case. Far above the Curie temperature, a critical behavior is observed [see especially Fig. 17(b)]. Closer to the Curie temperature, d_{33} saturates. This is equivalent to a

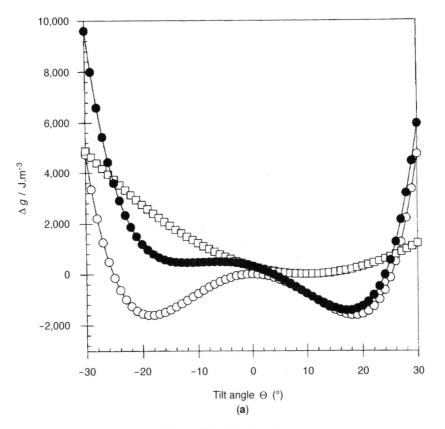

Figure 18. (*Continued*)

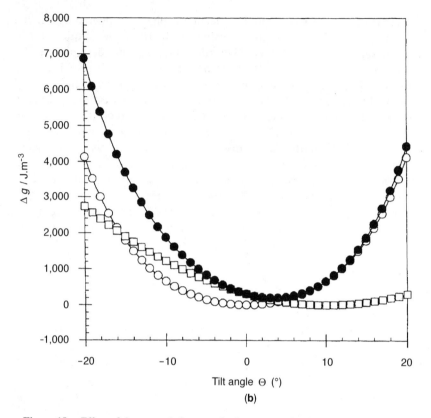

Figure 18.   Effect of the network force on the free energy density [12]. (a) 2K below the phase transition, S_C^* phase; (b) 2K above the phase transition, S_A phase, where ($\bigcirc$) is calculated potential, ($\square$) force due to the network and ($\bullet$) superposition of both.

measurement of the electroclinic effect by using strong electric fields, so that the regime of linear response is left. Measurements on systems with a lower cross-linking density would thus be necessary to observe the critical behavior of d_{33} in the smectic A phase more clearly.

VII. CONCLUSION

Ferroelectric LC-elastomers represent an interesting class of material insofar as they join in the ordering of liquid crystalline ferroelectric phases, together with the rubber elasticity of polymer networks. A switching of the electric polarization leads to a deformation of the polymer network, equivalent to streching a spring, and creates a stress in the network of polymer chains. The interaction of mesogens and network can be varied by using different topologies of netpoints: A cross-linking is carried out either within the siloxane sublayers (leading to fast-switching elastomers) or between the siloxane sublayers (resulting in an elastomer that favors the ferroelectric switching state in which the cross-linking reaction took place). As the polar orientation of the chiral smectic C^* phase is fixed in the elastomer, electromechanical measurements show a piezoelectric effect in the smectic C^* phase as well as in the smectic A^* phase. This last observation indicates a mechanoclinic effect, by which a definite tilt angle is induced within the cross-linked system.

Acknowledgment

The types of work presented in this summary article were made possible only through the close cooperation of several groups of researchers from the areas of physics and physical chemistry. We want to give special thanks to the groups of F. Kremer (Leipzig), F. Giesselmann and P. Zugenmaier (Clausthal-Zellerfeld), and J. K. Vij (Dublin).

References

1. M. G. Broadhurst and G. T. Davis, in: *Medical Applications of Piezoelectric Polymers*, Gordon and Breach Science Publishers S.A., Amsterdam, 1988.

2. A. Hult, F. Sahlén, Trollsås, S. T. Lagerwall, D. Hermann, L. Komitov, P. Rudquist, and B. Stebler, *Liq. Cryst.* **20**, 23–28 (1996).

3. R. Zentel, *Angew. Chem. Adv. Mater.* **101**, 1437 (1989).

4. M. Brehmer and R. Zentel, *Mol. Cryst. Liq. Cryst.* **243**, 353–376 (1994).

5. F. J. Davis, *J. Mat. Chem.* **3**, 551–562 (1993).

6. R. A. M. Hikmet, H. M. Boots, and M. Michielsen, *Liq. Cryst.* **19**, 65–76 (1995).

7. G. R. Mitchell, P. M. S. Roberts, K. Ahn, F. J. Davis, C. Hasson, H. Hischmann, and J. A. Pople, *Macromol. Symp.* **117**, 21–31 (1997).

8. G. C. L. Wong, W. H. de Jeu, H. Shao, K. S. Liang, and R. Zentel, *Nature* **389**, 576–579 (1997).

9. R. Zentel, *Liq. Cryst.* **1**, 589 (1986).

10. N. R. Barnes, F. J. Davis, and G. R. Mitchell, *Mol. Cryst. Liq. Cryst.* **13**, 168 (1989).

11. M. Brehmer, R. Zentel, K. Siemensmeyer, and G. Wagenblast, *Macromol. Chem. Phys.* **195**, 1891–1904 (1994).

12. M. Brehmer, R. Zentel, F. Giesselmann, R. Germer, and P. Zugenmaier, *Liq. Cryst.* **21**, 589–596 (1996).

13. E. Gebhard, M. Brehmer, R. Zentel, J. Reibel, G. Decher, H. M. Brodowsky, and F. Kremer, in: *The Wiley Polymer Networks Group Review*, Vol. 1 K. te Nijenhuis, W. Mijs (eds.). John Wiley & Sons, U.K. Ltd., Chichester, 1997.

14. H. Poths and R. Zentel, *Liq. Cryst.* **16**, 749–767 (1994).

15. L.-C. Chien and L. G. Cada, *Macromolecules* **27**, 3721 (1994).

16. S. Diele, S. Oelsner, F. Kuschel, B. Hisgen, H. Ringsdorf, and R. Zentel, *Makromol. Chem.* **188**, 1993 (1987).

17. H. Poths, E. Wischerhoff, R. Zentel, A. Schönfeld, G. Henn, and F. Kremer, *Liq. Cryst.* **18**, 811–818 (1995).

18. A. Kocot, R. Wrzalik, J. K. Vij, M. Brehmer, and R. Zentel, *Phys. Rev.* B **50**, 16346–16357 (1994); A. Kocot, R. Wrzalik, J. K. Vij, and R. Zentel, *J. Appl. Phys.* **75**, 728–733 (1994).

19. J. Küpfer and H. Finkelmann, *Macromol. Chem. Phys.* **195**, 1353–1367 (1994).

20. K.-H. Hanus, H. Kapitza, F. Kremer, G. Reckert, S. Vallerien, and R. Zentel, *Polymer Preprints* **30**, 493 (1989).

21. R. Zentel, *Liq. Cryst.* **3**, 531 (1989).

22. R. Zentel, G. Reckert, S. Bualek, and H. Kapitza, *Makromol. Chem.* **190**, 2869 (1989).

23. J. Küpfer and H. Finkelmann, *Makromol. Chem. Rapid Commun.* **12**, 717 (1991).

24. L. R. G. Treloar, *The Physics of Rubber Elasticity*, Clarendon Press, Oxford, U.K. 1975.

25. M. Brehmer and R. Zentel, *Macromol. Chem. Rapid Commun.* **16**, 659–662 (1995).

26. H. M. Brodowsky, U.-C. Boehnke, F. Kremer, E. Gebhard, and R. Zentel, *Langmuir* **13**, 5378–5382 (1997).

27. I. Benné, K. Semmler, and H. Finkelmann, *Macromolecules* **28**, 1854–1858 (1995).

28. R. A. M. Hikmet and J. Lub, *Prog. Polym. Sci.* **21**, 1165–1209 (1996).

29. M. Mauzac, H. T. Nguygen, F. G. Tournilhac, and S. V. Yablonsky, *Chem. Phys. Lett.* **240**, 461 (1995).

30. S. U. Vallerien, F. Kremer, E. W. Fischer, H. Kapitza, R. Zentel, and H. Poths, *Makromol. Chem., Rapid Commun.* **11**, 593 (1990).

Additional Reading

H. M. Brodowsky, U.-C. Boehnke, F. Kremer, E. Gebhard, and R. Zentel, *Langmuir* **15**, 274–278 (1999).

E. Gebhard and R. Zentel, *Macromol. Rapid Commun.* **19**, 341–344 (1998).

E. Gebhard and R. Zentel, *Liq. Cryst.* **26**, 299–302 (1999).

S. Shilov, E. Gebhard, H. Skupin, R. Zentel, and F. Kremer, *Macromolecules* **32**, 1570–1575 (1999).

STRUCTURE, MOBILITY, AND PIEZOELECTRICITY IN FERROELECTRIC LIQUID CRYSTALLINE ELASTOMERS

F. KREMER, H. SKUPIN, W. LEHMANN, AND L. HARTMANN

*Fakultät für Physik und Geowissenschaften, Universität Leipzig,
Leipzig, Germany*

P. STEIN AND H. FINKELMANN

*Institut für Makromolekulare Chemie, Albert-Ludwigs-Universität Freiburg,
Freiburg i. Br., Germany*

CONTENTS

I. INTRODUCTION

In ferroelectric liquid crystalline elastomers (FLCE) the properties of ferroelectric liquid crystals are combined with that of polymeric networks.

Advances in Liquid Crystals: A Special Volume of Advances in Chemical Physics, Volume 113,
edited by Jagdish K. Vij. Series Editors I. Prigogine and Stuart A. Rice.
ISBN 0-471-18083-1. © 2000 John Wiley & Sons, Inc.

With the use of refined synthetic techniques [1–11], it is possible to align the chiral mesogenic groups in FLCE on a macroscopic scale ($\sim \text{cm}^2$). By employing that set of methods, the development of a new class of materials has been achieved having an extraordinary profile of features, among them the following: (1) FLCE are piezoelectric because of the ferroelectric order of the chiral mesogens and because of the fact that the polymeric network prevents flow. (2) Due to a photochemically or thermally induced cross-linking reaction the viscoelastic properties of FLCE can be adjusted over a wide range. (3) The coupling between the elastomeric network and the (polar) chiral mesogens can be modified by chemical engineering, hence offering the possibility to tailor the piezoelectric as well as the viscoelastic properties of FLCE. (4) It is possible to prepare thick ($\sim 100\,\mu\text{m}$), thin ($\sim \mu\text{m}$), and ultrathin ($\sim \text{nm}$) films of macroscopically oriented FLCE, as self-supporting or as transferred samples. With this combination of properties FLCE have made a strong technological impact as basic materials for sensors and soft actuators in the area of microsystems technology. The present review article is focused on the piezoelectric effect in FLCE and its molecular interpretation. With that aim in mind, several experimental approaches are pursued.

II. EXPERIMENTAL

Macroscopically oriented samples of FLCE can be obtained essentially by two different methods. In the first of these, a thin ($\sim \mu\text{m}$) or ultrathin ($\sim \text{nm}$) self-supporting film of the uncross-linked ferroelectric liquid crystalline copolymer is prepared by enlarging its surface area slowly [8–14]. This is achieved by pulling apart movable steel blades surrounding the free-standing ferroelectric liquid crystalline copolymer film. By use of a photoinduced cross-linking reaction, UV-irradiation of a certain intensity and duration results in transformation into an FLC elastomer. (Samples of this type, as well as those oriented in surface stabilized LC cells [15–23], are not being discussed in this article.)

In the second method, thick ($\sim 100\,\mu\text{m}$), self-supporting layers of FLCE can be also produced by cross-linking the siloxane backbones under application of two successive deformations [1–3]. This yields a single crystal FLCE, as proved by X-ray diffraction. The chemical composition of the samples discussed in this article is shown in Figure 1(a). Two different ratios in the coelastomers lead to different phase sequences as examined by X-ray diffraction, polarization microscopy of the uncross-linked polymer, and by differential scanning calorimetry (DSC). Since the FLC copolymer has been uniformly aligned during the cross-linking by mechanical stress, the ferroelectric order is stabilized by the elastic network. Hence, the polar

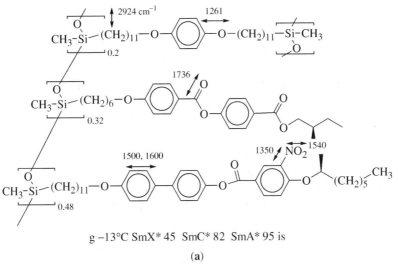

g −13°C SmX* 45 SmC* 82 SmA* 95 is

(a)

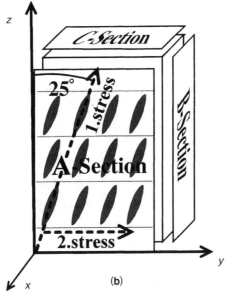

(b)

Figure 1. (a) Chemical composition of the FLCE under study. Two different chiral mesogenic subunits are attached in varying ratios (40:60 ≅ coelastomer 40/60; 50:50 ≅ coelastomer 50/50) to 80% of the Si–H groups of the polysiloxane backbone. The copolymer has been thermally cross-linked to 20% of the Si–H groups under application of two successive uniaxial deformations. (b) Scheme of the orientation of "single-crystal" FLCE. The direction of two successive mechanical stresses which were applied during the thermal cross-linking is indicated. Furthermore, the orientation of the microtomized sections (thickness ∼ 5μm) for the time resolved FTIR studies is shown.

axis of the system is defined and preserved by the elastic forces. Furthermore, the network prevents flow. In Figure 1(b) the orientation of the "single-crystal" FLCE and the directions of the two successive mechanical stresses are shown.

A.   Measurements of the (Direct and Inverse) Electromechanical Effect in FLCE

For the measurement of the *direct* electromechanical effect (Fig. 2), a periodic deformation is applied to the sample by means of a hysteresis-compensated Piezo-Actuator. The resulting AC-voltage at the sample electrodes is measured with phase-sensitive detection as a function of the amplitude and the frequency of the mechanical excursion. The set-up has been calibrated using an 80-μm-thick foil of poly(vinylidene fluoride)

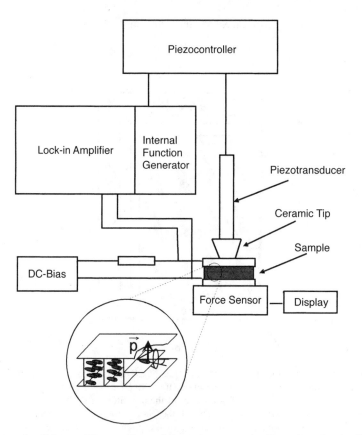

Figure 2.   Scheme of the experimental set-up to measure the direct piezoelectric effect in FLCE.

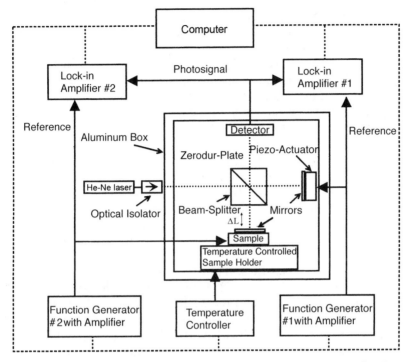

Figure 3. Scheme of the experimental set-up for the dynamic measurement of the inverse electromechanical effect based on a Michelson interferometer. The whole system has a resolution of $\Delta L/2 = \pm 100 \, \text{fm}$ [E].

(PVDF). The piezoelectric modulus of this standard was measured as $d_{33} = 24.2 \, \text{pC/N}$ (in the inverse effect) using the very accurate interferometer set-up (literature value: d_{33} (PVDF)$=30 \, \text{pC/N}$ [24]). In order to maintain permanent electrode contact in the measurements of the direct piezoelectric effect, the FLCE sample has to be prestressed. This has a direct influence on the spontaneous polarization P_s of the sample. In summary, the absolute error in the measurement of the direct piezoeffect is estimated to about $\pm 10\%$.

The *inverse* electromechanical effect is measured by means of an ultrastable Michelson interferometer (Fig. 3) having a resolution $\leq 100 \, \text{fm}$ in dynamical measurements [25]. A piece of the ~ 300-μm-thick sample sheet is contacted with 0.8-μm-thick aluminium foil electrodes, and a lightweight (5.6 mg) mirror is placed on top of it and mounted in one of the two interferometer arms. Phase-sensitive detection allows for a quantitative analysis of the (periodic) change in the sample thickness under the influence of AC electric and superimposed DC bias fields in the first and the second harmonic. By that the piezoelectric response (1st harmonic) can be distinguished from electrostriction and attraction of the capacitor plates (2nd harmonic). The DC bias-dependent contribution influences the 2nd harmonic effects as well [25]. From the relationship between the displacement amplitude $\Delta L/2$ and the applied AC amplitude $\hat{U}$ the linear and the quadratic fit coefficients $K_1(\Delta L/2 = K_1\hat{U})$ and $K_2(\Delta L/2 = K_2\hat{U}^2)$ can be derived. The contribution of the DC field in the second harmonic response can be separated by the following approach:

$$\Delta L/2 = aU(t) + bU^2(t) \tag{1}$$

With $U(t) = U_{DC} + U_{AC}\cos(\omega t)$ leading to the following equation (terms independent of ω are not detected in the phase-sensitive detection):

$$\Delta L/2 = (a + 2bU_{DC})U_{AC}\cos(\omega t) + b/2U_{AC}^2\cos(2\omega t) \tag{2}$$

Hence, the corrected coefficients a (linear) and b (quadratic) can be deduced as $K_1 = a + 2bU_{DC}$ and $K_2 = b/2$. It turns out that the true linear coefficient a does not differ much from K_1 because b is small. Therefore the coefficient b is neglected and K_1 is taken as the piezoelectric modulus d_{33} of the sample in nm/V ($\cong$ nC/N), where the change in sample thickness is parallel to the electric field. The whole interferometer is calibrated to the laser wavelength (632.8 nm) and the relative accuracy in the determination of the absolute value of the inverse piezoelectric coefficient is estimated to $\pm 10\%$.

B. Time-Resolved FTIR-Spectroscopy

In order to analyze structure and mobility in single-crystal FLCE, time-resolved Fourier Transform Infrared Spectroscopy (FTIR) with polarized light is employed (Fig. 4). Due to its specifity, FTIR-spectroscopy enables the determination of the orientation, the order parameter, and the extension and time-scale of the motion for the different molecular moieties. A Bio-Rad FTIR-spectrometer (FTS 6000) coupled to an IR microscope (UMA 500) with a measurement spot of $\leq$ 100 µm has been used to record IR spectra at different polarizations of the IR-beam. Microtomized sections [Fig. 2(b)] of the "single-crystal" FLCE are placed between two CaF$_2$ windows covered with indium tin oxide (ITO) electrodes. These windows and electrodes are transparent for IR as well as for visible light. The windows were separated by PET spacers of 5–8 µm in thickness to avoid electrical short circuits and mechanical stress on the sample. The measurement cell is mounted into a temperature-controlled sample holder (temperature range: 20–100°C; accuracy: ± 0.05°C). The static absorbance spectra are measured for a set of

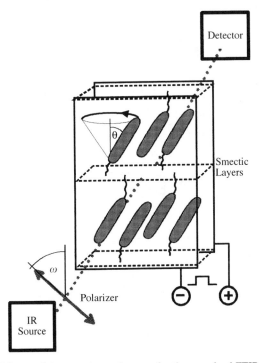

Figure 4. Scheme of the experimental set-up for time resolved FTIR measurement with polarized light. The orientation of the mesogens with respect to the incident IR-light and to the applied external electric field is shown.

polarizer directions ω from $0°$ to $180°$ with steps of $10°$. Absorbance spectra were recorded between $400\,\text{cm}^{-1}$ to $4000\,\text{cm}^{-1}$. For the time resolved FTIR measurements the step-scan technique enables a resolution in time of $5\,\mu s$. For further details see Ref. [19–21,26].

III. RESULTS AND DISCUSSION

A. The Direct and the Inverse Piezoelectric Effect in FLCE

The direct piezoelectric effect shows—as expected—a pronounced temperature dependence [Fig. 5(a)]; with increasing temperature the mobility of the (polar) mesogenic groups and the piezoelectricity increases. But it turns out that the effect has a strong hysteresis [Fig. 5(b)]: From the first heating run of a freshly prepared FLCE to the second heating cycle, a big difference is observed. It reflects the fact that in the course of the cross-

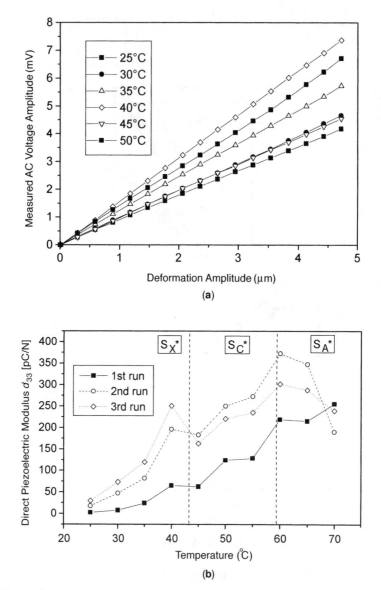

Figure 5. (a) Measured piezoelectric voltage vs. amplitude of the mechanical deformation of a fresh sample. The sample temperature was varied between 25 and 50°C as indicated. The sample thickness was 300 μm, and the frequency of the mechanical deformation was 13 Hz. Sample: Coelastomer 50/50, phase sequence: g13SmX*43SmC*59 SmA*75 is. (b) Hysteresis of the observed piezoelectricity in coelastomer 50/50: Piezoelectric moduli d_{33} vs. temperature for different heating cycles of the sample.

linking reaction during the twofold successive stretching of the sample a nonequilibrium state is "frozen." After several heating and cooling cycles the sample equilibrates and shows a maximum in its piezoelectricity near the phase transition temperature SmC*–SmA*. This maximum is interpreted as caused by the counterbalance between the increased mobility of the mesogens at elevated temperatures, which amplifies the piezoelectric effect and the destruction of the ferroelectric order with increasing temperature.

The direct effect can be enhanced or decreased by a superimposed electric DC-bias field (Fig. 6). This effect becomes stronger, if the phase transition SmC*–SmA* is approached. The fact, that at 65°C a minimum at about 1100 V/cm bias field is observed, reflects a still existing ferroelectric order in the sample. This field strength of 1100 V/cm corresponds to a coercive electric field, which compensates for the spontaneous polarization in the sample. At a temperature of 70°C the sample shows a true paraelectric DC-bias field dependence. It is remarkable that a bias field $\leq$ 3000 V/cm is sufficient to strongly enhance the piezoelectricity in the material.

The measurement of the *inverse* piezoelectric effect is carried out with the interferometric set-up described above. From the experiment [Fig. 7(a)] the piezoelectric modulus d_{33} can be directly deduced (Table I). At room temperature the piezoelectric moduli of the FLCE under study and of PVDF are comparable. At a temperature at 63°C the piezoeffect in FLCE is strongly enhanced. Similarly to the direct piezoelectric effect, the inverse piezoelectric effect shows a pronounced dependence on a superimposed

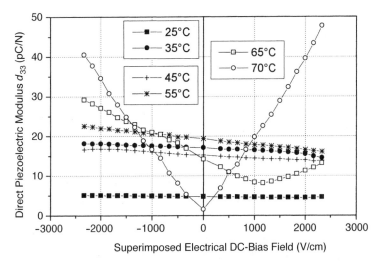

Figure 6.   DC-Bias field dependence of the direct piezoelectric modulus for the sample coelastomer 50/50 at different temperatures, open symbols correspond to the SmA*-phase.

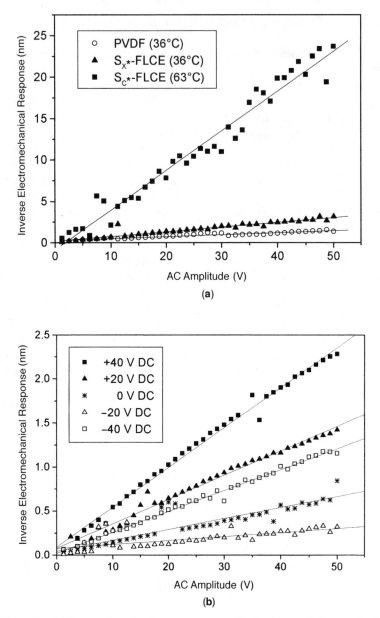

Figure 7. (a) Inverse piezoelectric response vs. amplitude of an applied external voltage. Sample: Coelastomer 40/60; sample thickness: 300 μm ±30 μm. Frequency of the external electric field: 13 Hz. Temperatures as indicated. The open symbol corresponds to measurements on polyvinylidene fluorid (PVDF) at 36°C. No external electric DC-bias field is superimposed. The solid lines are least squares linear fits. (b) Inverse piezoelectric responses vs. amplitude of an applied external voltage at different superimposed electric DC-bias voltages as indicated. Sample temperature: 25°C. Otherwise, the same as (a). The solid lines are least squares linear fits.

TABLE I

Piezoelectric Modulus d_{33} (as Deduced from Measurements of the Inverse Piezoelectric Effect for FLCE (at Two Different Temperatures) and for Poly(vinylidene fluoride) (at 36°C)

Substance	Piezoelectric Modulus d_{33} (pC/N)
PVDF (36°C)	24.2
Coelastomer 40/60 (36°C)	59.7
Coelastomer 40/60 (63°C)	479.5

external electric DC-bias field [Fig. 7(b)]. In complete analogy to the direct piezoelectric effect (Fig. 6), the DC-bias field dependence of the inverse effect is strongly temperature-dependent, reflecting the transition from a paraelectric (SmA*-phase) to a ferroelectric behavior [Fig. 8(a,b)]. The minima in Figure 8(b) define the coercive field strength which is (at a certain temperature) necessary to suppress the spontaneous polarization. If the phase transition SmC*–SmA* is approached, the coercive field strength becomes zero. From an estimation of the dielectric susceptibility $\chi' = (\varepsilon' - 1)$, the spontaneous polarization of the FLCE can be deduced according to:

$$P_S = \varepsilon_0 \chi' E \tag{3}$$

where ε_0 is the permittivity of free space. Assuming $\chi' \sim 13$ and with a coercive field $E \sim 1000$ V/cm, a value for P_S of ~ 1 nC/cm^2 is found. This calculation is very rough, because the DC-bias field dependence of χ' is not known for this sample, but the value of 1 nC/cm^2 is in accordance with recent pyroelectric measurements [27].

Summarizing the piezoelectric measurements, one has to state that the observed effects in FLCE are, in terms of its strength, fully comparable with the "classical" polymer PVDF. For FLCE a strong temperature dependence is observed. By use of a superimposed electric DC-bias field < 5000 V/cm the piezoelectric effect can be amplified by more than one order of magnitude.

B. Structure and Mobility of FLCE as Studied by (Time-Resolved) FTIR-Spectroscopy

The evolution of IR-absorbance spectra with time (Fig. 9) enables the detailed study of the response of a molecular system to an external electric field. Taking advantage of the specifity of IR-spectroscopy makes it possible to carry out this analysis for the different molecular moieties separately. By measuring the absorbance of a microtomized A-section [Fig. 10(a)] no reorientation is found for external fields $\leq 2.5 \cdot 10^5$ V/cm [Fig. 10(b)]. In

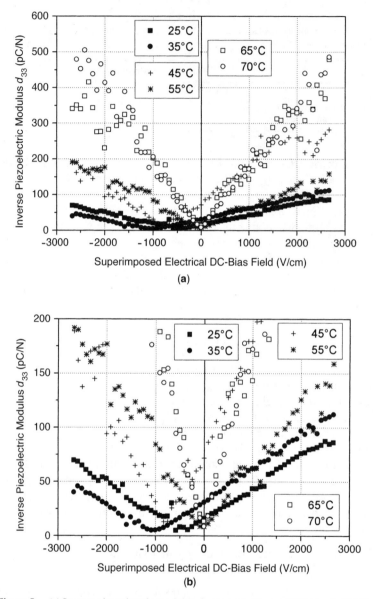

Figure 8. (a) Inverse piezoelectric modulus d_{33} vs. superimposed DC-bias field (V/cm) at different temperatures as indicated. Sample: Coelastomer 50/50. Amplitude of the AC-voltage 40 V. Error bars are omitted; relative accuracy $\pm 10\%$. (b) enlargement of (a).

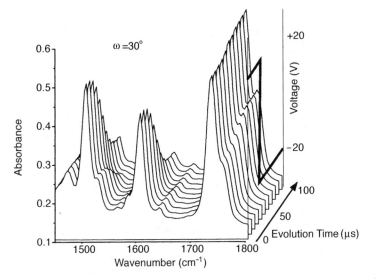

Figure 9. Waterfall plot of absorbance vs. wavenumber in its response to a jump of the applied external field as indicated in the third-scale "evolution time." The different molecular moieties react differently to the change in the external electric field. While both phenyl stretching bands at $1500\,cm^{-1}$ and $1609\,cm^{-1}$, decrease in their absorbance the carbonyl band at $1736\,cm^{-1}$ increases. Position of the polarizer with respect to the z-axis: 30°.

contrast, for a microtomized B-section [Fig. 11(a)] symmetric rearrangements of the director measured by reorientation of the phenyl groups are observed [Fig. 11(b)]. From the polar plot for $E = 0$ V/cm and $E = \pm 2.5 \cdot 10^5$ V/cm, the excursion of the angular motion—as projected in the plane perpendicular to the propagation direction of the IR beam—can be directly deduced as $\pm 15°$. How can the different response for the A- and B-section be comprehended? For the A-section the external electric field and the polarization vector P_S are oriented parallel to each other and the applied electric field cannot overcome the restoring forces of the network. In contrast, for the B-section the external field E is perpendicular to the polarization vector P_S. This allows for a torque to be exerted on the mesogens, causing a motion on a cone, as described in Figure 12. The different molecular moieties respond to the external electric field with rearrangements of different angular excursion [Fig. 13(a)]. By normalizing these plots, one can show that all molecular units move in phase with each other [Fig. 13(b)] at least on the time scale of this experiment ($\sim$ ms).

Summarizing the FTIR experiments, one can state that this method delivers refined information concerning structure and mobility of FLCE.

F. KREMER ET AL.

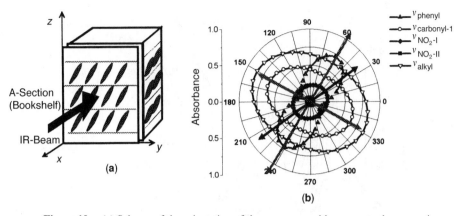

Figure 10. (a) Scheme of the orientation of the mesogens with respect to the measuring IR beam in the A-section. External electrical fields $\leq 2.5 \cdot 10^5$ V/cm were applied in X-direction. (b) Polar plot of the absorbance for the two phenyl-stretching vibrations I ($1500 \, \text{cm}^{-1}$) and II ($1609 \, \text{cm}^{-1}$), the carbonyl stretching vibration ($1736 \, \text{cm}^{-1}$), and the nitro-stretching vibration I ($1350 \, \text{cm}^{-1}$) and II ($1540 \, \text{cm}^{-1}$) for a microtomized A-section.

Details of the molecular structure and the molecular motion in response to an external electric field can be analyzed. Further details are discussed in Ref. [26].

C. Molecular Interpretation of the Observed Piezoelectric Effect in the FLCE Under Study

The fact that the amplitudes of the observed inverse piezoelectric effect can be as large as ± 50 nm excludes ferroelectric switching on the tilt cone and electroclinic tilt variation as the direct origin of the piezoresponses. But it is indicated that motions of the smectic layers as a whole are responsible for the observed effects: The spontaneous polarization in the tilted (layer tilt angle β, s: Fig. 14) layer is decomposed in its vector components orthogonal $\mathbf{P}_\perp$ and parallel $\mathbf{P}_\parallel$ to the applied external AC electric field and leads to a torque $\mathbf{P}_\perp \times \mathbf{E}$ that increases or reduces the layer tilt angle β, depending on the direction of the field. For electric fields in up direction left-tilted layers are turned right-handed, and right-tilted layers are turned left-handed. Hence both layer orientations are turned towards a more upright position (or toward a more tilted position for electric fields in down direction), thus changing the sample thickness. This interpretation is also backed by X-ray studies measuring the distribution of the layer inclination [28].

It remains to be explained how the superimposed DC-bias field can enhance or–in the case of the SmA-phase—induce the observed piezo-

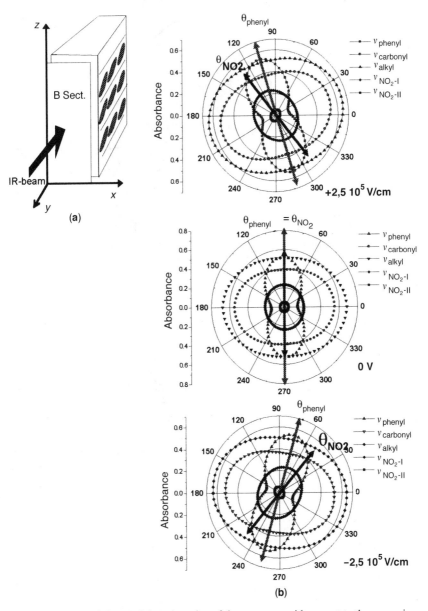

Figure 11. (a) Scheme of the orientation of the mesogens with respect to the measuring IR beam in the B-section. External electrical fields $\leq 2.5 \cdot 10^5$ V/cm were applied in the Y-direction. (b) Polar plot of the bands described in Figure 10(b) under the influence of an external electric field of $\pm 2.5 \cdot 10^5$ V/cm and 0 V/cm.

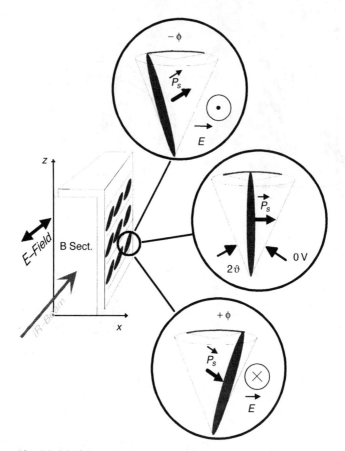

Figure 12.　Model of the molecular response of the mesogens to the external electric field for a microtomized B-section.

electricity. As proved by dielectric spectroscopy [29] on low molecular weight FLC a superimposed DC bias field can induce a rotational bias, i.e., it imposes a preferred orientation to the distribution of the lateral dipole moments around the long molecular axis. As shown [30], in molecular dynamics simulations as well this effect has a low activation energy and takes place for electric fields as small as 1000 V/cm. Hence the superimposed DC field amplifies (in the SmC*-phase) the net polarization vector of a smectic layer or it induces (in the SmA*-phase) such a polarization vector. This effect becomes more pronounced if the molecular mobility of the mesogens is increased at elevated temperatures.

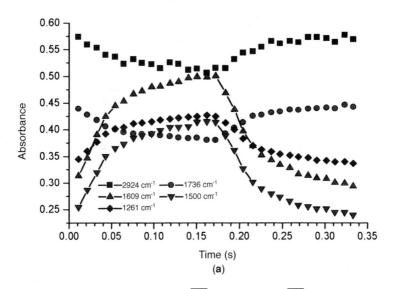

$$A_N = (A(t) - A_f) / (A_i - A_f), \text{where } A_i = \overline{[A(t)]}_{t \geq 300\text{ms}} \text{ and } A_f = \overline{[A(t)]}_{150\text{ms} \leq t \leq 180\text{ms}}$$

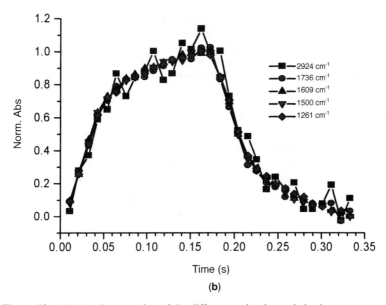

Figure 13. (a) Angular excursion of the different molecular moieties in response to an external electric field. (b) Normalized plot of the data from (a) showing that the different molecular moieties rearrange in phase with each other.

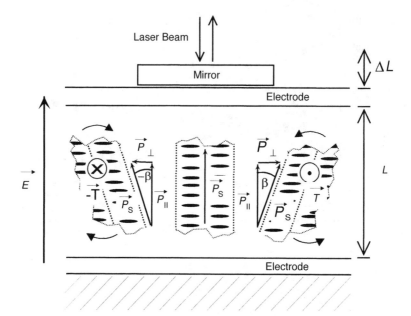

Figure 14. Scheme for the molecular interpretation for the observed piezoelectric effect in the FLCE under study (Coelastomer 40/60, resp. Coelastomer 50/50). In the proposed model the external electric AC field interacts with the polarization vector $\vec{P}_S$ and causes a torque $\vec{P}_S \times \vec{E}$. (The torques are indicated for the electric AC field in up direction). By that means, the angle of the layer inclination and hence the thickness of the sample are changed.

IV. CONCLUSION

The piezoelectric effect in FLCE is experimentally analyzed in detail. To comprehend the observed findings on a molecular level time-resolved FTIR-spectroscopy is employed. This results in a microscopic interpretation of the macroscopic piezoelectricity. By that means, new routes for the optimization of this extremely interesting class of materials can be envisaged.

Acknowledgment

Support by the German Science Foundation DFG in the framework of the Innovationskolleg "Phänomene an den Miniaturisierungsgrenzen" and the Sonderforschungsbereich SFB 294 "Moleküle in Wechselwirkung mit Grenzflächen" is gratefully acknowledged.

References

1. K. Semmler and H. Finkelmann, *Polym. Adv. Technol.* **5**, 231 (1994).
2. K. Semmler and H. Finkelmann, *Macromol. Chem. Phys.* **196**, 3197 (1995).

3. T. Eckert, H. Finkelmann, M. Keck, W. Lehmann, and F. Kremer, *Macromol. Chem. Rapid Commun.* **17**, 767 (1996).

4. W. Meier and H. Finkelmann, *Macromol. Chem. Rapid Comm.* **11**, 599 (1990).

5. S. U. Vallerien, F. Kremer, E. W. Fischer, H. Kapitza, R. Zentel, and H. Poths, *Macromol. Chem. Rapid Comm.* **11**, 593 (1990).

6. W. Meier, H. Finkelmann, *Macromolecules* **26**, 1811 (1993).

7. R. Zentel, *Angew. Chem. Adv. Mater.* **101**, 1437 (1989).

8. J. Reibel, M. Brehmer, R. Zentel, and G. Decher, *Adv. Mater.* **7**, 849 (1995).

9. E. Gebhard, M. Brehmer, R. Zentel, J. Reibel, G. Decher, H. M. Brodowsky, and F. Kremer, in K. Nijenhuis, W. Mijs (eds.) *The Wiley Polymer Networks Group Review*, Vol. I., John Wiley and Sons, Chichester, U.K., 1997.

10. G. Decher, J. Maclennan, and U. Sohling, *Adv. Mater.* **3**, 617 (1995).

11. J. Maclennan, G. Decher, and U. Sohling, *Appl. Phys. Lett.* **59**, 917 (1991).

12. H. M. Brodowsky, U.-C. Boehnke, F. Kremer, E. Gebhard, and R. Zentel, *Langmuir* **13**, 5378 (1997).

13. H. M. Brodowsky, U.-C. Boehnke, F. Kremer, E. Gebhard, and R. Zentel, *Langmuir* **15**, 274 (1999).

14. H. Skupin, F. Kremer, J. Prigann, S. V. Shilov, K. Skarp, and B. Helgee, in preparation.

15. M. Brehmer and R. Zentel, *Macromol. Chemie Rapid. Commun.* **16**, 659 (1995).

16. M. Brehmer, R. Zentel, G. Wagenblast, and K. Siemensmeyer, *Macromol. Chem. Phys* **195**, 1891 (1994).

17. A. Kocot, R. Wrzalik, J. K. Vij, M. Brehmer, and R. Zentel *Phys. Rev. B* **50**, 16346 (1994).

18. H. Poths and R. Zentel, *Liq. Cryst.* **16**, 749 (1994).

19. S. V. Shilov, H. Skupin, F. Kremer, E. Gebhard, and R. Zentel, *Liquid Crystals* **22**, 203 (1997).

20. S. V. Shilov, H. Skupin, F. Kremer, T. Wittig, R. Zentel, *Phys. Rev. Lett.* **79**, 1686 (1997).

21. S. V. Shilov, E. Gebhard, H. Skupin, R. Zentel, and F. Kremer *Macromolecules*, **32**, 1570 (1999).

22. A. Jákli and A. Saupe, *Liq. Cryst.* **9**, 519 (1991).

23. A. Jákli and A. Saupe, *Phys. Rev. E Rapid Commun.* **53**, R5580 (1996).

24. A. M. Glass, in T. T. Wang, J. M. Herbert, and A. M. Glass, eds, *The Applications of Ferroelectric Polymers*, Blackie, Glasgow, Scotland (1998).

25. W. Lehmann, P. Gattinger, M. Keck, F. Kremer, P. Stein, T. Eckert, H. Finkelmann, *Ferroelectrics* **208–209**, 373 (1998).

26. H. Skupin, F. Kremer, S. V. Shilov, P. Stein, and H. Finkelmann, *Macromolecules* **32**, 3746 (1999).

27. W. Lehmann, N. Leistner, L. Hartmann, D. Geschke, F. Kremer, P. Stein, and H. Finkelmann, *Mol. Cryst, Liq. Cryst.* **328**, 437 (1999).

28. F. Kremer, W. Lehmann, H. Skupin, L. Hartmann, P. Stein, and H. Finkelmann, *Polym. Adv. Technol.* **9**, 672 (1998).

29. A. Schönfeld, F. Kremer, *Ber. Bunsenges. Phys. Chem.* **97**, 1237 (1993).

30. M. Loos-Wildenauer, S. Kunz, A. Yakimanski, I. Voigt-Martin, E. Wischerhoff, R. Zentel, C. Tschierske, and M. Müller, *Adv. Mat.* **7(2)**, 170 (1995).

ORIENTATIONAL EFFECTS IN FERROELECTRIC AND ANTIFERROELECTRIC LIQUID CRYSTALS USING INFRARED SPECTROSCOPY

A. KOCOT

Institute of Physics, University of Silesia, Katowice, Poland

J. K. VIJ AND T. S. PEROVA

Department of Electronic and Electrical Engineering, Trinity College, University of Dublin, Ireland

CONTENTS

Advances in Liquid Crystals: A Special Volume of Advances in Chemical Physics, Volume 113, edited by Jagdish K. Vij. Series Editors I. Prigogine and Stuart A. Rice.
ISBN 0-471-18083-1. © 2000 John Wiley & Sons, Inc.

I. INTRODUCTION

Mayer's [1,2] prediction of the mechanism for the emergence of the
ferroelectricity in liquid crystals (LC), based on symmetry arguments and
the molecular structure, have now been confirmed experimentally. The
liquid crystalline phases, which appear successively on lowering temperature from the isotropic, nematic, and SmA phases, are paraelectric owing to
the isotropic rotation around the long molecular axis and the head and tail
equivalence. In the chiral SmC* phase, however symmetry is significantly
reduced, to C_2, which allows the existence of the chirality induced improper
ferroelectricity. Microscopically, the molecules do rotate, but this rotation is
now biased, hence a net polarization emerges in the SmC* phase. In addition
to the classical ferroelectric SmC* phase, a number of SmC*-like tilted
phases have also been observed in liquid crystals. These are the antiferroelectric SmC_A^*, SmC_α^*, and other intermediate ferrielectric phases. Since the
discovery of a number of phases, extensive investigations for the tilted
phases have been carried out to clarify their structure and to find the origin
of their emergence.

    Fourier transform infrared (FTIR) spectroscopy has proven itself to be
one of the most useful techniques for investigating liquid crystals. This
method is commonly used for obtaining information about the order
parameter, for studying the conformational changes that may occur, the
dynamics of molecules, and of the molecular segments. More recently, this
method was successfully used for obtaining microscopic proof of the biased
rotation of molecules about their long axes in both ferroelectric liquid
crystals (FLCs) and antiferroelectric liquid crystals (AFLCs). Again using
polarized FTIR, it has been found that in an antiferroelectric phase the
molecules in alternate layers are tilted in opposite directions but with the
same tilt angle. Fourier time-resolved spectroscopy has been used to observe
the dynamics of molecules, following the field-induced switching. It is,
therefore, desirable to review some of the work reported in the literature
during the last decade using polarized FTIR and time-resolved infrared (IR)
spectroscopy.

The middle IR range, in the wavenumber range of $400-4000\,cm^{-1}$, is found to be most useful, because an interaction of light with the molecular vibration in this range dominates the spectra. The absorption peaks that are observed in the spectra belong to the normal vibration of a repeating unit of the sample: a molecule or a unit cell for the case of crystals. In these normal vibrations, all atoms of the unit, whether a molecule or a unit cell itself, take part and give rise to the spectra. However, for the organic molecules, most of the strongest absorption peaks are assignable to the vibration modes of well-defined groups of the molecule, such as the benzene ring or a CH_2 or $C=O$ bond. It is well known that the gross features of the IR spectra are determined by the molecule itself and in particular the groups and bonds that it is built from; nevertheless, the intermolecular interactions and the symmetry of the environments of a certain bond or groups do have a relatively small but measurable effect on the corresponding spectra. A splitting of the bands may occur owing to a symmetry lowering or from the intermolecular interactions. These interactions and the shape of the molecule play a dominant role not only in establishing the liquid crystalline phases but also in determining the transition temperatures. The reason for the existence of the liquid crystalline phases can, therefore, in principle, be determined from the IR spectra. From the observation of a shift in the frequency of the bands and from a change in their intensity at the transitions, we can determine the particular bonds or groups of atoms that play the most significant role in the phase transformation.

II. ORIENTATIONAL ORDER PARAMETERS AND THEIR DETERMINATION BY VIBRATIONAL SPECTROSCOPY

Consider an almost perfectly oriented domain of a liquid crystal sample; the laboratory frame of reference is determined by the anisotropy of the material. The orientation of the molecule in the laboratory frame of reference can be described by the three Euler angles: θ, φ, and ψ (Fig. 1). Denoting the unit vectors of the laboratory frame by e_X, e_Y, and e_Z and those of the molecular frame of reference by e_x, e_y, and e_z, the following formulas relate the unit vectors and the Euler angles [3,4].

In Figure 1, θ is the angle between e_Z and e_z

$$\cos\theta = (e_Z \cdot e_z) \tag{1}$$

n, the normal vector to the plane defined by e_Z and e_z, makes an angle φ with e_X

$$n = [e_Z \times e_z]/\sin\theta$$
$$(n \cdot e_x) = \cos\varphi \tag{2}$$

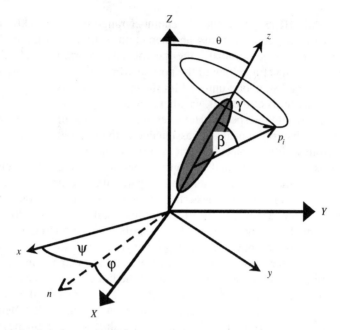

Figure 1. Laboratory (X, Y, Z) and molecular frames (x, y, z) of reference. Angles identified in the text. n is normal to the plane defined by e_Z and e_z.

θ and φ define the z axis in the molecular frame of reference: Usually the axis of highest symmetry of the molecule is chosen as z axis. For rod-like molecules, this is the molecular long axis. ψ describes the rotation of the molecule around the axis Z: the angle that e_x makes with n.

$$(n \cdot e_x) = \cos \psi \tag{3}$$

Any vector defined with respect to the molecular frame of reference can be transformed into the laboratory frame of reference by the transformation matrix M

$$(p_x, p_y, p_z) \underset{M}{\Rightarrow} (p_x, p_y, p_z)$$

$$M = \begin{bmatrix} \cos \varphi \cos \psi - \cos \theta \sin \varphi \sin \psi & \cos \varphi \sin \psi + \cos \theta \sin \varphi \cos \psi & \sin \theta \sin \varphi \\ -\sin \varphi \cos \psi - \cos \theta \cos \varphi \sin \psi & -\sin \varphi \sin \psi + \cos \theta \cos \varphi \cos \psi & \sin \theta \cos \varphi \\ \sin \theta \sin \psi & -\sin \theta \cos \psi & \cos \theta \end{bmatrix} \tag{4}$$

The probability of a particular orientation of a molecule is described by the orientational distribution function $f(\theta, \varphi, \psi)$. When the molecule is rod shaped or the orientation around its axis is equally probable, f does not depend on ψ. For SmA or a nematic phase, the material is uniaxial, that is the distribution function is independent of φ. If both conditions are satisfied, the orientational order in the material is described by a single-order parameter

$$S = 0.5\langle 3\cos^2\theta - 1 \rangle \tag{5}$$

where the angle brackets imply the ensemble average for all molecules in the system. It can happen that the rotation around the long molecular axis is not free. The distribution function then depends on ψ, and the axes x and y have different average inclinations to the Z axis. This biaxiality of the molecule is characterized by a parameter called dispersion and denoted by D [3,4]:

$$D = \tfrac{3}{2}\langle (e_{xZ})^2 - (e_{yZ})^2 \rangle = \tfrac{3}{2}\langle \sin^2\theta \cos 2\psi \rangle \tag{6}$$

A. IR Dichroism

The study of the IR dichroism of the absorption bands that belong to identified groups in the molecule, groups for which direction of the vibrational transition dipole moment is known, can give reliable information about the structure and alignment of liquid crystals [5]. The vibrational motions of a molecule can be described as a linear combinations of $3N-6$ normal modes, where N is the number of atoms in the molecule. In a normal mode, the vibration is accompanied by a vibrating dipole moment

$$\Delta m = (\delta m/\delta Q)Q = pQ \tag{7}$$

where Q is the normal coordinate of vibration and $(\delta m/\delta Q)$ is called the transition moment belonging to the normal vibration Q and is taken at $Q = 0$. When interacting with light, the energy of this interaction is

$$(\Delta m \cdot E) = (p \cdot E)Q \tag{8}$$

where E is the electric field vector in the light beam. The solution of the equation of motion for the normal coordinate Q in a harmonic electric field is a forced oscillation, the amplitude of which is proportional to $(p \cdot E)$

$$Q = g(\nu)(p \cdot E) \tag{9}$$

where $g(v)$ is a complex function of the frequency v of the IR radiation. Any vibration mode of an individual molecule contributes to the dipole moment of the sample by:

$$\Delta m = p(p \cdot E) g(v) \qquad (10)$$

This contribution depends on the orientation of the individual molecule with respect to the electric field. Taking all the molecules in a unit volume into account, we get the contribution of the normal vibration to the polarization of the material.

$$P = Ng(v)\langle p(p \cdot E)\rangle \qquad (11)$$

where N is the number of molecules in the unit volume and the angle brackets imply the average for all possible orientations of a molecule

$$\langle p(p \cdot E)\rangle = \int\int\int p(p \cdot E) f(\theta, \psi, \varphi) \, d\varphi \, d\psi \qquad (12)$$

The electric displacement vector is

$$D = \varepsilon_0 E + P = \varepsilon E \qquad (13)$$

the normal mode with transition moment p contributes by $N\langle p_k p_l\rangle g(v)$ to the $\varepsilon_{k,l}$ component of ε, the tensor of complex permittivity.

B.   Uniaxial Phase

If we chose the Z axis of the laboratory frame of reference so as it to be parallel to the director of the sample, then the permittivity tensor becomes diagonal and $\varepsilon_{XX} = \varepsilon_{YY} \neq \varepsilon_{ZZ}$ if the material is uniaxial. If the sample is oriented either homogeneously or homeotropically, the Z axis is parallel to the sample surface in the first case and normal to it in the second case. Let the light beam be incident normally on the sample and the electric field parallel to one of the optical axes. In this case, the light beam traverses normally through the sample and behaves as an ordinary ray. Propagation and absorption are determined by the corresponding diagonal element of the permittivity tensor ε, that is by ε_{YY}, if E is perpendicular to the director and by ε_{ZZ}, if E is parallel to it. The absorption is governed by the imaginary part of the permittivity, which, in turn, is proportional to $\langle p_Y^2\rangle$ and $\langle p_Z^2\rangle$ respectively [4,6].

The dichroic ratio N of an absorption peak is defined as the ratio of the peak intensity measured with the light polarized parallel to the director to the band intensity when the light is polarized perpendicular to it. The band intensity can be taken as the area of the band in the absorbance spectrum or as the peak intensity. The normal modes reflect the symmetry of the unit they belong to: either a molecule or a unit cell in a crystal, but these normal modes of vibration are built up from the vibrations of smaller groups of atoms, like those of the benzene ring, the alkyl group, or the C=O bond. For the normal modes of the molecule, in which the vibrational energy is concentrated in a particular kind of unit, the absorption bands we observe in the spectra can be assigned to the vibrating bond or a group rather than to a vibrational normal mode of the whole molecule. Accordingly, the direction of the transition moment of the vibration is determined by the orientation of this group of atoms. Assume that the direction of the transition moment is given by the polar angles β and γ in the molecular frame of reference; that is, its components are

$$p = |p|(\sin \beta \cos \gamma, \sin \beta \sin \gamma, \cos \beta)$$

To calculate how the measurable dichroic ratio is determined by the direction of the transition moment and by the order parameters S and D, the vector p must be transformed from the molecular system into the laboratory frame of reference using the transformation matrix M, and then the average of the components must be determined according to the text of Blinov [7]. Assuming that the distribution function f does not depend on φ (the material is uniaxial) we get

$$\langle p_Y^2 \rangle = \langle p_X^2 \rangle = 0.5\, p^2 \{ \sin^2 \beta [1 - \langle \sin^2 \theta \sin^2 (\phi - \gamma) \rangle] + \cos^2 \beta \langle \sin^2 \theta \rangle \}$$
$$\langle p_Z^2 \rangle = p^2 \{ \sin^2 \beta \langle \sin^2 \theta \sin^2 (\phi - \gamma) \rangle + \cos^2 \beta \langle \cos^2 \theta \rangle \} \tag{14}$$

Assuming at least C_2 symmetry for the distribution function with respect to ψ, the averages $\langle p_X^2 \rangle$ and $\langle p_Z^2 \rangle$ will be as follows:

$$\langle p_Y^2 \rangle = \frac{p^2}{6} \{ 2(1 - S) + \sin^2 \beta \, (3S + D \cos 2\gamma) \}$$
$$\langle p_Z^2 \rangle = \frac{p^2}{3} \{ 2S + 1 - \sin^2 \beta \, (3S + D \cos 2\gamma) \} \tag{15}$$

For homeotropic alignment, the absorbance is proportional to $\langle p_Y^2 \rangle$; in any direction in the plane of the sample. When the material is warmed to the isotropic state, S and D become 0, hence $\langle p_{Y\mathrm{iso}}^2 \rangle = p^2/3$. The dichroic ratio, R, can be defined as

$$R = \langle p_Y^2 \rangle / \langle p_{Y\mathrm{iso}}^2 \rangle = 1 - S + \sin^2 \beta \, (1.5\, S + 0.5\, D \cos 2\gamma) \tag{16}$$

Because absorbance components are proportional to the corresponding $\langle p_i^2 \rangle (i = X, Y, Z)$, the dichroism is experimentaly described by the ratio of the absorbances. For homogeneous alignment, the dichroic ratio R is defined as

$$R = \langle p_Z^2 \rangle / \langle p_Y^2 \rangle = \frac{(1 - S) + 0.5 \sin^2 \beta \, (3S + D \cos 2\gamma)}{2S + 1 - \sin^2 \beta \, (3S + D \cos 2\gamma)} \tag{17}$$

which for $D = 0$ corresponds to the formula obtained by Neff et al. [8]:

$$R = \frac{\frac{1}{2} \sin^2 \beta \, (1 - \langle \cos^2 \theta \rangle) + \cos^2 \beta \, \langle \cos^2 \theta \rangle}{\frac{1}{4} \sin^2 \beta (1 + \langle \cos^2 \theta \rangle) + \frac{1}{2} \cos^2 \beta \, \langle 1 - \cos^2 \theta \rangle} \tag{18}$$

The apparent order parameter, defined as S_a [5], is

$$S_a = (R - 1)/(R + 2) = S - 0.5 \sin^2 \beta \, (3S + D \cos 2\gamma) \tag{19}$$

For $D = 0$ this is equivalent to a well-known relation between S_a/S and the angle β of the transition moment with respect to the molecular long axis [4,8,9]. From the definition of the Euler angles, it follows that the reference direction of the measurements is fixed by the choice of θ and φ. The rotation is not free if the e_x axis of the molecule tends to incline at a certain angle to the Z, z plane. But such an inclination from both sides of the plane is equally probable, which means that $f(\theta, \psi)$ is symmetric in ψ. If the molecule possesses a plane of symmetry and e_x is chosen normal to that plane, $\psi = 0$ means that the plane of symmetry coincides with the Z, z plane. If the molecule has no symmetry, we can imagine that the molecule with its mirror image makes a symmetrical unit, and the symmetry $f(\theta, \psi)$ still holds. Assume that two mutually perpendicular vibrations are found in the plane normal to the axis z of the molecule or unit but that neither of them lies in the hypothetical mirror plane that determines the axes of the molecular frame of reference. Let the polar angles of both mutually perpendicular transition moments be

$$\beta_1 = 90°, \; \gamma_1 = \gamma_0, \; \beta_2 = 90°, \; \gamma_2 = \gamma_0 + 90° \tag{20}$$

Using Eq. (19), the order parameter S and $D \cos 2\gamma$ are obtained from the apparent order parameters S_{a1} and S_{a2} as

$$S = -S_{a1} - S_{a2}, \; D \cos 2\gamma_0 = -S_{a2} - S_{a1} = D' \tag{21}$$

For the other vibrations, S_a/S is a linear function of D'/S if the geometry of the molecule is unchanged during the temperature cycle

$$\frac{S_a}{S} = (1 - 1.5 \sin^2 \beta) - \frac{D'}{2S} \sin^2 \beta \frac{\cos 2\gamma}{\cos 2\gamma_0} \qquad (22)$$

From the intercept of this straight line, angle β from the slope $\cos 2\gamma/\cos 2\gamma_0$, is obtained.

Equations (21) and (22) have been used to estimate both order parameters S and D for two samples of phenyl benzoate side group–methacrylate main chain polymeric liquid crystal in the achiral SmA phase [4] (Fig. 2). Both samples were aligned homogeneously, the first sample by polyvinyl alcohol (PVA), the second using a magnetic field. For this molecule, neither of the longitudinal vibrations proved to be really parallel to the molecular axis z. Therefore, two mutually perpendicular transverse vibrations were chosen; $\beta = 90°$ was assumed for both. One was the transverse β_{CH} vibration at 1110 cm^{-1}, and the other was chosen from the two out-of-plane benzene ring vibrations at 840 cm^{-1} for the PVA-aligned sample and 764 cm^{-1} for the magnetically aligned sample. The factor $\cos 2\gamma_0$ in Eq. (21) was estimated to be close to unity; hence D' can be replaced by D. The resulting

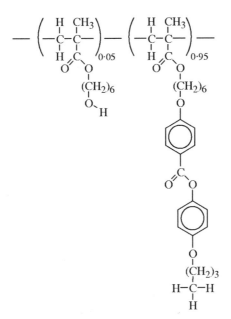

Figure 2. Chemical formula of phenyl benzoate side group–methacrylate main chain polymeric liquid crystal.

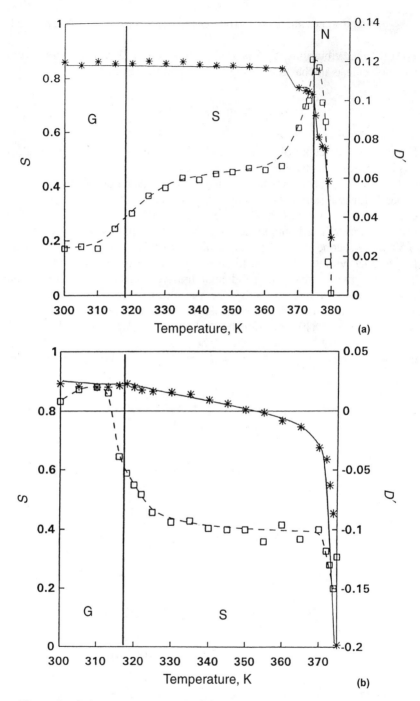

Figure 3. Order parameters S and D' for (a) the PVA-aligned sample and (b) the magnetically aligned sample. $*$, S; □, D'.

values of S and D for the material with the formula shown in Figure 2 vs. temperature are plotted in Figure. 3. The value $S \approx 0.85$ is quite typical for the liquid crystals in achiral SmA phase. The difference between the two samples can be attributed to the presence or absence of the alignment layer. Thus the order parameter S of the PVA-aligned sample decreased very slowly in the SmA phase, followed by an abrupt fall in SmA–N phase transition. The magnetically aligned sample shows a more gradual reduction, and S drops to zero at the SmA–N phase transition. The obtained D value, ~ 0.02, is rather low; nevertheless, it is clear that it has opposite signs for the two samples. This means the inclination of the planes of the benzene rings in the Z, z plane are different for the two samples.

III. MOLECULAR ORDER IN THE SmC* PHASE

In the uniaxial SmA phase, the distribution function depends just on θ, but in the SmC* phase, Zgonik et al. [10] simply considered the dependence of the distribution function on the polar angle θ and on the azimuthal angle γ of the dipole p in the laboratory system of reference. The orientational distribution function has been expanded in a series of spherical harmonics, consistent with the local symmetry of the SmC* phase (Fig. 4).

$$f(\theta, \gamma) = \frac{1}{4\pi}\left[1 + 3A\sin\theta\sin\gamma + 5BP_2(\cos\theta) + \frac{15}{4}\sin 2\theta\right.$$
$$\left.(C\cos\gamma + E\sin\gamma) + \frac{15}{4}\sin^2\theta(D\cos 2\gamma + F\sin 2\gamma)\right] \quad (23)$$

where A, B, C, D, E, and F are the expansion coefficients, which can serve as order parameters each with the maximum value of 1 for completely ordered systems. A measures the polar order and can be nonzero in the ferroelectric phase. If the dipole p is chosen so that it is aligned along the molecular long axis, B is the second rank order parameter, usually denoted by S, $S = \langle P_2(\cos\theta)\rangle$. In general, the coefficient B measures the order of the selected dipole relative to the Z axis. A maximum value of unity corresponds to the case when all of the dipoles are parallel to the Z axis, and minimum value of -0.5 occurs when dipoles are orthogonal to Z; $B = 0$, is found when the dipoles are disordered around the Z axis. The remaining coefficients C, D, E, and F are the four parameters that describe the quadrupolar order. Zgonik et al. [10] calculated the contribution to the absorption coefficient for the arbitrarily oriented dipole and the incident field E in experimentally accessible principal planes of the laboratory coordinate system. The components of p can be written $p = /p/(\sin\theta\cos\gamma,$

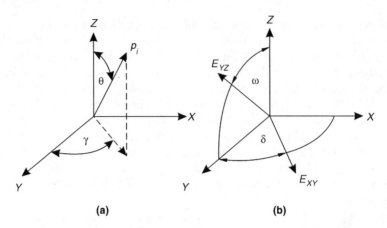

Figure 4. (a) Orientation of the induced molecular dipole moment p_i in the laboratory frame with the Z axis normal to the smectic layers and the X axis parallel to the ferroelectric polarization in the SmC* phase. (b) The angles ω and δ, which measure the orientation of the optical electric field in the $X - Y$ and $X - Z$ planes, respectively. (Redrawn with permission from [10].)

$\sin\theta \sin\gamma,\ \cos\theta$), and the components of the incident electric field polarized in the X-Z plane are $/E/(\sin\omega, 0, \cos\omega)$, where ω is the angle between the Z axis and the electric field direction. The contribution $(\boldsymbol{E}\cdot\boldsymbol{p})^2$ is as follows:

$$\left(\frac{\boldsymbol{E}(\omega)\cdot\boldsymbol{p}}{|E\|p|}\right)^2 = \frac{1}{3} + \frac{2}{3}\,P_2(\cos\theta) + \left[\frac{1}{2}\sin^2\theta\cos 2\gamma - P_2(\cos\theta)\right]\sin^2\omega$$
$$+ \frac{1}{2}\sin 2\theta\cos\gamma\sin 2\omega \tag{24}$$

An analogous expression for the case of the optical electric field in the X-Y plane, $/E/(\cos\delta, \sin\delta, 0)$, where δ is the angle between the Y axis and optical electric field direction E_{XY} is

$$\left(\frac{\boldsymbol{E}(\delta)\cdot\boldsymbol{p}}{|E\|p|}\right)^2 = \frac{1}{3} - \frac{1}{3}\,P_2(\cos\theta) + \frac{1}{2}\sin^2\theta\cos 2\gamma\cos 2\delta$$
$$+ \frac{1}{2}\sin^2\theta\sin 2\gamma\sin 2\delta \tag{25}$$

To calculate the measured absorption coefficient, the absorption of a single molecule in Eq. (24) or (25) must be multiplied by the distribution function in Eq. (23) and then integrated over $d\gamma$ and $d(\cos\beta)$. The resultant absorption coefficients normalized to the absorption in the isotropic phase are

$$\frac{A(\omega)}{A_{\text{iso}}} = 1 - 2B + \left[\frac{3}{2}D - 3B\right]\sin^2\omega + \frac{3}{2}C\sin 2\omega \qquad (26)$$

and

$$\frac{A(\delta)}{A_{\text{iso}}} = 1 - B + \frac{3}{2}D\cos 2\delta + \frac{3}{2}F\sin 2\delta \qquad (27)$$

Here, only second order coefficients contribute to the IR absorption. No information is obtained from IR experiments on the A coefficients, which allows one to distinguish the ferroelectric from the nonferroelectric SmC phase. To avoid the problem of a birefringent medium, the incident beam should be polarized in one of the two eigen polarization directions for the selected directions of propagation.

The order parameter is determined in both the SmA* phase and the ferroelectric SmC* phase for 4'-(2S,3S)-2''-chloro-3''-methylpentyloxyphenyl-4-decyloxythiobenzoate, ($\overline{10}.S.$CIIsoleu) [10] (Fig. 5). The measurements in homeotropically oriented samples were initially used to determine the parameter B for selected dipoles. The absorption was found not to be dependent on the angle δ; therefore, the average of the absorption spectra over δ was obtained. The parameter B was calculated from the absorption spectra as $B = 1 - N$, where N being the dichroism ratio. Figure 6 shows the temperature dependence of parameter B for seven different dipoles. As seen in the figure, these parameters do not change much at T_{AC} and stay nearly constant through the SmC* phase. To determine the other two coefficients, C and D, measurements in a cell with bookshelf orientation were performed. In the SmA* phase, the results confirmed previously obtained values for B. Their absolute values were a few percent lower, and this difference was attributed to a poorer sample orientation. Owing to the less reliable sample orientation quality, only the strongest band, centered at 1600 cm^{-1}, was investigated, and no precise temperature dependence was established. At the corresponding saturated value (10 K below T_{AC}), the parameter C was found to be 0.4 ± 0.1, and the absolute value of D is < 0.05.

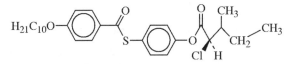

Figure 5. The chemical formula of the $\overline{10}\,S.$CIsoleu liquid crystal molecule. (Redrawn with permission from [10].)

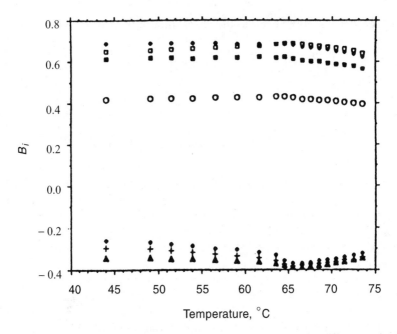

Figure 6. Temperature dependence of the order parameter B_i for different molecular dipole moments associated with different molecular vibrations. Peak positions are marked. Phase transition on cooling are $T(\text{SmA*-I}) = 74°C$ and $T(\text{SmC*–SmA*}) = 66°C$. $+$, 1762 cm^{-1}; $\bullet$, 1672 cm^{-1}; $\square$, 1600 cm^{-1}; $\blacklozenge$, 1163 cm^{-1}; $\blacksquare$, 1015 cm^{-1}; $\bigcirc$, 900 cm^{-1}; $\blacktriangle$ 839 cm^{-1}. (Redrawn with permission from [10].)

Similar observations are reported for the FLC mixture ZLI 3654 [11] (Fig. 7). The IR dichroism for the sample in homeotropic and homogeneous orientations has been studied throughout the SmA and SmC* phases. A number of vibrational bands were selected for calculationing the order parameter: 1607, 1498, and 1432 cm^{-1} phenyl ring C–C stretching

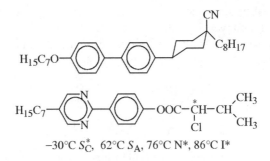

$-30°C\ S_C^*,\ 62°C\ S_A,\ 76°C\ N^*,\ 86°C\ I^*$

Figure 7. Structure of the ZLI-3654 mixture and its transition temperatures [11].

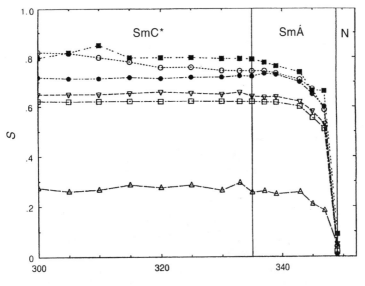

Figure 8. Temperature dependence of the order parameter for ZLI-3654 sample obtained for different dipole moments [11]. $\triangle$, CH_2 stretch (2800–3000 cm^{-1}); $\blacksquare$, C=O stretch (1771 cm^{-1}); $\square$, C–C stretch in plane (1607 cm^{-1}); $\triangledown$, C–C stretch in plane (1498 cm^{-1}); $\bullet$, C–C stretch in plane (1432 cm^{-1}); $\bigcirc$, C–H stretch out of plane (860–780 cm^{-1}).

vibrations; the group of bands 860–780 cm^{-1} connected with C–H aromatic out-of-plane vibrations; and vibrations located in the tails CH_2 and CH_3 in the range of 2800–3000 cm^{-1}. The band centered at 650 cm^{-1} corresponds to the C–Cl stretching band, and the other at 1771 cm^{-1} corresponds to the C=O stretching vibrations. These are the two vibrations located near the chiral center. Figure 8 shows the order parameter calculated for the above bands in the SmA and SmC* phases. The order parameters that correspond to the long molecular axis were calculated by ignoring the higher terms (C, D, E, and F) in the distribution function. The temperature dependence of the order parameter was similar for all bands.

We do not observe any step or even a decrease in the value of the order parameter at the SmA-SmC* phase transition. This can be explained by a specific ordering of molecules at the SmA-SmC* phase transition [11]. The orientation of a single molecule relative to the smectic layer normal does not significantly change at the phase transition. Molecular long axes are disordered on the cone in the SmA phase and start to become correlated in between layers on entering SmC* phase. The tilt angle of the molecules, therefore, does not change at the phase transition. This can be considered as a disordered tilted SmA* phase.

For the group of transversal dipoles—that is, dipoles normal to the long molecular axis—the order parameter was found to be significantly higher than for the group of parallel dipoles. This indirectly shows that the rotation around the long molecular axis is not isotropic. It seems that the distribution function is dependent on γ in the SmA* phase as well.

IV.   EFFECT OF THE ELECTRIC FIELD IN THE SmC* PHASE: DETERMINATION OF THE MOLECULAR TILT ANGLE

The experimental arrangement for a study of the effect of the electric field on IR absorbance of FLC material is shown in Figure 9 [11]. In the SmA* phase, the absorbance of the bands with transition dipole moments along the long molecular axis is maximal when the polarizer is parallel to the smectic layer normal, and consequently the absorbance of the perpendicular bands has maxima rotated by 90°. When an electric field is applied to the sample in the SmC* phase, the maxima of absorbance rotate through approximately the optical tilt angle either clockwise or counterclockwise (with respect to their position in the SmA* phase), depending on the polarity of the electric field. Figure 10 shows the temperature dependence of the tilt angle $\Delta\omega$ observed for the phenyl stretching dipoles of ZLI-3654 [11]. In the range of temperatures $< 329K$, tilt angles observed for the various bands are significantly lower than the optical tilt angles. This is due to the reason that the IR cell was in the chevron structure whereas the optical cell was in the

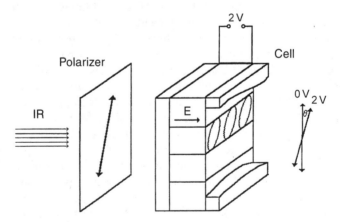

Figure 9.   Experimental setup for the study of the electric field–induced tilt angle [11].

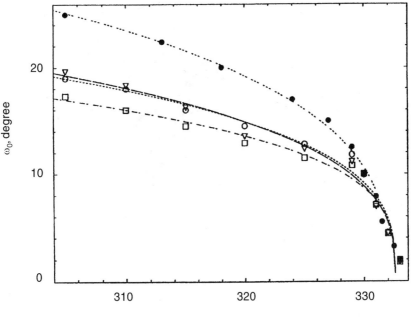

Figure 10. A comparison of the temperature dependence of the molecular tilt angle for different bands [11]. ●, optical tilt angle; □, FTIR 1430 cm^{-1}; ▽, FTIR 1585 cm^{-1}; ○, FTIR 1600 cm^{-1}.

bookshelf structure. The angular dependence of the peak absorbance and the effect of the electric field on this dependence can conveniently be represented in the form of polar plots (circular graphs) [12] (Fig. 11). A distance from the center represents the peak absorbance at the polarizer position ω. The results of the absorbance profile vs. the polarization angle for the ZLI-3654 sample are given in Figure 12. As can be seen from these figures, the maximum absorbances for different bands undergo quite different angles of rotation after the reversal of the polarity of the DC field. To obtain a precise value for the rotation angle of the particular dipole moment, the absorbance profile is fitted to the following formula [13,14]:

$$A(\omega) = -\log\left(10^{-A_{\parallel}} + \left(10^{-A_{\perp}} - 10^{-A_{\parallel}}\right) \cdot \sin^2(\omega - \omega_0)\right) \qquad (28)$$

where $A_{\parallel}$ and $A_{\perp}$ are the maximal and minimal values of the absorbance, ω is a polarizer angle, and ω_0 is the angle for which a maximal absorbance is obtained. For the ZLI sample, 20 V of DC voltage is sufficient to obtain the

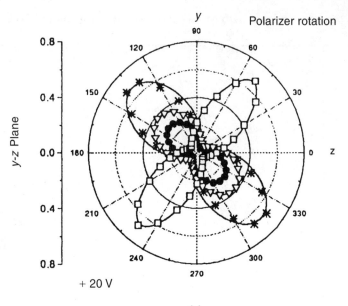

(a)

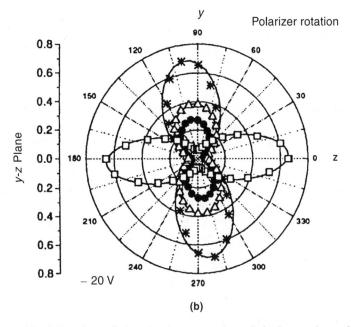

(b)

Figure 11. Polar plots of the absorbance vs. the polarization angle ω for four representative bands. Symbols represent the experimental data; solid lines are fits to Eq. (28). ●, $-653\ \mathrm{cm}^{-1}$; ∗, $-798\ \mathrm{cm}^{-1}$; ▽, $-1772\ \mathrm{cm}^{-1}$; □, $-1607\ \mathrm{cm}^{-1}$. The last two polar plots are multiplied by 1.3 and 0.5, respectively.

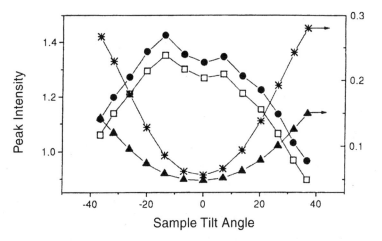

Figure 12.   The dependence of the peak intensity on the sample rotation angle for $+20$ V across the cell. Wave numbers: □, 1584 cm^{-1}; ●, 1608 cm^{-1}; ▲, 654 cm^{-1}; *, 798 cm^{-1}.

saturated value of the switching angle. The values obtained are as follows: 21.8° for 1772 cm^{-1}, 25° for 1607 cm^{-1}, 24° for 1584 cm^{-1}, 20.5° for 1543 cm^{-1}, 20° for 822 cm^{-1}, 19° for 798 cm^{-1}, 23.2° for 722 cm^{-1}, and 18.6° for 653 cm^{-1}. It must be noted that the angles for the dipoles parallel to the phenyl ring are significantly higher than those for the perpendicular dipoles. These results for switching angles can be explained in terms of the nonisotropic rotation around the long molecular axis and will be discussed in detail later. Such a nonisotropic rotation can be considered a reason for the existence of ferroelectricity for the chiral systems. To study the distribution of the transverse dipoles around the long molecular axis, it is better to measure the absorbance profile in the plane perpendicular to the optical axis i.e., in (X-Y plane). For a homogeneously aligned sample, this can be realized by tilting the sample out of the normal angle of incidence [15]. Figures 12 and 13 show the distribution of the absorbance for several tranverse dipoles vs. the angle of the sample tilt (in the plane consisting of layer normal and the DC field direction). Although the angle through which the cell is rotated is limited ($\sim -60°$ to $+60°$) in this method, one can directly see a noncylindrical symmetry of the distribution of the transversal dipoles. The CH$_2$ rocking dipole in the tail lies in the window plane, the phenyl out-of-plane dipole is inclined at a small angle to the window, and the other dipoles are highly inclined to the plane of the windows. In particular the dipole of the Cl–C chiral group is rather close to the direction of the electric field.

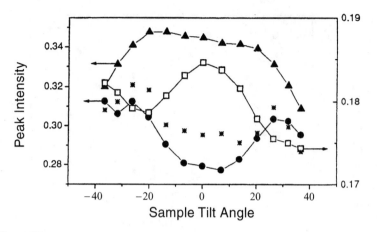

Figure 13. The dependence of the peak intensity on the sample rotation angle for -20 V across the cell. Wave numbers: $\bullet$, -653 cm^{-1}; $\blacktriangle$, -798 cm^{-1}; $\square$, -722 cm^{-1}; $*$, -1109 cm^{-1}.

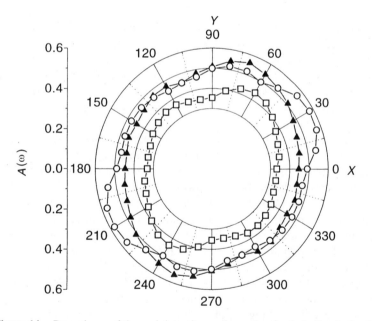

Figure 14. Dependence of the peak intensity on the angle of polarization in the $X - Y$ plane under DC field (X and Y axes correspond to p and s polarizations in ATR experiments, respectively). $\blacktriangle$, -653 cm^{-1}; C–Cl stretching dipole at the chiral center; $\bigcirc$, -798 cm^{-1}; C–H aromatic out-of-plane deformation; $\square$, -1109 cm^{-1}; C–H aromatic in-plane deformation.

By using the attenuated total reflection (ATR) method, there is no such limitation for the angle; hence the full angular dependence for the absorbance profile can be reproduced (despite some difficulties encountered in the precise calculation of the depth of penetration of the IR beam in this method). Figure 14 shows the angular dependence of the absorbance for ZLI-3654 in the plane normal to the director under DC field [16]. It is clear that the distribution of the transition dipole moments around the long molecular axis is not uniform. The preferable direction of the chiral Cl–C and phenyl transversal dipoles (in the phenyl plane) is rather close to the X axis (the X axis is in the direction of DC field), whereas the phenyl C-H out-of-plane dipole is close to the Y axis. This is direct evidence that the rotational motion of the molecule is biased around its long molecular axis.

V. MOLECULAR CONFORMATION AND HINDERED ROTATION IN THE FERROELECTRIC AND ANTIFERROELECTRIC PHASES

Kim et al. [17] studied several antiferroelectric liquid crystals, the results shown in Figure 15 suggest an averaged alkyl chain orientation parallel to the long molecular axis and the hindrance of rotation of the carbonyl groups. Miyachi et al. [18] reported results for MHPOBC, a prototype antiferroelectric liquid crystal. They showed that the hindered rotational motion of the C=O group near the chiral center in the SmC_A^* phase produces spontaneous polarization parallel to the Y axis, P_Y, which is much greater than P_X. It is important to note that the C=O group in question is not in the core but is not at the end of the molecule either.

A. Homogeneously Aligned Cells

1. SmA* Phase

The experimental setup for the FTIR studies by Kim et al. [17] is shown in Figure 16. Figure 17 shows the polarized IR spectra as a function of the polarizer rotation angle, ω, in SmA* of a homogeneously aligned TFMHPODB cell. Here, ω is the angle between the smectic layer normal and the polarization direction of an incident radiation. The assignment of several absorption peaks for different AFLCs are listed in Table I. Angular dependencies of the absorbance peaks can generally be classified under three types.

1. The phenyl stretching bands at 1604 and 1504 cm^{-1} attain their maxima when the polarization direction of the incident IR beam coincides with the smectic layer normal, ω = 0° and ω = 180°. The peaks show quite large dichroisms (on the order of 5–10) because the transition dipole

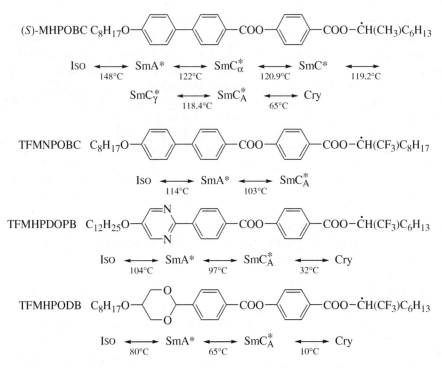

(S)-MHPOBC $C_8H_{17}O$—⬡—⬡—COO—⬡—COO—$\overset{\cdot}{C}H(CH_3)C_6H_{13}$

Iso $\xleftrightarrow{148°C}$ SmA* $\xleftrightarrow{122°C}$ SmC$_\alpha^*$ $\xleftrightarrow{120.9°C}$ SmC* $\xleftrightarrow{119.2°C}$

SmC$_\gamma^*$ $\xleftrightarrow{118.4°C}$ SmC$_A^*$ $\xleftrightarrow{65°C}$ Cry

TFMNPOBC $C_8H_{17}O$—⬡—⬡—COO—⬡—COO—$\overset{\cdot}{C}H(CF_3)C_8H_{17}$

Iso $\xleftrightarrow{114°C}$ SmA* $\xleftrightarrow{103°C}$ SmC$_A^*$

TFMHPDOPB $C_{12}H_{25}O$—⬡(N,N)—⬡—COO—⬡—COO—$\overset{\cdot}{C}H(CF_3)C_6H_{13}$

Iso $\xleftrightarrow{104°C}$ SmA* $\xleftrightarrow{97°C}$ SmC$_A^*$ $\xleftrightarrow{32°C}$ Cry

TFMHPODB $C_8H_{17}O$—⬡(O,O)—⬡—COO—⬡—COO—$\overset{\cdot}{C}H(CF_3)C_6H_{13}$

Iso $\xleftrightarrow{80°C}$ SmA* $\xleftrightarrow{65°C}$ SmC$_A^*$ $\xleftrightarrow{10°C}$ Cry

Figure 15. Structural formula for various compounds. (Redrawn with permisssion from [17].)

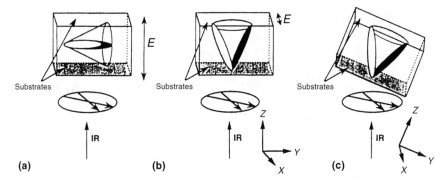

Figure 16. Geometries used in polarized IR measurements for (a) homogeneously aligned samples, (b) homeotropically aligned samples and (c) the oblique incidence of the radiation (the sample is rotated on the Y axis). (Redrawn with permisssion from [17].)

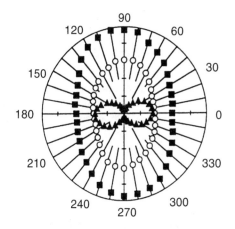

Figure 17. Polar plot of the absorbance vs. the polarizer rotation angle ω for three representative stretching peaks in the SmA phase of homogeneously aligned TFMHPODB. ▲, phenyl ring; ◯, CH₂; ■, C=O. (Redrawn with permisssion from [17].)

TABLE I
Band assignments of Several Peaks in the IR Spectrum of TFMHPODB

Observed Frequencies (cm⁻¹)	Band
2960	CH_3 degenerate asymmetric stretching
2929	CH_2 asymmetric stretching
2856	CH_2 symmetric stretching
1743	C=O stretching
1604,1504	Phenyl ring C–C stretching
1465	CH_2 scissor vibration
1260	Asymmetric C–O stretching
1180	Aromatic CH in-plane deformation

Reprinted with permission from [17].

moment for these bands is parallel to the long molecular axis and the stretching directions in these phenyl rings coincide with each other.

2. The CH_2 peaks at 2929, 2856, and 1465 cm⁻¹ and the phenyl C–H out-of-plane band reach their maxima at ω = 90° and ω = 270°, which is consistent with the expected direction of the transition dipole moment. For the phenyl C–H, the transition dipole is perpendicular to the phenyl para axis, and for CH_2 peaks, the transition dipole moment is perpendicular to the alkyl chain. The CH_2 peaks show relatively little absorbance variation with the polarizer rotation angle ω, because the disorder of the aliphatic chains. The C=O stretching peak also belongs in this group. Although the transition

dipole moment of the C=O vibration cannot be considered as perpendicular to the molecular long axis, the absorbance maximum appears at $\omega = 90°$ and $\omega = 270°$ because of the free rotation and the head and tail equivalence of the molecular orientation.

3. The absorbance profile of a number of peaks scarcely depends on ω, for example, the CH_3 stretching peaks and some of the C=O peaks. The isotropic character of the CH_3 peak may occur because there are usually two CH_3 groups at the ends of the flexible carbon chains that are strongly fluctuating. In case of the chiral molecules, there is another CH_3 group that rotates freely around the long molecular axis.

2. Helicoidal SmC_A^* and Unwound, Uniform SmC^* Phases

In the helicoidal SmC_A^* and SmC^* phases at zero electric field [Fig 18(b), Fig. 19(b)], the maxima position for the stretching peak of the phenyl ring is close to $\omega = 0°$, and for the C–H out of plane it is $\omega = 90°$. This is the same as it was in the SmA* phase, except that the maximum absorbance decreases and the minimum of absorbance increases by about 15% compared with values in the SmA* phase, owing to the helicity and the tilting of the molecular axis.

When an electric field that is greater than the threshold is applied in the SmC_A^* or SmC^* phase, all profiles usually rotate through an angle almost equal to the optical tilt angle; the sense of rotation depends on the field direction, as shown in Figure 18(a) and (c) for TFMHPODB and Figure

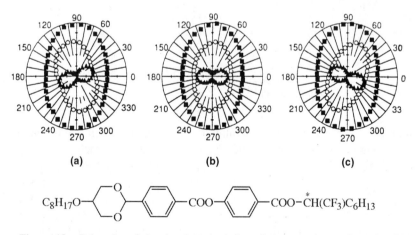

Figure 18. Polar plot of the absorbance vs. the polarizer rotation angle ω for three representative peaks in SmC_A^* of homogeneously aligned TFMHPODB under the application of DC electric fields. (a) $+10$ V/μm; (b) 0 V/μm; (c) -10 V/μm ▲, phenyl ring; ○, CH_2; ■, C=O. (Redrawn with permisssion from [17].)

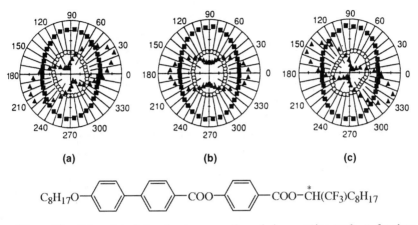

Figure 19. Polar plot of the absorbance vs the polarizer rotation angle ω for three representative peaks in SmC_A^* of homogeneously aligned TFMNPOBC under the application of DC electric fields. (a) $+15$ V/μm; (b) 0 V/μm; and (c) -15 V/μm. ▲, phenyl ring; $\bigcirc$, CH$_2$; ■, C=O. (Redrawn with permisssion from [17].)

19(a) and 19(c) for TFMNPOBC. These changes of the ω_{max} imply the field-induced phase transition occurs from helicoidal SmC^* or SmC_A^* to an unwound and uniform SmC^*. As a result of the helix's unwinding, the absorbance is brought to almost the same value as in the SmA* phase.

It should be noted that although the character of the angular dependence remains similar to that in the SmA* phase, most of the peaks (CH$_2$ and C–H out of plane, C=O) rotate by a lower angle than does the phenyl peak. Furthermore, the absorbance profiles are not entirely symmetric with respect to the ω_{max} of the phenyl peak. For the C=O band, the peak rotates much less than does the phenyl ring or may even show no rotation. Such behavior of the absorbance profiles clearly suggests the hindered rotation of the molecules around the long molecular axis.

B. Homeotropically Aligned Cells

For homeotropically aligned cells, the IR radiation is incident normal to the substrate along the smectic layer normal. The polarization rotation angle is measured with reference to the direction of the applied electric field. For an undisturbed helicoidal structure, when the electric field is not present, the absorbancies of the phenyl peak at 1604 cm^{-1}, the CH$_2$ symmetric stretching peak at 2856 cm^{-1}, and the two C=O peaks at 1739 and 1720 cm^{-1} do not vary with ω [Fig. 20(a)]. By applying an electric field greater than the threshold, weak angular dependence appears for the phenyl peak, with the maximum at $\omega = 90°$ and $\omega = 270°$; that is, the polarization

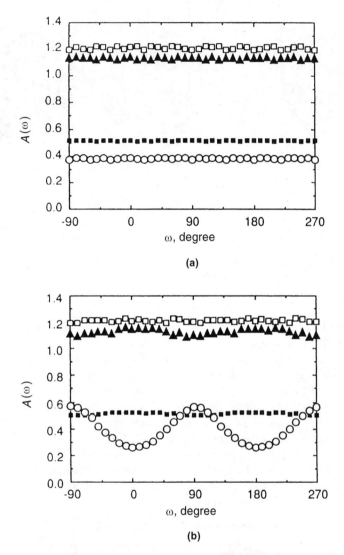

Figure 20. Angular dependences of several absorbance peaks for normal incidence in SmC* of homeotropically aligned MHPOBC at (a) $E = 0$ V/mm and (b) $E = 75$ V/mm. The electric field direction is taken as $\omega = 0°$. □, C=O in the core part; ▲, C=O in the chiral part; ■, CH$_2$ symetric stretching; ○, phenyl ring stretching. (Redrawn with permission from [17].)

direction appears perpendicularly to the electric field [Fig. 20(b)]. For other peaks, even in the unwound SmC* phase, the absorbance only weakly depends on the polarization rotation angle. For example, the 1720 cm^{-1} peak, which corresponds the chiral C=O group, shows angular dependence with maximum at $\omega = 0°$ and $\omega = 180°$, which is 90° out of phase with phenyl peak. A similar absorbance profile is also observed for the CH_2 peak at 2856 cm^{-1}. To confirm that the weak angular dependence of the chiral C=O and CH_2 peaks, really corresponds to the nearly isotropic distribution of the transition dipole moments around the long molecular axis, the polarization IR spectra were measured using IR radiation obliquely incident along the long molecular axis [Fig. 16(c)]. To obtain this geometry, a homeotropically aligned cell was rotated on an axis parallel to the applied electric field by a tilt angle. The resultant angular dependence of the absorbance for several peaks under positive and negative electric fields are given in Figure 21. For the positive electric field, the phenyl peak naturally shows a weak angular dependence; the CH_2 peak also shows a small 90° out-of-phase angular dependence. In the negative electric field, however, the propagation direction of the incident radiation is parallel to the average stretching direction of the phenyl ring (i.e., the long molecular axis), so the absorbance of the phenyl peak is weak and does not depend on ω. This independence of ω ensures that the cell is oriented properly with respect to the incident radiation; that is, that the incident radiation propagates along the average direction of the long molecular axis. Even in this geometry, the angular dependence of the 1720 cm^{-1} chiral C=O peak is still present and directly shows the noncylindrical symmetry of the transverse dipole rotation.

C.   Homeotropically Oriented Free-Standing Film

The MHPOBC sample in Fukuda and his co-workers' [18] experiment was partially racemized to simplify the phase sequence and elongate the pitch of a helicoidal structure in the SmC_A^* phase, so its unwinding occurs at a relatively low electric field (1500 Vmm^{-1} at 800 Hz). A free-standing film of MHPOBC, formed in a frame depicted in Figure 22, was mounted in an oven in between two SrF_2 windows. The measuring geometry is shown in the figure. The rotation angle is defined as zero when the polarization direction of the incident IR radiation coincides with the tilt plane normal. The angular dependence of the phenyl ring stretching peak in the SmC_A^* phase is quite similar to that in SmC^* phase (Fig. 23). This indicates that the average stretching direction of the phenyl rings is parallel to the long molecular axis and that its orientational order is fairly high. On the contrary, the angular dependence of the C=O stretching peaks in SmC_A^* is different from that in the SmC^* phase (Fig. 21) [17]. The dependence is characteristic and in phase with that of phenyl ring stretching peak, whereas it is out of phase in SmC^*. Therefore, the rotational motion of the C=O group is expected to be of a substantially different character in the SmC_A^* phase than in the SmC^* phase.

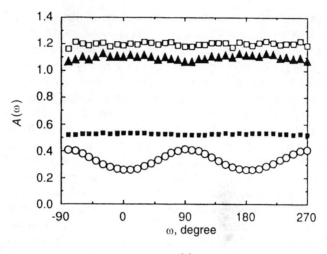

(a)

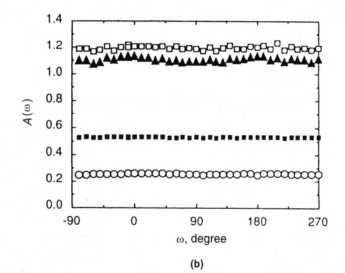

(b)

Figure 21. Angular dependences of several absorbance peaks for oblique incidence in SmC* of homeotropically aligned MHPOBC at (a) $E = +75$ V/mm and (b) $E = -75$ V/mm. The electric field direction is taken as $\omega = 0°$. □, C=O in the core part; ▲, C=O in the chiral part; ■, CH$_2$ symetric stretching; ○, phenyl ring stretching. (Redrawn with permisssion from [17].)

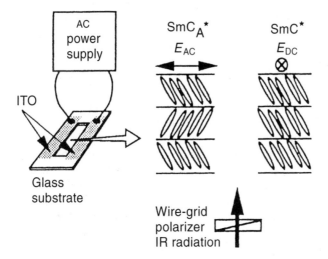

Figure 22. Free-standing film prepared in a frame, in which the smectic layer is formed parallel to the interface and the measuring geometry is in the SmC$_A^*$ and SmC* phases. (Redrawn with permisssion from [18].)

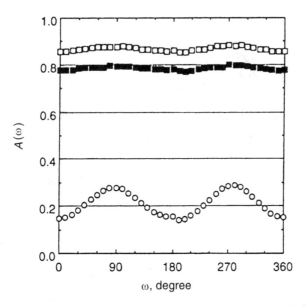

Figure 23. Absorbance as a function of the polarizer rotation angle in SmC$_A^*$ unwound by applying an 800-Hz, 1400-Vmm^{-1} electric field for the MHPOBC free-standing film. □, core C=O; ■, chiral C=O; ○, phenyl ring. (Redrawn with permisssion from [18].)

D. Rotation of Carbonyl Groups in the SmA, SmC*, and SmC$_A$* Phases

It is quite common to assume that C=O groups in the SmA* phase rotate freely around the long molecular axes and that these axes fluctuate around the smectic layer normal with the order parameter S [11]. It has been shown that the absorbance can qualitatively be well reproduced in the SmA* phase. However, in the SmC* phase, it is not possible to explain the angular dependence of the peaks' absorbance by assuming the free rotation of the C=O groups. If they rotate freely around the long molecular axis, the angular dependence of the C=O peaks should be symmetrical with respect to the director (average position of the phenyl peak). This is not the case, not only in the chiral C=O group but also in the core C=O group, as is clear in Figures 18 and 19. The most probable orientation of the C=O group is shown in Figure 24, which simply explains why this orientation is almost in the smectic layer plane. Although the orientation of each individual chiral C=O group is not in the plane containing the C$_2$ axis, the spontaneous polarization appears along the C$_2$ axis because of the head and tail equivalence of the molecule.

To describe the hindered rotational motion, four molecular configurations must be considered, with the chiral C=O in the top or bottom and tilted to the right or left (Fig. 24). In the SmC* phase, the orientation of the

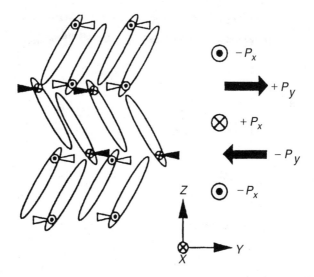

Figure 24. Most probable orientation of chiral carbonyl groups in the SmC$_A$* phase. (Redrawn with permisssion from [18].)

molecular system is chosen; $\psi = \psi_0$ (with respect to the tilt plane normal) and $\theta = \theta_0$ and $\theta = 180 + \theta_0$, due to the head and tail equivalence. In SmC$_A^*$ phase, four equivalent configurations exist because the X and Y axes are the twofold symmetry axes. To a first approximation, the hindered rotational motion may be described by

$$f(\psi) = \frac{1}{4}\{f_{t+}(\psi) + f_{b+}(\psi) + f_{t-}(\psi) + f_{b-}(\psi)\} \qquad (29)$$

where

$$f(\psi) = \frac{1}{2\pi}\{1 + a\cos(\psi - \psi_0)\}$$

and the remaining three terms are the corresponding distribution functions, with the other biased directions in the laboratory frame. Here a is the degree of biasing and t, b, $+$, $-$ refer to top, bottom, left, and right, respectively. By neglecting the fluctuation of the long molecular axis, the absorbance vs. polarizer rotation angle can be simulated for homeotropically oriented sample in the SmC* and SmC$_A^*$ phases. The normalized absorbance $A(\omega)$, is given by

$$A(\omega) = \frac{1}{4}\int_0^{2\pi} f_{t+}(\psi)\{\sin\omega(-\cos\beta + \sin\beta\sin\psi)\cos\theta$$

$$+ \cos\omega(\sin\beta\cos\psi)\}^2 d\psi + \text{(three corresponding terms)} \qquad (30)$$

where $\omega = 0$ corresponds to the tilt plane normal. The simulated results depend on the degree of biasing a and the angle of biasing ψ_0 (Fig. 25a). The degree of polarization $D(\psi_0) = 2[A(0°) - A(90°)]/[A(0°) + A(90°)]$ is plotted as a function of the biasing angle in Figure 25b; note that the positive and negative signs of $D(\psi_0)$ correspond to out-of-phase and in-phase dependence with respect to the phenyl ring stretching peak, respectively. The conspicuous in-phase angular dependence in SmC* (Fig. 23) clearly indicates that ψ_0 in SmC$_A^*$ is different from that in SmC* (Fig. 21), because the angular dependence is out of phase in SmC*. The changeover from in-phase to out-of-phase occurs with the parameters $a = 0.2$ and $\beta = 25°$.

For homogeneously aligned cells the normalized absorbance $A(\omega)$ is given by

$$A(\omega') = \frac{1}{2}\int_0^{2\pi} f_{t+}(\psi)(\sin\omega'\sin\beta\sin\psi + \cos\omega'\cos\beta)^2 d\psi$$

$$+ \text{(a corresponding term with } f_{b+}(\psi)) \qquad (31)$$

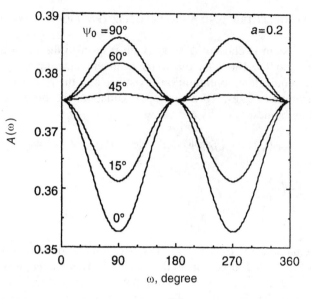

(a)

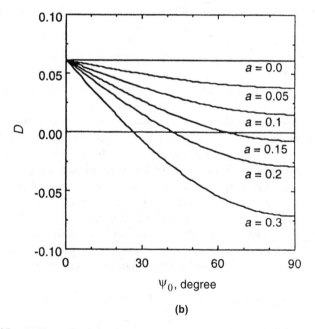

(b)

Figure 25. (a) Normalized absorbance vs. polarizer rotation angle $A(\omega)$ calculated in the SmC* and SmC$_A^*$ phases for the radiation incident along the smectic layer normal by using the distribution function given in Eq. (29). The molecular tilt angle and the degree of biasing are assumed to be $\theta = 25°$ and $a = 0.2$, respectively; and $\omega = 0°$ is the tilt plane normal. (b) Degree of polarization vs. most probable orientation $D(\psi_0)$ for several values of a. (Redrawn with permisssion from [18].)

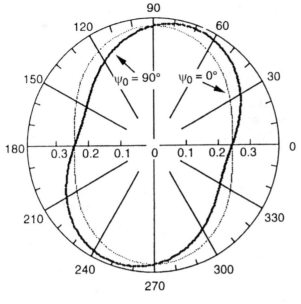

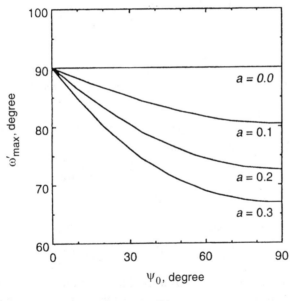

(b)

Figure 26. (a) Normalized absorbance vs. the polarization rotation angle $A(\omega)$ calculated in SmC* for the IR radiation incident along the tilt plane normal by using the distribution function given in Eq. (29). Here, $a = 0.2$, $\omega' = \omega - \theta$, and $\omega = 0$ is the layer normal. (b) Maximum angle vs. the most probable orientation for several values of the parameter a. (Redrawn with permisssion from [18].)

where $\omega' = \omega - \theta$ is used for convenience. Figure 26 shows some calculated results for normalized absorbance $A(\omega')$ when $a = 0.2$ and for the expected maximum angle $\omega_{max}(\psi_0)$ vs. the most probable orientation. By comparing the experimental data with the prediction of the model, the following conclusion is drawn [18]: The C=O group in the core part appears to lie in the tilt plane in both phases, whereas the chiral C=O group has a tendency to lie in the tilt plane in SmC_A^* and takes a considerably upright position in SmC^*.

VI. BIASED ROTATION AROUND THE LONG AXIS—CALCULATION OF IR ABSORBANCE COMPONENTS

It can be shown that the orientational distribution functions of the transverse dipoles can be quantitatively analyzed through a detailed analysis of the absorbance profiles. In the case of unwound SmC^* phase, a simplification can be introduced if one considers molecular fluctuation in the tilt plane (soft mode) but neglect thermal fluctuations in the azimuthal direction under an electric field [13]. Let the angle ψ be measured from the normal to the tilt plane (the usual C_2 axis in the SmC^* phase) to the position of the transition dipole moment on its rotation cone around the molecular long axis.

The absorbance components A_Z, A_Y, and A_{YZ} in the laboratory system can be expressed in terms of $\langle \sin \psi \rangle$ and $\langle \sin^2 \psi \rangle$ as follows [13]:

$$A_Z = k\{\cos^2 \beta \langle \cos^2 \theta \rangle - 2 \sin \beta \cos \beta \langle \sin \theta \cos \theta \rangle \langle \sin \psi \rangle$$
$$+ sin^2 \beta \langle \sin^2 \theta \rangle \langle \sin^2 \psi \rangle \}$$

$$A_Y = k(\cos^2 \beta \langle \sin^2 \theta \rangle + 2 \sin \beta \cos \beta \langle \sin \theta \cos \theta \rangle \langle \sin \psi \rangle$$
$$+ \sin^2 \beta \langle \cos^2 \theta \rangle \langle \sin^2 \psi \rangle)$$

$$A_{YZ} = k(\langle \sin \theta \cos \theta \rangle [\cos^2 \beta - \sin^2 \beta \langle \sin^2 \psi \rangle]$$
$$+ \sin \beta \cos \beta [\langle \cos^2 \theta \rangle - \langle \sin^2 \theta \rangle] \langle \sin \psi \rangle) \quad (32)$$

The polar angles β of the bands under investigation are widely reported in the literature [18,19]. Because θ describes the orientation of the molecular long axis (z axis) in the laboratory system, the averages $\langle \cos^2 \theta \rangle$ and $\langle \sin \theta \cos \theta \rangle$ provide information about the distributions of the long molecular axis. To reduce the number of unknown variables in Eq. (32), the averages $\langle \sin \psi \rangle$ and $\langle \sin^2 \psi \rangle$ can be expressed in terms of the following approximation for the azimuthal distribution function $f(\psi)$ [13]:

$$f(\psi) = \frac{1}{2\pi} + a\cos(\psi - \psi_0) + b\cos(2(\psi - \psi_0))$$

$$\langle \sin\psi \rangle = \int_0^{2\pi} \sin(\psi)f(\psi)\,d\psi = \pi a\sin(\psi_0) \qquad (33)$$

$$\langle \sin^2\psi \rangle = \int_0^{2\pi} \sin^2(\psi)f(\psi)\,d\psi = \frac{1}{2}(1 - \pi b\cos(2\psi_0))$$

where a and b indicate the degree of polar and quadrupolar biasing and ψ_0 is the angle of biasing. The overall distribution for all molecules is generally a linear combination of $[f(\psi) + f(-\psi)]/2$ and $[f(\pi - \psi) + f(\psi - \pi)]/2$. For the positive polarization, the first term is predominant whereas for negative, the second one is predominant.

Jang et al. [13] quantitatively investigated the orientational distribution of C=O groups obtained through detailed analyses of polarized FTIR in various FLC materials. The structural formulas and phase sequences of materials are given in Figure 27. Several absorption peaks have been observed in the spectra, but the attention has been focused mainly on the CH_2 asymmetric stretching (~ 2920 cm^{-1}), C=O stretching ($k1$, ~ 1760 cm^{-1}, $k2$, ~ 1735

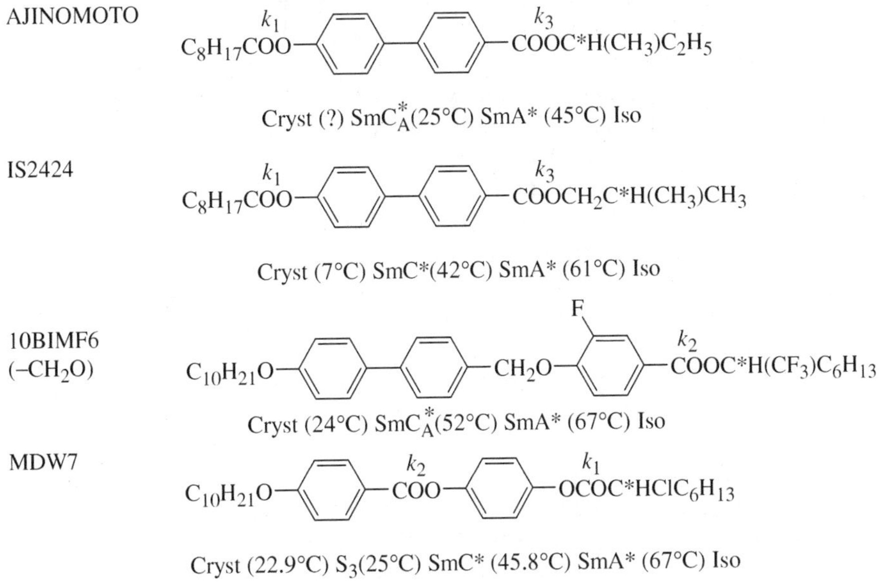

AJINOMOTO

k_1

$C_8H_{17}COO$—⬡⬡—$COOC^*H(CH_3)C_2H_5$ k_3

Cryst (?) SmC_A^*(25°C) SmA* (45°C) Iso

IS2424

k_1

$C_8H_{17}COO$—⬡⬡—$COOCH_2C^*H(CH_3)CH_3$ k_3

Cryst (7°C) SmC*(42°C) SmA* (61°C) Iso

10BIMF6
(–CH$_2$O)

F

$C_{10}H_{21}O$—⬡⬡—CH_2O—⬡—$COOC^*H(CF_3)C_6H_{13}$ k_2

Cryst (24°C) SmC_A^*(52°C) SmA* (67°C) Iso

MDW7

k_2 k_1

$C_{10}H_{21}O$—⬡—COO—⬡—$OCOC^*HClC_6H_{13}$

Cryst (22.9°C) S$_3$(25°C) SmC* (45.8°C) SmA* (67°C) Iso

Figure 27. Structural formulas and phase sequences of various FLC samples. (Redrawn with permission from [13].)

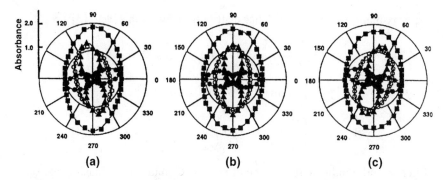

Figure 28. Absorbance vs. polarizer rotation angle for four representative stretching peaks in the SmA* and field-induced SmC* phases of a homogeneously aligned AJINOM-OTO sample. (a) SmC_A^* ($-E$), (b) SmA*. (c) SmC_A^* ($+E$). ●, phenyl ring; ○, CH_2; ■ C=O near the chiral center; ▲, C=O far from the chiral center. (Redrawn with permission from [13].)

cm^{-1}, $k3$, ~ 1715 cm^{-1}; see the chemical structure, and phenyl ring stretching (~ 1600 cm^{-1}) peaks, because they are well separated from the other absorption peaks and their transition moment directions are also well defined. The absorbance was measured in the SmA* phase and under the applied field in the SmC* and SmC_A^* phases. Figure 28 shows the polar plot of the absorbance for the AJINOMOTO sample. Note that the two carbonyls, one attached near the chiral center and the other far from chiral center, show different angular dependencies: The absorbance variation of the chiral C=O vs. ω is not particularly large ($R = 1.63$), whereas the other C=O peak has a rather large dichroism ratio ($R = 6.25$). The angles of rotation of the absorbance profiles after application of the electric field were investigated for several materials; the results are listed in Table II.

In the materials IS2424 and MDW7, both of the C=O profiles rotate in the same direction as the phenyl profile, but their angles are lower than the phenyl profile. Surprisingly, in the material 10BIMF6 the C=O profile rotates in the opposite direction to the phenyl ring profile. Compared to the optical tilt measurements, the phenyl stretching transition dipole moment nearly coincides with the molecular long axis [17]. Thus an asymmetry of the C=O absorbance profile with respect to the phenyl profile clearly shows that the rotation of the C=O group is biased. Values for $\langle \cos^2 \theta \rangle$ and $\langle \sin \theta \cos \theta \rangle$ were obtained from the phenyl profile in the SmC* phase by assuming its polar angle is $\beta \approx 0°$; k is obtained from the C=O profile in the SmA* phase. By substituting Eq. (33) for $\langle \sin^2 \psi \rangle$ and $\langle \sin \psi \rangle$ into Eq. (32) and fitting to the experimental C=O absorbance profile, four unknown parameters (a, b, ψ_0, and $\beta_{C=O}$) must be determined. The polar angle of the C=O transition moment with respect to the phenyl ring has been reported

TABLE II
Angular shift of the peak profiles $\Delta\omega_0$ with an applied electric field

Liquid Crystal	Phase	(1)[a] ω_0 (phenyl)	(2) ω_0 (alkyl)	(3) ω_0 (k_1 C=O)	(4) ω_0 (k_2 C=O)	(5) ω_0 (k_3 C=O)
	SmA*	0.00	0.49	0.15	—	1.78
AJINOMOTO	SmC$_A^*$,+E	−15.86	−15.44	−13.68	—	−0.65
	SmC$_A^*$,−E	13.97	14.61	11.99	—	0.55
	SmA*	0.00	0.51	0.23	—	0.66
IS2424	SmC*,+E	18.29	16.84	13.95	—	2.58
	SmC*,−E	−17.63	−18.06	−13.74	—	−2.10
	SmA*	0.00	0.59	0.04	1.05	—
MDW7	SmC*,+E	19.82	15.95	13.65	13.66	—
	SmC*,−E	−19.83	−16.31	−13.83	−13.68	—
	SmA*	0.00	0.74	—	1.09	—
10BIMF6	SmC$_A^*$,+E	−17.53	−17.58	—	12.78	—
	SmC$_A^*$,−E	17.72	16.29	—	−15.21	—

[a] Shift angles (degrees) of the bands.
Reprinted with premission from [13].

for several ester compounds as being $\sim 60°$. This is based on X-ray analysis [19] and molecular orbital calculations [20]. In fact, the best result was obtained for an angle of $61.7°$ for the low dichroic C=O profiles. However, researchers could not fit the high dichroic C=O profile well without increasing β to $78.9°$. Jang et al. [13] noted that all C=O groups that show large dichroic ratio are not directly attached to the phenyl ring but are connected through ether (–O–) linkages; whereas the C=O groups that show low dichroic ratio are connected directly to the phenyl ring. The authors assumed that when the polar group is separated from the phenyl core, the angle $\beta_{C=O}$ can substantially differ from $60°$.

Figure 29 shows the fitted experimental data for the low and high dichroic carbonyl groups of the AJINOMOTO compound. The fitting parameters are $a = -0.049$, $b = 0.039$ for the high dichroic C=O group and $a = -0.039$, $b = 0.019$ for low dichroic C=O group. The averages $\langle \sin^2 \psi \rangle$ and $\langle \sin \psi \rangle$ show similar values (0.492 and −0.105 for high dichroic C=O and 0.504 and −0.125 for low dichroic C=O groups, respectively), so the rotational distribution functions generally look similar to those shown in Figure 30. The most probable orientation of the C=O groups is not in the plane containing the twofold C_2 axis. The molecule on the average spends more time around $\psi \approx \pm 40°$ and $\psi \approx \pm 220°$ for positive and negative electric fields, respectively. Because the rotation is biased not in a unique direction but in two equivalent directions (owing to the head-and-tail equivalence), the spontaneous polarization does appear along C_2 axis.

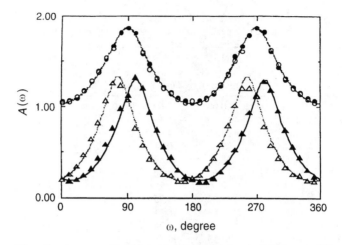

Figure 29. Absorbance profiles $A(\omega)$ as a function of electric field for two different C=O groups in the SmC* phase of the AJINOMOTO sample. Data are shown for 1716 cm^{-1} ($\bullet$, $-E$; $\bigcirc$, $+E$) and 1756 cm^{-1} ($\blacktriangle$, $-E$; $\triangle$, $+E$). Solid and dotted lines are fit to experimental data using Eq. (28). (Redrawn with permission from [13].)

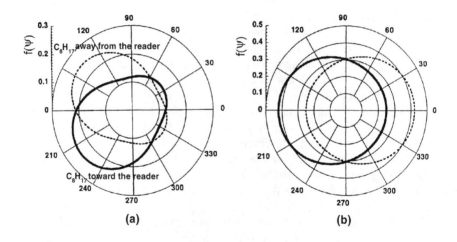

(a)   (b)

Figure 30. Orientational distribution function $f(\psi)$ of the 1756 cm^{-1} C=O group of the AJINOMOTO sample in an electric field. (a) The *solid curve* $f(\psi)$ represents a possible distribution of molecules with the C_8H_{17} end towards the reader. The dashed curve $f(\psi)$ represents a possible distribution of molecules with the C8H17 end away from the reader. (b) The overall distribution for all molecules is a linear combination of $[f(\psi) + f(-\psi)]/2$ (*solid curve*) and $[f(\psi - \pi) - f(-\psi + \pi)]/2$ (*dashed curve*). The positive polarization of this material indicates that the solid distribution is predominant. (Redrawn with permission from [13].)

Cr 388 SmC$_A$* 371.7 Fi1 374.2 Fi2 376.2 SmC* 392 SmA* 403 I

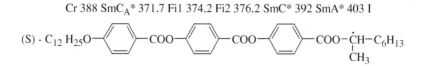

Figure 31.   Chemical formula of the AFLC compound, (S)12 OBBB 8. (Redrawn with permission from [21].)

A similar calculation has been carried out for the AFLC sample (S) 12 OBBB 8 [21] of the structure given in Figure 31, which contains three carbonyl groups in the core [22]. Results for the angular shifts of different bands are given in Table III. Because the molecule has two carbonyl linkages in between the phenyl rings, the phenyl para axis, is inclined to the molecular axis, and the assumption $\beta \approx 0°$ for phenyl stretching dipole cannot be used. The absorbance profiles are shown in Figure 32. The profiles for the three bands were used simultaneously for calculating the rotational distribution around the long molecular axis: phenyl ring stretching peak (~ 1600 cm^{-1}), phenyl out-of-plane deformation (764 cm^{-1}), and C=O stretching (~ 1718 cm^{-1}). Factors k for each band were obtained from the C=O profile in the SmA* phase using the different formulas for the absorbance components [21] for the fluctuations of the long molecular axis, which are not restricted to the tilt plane in the SmA* phase.

$$A_Y = 0.5\,k \left\{ \sin^2 \beta [1 - \langle \sin^2 \theta \rangle \langle \sin^2 \psi \rangle] + \cos^2 \beta \langle \sin^2 \theta \rangle \right\}$$
$$A_Z = k \left\{ \sin^2 \beta \langle \sin^2 \theta \rangle \langle \sin^2 \psi \rangle + \cos^2 \beta \langle \cos^2 \theta \rangle \right\} \tag{34}$$

The A_Y and A_Z components are proportional to $\langle p_Y^2 \rangle$ and $\langle p_Z^2 \rangle$ in Eq. (14), respectively. Possible molecular conformations have been carefully studied using quantum orbital calculations. The torsional angles of C_k-O-C_{ph}-C_{ph} and C_{ph}-C_{ph}-C_k=O were found to be about $55°$ and $5°$, respectively (k=carbonyl; ph=phenyl). These imply the angle between the carboxyl group and the middle phenyl ring and between the end phenyl group and the carboxyl groups respectively. These agree well with the values reported for similar structures [23,24]. Hence the noncoplanar structure of the core with the phenyl rings rotated by $50°$ to $60°$ with respect to each other were considered. The polar angles of the dipoles have been taken: $\beta = 24.4°$ for the phenyl stretching vibration (1600 cm^{-1}), $\beta = 102°$ for the C–C out-of-plane of vibration (764 cm^{-1}) and $\beta = 61.4°$ for the carbonyl C=O stretching vibration.

In the SmA* phase, an assumption is commonly made that molecules are rotating freely around their long axes [4,8,9] ($\langle \sin \psi \rangle = 0$ and $\langle \sin^2 \psi \rangle = 0.5$). In such a case, the transition dipole moment is uniformly

TABLE III
Angular Shifts $\Delta\omega_0$ (degrees) of Peak Profiles with an Applied Electric Field

		Band Frequency, cm^{-1}						
T, K	Phase	Phenyl 1600	Phenyl 764	CH_2 symmetric 2928	CH_2 asymmetric 2928	CH_3 symmetric 1380	C=O core 1740	C=O chiral 1718
345	SmC_A^{*a} F(+)	−30.57	−25.18	−22.58	−26.75	−30.19	−23.34	−14.05
	$F(-)$	31.30	24.69	22.08	26.56	29.43	21.94	15.67
380	SmC^{*b} F(+)	−25.70	−20.31	−18.61	−23.17	−24.95	−18.86	−13.63
	$F(-)$	25.29	20.85	17.95	23.94	25.11	19.26	13.11

[a]$E = 12$ V/μm.
[b]$E = 6$ V/μm.
Reprinted with permission from [21].

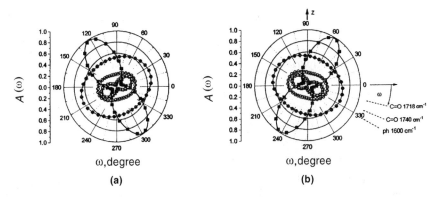

Figure 32. Absorbance profiles $A(\omega)$ vs. the polarizer rotation angle ω for different transition dipole moments of the molecular core at $T = 345$ K and $E = -12$ V/μm. ■, 1600 cm^{-1} in-plane phenyl; □, 764 cm^{-1} out-of-plane phenyl; ○, 1718 cm^{-1} C=O near the chiral center; ⊕ 1740 cm^{-1} C=O far from the chiral center, *solid lines* and *dashed lines* calculated from parameters shown in Table 3. (Redrawn with permission from [21].)

distributed around the long molecular axis, and the order parameter S for different transition dipole moments can be simply determined using Eqs. (18) and (19). It was found, however, that S for each band obtained in this way is not the same for each band, and the differences are much greater than experimental error. Therefore, we conclude that the rotation around the long molecular axis is not free; that is, $\langle \sin^2 \psi \rangle \neq 0.5$. It is reasonable to assume that the azimuthal angles ψ of the C–C phenyl stretching dipole and the out-of-plane (764 cm^{-1}) vibration dipole moment differ by 90°. The calculated averages for the long molecular axis fluctuations were $\langle \cos^2 \theta \rangle = 0.74$ and $\langle \sin^2 \psi \rangle = 0.48$, using the experimental absorbance for the C–C phenyl stretching peak and Eq. (34). The same value of $\langle \sin^2 \psi \rangle = 0.48$ has also been obtained for the carbonyl dipole near the chiral carbon (Table IV). In the next step, k factors for each band were obtained from the absorbance components in the SmA* phase using Eq. (34).

TABLE IV
Calculated Bias Parameters

T, K	Phase		a	b	$|\psi_0|$	$|\langle \sin\psi \rangle|$	$\langle \sin^2\psi \rangle$
345	SmC$_A^*$	$F(+)$	0.039	0.052	41°	0.081	0.486
		$F(-)$	0.036	0.065	39°	0.072	0.479
380	SmC*	$F(+)$	0.031	0.055	38°	0.050	0.478
		$F(-)$	0.032	0.047	37°	0.049	0.479
395	SmA*						0.480

Reprinted with permission from [21].

The possible nonisotropic rotation of the molecular short axis in the SmA* phase was determined through FLC dynamic studies by Lalanne et al. [25]. They observed a pretransitional slowing down far above the SmC*–SmA* phase transition temperature using a degenerate four-wave mixing technique on chiral compounds. The effect appears to be directly linked to the ψ-privileged direction evidenced by IR experiments.

To solve this problem in unwounded the SmC_A^* and SmC* phases, a calculation was carried out for the three transition dipole moments located in the molecular core: the phenyl stretching vibration (1600 cm^{-1}), the C–C out of plane of vibration (764 cm^{-1}), and the carbonyl C=O stretching vibration near the chiral center. The molecular core was assumed to be rigid; hence the biasing parameters for the constituents of the core were calculated as an entity. The Eqs. (32) and (34) are generally used, replacing ψ with $\psi - \gamma$. The angle γ is the azimuthal angle of a dipole in the molecular frame of reference (for the chiral C=O dipole, γ is chosen to be 0 so ψ directly measures the position of C=O dipole as it was before). The averages $\langle \sin \psi \rangle$ and $\langle \sin^2 \psi \rangle$ were substituted in Eq. (32) using the approximation for the azimuthal distribution function shown in Eq. (33). It was reasonable to assume that the transition dipole moments under consideration must have the same degree of biasing (a and b) but have different azimuthal angles γ in the molecular frame of reference. The azimuthal angles, however, should directly correspond to the torsional angles between components of the molecular core. The overall distribution for all molecules was taken as a linear combination of $[f(\psi) + f(-\psi)]/2$ and $[f(\pi - \psi) + f(\psi - \pi)]/2$.

The results of the calculations are listed in Table IV in terms of the rotational bias parameters. Values of $\langle \sin \psi \rangle$, $\langle \sin^2 \psi \rangle$ and ψ_0 are related to the carbonyl dipole near the chiral carbon for a positive electric field. For a negative field, the angle ψ_0 becomes $\psi_0 + 180°$. In these cases, the calculated parameters reproduce the experimental profiles extremely well (see Fig. 32).

The fitting parameters: $\langle \cos^2 \theta \rangle$, $\langle \sin \theta \cos \theta \rangle$, a, and b, can eventually be used for reproducing the absorbance profiles of the remaining carbonyl group centered at 1740 cm^{-1} and the symmetrical stretching CH$_3$ band at 1380 cm^{-1}. The absorbance profile of the 1740 cm^{-1} band (Fig. 32) was found to be a superposition of the two profiles, one with an angle $\omega_0 = 13.5°$, the second with an angle $\omega_0 = 9°$ in the SmC_A^* phase. ω_0 is defined in Eq. (28). This reflects the fact that one of the C=O dipole moments is rotated with respect to the second (around the long molecular axis) by a torsional angle of $\sim 55°$ so these have different azimuthal angles γ in the molecular frame of reference. The calculated profile reproduces the experimental data quite well. For the chiral CH$_3$ group, the polar angle β was taken to be 109°, and the best fit to the experimental data is obtained for a γ of 258°. This is quite a reasonable value for the torsional angle between the carbonyl group and CH_3 group, which agrees well with the most probable conformation of such a molecular segment [26].

The results show a slightly higher angle and degree of biasing for the C=O transition dipole moment in the SmC_A^* phase than in SmC^*, although the angles of biasing are similar to those reported for the AJINOMOTO compound [13]. The degree of polar biasing (factor a) was found to be significantly lower than factor b for quadrupolar biasing of the distribution function in contrast to the result for the AJINOMOTO compound.

VII. DYNAMICS OF THE ELECTRIC FIELD–INDUCED MOLECULAR REORIENTATION OF THE FLCS

FTIR time-resolved spectroscopy (FTIR TRS) with simultaneous temporal and spectral resolution has emerged to be a powerful technique for elucidating dynamic, segmental information concerning both the structural and orientational changes of molecules in transient systems [27]. The FTIR TRS technique applied to the dynamic analysis of liquid crystals has revealed new information that enables us to understand the detailed submolecular mechanisms of the electrically induced liquid crystal reorientation. FTIR TRS may be categorized into two types: time domain and frequency domain. The former is commonly referred to as "strobo-scopic," or "transient" spectroscopy. The output from these experiments is a stack of time-resolved spectra; the structural and the orientational changes of the sample can be analyzed in terms of the spectral changes. In the frequency domain, a frequency correlation analysis is applied to the time-resolved spectra to yield a spectrum defined by two independent IR wavenumbers [28]. Thus information about the orientational dynamic correlation between the different parts of the molecule can be obtained. Although FTIR TRS measurements were first carried out by Sakai and co-workers [29,30] using a step-scan spectrometer; until recently, this technique was overshadowed by continuous-scan instruments. The recent improvement in the step-scan technique, however has resulted in a time resolution to the submicrosecond level. The most important advantage of step-scan interferometry lies in the fact that the optical retardation of the interferometer is held constant during the sampling of interferogram elements; consequently, the spectral multiplexing is decoupled from the time dependence of the data collection. Interferogram elements are then time sorted and rearranged to obtain complete interferograms.

The step-scan vibrational spectroscopy with polarized light was used to study the dynamics of molecular orientation and conformational changes during large-scale externally induced change accompanying electro-optic switching in the nematic, chiral SmA*, antiferroelectric, ferrielectric, and ferroelectric phases of the nonpolymeric and polymeric liquid crystals [17, 31–53]. Unlike the nonpolymeric LCs, the ferroelectric LC polymers in SmC* phase show pronounced differences in tilt angles of the mesogen, spacers, and backbone.

A. Dynamics of the Molecular Reorientation in a Chiral SmA* Phase

In a system of uniaxial symmetry, the measurements of the dichroism ratio can yield information about the molecular organization. With a transition to a lower symmetry, such as that of monoclinic field-induced or temperature-induced SmC* phase, however, dichroism changes may arise from either (or both) director reorientation or symmetry changes. Unraveling these questions requires measurements and analysis of the overall pattern of the absorbance vs. polarizer orientation.

Experiments were carried out at a temperature of 25°C on the mono-domains sample of the compound W317 [31] in the bookshelf geometry. W317 has the electronegative polar substituent, $-NO_2$, on the core ring with *ortho* to the chiral tail to give high ferroelectric polarization and fast electroclinic switching. Square wave excitation with an amplitude of 160 V applied across the sample produced a director reorientation between $+18°$ and $-18°$ with respect to the layer normal with a response time of 10 to 20 μs as determined via the rotation of the optic axis in visible range. The IR spectra obtained vs. the polarizer angle ω and peak absorbance $A(\omega)$ for the principal IR bands were determined and analyzed. A polar plot of the absorbance with $E = 0$, Fig. 33, clearly shows C_2mm symmetry about $\omega = 0$, with the mirror planes along the layer normal Z ($\omega = 0°$) and the layer direction ($\omega = 90°$), as expected for SmA* phase resulting from isotropic averaging of an orientation around n parallel to Z (Fig. 33). Figure 34 shows the effect of applying an electric field and displays temporal evolution of W317 absorbance profiles during a transient rotation of n induced by switching the field from -45 to $+45$ V/μm at $t = 0$. Although each band has the C_2mm symmetry form of $A(\omega)$, the overall structure of these profiles exhibits the lower C_2 symmetry required by the monoclinic field–induced SmC* structure, resulting from a tilt of n. It is clear that the core bands have rotated through a larger angle than have the tail bands, indicating that in equilibrium under the applied field the tails are less tilted than the cores relative to the layer normal Z (Fig. 34, *dotted line*). The *solid lines* and *dashed lines* in the figure indicate the mean orientation of the core $\omega_0(t)_{core}$ and tail groups $\omega_0(t)_{tail}$, respectively at $t = -2$ and 300 μs, clearly

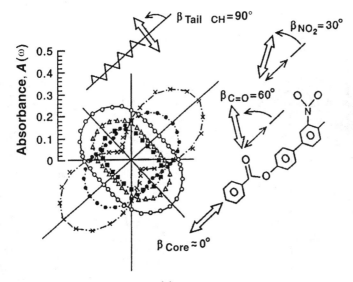

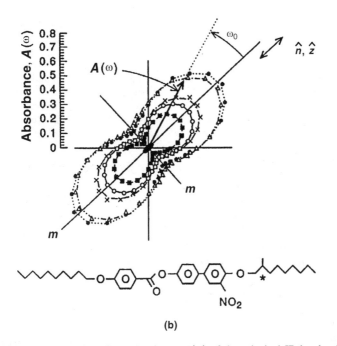

(b)

Figure 33. Polar plot of the absorbance $A(\omega)$ of the principal IR bands of W317 in a SmA* monodomain with a zero electric field applied. The main dipoles are indicated by the open arrows ($\Leftrightarrow$). Plot exhibit C_2mm symmetry with a mirror line along the director *n* (layer normal *z*), a result of the SmA* uniaxial symmetry. (a) $\bigcirc$, tail CH, 2927 cm^{-1}; ■, tail CH, 2856 cm^{-1}; $\triangle$, C=O, 1732 cm^{-1}; ×, core, 1606 cm^{-1}; ●, NO$_2$, 1535 cm^{-1}. (b) core: $\bigcirc$, 1512 cm^{-1}; ■, 1489 cm^{-1}; $\triangle$, 1259 cm^{-1}; ×, 1211 cm^{-1}; ●, 1169 cm^{-1}. (Redrawn with permission from [31].)

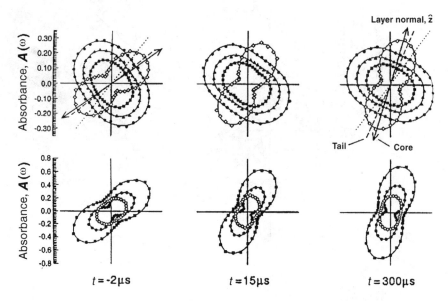

Figure 34. The polar absorbance patterns $A(\omega, t)$ before and during the dynamic response of the W317 cell shown in Figure 33 to an applied voltage step from -80 to $+80$ V at $t = 0$. The symmetry reduction accompanying applied electric field is evident in the *upper row*, which shows that the net rotation of the core and tail during switching is through different net angles. The *solid lines* and *dashed lines* indicate the mean orientation $\omega_0(t)$ of the core and tail groups, respectively. *Upper row*; ●, tail CH, 2927 cm^{-1}; ■, tail CH, 2856 cm^{-1}; ▲, C= O, 1732 cm^{-1}; ◇, core, 1489 cm^{-1}. *Lower row*: ●, core, 1606 cm^{-1}; ■, 1169 cm^{-1}; ○, NO$_2$, 1535 cm^{-1}. (Redrawn with permission from [31].)

showing the larger tilt of the cores and the overall C$_2$ symmetry of the patterns.

The resulting temporal evolution of the mean orientation for W317 are shown on two time scales in Figure 35 as a plot of $\omega_0(t)$ and as a fractional orientation function $f(t)$. The plot for $f(t)$ of W317 shows that just after field switching the tail reorientation rate is slower than that for the core, with $\omega_0(t)_{\text{tail}}$ changing from being smaller than the $\omega_0(t)_{\text{core}}$ to being larger than it during the first 5 μs after field switching. The reorientation of the C=O and NO$_2$ groups are particularly interesting. The C=O bands overshoot in $f(t)$, eventually reorienting through $\Delta\omega_0(t)_{\text{C=O}} = 29°$, a smaller angle than the core; the NO$_2$ band reorients through $\Delta\omega_0(t)_{\text{NO}_2} = 42°$, a larger angle than the core. Such difference in $\Delta\omega_0(t)$ must arise as a result of anisotropy of reorientation about n. The overshoot on field switching indicates that the process of equilibration involves two steps: a director reorientation, which leaves a nonequilibrium orientation distribution around the long axis, and equilibration around the long axis.

B.   Reorientation Dynamics of FLC in the SmC_A^* and SmC^* Phases

1.   Synchronous Reorientation in TFMHPODB

The possibilities of observing conformational and orientational changes during tristable switching: the electric field–induced phase transition

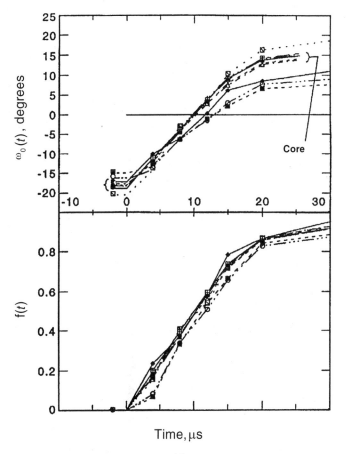

(a)

Figure 35. Time dependence of the mean orientation $\omega_0(t)$ and the fractional $f(t)$ change of $\omega_0(t)$ for the principal bands in W317 at (a) short and (b) long time scales. Inset, extremes $\psi = 90°$ and $\psi = -90°$ of the C=O orientation about the molecular long axis and the two-stage reorientation process that would yield a transient overshoot in the apparent C=O orientation. $\bigcirc$, tail CH, 2927 cm^{-1}; $\blacksquare$, tail CH, 2856 cm^{-1}; $\blacklozenge$; C=O, 1732 cm^{-1}; $\times$, 1606 cm^{-1}; $\square$, NO$_2$, 1535 cm^{-1}; $\boxminus$, 1512 cm^{-1}; $\triangle$, 1489 cm^{-1}; $\blacklozenge$; 1259 cm^{-1}; $\boxplus$, 1211 cm^{-1}; $\blacktriangle$, 1169 cm^{-1}. (Redrawn with permission from [31].) In the figure, γ in the notation here implies ψ.

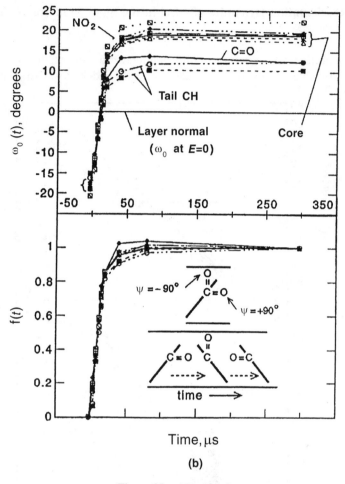

Figure 35. (*Continued*)

between SmC$_A^*$ and SmC*, have been explored by observing time-resolved IR spectra for homogeneously aligned samples of TFMHPODB [17]. Figure 36 compares the peak absorbance plot as a function of time with the corresponding optical response. A pulsed electric field was applied for 40 μs, and the spectra were measured at a time interval of 1.5 μs. The polarizer was set at 45° to the layer normal to obtain large intensity changes. The C=O and CH$_3$ stretching peaks do not show any absorbance changes during switching. The absorbances of phenyl and CH$_2$ peaks start to change immediately after the application of the electric field and do not change after

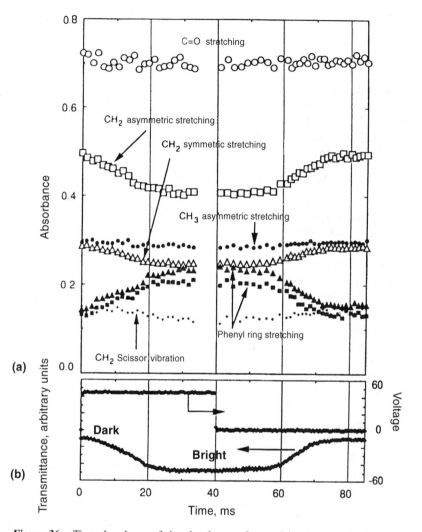

Figure 36. Time depedence of the absorbance of several bands and with the optical response in SmC$_A^*$ of a homogeneously aligned cell containg TFMHPODB. (Redrawn with permission from [17].)

a lapse of ~25 µs. The relaxation process from SmC* to SmC$_A^*$, does not occur immediately after the field is turned off; it requires an induction period before the actual changes begin to occur. The induction period is approximately 15 µs, and the transition is completed in about 35 µs after the electric field was removed. These absorbance variations are completely consistent with the optical response. No differences in reorientation rates have been found between the alkyl and phenyl groups; therefore, it is throught that these groups in AFLCs reorient simultaneously on the microsecond time scale. Similar conclusions have been reached for the FLC dimers and polymers (see section VIII). Such a conclusion is contrary to several reports for the nematic liquid crystals, in which the alkyl chain is reported to respond more quickly than the phenyl part [28,30,31] and for some FLCs and AFLCs in which the response of the alkyl chains is delayed with respect to the core [32,52].

2. Delayed Response of the Alkyl Chain During Dynamical Switching in VOH8 and MHDBOB

To our knowledge, only Ozaki and his co-workers have so far, reported a delayed response for the alkyl chain in the ferroelectric liquid crystal VOH8 [53] and the antiferroelectric liquid crystal MHDBOB [54]. The structures are given in Figure 37.

The following bands are used to characterise the behavior of different molecular moieties: for VOH8, the comparatively isolated bands at 2854 cm^{-1} (CH$_2$ symmetric stretching) and 2962 cm^{-1} (CH$_3$ asymmetric stretching); for the alkyl chain, the bands at 1771 cm^{-1} ($\nu_{C=O\,free}$) and 1753 cm^{-1} ($\nu_{C=O\,H\text{-bonded}}$) for the chiral part and the bands at 1510 cm^{-1} ($\nu_{C=C}$), 1621 cm^{-1} ($\nu_{C=N}$) and 983 cm^{-1} (phenyl ring C–H in-plane deformation) for the mesogen part. A complete assignment of bands is shown in Table V. The polarized IR spectra of the sample were measured in the SmC* phase (at 83°C and 75°C) under DC voltages with positive and negative polarity for monitoring the relative orientation of the alkyl, mesogen, and chiral segments. Absorbance as a function of the angle of polarization $A(\omega)$ clearly shows that the maxima of the band absorbance of the three segments is obtained for different angles of polarization. The observed rotation of the

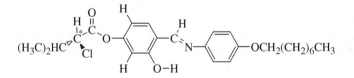

$$k \underset{43.0}{\overset{76.0}{\rightleftharpoons}} \text{SmC*} \underset{89.5}{\overset{90.3}{\rightleftharpoons}} \text{SmA*} \underset{104.6}{\overset{106.5°C}{\rightleftharpoons}} I$$

Figure 37. Structure of the VOH8 ferroelectric liquid crystal and phase transition temperatures. (Redrawn with permission from [53].)

TABLE V

Dichroic Ratio and Vibrational Band Assignments for the Relevant Peaks in the *IR* Spectra of Various and Dichroic Ratios

Wave number, cm^{-1a}	$D(A_{\parallel}/A_{\perp})$	Assignment
2962 (m)	1.0	CH$_3$, asymmetric stretching
2929 (s)	0.5	CH$_2$, antisymmetric stretching
2901 (sh)	1.0	CH, stretching
2872 (m)	1.0	CH$_3$, symmetric stretching
2854 (m)	0.8	CH$_2$, symmetric stretching
1771 (m)	0.2	C=O, stretching (free)
1753 (w)	0.4	C=O, stretching (hydrogen bonded)
1621 (s)	3.0	C=N, stretching
1577 (m)	2.5	Phenyl ring C=C, stretching
1510 (s)	4.8	Phenyl ring C=C, stretching
1469 (m)		CH$_2$, deformation
1292 (m)	5.7	C–O, stretching
1274 (m)	1.2	C–OH, stretching
1249 (m,sh)	5.4	C–O, stretching
1186 (m)	4.0	Phenyl ring C–C, deformation
1146 (m)	3.0	Chain C–C stretching
1117 (m)	3.2	Phenyl ring C–C, deformation
1077 (w,sh)	4.5	Phenyl ring C–C, deformation
983 (m)	4.3	Phenyl ring C–H in plane, deformation
883 (w)	0.4	Phenyl ring C–H out of plane, deformation
831 (w)	0.2	Phenyl ring C–H out of plane, deformation

^{a} *w*, weak; *m*, medium; *s*, strong; *sh*, shoulder.

Reprinted with permission from [53].

absorbance maxima between plus and minus polarities are as follows: for mesogen, $\Delta\omega = 58°$ at 75°C (7 V) and $\Delta\omega = 53°$ at 83°C (5 V), respectively. On the other hand, the absorbance maximum of CH$_2$ symmetric stretching band rotates by $\Delta\omega = 48°$ at 75°C (7 V) and $\Delta\omega = 44°$ at 83°C (5 V), respectively. Particularly interesting is the behavior of the C=O peaks, which show maxima for the different polarization angles for positive and negative voltages. Such results are significant and provide an unambiguous evidence for the biased rotation of the chiral group and noncolinearity of the alkyl and chiral chains with respect to the mesogen and the molecular long axis.

To explore the segmental mobility and the dynamical behavior during switching from one to the other surface-stabilized state, time-resolved FTIR spectra were measured as a function of delayed time from 0 to 100 μs with respect to the step voltage application. For quantitative comparison of the reorientation rate, data were presented as the normalized absorbance

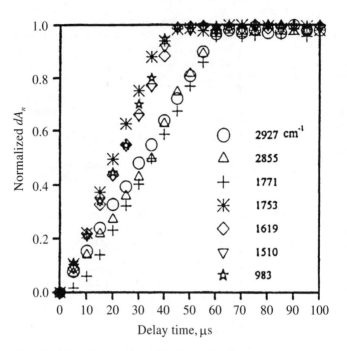

Figure 38. Time dependence of normalized intensity changes (dA_n) vs. delay time for the representative IR bands of the mesogen, alkyl chain, and chiral chain segments of VOH8 in the SmC* phase under an electric field of ± 7 V and 5 kHz repetition rate. $\bigcirc$, 2927 cm^{-1}; $\triangle$, 2855 cm^{-1}; +, 1771 cm^{-1}; *; 1753 cm^{-1}; $\diamondsuit$, 1619 cm^{-1}; ∇, 1510 cm^{-1}; $\star$, 983 cm^{-1}. (Redrawn with permission from [53].)

changes dA_n

$$dA_n = \frac{[A(t) - A_1]}{[A_1 - A_2]} \tag{35}$$

where $A(t)$ is the peak absorbance at time t and A_1 and A_2 are the peak absorbances for the two stabilized states, respectively. Figure 38 provides information about the fractional reorientation of the segments of the molecule at a particular time. It is clear that the reorientation from one to the other state starts immediately on a reversal of the polarity of the applied field; and after a certain time, the molecule attains the position of the second stabilized state; however, the different segments reorient at distinct and different times. Particularly, the hydrogen-bonded C=O group moves with the mesogen at a faster rate than does the alkyl chain. The results are summarized in Tables VI and VII.

TABLE VI
Electric Field and Temperature Dependence of Reorientation Times (μs) of Different Segments for VOH8

Segment	Temperature 83°C			Temperature 75°C	
	$\pm 3V^a$	$\pm 5V$	± 7 V	± 5 V	± 7 V
$\nu_{C=O}$ free and alkyl chain	75	65	62	70	64
$\nu_{C=O}$ H bonded and mesogen	65	50	45	59	53

[a] Applied voltage.
Reprinted with permission from [53].

TABLE VII
Electric Field and Temperature Dependence of Reorientation Time (μs) of Different Segment for MHDBOB.

Segment	Temperature 109°C		Temperature 99°C	
	± 20 V τ^a	± 15 V τ	± 20 V τ	± 15 V τ
C=O group and alkyl chains	64	68	67	71
Core	33	45	48	56

[a] Applied voltage.
Reprinted with permission from [54].

Similarly for an antiferroelectric liquid crystal, MHDBOB, three types of experiments have been performed [54] to obtain information on the dynamics of the molecules. Measurements of the absorbance changes of the representative IR bands as a function of polarisation angle at 109°C under voltages 0, 15 and −15 V were made to determine the mutual arrangement of different molecular segments in their equilibrium state in the SmC_A^* phase (0 V), and field-induced, unwound SmC^* phase (15 and −15 V). Time-resolved IR measurements were made on an application of a rectangular wave electric field of ± 20 V, 5 kHz repetition rate as a function of a delay time ranging from 0 to 100 μs to monitor the dynamical behaviour during switching and to obtain the reorientation times of the different molecular segments.

The observed rotation of the absorption maxima of these bands between the positive and negative polarity of the electric field gives a tilt angle $\omega_0 = \Delta\omega/2 = 45°$ for the mesogen axis. Tilt angles for the C=O group and the alkyl chain modes are 27.5° and 25.5°, respectively. The polarization dependence of the C=O peak is typical in the sense that it is not symmetrical with respect to the peak position of the phenyl group. This provides clear evidence for a hindered rotation of the C=O group, and its biased orientation in a specific direction relative to the tilt plane normal. Moreover, the averaged axis of the alkyl chain does not coincide with the average mesogen axis in agreement with the zig-zag model [55] for FLCs.

Figure 39 shows the normalized absorbance changes dA_n, Eq. (35), vs. delay time during the electro-optical switching for the characteristic bands at 109°C and a ±20 V square electric field. It is clear that different segments of the molecule require different times for a completion of the reorientation. The core moves faster and takes ∼33 μs to complete the reorientation, whereas the alkyl chain and the C=O groups take ∼64 μs for the same process. Tables VI and VII summarize the results for different field strengths and temperature. Figure 40 shows a plot of the mean orientation angle $\omega_0(t)$ as a function of the delay time for the representative bands. The most

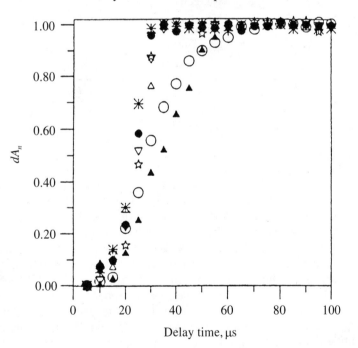

Figure 39. Dependence of the normalized intensity changes (dA_n) vs. delay time for the representative IR bands of the core, achiral, and chiral chain segments of MHDBOB in the SmC$_A^*$ phase on the application of a square wave electric field of ±20 V with a 5-kHz repetition rate. ▲, 2856 cm^{-1}; ◯, 1717 cm^{-1}; ▽, − 1607 cm^{-1}; *, 1499 cm^{-1}; △, 1274 cm^{-1}; ● 1243 cm^{-1}; ☆, 1174 cm^{-1}. The different reorientation times of the core and alkyl chains are evident from this figure. (Redrawn with permission from [54].)

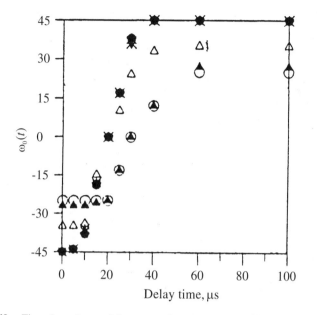

Figure 40. Time dependence of the mean orientation angle $\omega_0(t)$ as a function of delay time τ for the representative IR bands of the core, achiral, and chiral chain segments of MHDBOB in the SmC$_A^*$ phase on the application of a square wave electric field of ± 20 V and with a 5-kHz repetition rate. ▲, 2856 cm^{-1}; ◯, 1717 cm^{-1}; ▽, -1607 cm^{-1}; ✳, 1499 cm^{-1}; △, 1274 cm^{-1}; ● 1243 cm^{-1}; ☆, 1174 cm^{-1}; (Redrawn with permission from [54].)

important feature in these plots is that the core responds immediately, but very slowly, on switching the electric field, whereas the alkyl chain requires an induction period of ~ 15 µs before responding to the electric field. The in-duction period is suggested to be a characteristic property of liquid crystals.

C. Double-Modulation FTIR Spectroscopy of an FLC Sample under an Alternating Electric Field

Reorientation dynamics of the FLC molecules in the SmC* phase has been studied in the frequency domain using a double-modulation FTIR spectro-scopy [56], which takes advantage of the difference in the modulation frequency of the applied electric field and the IR beam. The principle of this method is to analyze the variation of the IR transmittance (T) induced in the sample by a sinusoidal electric field. The signal is sensitive only to the anisotropic properties of FLC produced by the electric field. This method combines the high-frequency modulation of the applied electric field with the beam intensity modulation produced by the interferometer at a lower frequency. Simultaneous acquisition on two channels of the high-frequency

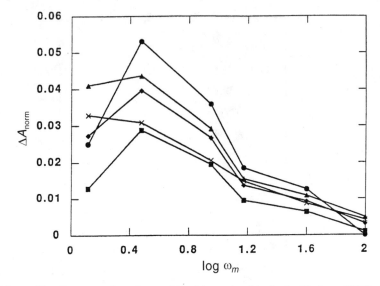

Figure 41. Frequency dependence of the ΔA_{norm} signal in the SmC* phase (62°C) of the $\overline{10}$S.CIsoleu liquid crystal for the five molecular vibrations on the application of a square wave electric field of ± 20 V at the polarizer angle of 40° ▲, 1762 cm^{-1}; ◆, 1672 cm^{-1}; ■, 1600 cm^{-1}; ×, 1163 cm^{-1}; ●, 900 cm^{-1}. (Redrawn with permission from [56].)

and low-frequency interferograms allows us to get a signal that depends only on the molecular group reorientation induced by the applied field.

The experiment was carried out on a $\overline{10}$ S CIIsoleu sample (structure given in Fig. 5). A fixed field of 2 V/µm is applied across the sample. Five characteristic dipoles were selected for analyzing the dynamics of FLC switching in the frequency range. The band centered at ~ 1600 cm^{-1} corresponds to the C–C stretching and the C–H bending in-plane vibrations are directed predominantly along the long molecular axis. The band centered at 900 cm^{-1} is the C–S stretching, and two C=O vibrations are centered at 1672 cm^{-1} (C=O between the phenyl rings) and 1762 cm^{-1} (ester group). Figure 41 shows the normalized differential signal ΔA_{norm} [normalized by the dichroic difference as in Eq. (35)] vs. logarithm of frequency. ΔA_{norm} for most of the bands (for instance those at 900, 1600, 1672, and 1762 cm^{-1}) exhibit a more or less pronounced maximum at ~ 3 kHz. In contrast, the normalized differential signal for some other bands, especially the band at 1163 cm^{-1}, is continuously decreasing with frequency in the range explored.

VIII. SEGMENTAL ORIENTATION AND THE MOBILITY OF FERROELECTRIC LC DIMERS AND POLYMERS

The synthesis of polymeric LCs that form chiral SmA and SmC* phases make it possible to combine the properties of ferroelectrics with those of polymers. At present a number of empirical facts about the influence of the chemical composition on collective dynamics of ferroelectric liquid crystals and ferroelectric liquid crystal polymers based on the electro-optical switching studies have been established. Because all orientation processes are controlled by the specific functional mobility, their study is a key to the understanding of the orientational phenomena induced by the electric field. With the aim of elucidating the reorientation mechanism, step-scan FTIR TRS has been applied to a study of the fast reorientation motion of the different molecular segments in a ferroelectric liquid crystal dimer (FLCD) [47] and ferroelectric liquid crystalline polymers (FLCP) [45,48–52] under the influence of an electric field.

A. FLCP

The chemical structure and the transition temperatures of FCLPs under investigation are presented in Figure 42 [46,47]. FLCP was sandwiched between two CaF$_2$ windows and a combination of shearing and alternating voltage was used to reduce the defects in the orienting film. For the first FLCP, several bands have been assigned, that characterize the mesogenic unit, spacer, tail, and polysiloxane backbone [Fig. 42(b); Table VIII) [46].

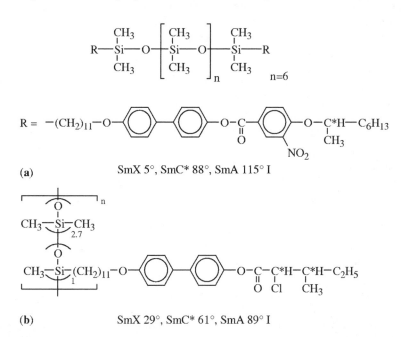

(a) SmX 5°, SmC* 88°, SmA 115° I

(b) SmX 29°, SmC* 61°, SmA 89° I

Figure 42. Chemical structure and the transitions temperatures for (a) FLCD and (b) FLCP. (Redrawn with permission from [46 and 47].)

TABLE VIII
Band Assignment of FLCPs Investigated

Wave number, cm^{-1}	Dichroic Ratio, R	Assignment
2963	1.05	$v(CH_3)$
2924	0.61	$v_{as}(CH_2)$
2855	0.63	$v_s(CH_2)$
1765	0.18	$v(C=O)$
1608	13.12	$v(C–C)_{ar}$
1499	21.61	$v(C–C)_{ar}$
1463	0.59	$v(CH_2)$
1261	1.29	$v_{as}(C–O–C) + \delta(Si–CH_3)$
1207	23.82	$v(C–O) + v(C–O–C)$
1168	12.22	$v(C–O) + v(C–O–C)$
1096	1.31	$v(C–O) + v(Si–O–Si)$

Reprinted with permission from [48].

The 1608, 1499, 1207, and 1168 cm^{-1} bands are characteristic of the mesogenic unit and have high dichroism. It is concluded that the transition dipole moments for these bands are nearly parallel to the mesogen, and the various mesogens in the polymer are almost parallel to each other. The 2924 and 2855 cm^{-1} bands originate from the asymmetric and symmetric CH$_2$ stretching vibrations, respectively. The majority of CH$_2$ groups are in the spacer (and only one is located in the tail), hence this group characterizes mainly the poly(methylene) spacer. The transition moment of these vibrations are nearly perpendicular to the poly(methylene) chain in all trans conformations. Owing to a conformational disorder of the spacer, the dichroism ratio for this band is lower than that for the bands of the mesogen unit. For characterization of the backbone, the 1261 cm^{-1} band for δ (Si–CH$_3$) is chosen. The transition dipole of this band is nearly perpendicular to the polysiloxane chain. Two groups are mainly responsible for the spontaneous polarization in the FLCP samples: the C–Cl group absorbance in the 800 to 600 cm^{-1} region and the carbonyl group that absorbs at 1765 cm^{-1}. Only the latter band is accessible in the transmission region of the CaF$_2$ windows.

The polar plot of the absorbance vs. polarization angle ω in the SmA* phase is presented in Figure 43. The 90° position of the polarizer corresponds to the shearing direction. As expected, the overall absorbance profile is symmetrical in the SmA* phase: The absorption maxima for 1499 and 1261 cm^{-1} bands are observed at 0°, and those for the 2924 and 1765 cm^{-1} bands are observed at 90°. Taking the angles between the transition moments for these bands and the corresponding molecular segments in consideration, the conclusion is as follows: The polymethylene spacers are

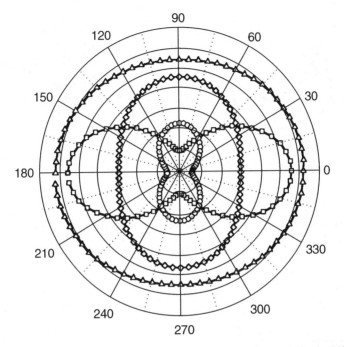

Figure 43. Polar plot of absorbance vs. the polarization rotation angle for the 1499 cm^{-1} ($\square$), 1765 cm^{-1} ($\bigcirc$), 1261 cm^{-1} ($\triangle$), and 2924 cm^{-1} ($\lozenge$) bands for the FLCP in the SmA* phase at 70°C (Redrawn with permission from [48].)

preferentially oriented in the direction of the long axis of the mesogenic groups, whereas the main chain segments are oriented perpendicular to it. The shearing of the sample in the SmA* phase orients the mesogenic groups and the spacers perpendicular to the shear direction and the backbone parallel to the shear direction.

In the SmC* phase, the optical switching between the two stable orientations of the optical axis were observed after the application of $+10$ and -10 V. For IR absorbance measurements, the polarizer position $\omega = 0$ was set to coincide with the position of the optical axis for $+10$ V. The polar plots of some absorption bands for sample at 35°C under $+10$ V are shown in the Figure 44. It is clear that the overall absorbance profile has lost symmetry (with respect to $\omega = 0$). There is direct evidence for the cylindrical symmetry of the distribution function to be broken in the ferroelectric phase. The graphs for positive and negative voltages are symmetric relative to the bisector of the angle between the absorption maxima for positive and negative voltages (Fig. 44). The temperature dependence of the apparent tilt angles is shown in Figure 45. The largest tilt

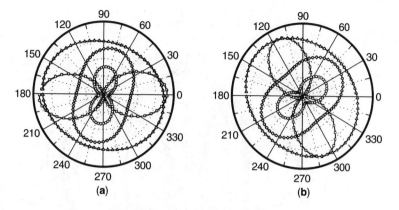

Figure 44. Polar plot of absorbance vs. the polarization rotation angle for the 1499 cm^{-1} ($\square$), 1765 cm^{-1} ($\bigcirc$), 1261 cm^{-1} ($\triangle$), and 2924 cm^{-1} ($\diamond$) bands for the FLCP in the SmC* phase at 35°C: (a), +10 V, (b), −10 V, (Redrawn with permission from [48].)

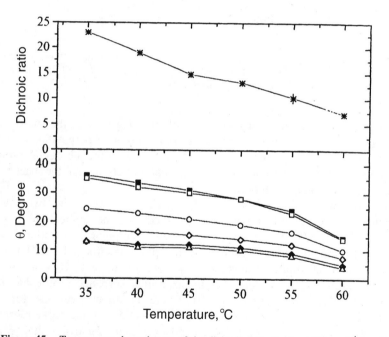

Figure 45. Temperature dependences of the dichroic ratio for the 1499 cm^{-1} band ($*$) and the apparent tilt angle calculated from the absorbances of the 1499 cm^{-1} ($\square$), 1608 cm^{-1} ($\blacktriangle$), 1765 cm^{-1} ($\bigcirc$), 2924 cm^{-1} ($\diamond$), 2855 cm^{-1} ($\blacklozenge$), and 1261 cm^{-1} ($\triangle$) bands. (Redrawn with permission from [48].)

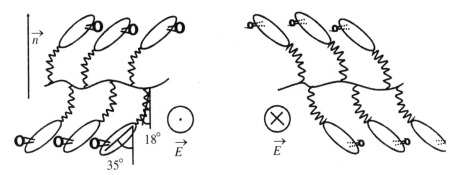

Figure 46. Model of the mutual arrangments of molecular segments at 35°C. (Redrawn with permission from [48].)

angle corresponds to the bands for the benzoate ring of the mesogenic unit, whereas the smallest tilt angle is related to the backbone. The IR data on tilt angles for the mesogenic unit (1608 and 1499 cm^{-1}) are in agreement with results from the electro-optical investigations of the FLCP [57]. A small shift of the carbonyl maximum position can be explained on the basis of the biased rotation around the long axis as was already attempted for the monomeric FLC. The proposed model, however, does not explain the differences in the ω_{max} and the tilt angles for bands corresponding to the spacer and those corresponding to the benzoate ring. The polar angle of the transition dipole for the 2924 and 2855 cm^{-1} bands is considered around 90°, hence no difference is expected for ω_{max} for these bands and ω_{max} of the bands for the mesogen. The detected difference directly corresponds to an inclination of the spacer with respect to axis of the mesogen, in agreement with the zigzag model [55]. The interpretation of the absorbance of the backbone band (1261 cm^{-1}) leads to conclusion that the backbone on average is oriented perpendicular to the smectic layer normal (Fig. 46).

B. Segmental Mobility of FLCP in the SmC* Phase

The spacer effectively decouples the movement of the mesogenic unit from the movement of the polymer backbone; hence the behavior of the mesogenic group, spacer, and main chain is expected to be different during a switching process. A time-resolved experiment was carried out at 50°C and a field strength of 30 V/μm; the switching process between the two states was observed for the polarizer orientation $\omega = 0$. The evolution of the absorbance bands corresponding to the mesogenic unit, spacer, and backbone are given in Figure 47. The orientation process starts immediately after a reversal of the polarity of the field and takes approximately 300 μs to reach the second stabilized state. For a qualitative comparison of the

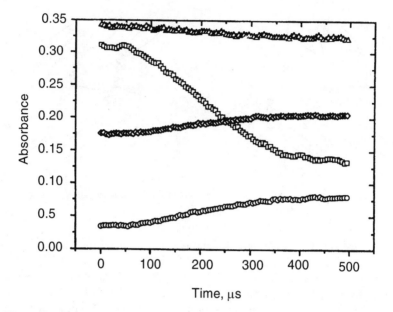

Figure 47. Time dependence of absorbance changes for the 1499 cm^{-1} (□), 1765 cm^{-1} (○), 1261 cm^{-1} (Δ), and 2924 cm^{-1} (◇) bands of FLCP. (Redrawn with permission from [48].)

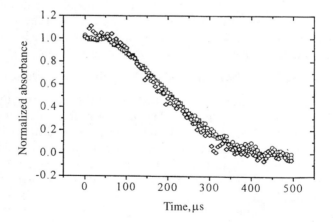

Figure 48. Normalized absorbance vs. time for the 1499 cm^{-1} (□), 1765 cm^{-1} (○), 1261 cm^{-1} (Δ), and 2924 cm^{-1} (◇) bands of FLCP. Redrawn with permission from [48].)

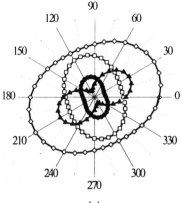

(a)

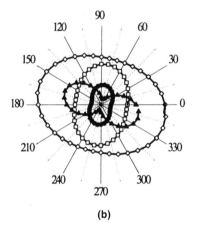

(b)

Figure 49. Time-resolved polar plot of the absorbance vs. polarization angle (a) before reversal of the field polarity and (b) after completion of the reorientation. □, 2924 cm^{-1}; ●, 1756 cm^{-1}; ▲, 1604 cm^{-1}; ○, 1261 cm^{-1}. (Redrawn with permission from [49].)

orientation rates for different molecular segments, the normalized intensity A_N was calculated using $A_N = (A(t) - A_1)/(A_1 - A_2)$, where $A(t)$ is the peak absorbance at time t; A_1 is the value before the reversal of the polarity, and A_2 is the corresponding absorbance after completion of the reorientation. The results show that the normalized time profiles for the selected bands are similar to each other (Fig. 48).

For another FLCP [49], the following bands were chosen to characterize the motion of the molecular segments: 2924 cm^{-1} (CH$_2$ polymethylene chain), 1737 cm^{-1}(C=O), 1500 cm^{-1} (C–C$_{ar}$ benzoic rings of the mesogen), and 1260 cm^{-1} (Si–CH$_3$ polysiloxane chain). The polar plots

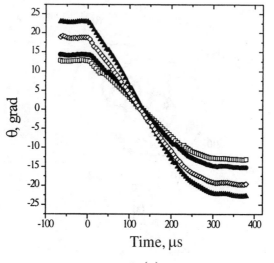

(a)

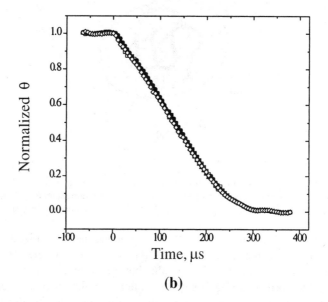

(b)

Figure 50. (a) Time dependence of apparent θ angles for different bands. (b) Time dependence of the normalized θ. □, 2924 cm^{-1}; ●, 1756 cm^{-1}; ▲, 1604 cm^{-1}; ◇, 1261 cm^{-1}. (Redrawn with permission from [49].)

of absorbance of selected bands vs. ω are presented in Figure 49 for the two polarities of the field. The observations are similar to those for the first polymer, for which the results have been reviewed. The overall structure of the absorbance profile is a consequence of the noncylindrical symmetry of the orientational distribution in SmC* phase, whereas differences between the 2924 cm^{-1} (spacer or flexible tail) and 1604 cm^{-1} bands correspond to a different tilt of the segments to the smectic layer normal. To follow the orientational dynamics of different molecular segments, the time-resolved experiment was carried out for different angles ω of the polarizer. The apparent tilt angles for different molecular moieties have been calculated from the absorbance profiles $A_\omega(t)$. The obtained results as a dependence of $\Delta\omega \equiv \theta$ and the normalized θ as a function of time are given in Figure 50. The apparent angles show significant differences during reorientation (except at the inversion point when all tilt angles change their sign). Nevertheless the normalized dependencies are perfectly superimposed. On the basis of these results, Shilov et al. [49] concluded that during the reorientational process between the two surface-stabilized states the molecular segments move synchronously on cones.

Acknowledgments

The authors thank Professor Noel A. Clark for fruitful discussions on his results on the polarized FTIR work. We also thank Professor F. Kremer for discussions on his time-resolved experiments. We thank E. Hild, R. Wrzalik, G. Kruk, and R. Zentel for their collaboration on the IR studies and S. Tstvekov for his assistance with the diagrams. A.K. thanks the Committee for Scientific Research for grant 2P03B 09817 and for instrumentation support. J.K.V. acknowledges the European Commission for its approval of research grants under the following programs: science, HCM, PECO, INTAS, and network, and Enterprise Ireland for their basic and international collaboration research awards.

References

1. R. B. Meyer, L. Liebert, L. Strzelecki, and P. Keller, *J. Phys. (Paris)* **36**, L69 (1975).

2. R. B. Mayer, *Mol. Cryst. Liq. Cryst.* **40**, 33 (1977).

3. W. H. de Jeu, *Physical Properties of Liquid Crystalline Materials*, Gordon & Breach, New York, 1980.

4. E. Hild, A. Kocot, J. K. Vij, and R. Zentel, *Liq. Cryst.* **16**, 783 (1994).

5. R. Kiefer and G. Baur, *Mol. Cryst. Liq. Cryst.* **174**, 101 (1989).

6. G. Turrell, *Infrared and Raman Spectra of Crystals* Academic Press, New York, 1972.

7. L. M. Blinov, *Electro-Optical and Magneto-Optical Properties of Liquid Crystals*, J. Wiley and Sons, New York.

8. V. D. Neff, L. W. Gulrich, and G. H. Brown, in *Proc. Int. Conf. Liq. Cryst. 1967*, G. H. Brown, G. J. Dienes, and M. M. Labes, Eds., Gordon & Breach, New York.

9. J. K. Vij, A. Kocot, G. Kruk, R. Wrzalik, and R. Zentel, *Mol. Cryst. Liq. Cryst.* **257**, 337 (1993).

10. M. Zgonik, M. Rey-Lafon, C. Destrade, and H. T. Nguyen, *J. Phys. Paris* **51**, 2015 (1990).

11. A. Kocot, G. Kruk, R. Wrzalik, and J. K. Vij, *Liq. Cryst.* **12**, 1005 (1992).

12. T. Perova, A. Kocot, J. K. Vij, and R. Zentel, *SPIE* **2731**, 107 (1996).

13. W. G. Jang, C. S. Park, J. E. Maclennan, K. H. Kim, and N. A. Clark, *Ferroelectrics* **180**, 213 (1996).

14. A. Kocot, R. Wrzalik, and J. K. Vij, *Liq. Cryst.* **21**, 147 (1996).

15. T. Perova, A. Kocot, and J. K. Vij, *Ferroelectrics* **214**, 83 (1998).

16. A Kocot, T. Perova, J. K. Vij, and R. Wrzalik, *Ferroelectrics* **214**, 1 (1998).

17. K. H. Kim, K. Ishikawa, H. Takezoe, and A. Fukuda, *Phys. Rev. E* **51**, 2166 (1995).

18. K. Miyachi, J. Matsushima, Y. Takanishi, K. Ishikawa, H. Takezoe, and A. Fukuda, *Phys. Rev. E* **52**, R2153 (1995).

19. J. P. Hummel and P. J. Flory, *Macromolecules* **13**, 479 (1980).

20. P. Coulter and A. Windle, *Macromolecules* **22**, 1129 (1989).

21. A. Kocot, R. Wrzalik, B. Orgasińska, T. Perova, J. K. Vij, and H. T. Nguyen, *Phys. Rev. E* **59**, 551 (1999).

22. J. W. O'Sullivan, J. K. Vij, and H. T. Nguyen, *Liq. Cryst.* **23**, 77 (1997).

23. K. Tishiro, J. Hao, M. Kobayashi, and T. Inoue, *J. Am. Chem. Soc.* **112**, 8273 (1990).

24. R. Centore and A. Tuzi, *Acta Cryst.* **C45**, 107 (1998).

25. J. R. Lalanne, J. Buchert, S. Kielich, *Modern Nonlinear Optics, Advances in Chemical Physics,* **LXXXV,** 159 (1993), (I. Prigogine and S. A. Rice (Eds.), J. Wiley 1993.

26. H. Toriumi, M. Yoshida, M. Mikami, M. Takeuchi, and A. Mochizuki, *J. Phys. Chem.* **100**, 1507 (1996).

27. T. Nakano, T. Yokoyama, and H. Toriumi, *Appl. Spectrosc.* **47**, 1354 (1993).

28. I. Noda, *J. Am. Chem. Soc.*, **111**, 8116 (1989) and *Appl. Spectrosc.* **44**, 550 (1990).

29. R. E. Murphy, F. H. Cook, and H.Sakai, *J. Opt. Soc. Am.* **65**, 600 (1975).

30. H. Sakai and R. E. Murphy, *Appl. Opt.* **17**, 1342 (1978).

31. F. Hide, N. A. Clark, K. Nito, A. Yasuda, and D. M. Walba, *Phys. Rev. Lett.* **75**, 2344 (1995).

32. H. Toriumi, H. Sugisawa, and H. Watanabe, *Jpn. J. Appl. Phys.* Part 2, **27**, L935 (1988).

33. V. G. Gregoriu, J. L. Chao, H. Toriumi, and R. A. Palmer, *Chem. Phys. Lett.* **179**, 491 (1991).

34. H. Sugisawa, H. Toriumi, and H. Watanabe, *Mol. Cryst. Liq. Cryst.* **214**, 11 (1992).

35. T. I. Urano and H. Hamaguchi, *Chem. Phys. Lett.* **195**, 287 (1992).

36. T. I. Urano and H. Hamaguchi, *Appl. Spectrosc.* **47**, 2108 (1993).

37. K. Masutani, H. Sugisawa, A. Yokota, Y. Furukawa, and M. Tasumi, *Appl. Spectrosc.* **46**, 560 (1992).

38. K. Masutani, A. Yokota, Y. Furukawa, M. Tasumi, and A. Yoshizawa, *Appl. Spectrosc.* **47**, 1370 (1993).

39. K. Kawasaki, H. Kidera, T. Sekiya, and S. Hachiya, *Ferroelectrics* **148**, 233 (1993).

40. M. Czarnecki, N. Katayama, Y. Ozaki, M. Satoh, K. Yoshio, T. Watanabe, and T. Yanagi, *Appl. Spectrosc.* **47**, 382 (1993).

41. N. Katayama, M. Czarnecki, Y. Ozaki, K. Murashiro, M. Kikuchi, S. Saito, and D. Demus, *Ferroelectrics* **147**, 441 (1993).

42. M. Czarnecki, N. Katayama, M. Satoh, T. Watanabe and Y. Ozaki, *J. Phys. Chem.* **99**, 14101 (1995).

43. N. Katayama, T. Sato, Y. Ozaki, K. Murashiro, M. Kikuchi, S. Saito, D. Demus, T. Yuzawa, and H. Hamaguchi, *Appl. Spectrosc.* **49**, 977 (1995).

44. S. V. Shilov, S. Okretic, and H. W. Siesler, *Vibr. Spectrosc.* **9**, 57 (1995).

45. S. V. Shilov, S. Okretic, H. W. Siesler, R. Zentel, and T. Öge, *Macromol. Chem. Rapid Commun.* **16**, 125 (1995).

46. S. V. Shilov, H. Skupin, F. Kremer, T. Wittig, and R. Zentel, *Macromol. Symp.* **119**, 261 (1997).

47. S. V. Shilov, H. Skupin, F. Kremer, T. Wittig, and R. Zentel, *Phys. Rev. Lett.* **79**, 1686 (1997).

48. S. V. Shilov, H. Skupin, F. Kremer, E. Gebhard and R. Zentel, *Liq. Cryst.* **22**, 203 (1997).

49. S. V. Shilov, H. Skupin, F. Kremer, K. Skarp, P. Stein, and H. Finkelmann, *SPIE* **3318**, 62 (1998).

50. A. Merenga, S. V. Shilov, F. Kremer, G. Mao, C. K. Ober, and M. Brehmer, *Macromolecules* **31**, 9008 (1998)

51. S. Shilov, E. Gebhard, H. Skupin, R. Zentel, and F. Kremer, *Macromolecules* **32**, 1570 (1999).

52. H. Skupin, F. Kremer, S. V. Shilov, P. Stein, and H. Finkelmann, *Macromolecules* **32**, 3746 (1999).

53. A. L. Verma, B. Zhao, S. M. Jiang, J. C. Sheng, and Y. Ozaki, *Phys. Rev. E* **56**, 3053 (1997).

54. A. L. Verma, B. Zhao, H. Terauchi, and Y. Ozaki, *Phys. Rev. E* **59**, 1868 (1999).

55. R. Bartolino, J. Doucet, and G. Durand, *Ann. Phys. Paris*, **3**, 389 (1978).

56. B. Desbat, M. Rey-Lafon, T. Buffetau, C. Destrade, J. J. Bonvent, and H. T. Nguyen, *Ferroelectrics* **129**, 43 (1992).

57. A. Kocot, R. Wrzalik, J. K. Vij, M. Brehmer, and R. Zentel, *Phys. Rev. B* **50**, 16346 (1994).

THE STRUCTURE AND PROPERTIES OF ANTIFERROELECTRIC LIQUID CRYSTALS

YU. P. PANARIN AND J. K. VIJ

*Department of Electronic and Electrical Engineering,
Trinity College, University of Dublin, Ireland*

CONTENTS

I. INTRODUCTION

The discovery of the antiferroelectricity in chiral tilted smectic liquid crystals (SmC*) has led to extensive theoretical and experimental investigations [1–4]. Antiferroelectric liquid crystals (AFLCs) exhibit interesting and complex physical phenomena and are the prime candidates for applications in the display technology [5]. Since this discovery, a number of ferrielectric phases have been observed in the temperature range between the ferroelectric and antiferroelectric phases [6]. As a result of extensive investigations made in the 1990s, the understanding of the phenomena has advanced considerably. New physical properties of AFLCs have been found.

Advances in Liquid Crystals: A Special Volume of Advances in Chemical Physics, Volume 113,
edited by Jagdish K. Vij. Series Editors I. Prigogine and Stuart A. Rice.
ISBN 0-471-18083-1.   © 2000 John Wiley & Sons, Inc.

Hundreds of new antiferroelectric compounds have been synthesized, several different antiferroelectric and ferrielectric subphases have also been discovered in addition to the basic SmC^* and SmC_A^* phases. The first video-rate passive-matrix AFLC display has been designed [5]. Several theoretical models for explaining the structure and physical properties of the AFLC have been developed. These developments are summarized in published reviews [7,8]. Nevertheless, since 1995 new important results have emerged. The interesting results include a discovery of thresholdless antiferroelectricity, the observations of some new field-induced ferrielectric subphases with weak order between the neighboring smectic layers, the understanding of the relaxation processes in the antiferroelectric phase by dielectric and electrooptic spectroscopy, and the improvement of theoretical models to explain most of the experimental results published so far. This review summarizes these new results. We will briefly discuss previous reviews [7,8] and then concentrate on new discoveries.

II. DISCOVERY OF ANTIFERROELECTRIC LIQUID CRYSTALS

The history of the discovery of antiferroelectric liquid crystals (AFLC) is interesting and rather intriguing; therefore, it is worth discussing the early research in this area. The best known work on AFLCs was presented at the first International Symposium on FLC in Arcachon (1987) by Fukuda et al. and published a year later [2]. They presented the electro-optic properties of the AFLC compound MHPOBC:

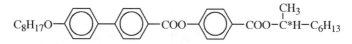

The homogeneously aligned cells show the so-called tristable switching; a double-hysteresis loop and the DC threshold of the apparent tilt angle are shown in Figure 1 [2]. In addition to the two switched states with structures similar to the usual SmC^*, another state that is stable in the absence of applied voltage is seen, which possesses neither apparent tilt angle nor net spontaneous polarization. No conclusion about antiferroelectricity in MHPOBC was made until the following year when Chandani et al. [3] concluded that both these phenomena arose from the existence of an antiferroelectric order in MHPOBC.

The first conoscopic observation of MHPOBC showed the existence of new phases [6]. The application of a DC electric field to the homeotropically aligned AFLC cells changes the helical structure (and, therefore, the conoscopic picture) of different phases in different ways. The existence of the several subphases in addition to SmC^* and SmC_A^* was reported. The

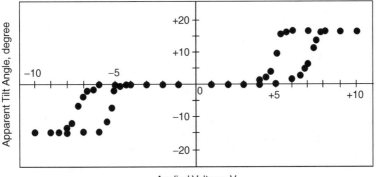

Figure 1. Hysteresis observed in an apparent tilt angle as a function of the applied voltage. (Redrawn with permission from [2].)

phases are termed SmC_α^*, SmC_β^*, and SmC_γ^* and are obtained with a decrease in temperature; SmC_β^* is the usual SmC^* phase, and SmC_γ^* is a ferrielectric phase. The appearance of these phases strongly depends on the optical purity of the compound (MHPOBC) and changes from $SmA^* - SmC_\alpha^* - SmC_\beta^* - SmC_\gamma^* - SmC_A^*$ in the pure enantiomer to $SmA^* - SmC_\beta^*$ (or SmC^*) $- SmC_A^*$ in the racemic mixture with unequal ratios of the two enantiomers. Several interesting findings on antiferroelectric liquid crystals were made in Japan. Antiferroelectricity in liquid crystals had been discovered earlier, but the attention to the discovery was not focused until Chandani et al. demonstrated tristable switching.

It is now well known that the SmO^* phase, investigated by Levelut et al. [9] in 1983, is actually the antiferroelectric liquid crystalline phase. X-ray investigations of the chiral and racemic compound M7TAC showed the extremely low smectic layer spacing (31 Å) comparised to the molecular length (48 Å). The similar observations of binary mixtures had been reported in an earlier paper [10]. The optical investigation of M7TAC, given by Galerne and Liebert [11], confirmed the herringbone structure (or chevron structure) for the SmO^* phase. Electro-optic investigations of the compound showed the existence of the third stable state (in the absence of applied electric field) between the two switched states [12]. Such a property is certainly antiferroelectric, but the authors did not draw any conclusions about the antiferroelectricity in the investigated sample.

The first discovery of AFLC was reported at the fourth Conference of Socialist Countries on Liquid Crystals in Tbilisi in 1981 [1]. The authors investigated the pyroelectric and electrooptic properties of the binary mixtures of SmC^* compounds doped with the nonmesogenic chiral dopants. They showed the existence of the DC threshold in the measured value of the pyroelectric signal. This phenomenon was explained by the existence of antiferroelectric order in the smectic structure and termed the antiferro-electric liquid crystalline phase. The layer structure of this phase that they described is now well accepted to be an antiferroelectric phase (Fig. 2).

It should be remarked that the term *antiferroelectric liquid crystalline phase* appeared in literature in 1981 [12] but was used for a bilayered SmA phase (SmA$_2$), which is not currently accepted as antiferroelectric.

III. THE STRUCTURE OF ANTIFERROELECTRIC LIQUID CRYSTALS AND THEORETICAL MODELS

The appearance of the antiferroelectric and ferrielectric subphases in the tilted chiral smectic liquid crystals can be understood to be a result of the competition between the antiferroelectric and ferroelectric interactions in adjacent smectic layers, which stabilize the SmC$_A^*$ and the SmC* phases. This competition produces various ferrielectric subphases with different sequences of antiferroelectric (A) and ferroelectric (F) orderings among the smectic layers. Different theoretical concepts have been given for explaining a variety of the ferrielectric subphases; these postulates are based mostly on the expanded Landau model [13–16] or on statistical models: the Ising model [17–19] and the axial next-nearest neighbor (ANNNI) model [20–22]. These models will be discussed latter in greater detail. We take the Ising model as the basis for explaining a variety of different ferrielectric phases in tilted chiral liquid crystals. Note that different theoretical models predict different structures of ferrielectric phases and experimental results support some of these models and contradict others. Therefore, we take the Ising model as an example for presenting the structure of the different anti-ferroelectric and ferrielectric phases in the first instance.

The one-dimensional Ising model has been used to explain some spatially modulated systems [17]. Latter it was adopted to AFLC, and nonrealistic spins were introduced into liquid crystals [18]. According to this model, the energy of the interactions is given by

$$E = -\sum HS_i + \frac{1}{2}\sum j(i-j)\frac{S_i+1}{2}\frac{S_j+1}{2} \qquad (1)$$

where spin $S_i = -1$ is for the antiferroelectric ordering between the two neighboring smectic layers, $S_i = 1$ is for ferroelectric ordering, $J(i-j)$ is the long-range repulsive interaction between the ferroelectric orderings (this energy term implies an unlimited number of interacting smectic layers), and $H = H_A - H_F$ is the chemical potential (where H_A and H_F, respectively, represent the stability of antiferroelectric and ferroelectric orderings between the two neighboring smectic layers). The first term in Eq. (1) stabilizes the ferroelectric ordering for positive values of H (i.e., $H > 0$) and

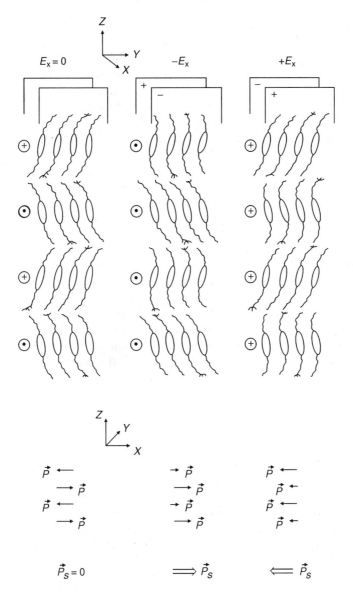

Figure 2. A model structure for an antiferroelectric compound and its field behavior. Z is the normal to the smectic layers. (Redrawn with permission from [1].)

the antiferroelectric ordering for $H < 0$; the second term always stabilizes antiferroelectric ordering with spin $S = -1$. Therefore, for negative H, the preferable structure is antiferroelectric SmC_A^* and for sufficiently large positive values of H, the structure is ferroelectric SmC^*. For the smaller positive values of H, a minimization of Eq. (1) gives an unlimited number of stable ferrielectric structures, which consist of the regularly arranged ferroelectric and antiferroelectric orderings. Such phases (sometimes termed subphases) can be specified by the irreducible rational parameter $q_T = F/(F + A)$, which means the fraction of ferroelectric ordering in the periodic ferrielectric structure $(F + A)$. A complete set of the ferrielectric subphases fills up the entire temperature range between SmC_A^* and SmC^* (Fig. 3), and these subphases are exhibited without any first-order transitions. Such a sequence is termed the Devil's staircase [17]. It can be shown that the ferrielectric subphases with $q_T = m/n$, where either m or n is even, possesses no net polarization and hence can be considered antiferroelectric ferri-electric subphases. These are ferrielectric because they consist of both ferroelectric or antiferroelectric orderings between neighboring smectic layers and antiferroelectric owing to an absence of the net polarization. Otherwise, the net polarization is equal to P_S/n (both n and m are odd). P_S is the spontaneous polarization in the ferroelectric SmC^* phase. The stability of the phases with $q_T = m/n$ quickly decreases with increasing n (see Fig. 3). Figure 4(a) shows some of the possible ferrielectric phases with the parameter q_T. If at a temperature T_A, the coefficient $H = H_A - H_F$ becomes positive and monotonically increases with increasing temperature, the parameter q_T of the phase sequence increases from 0 to 1. The smoother the temperature dependence of the parameter H, the larger the number of ferrielectric subphases observed between SmC_A^* and SmC^*. Practically,

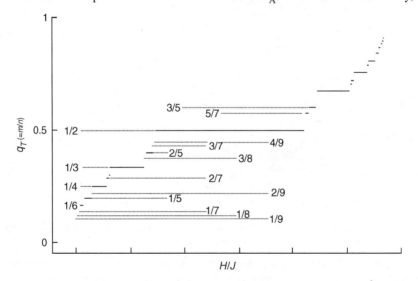

Figure 3. Stability range $q(T)$ based on the assumption of $j(i - j) = J/|i - j|^2$, where J is a constant. Note that $q(T) = m/n$ is a width independent of m and that $q(T)$ is insensitive to $j(i - j)$, provided that it is convex and repulsive. (Redrawn with permission from [17].)

however, the full sequence of the antiferroelectric liquid crystals is SmC_A^* ($q_T = 0$), spr3, SmC_γ^* ($q_T = 1/3$), spr2, AF ($q_T = 1/2$), spr1, SmC^* ($q_T = 1$), SmC_α^*, SmA* [23,24]. The phases with the simplest structures such as SmC_A^*, SmC_γ^*, AF, and SmC^* are thermodynamically stable, whereas the intermediate subphase regions (spr1–spr3) seem to be more complicated and were observed and investigated rather recently.

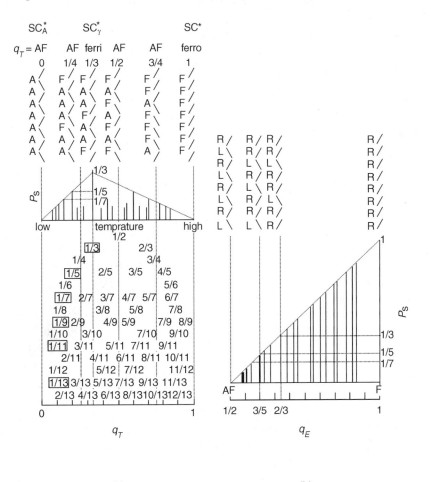

Figure 4. Two types of staircases. (a) The temperature-induced staircase produced by assigning the spin variable S_i to the antiferroelectric or ferroelectric ordering, A or F: P_s (or θ) plotted as a function of q_T, where q_T is a fraction of F, and is given with the Farey sequence and some structures specified by q_T. (b) The field-induced staircase produced by assigning S_i to the tilting sense R or L: P_s (or θ) plotted as a function of q_E, where q_E is a fraction of R. Some structures specified by simple q_E are shown. (Redrawn with permission from [8].)

In addition to the structural changes with temperature (the so-called temperature Devil's staircase), there exists the electrical field–induced Devil's staircase, which leads to the structural changes with applied electric field. Molecular structures in this model are described by the parameter $q_E = L/(R+L)$, where L and R are the number of left and right tilted smectic layers in the period of the structure. The application of the electric field can change the molecular structure, thus increasing q_E, as shown in Figure 4(b) [19].

The Ising model satisfactorily describes the structural changes with temperature and electric field. Nevertheless, there are definite concerns as to the validity of using the long-range repulsive interactions between the nonrealistic spins for stabilizing the ferrielectric phases. In addition, the Ising model ignores the disordering effect of the thermal fluctuations, the symmetry restrictions, and the effect of the external electric field. These important factors were taken in consideration in the ANNNI model [20]. In the ANNNI model with the third-neighbor interactions adopted for AFLC, the Hamiltonian is given by [21,22]

$$H = -J_1 \sum_{(i,j)} \sigma_i \sigma_j - J_2 \sum_{(i,j')}{}^A \sigma_i \sigma_{j'} - J_3 \sum_{(i,j'')}{}^A \sigma_i \sigma_{j''} \qquad (2)$$

where the Ising spin σ_i takes values ± 1, the first summation Σ is taken over all the nearest neighboring pairs (i, j), and other summations Σ^A are taken only over the second and the third neighboring pairs, (i, j') and (i, j''), respectively, of the axial direction parallel to the layer normal. J_1, J_2 and J_3 are the parameters of the energy terms, and J_2 is always negative.

The minimization of the Hamiltonian gives four ground states with wave number $q = 0, 1/4, 1/3, 1/2$, which corresponds to SmC*, AF, SmC$_\gamma^*$, and SmC$_A^*$ respectively. The wave number q is this model, defined as the reciprocal of the period of the basic repeatable unit, is related to the q_T parameter (in the Ising model) by $q = (1 - q_T)/2$. These phases correspond to the four most stable states in the Ising model.

The effect of the thermal fluctuations on the ferrielectric structure was taking into account by applying the symmetry breaking field corresponding to the order of ferrielectric phases through the thermodynamic potential written [22, 25]

$$\Phi_X(\sigma_X) = (-J + E_X)\sigma_X^2 + \frac{kT}{2}\{(1 + \sigma_X)\ln(1 + \sigma_X) + (1 - \sigma_X)\ln(1 - \sigma_X)\}$$

$$= \left(\frac{kT}{2} - J + E_X\right)\sigma_X^2 + \frac{kT}{12}\sigma_X^4 + O(\sigma_X^6) \qquad (3)$$

where the index X denotes the different phases: SmC*, AF, SmC$_\gamma^*$, and SmC$_A^*$, E_X for these phases are $E_F = -J_1 - J_2 - J_3$; $E_{AF} = J_2$;

$E_\gamma = J_1/3 - J_2/3 - J_3$; and $E_A = J_1 - J_2 + J_3$. σ_X is the average σ_i in the phase X, J is a positive constant and accounts for the interactions among the intralayer spins, k is the Boltzmann constant, and T is the temperature. In other words, for sufficiently low temperatures, the model shows the existence of four ground phases: SmC*, AF, SmC$_\gamma^*$ and SmC$_A^*$. The parameter J_1 is considered to be temperature dependent: $J_1(T) = J_0 + TS$, where J_0 accounts for the dipole–dipole interactions. Figure 5(a) shows the possible successive phase transitions with temperature. It was shown that at higher temperatures an additional ferrielectric subphase does exist in addition to the four basic phases [21]. The q parameter describing these

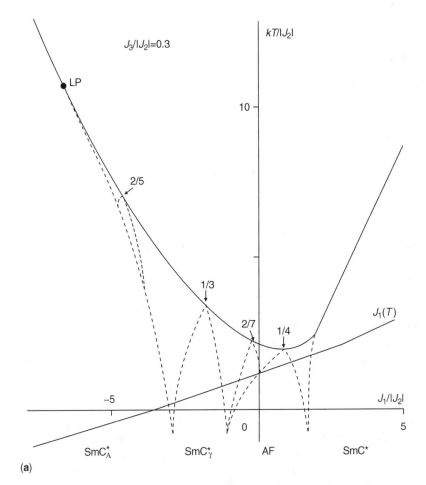

(a)

Figure 5. (a) The phase diagram and locus of $J_1(T)$. (Redrawn with permission from [22].) (b) Phase diagram of E vs. J_1 plane, where $T = 7.0$. (Redrawn with the permission from [26].)

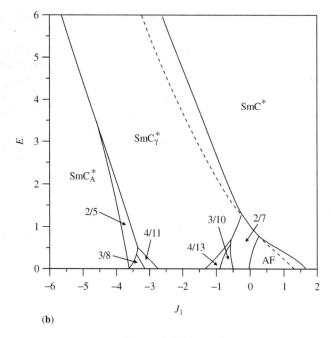

Figure 5. (*Continued*)

subphases decreases from $\frac{1}{2}$ to 0 (which corresponds to an increase of q_T from 0 to 1 in the Ising model) when temperature increases, forming a phase sequence that is similar to the temperature Devil's staircase. This model predicts the existence of disordered antiferroelectric phases for the case of $kT \gg J_i$.

Later Yamashita and Tanaka [26] developed the ANNNI model by introducing the interactions of the electric field with the local spontaneous polarization to the Hamiltonian (Φ_x) and showed the mechanism of the successive phase transitions under the electric field (electric field–induced Devil's staircase) [Fig. 5(b)].

In summary, these models give similar structures of ferrielectric phases, in which the direction of the local spontaneous polarization in the neighboring smectic layers must be almost parallel or antiparallel to each other.

The theoretical models based on the expanded Landau model [13–16] lead to different structures for the various ferrielectric phases. These models consist of a Landau expansion in terms of the ferroelectric and antiferroelectric order parameters: $\xi_f = (\xi_1 + \xi_2)/2$ and $\xi_a = (\xi_1 - \xi_2)/2$, where $\xi_i = (n_{ix}n_{iz}, n_{iy}n_{iz})$ is a two-component tilt vector (*C* director) in even and odd smectic layers and *n* is the molecular director. The Landau free energy expansion is described as [13,14]:

$$g(z) = \frac{1}{2}\alpha_a\xi_a^2 + \frac{1}{2}\alpha_f\xi_f^2 + \frac{1}{4}\beta_a\xi_a^4 + \frac{1}{4}\beta_f\xi_f^4 + \frac{1}{2}\gamma_1\xi_a^2\xi_f^2 + \frac{1}{2}\gamma_2(\xi_a\xi_f)^2$$
$$+ \delta_a\left(\xi_{ax}\frac{\partial\xi_{ay}}{\partial z} - \xi_{ay}\frac{\partial\xi_{ax}}{\partial z}\right) + \delta_f\left(\xi_{fx}\frac{\partial\xi_{fy}}{\partial z} - \xi_{fy}\frac{\partial\xi_{fx}}{\partial z}\right) \quad (4)$$
$$+ \frac{1}{2}k_a\left(\frac{\partial\xi_a}{\partial z}\right)^2 + \frac{1}{2}k_f\left(\frac{\partial\xi_f}{\partial z}\right)^2$$

where α and β are the coefficients for the expansion up to the fourth order in magnitude of ξ_a and ξ_f, δ is the Lifshitz term, which appears as a result of the chirality of the molecules, k is the elastic constant, and γ_2 is the coupling term between the two order parameters.

Assuming that only $\alpha_a = a(T - T_{a,0})$ and $\alpha_f = a(T - T_{f,0})$ $(T_{a,0} < T_{f,0})$ are temperature dependent, a minimization of free energy [Eq. (4)] with respect to the three parameters θ_a, θ_f, and q gives four types of solutions for the different phases:

$$\text{SmA}^* : \theta_a = 0, \ \theta_f = 0$$
$$\text{SmC}^* : \theta_a = 0, \ \theta_f \neq 0$$
$$\text{SmC}_\gamma^* : \theta_a \neq 0, \ \theta_f \neq 0$$
$$\text{SmC}_A^* : \theta_a \neq 0, \ \theta_f = 0$$

where $\theta_a = (\theta_1 - \theta_2)/2$, $\theta_a = (\theta_1 + \theta_2)/2$, and θ_i is the molecular tilt angle in the ith smectic layer.

As temperature is lowered, the SmA* phase transforms to the ferroelectric SmC* phase and then to the ferrielectric SmC$_\gamma^*$ phase. In this phase the ferroelectric order parameter (ξ_f) decreases from ξ to 0 when temperature decreases, whereas the antiferroelectric order parameter (ξ_a) increases from 0 to ξ. In other words, the angle between the tilt vectors in the neighboring smectic layers changes from 0 to π on cooling through the SmC$_\gamma^*$ phase [Fig. 6(a)]. This model also predicts the existence of incommensurate phases when the coupling term between the two order parameters γ_2 is not strong enough.

This model was later improved by taking the symmetry considerations and the interactions between the nearest and the next-nearest smectic layers into account [27,28]. Both of these models predict the existence of periodical structures with the n-layer period, in which the angle between the tilt directions in the neighboring smectic layers $\Delta\varphi$ is equal to $2\pi/n$ [Fig. 6(b)]. Because the C directors in the neighboring layers are directed to each other by the angle, $2\pi/n$; hence the models are known as the clock models.

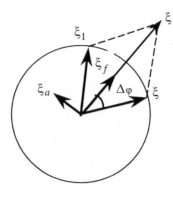

(a)

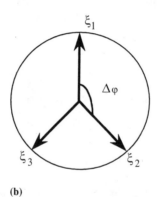

(b)

Figure 6. The phase in the azimuthal angle $\Delta\varphi$ between the tilt direction in neighboring smectic layers in the (a) bilayered model and (b) clock model.

The structure of antiferroelectric SmC_A^* phase ($n = 2$) predicted by the clock model is the same as that derived from Ising and ANNNI models, whereas the structures for SmC_γ^* ($n = 3$) and AF ($n = 4$) phases are different from them.

The similar clock structures follow from the short pitch mode (SPM) model [29,30] . This approach was related to the assumed structures of such smectics possessing a complicated spatial director distribution $\varphi(z)$ along the smectic layer normal. This distribution is in fact, the superposition of the short pitch modes responsible for the clock-like structure and long pitch mode (LPM), which forms the helix.

In summary, note that the structure of antiferroelectric SmC_A^* phase derived from different models is the same, whereas the structures of ferrielectric (three-layered) SmC_γ^* phase and antiferroelectric (four-layered) AF phase based on different models are different.

IV. THE METHODS FOR THE PHASE IDENTIFICATION OF AFLCs

Although the existence of a variety of ferrielectric and antiferroelectric liquid crystalline phases is well established, the precise structure of these phases is still under discussion. The structure of the ferrielectric phases, deduced from the Ising model, although initially accepted, is currently being revised [13–16,27–30]. The complexity of the antiferroelectric and fer-rielectric structures makes it necessary to use the experimental methods for their identification. In fact there is not a single method, that can give precise structure in different antiferroelectric and ferrielectric phases. In practice, several different complementary methods are being used to find the molecular structure. The most important experimental techniques are calorimetry, conoscopic observations, measurements of the apparent tilt angle, induced polarization, X rays, dielectric and IR spectroscopy.

Calorimetry is usually effective in detecting the phase transitions between the different liquid crystalline phases and has been widely used in AFLCs [31–35]. For ferrielectric phase transitions, the different calorimetric tech-niques give different results. A comparison of these techniques for detecting the phase transitions in MHPOBC is given in Figure 7. Although the enthalpy changes at the phase transitions are extremely low (few $\mu J\ s^{-1}g^{-1}$) [12], differential scanning calorimetry (DSC), relaxation calorimetry, and adiabatic calorimetry enable detection of most phase transitions in MHPOBC. AC calorimetry (ACC), which is usually considered to be the most sensitive method, is not effective for AFLCs because of the large temperature hysteresis between the phase transitions and the long relaxation times [8,21]. Although calorimetry is an effective method for detecting the phase transitions, it does not provide the structure of the subphases.

Conoscopy is an important technique that provides proof for the existence of the ferrielectric SmC_γ^* phase and gives information about its possible structure. The first observations of conoscopic patterns of the homeo-tropically aligned AFLC cell under DC voltage shows different responses of liquid crystals at different temperatures which is explained through the existence of different molecular structures in different phases [6] (Fig. 8). These structures are consistent with the Ising model. This technique was later employed to investigate the structure of more complicated ferrielectric phases [36–40]. Nevertheless, it provides only indirect information about the molecular structures in the neighboring smectic layers.

X-ray diffraction is normally a powerful method in crystallography for the identification of different structures. For determining the phase tran-

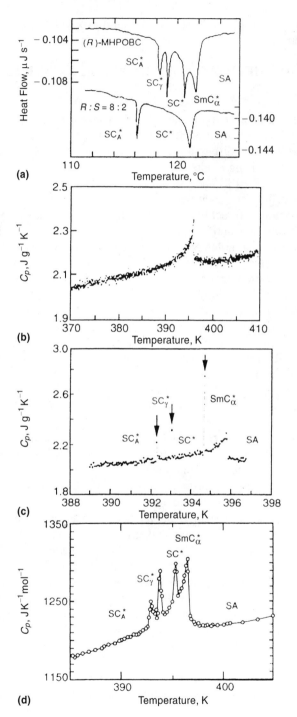

(a)

(b)

(c)

(d)

sitions in AFLCs, this method (at least in its conventional use) is not as effective owing to the following reasons. During the phase transitions from SmA* to SmC_A^* through the different ferrielectric phases, the molecular structures change by the distribution of the azimuthal angle φ, whereas the smectic layer spacing smoothly decreases with temperature and shows no jumps [12,41–44]. In other words, conventional X-ray diffraction does not distinguish between the molecules in smectic layers tilted in different directions (or having different azimuthal angles). Nevertheless, a comparison of the optical tilt angle (defined from the optical switching) with the steric tilt, defined from X-ray diffraction, reveals some interesting results about the possible molecular structure in ferrielectric phases [43]. Usually, in the tilted smectic phases, the optical tilt angle is higher than the steric tilt. Recent investigations show that the optical tilt angle is lower than steric tilt in the antiferroelectric liquid crystalline sample 12OBP1M7 (named AS573 in some previous publications) at temperatures 0 to 8°C below the SmA* phase. For this reason, the phase below SmA phase was assigned to SmC_β^* by these authors. This result has been interpreted as a coexistence of two different molecular conformers in the SmC_β^* phase with two different tilt angles and smectic layer spacings. With a change in temperature, one of these two conformations becomes more dominant than the second causing a change in the average polarization. However, authors of Ref. 43 interpret SC_β^* as a special ferroelectric phase but consisting of two different conformers. The result can also be interpreted differently as has been pointed out in our publications. The possibility of such ferrielectric and not ferroelectric structures was predicted by Lorman et al. [16].

Considerable progress in using X-ray diffraction for determining ferrielectric structures was recently made by Mach et al. [45] through the application of resonant X-ray diffraction in the AFLC compound 10OTBBB1M7, which contains the sulfur atom within the rigid core part of the molecule. By working in the energy range at sulfur's K absorption edge ($\approx 2.5\,\mathrm{keV}$), the scattered X-ray intensity becomes sensitive to the azimuthal orientation of the molecules in the smectic layers. Figure 9 shows the X-ray intensity scan for 250 layers of a thick free-standing film of 10OTBBB1M7 in different tilted phases. From this figure, one can easily find the double-layer periodicity in the SmC_A^* phase, the three-layer periodicity in the SmC_{FI1} (or SmC_γ^*) phase, and the four-layer periodicity in the SmC_{FI2} (or AF) phase. Careful analysis of the relative intensity of scattering peaks based on the structure factor calculations shows that the angle between the tilt directions in any pair of neighboring smectic

Figure 7. Thermograms obtained by (a) DSC, (b) ACC, (c) relaxation calorimetry, and (d) adiabatic calorimetry measurements of MHPOBC. (Redrawn with permission from [8].) In the figure, SmC_a^* stands for SmC_α^*.

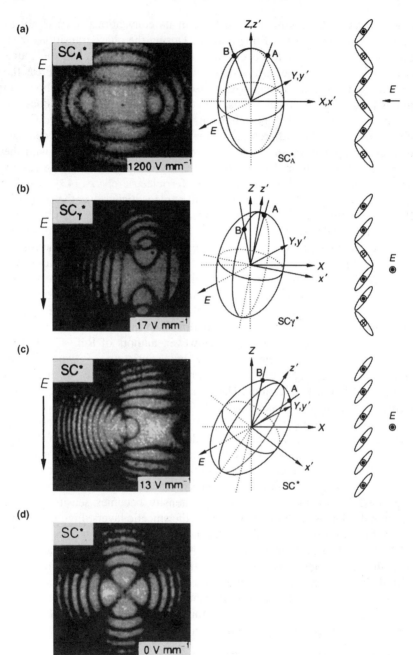

Figure 8. Typical conoscopic figures for MHPOBC. (a) SmC_A^* at 1200 V mm^{-1}; (b) SmC_γ^* at 17 V mm^{-1}; (c) SmC^* at zero field. (Redrawn with the permission from [8].)

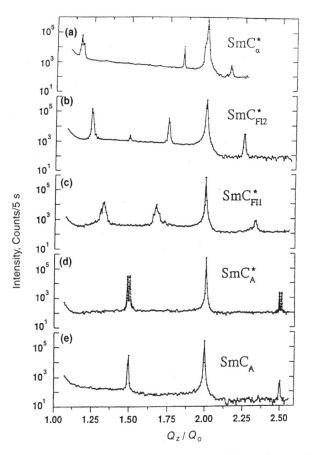

Figure 9. X-ray intensity scans in the indicated phases of (R)-enantiomer (plots a–d) and racemic 10OTBBB1M7 (plot e). The temperatures corresponding to the curves are 199.0, 115.4, 113.2, 109.3, and 107.0, in °C respectively. (Redrawn with permission from [45].)

layers is $\Delta\varphi = 2\pi/n$, where n is number of smectic layers in a periodic unit [46]. This is consistent with the structures, suggested by the clock models [28–30].

Recently Raman light scattering [47] and Fourier transform infrared (FTIR) spectroscopy [48–51] have been employed for investigating the molecular conformation and the hindered rotation of different molecular groups in AFLC. The results of FTIR spectroscopy shows that the rotation of both chiral and core carbonyl C=O groups around the long molecular axis is hindered in tilted smectic phases [48–51]. The investigation of the switching mechanism under application of electric field by time-resolved FTIR spectroscopy [48] shows that all the molecular segments reorient simultaneously, at least in the microsecond range. These methods do not provide direct information about the structure of ferrielectric phases; however, a careful analysis of the experimental results enables a validation of the structure of some phases, such as SmC_A^* and even SmC_γ^* [47].

In addition to the various experimental techniques mentioned above, several analytical methods have been widely used for studying AFLCs. These are the measurements of the apparent tilt angle, induced polarization, and dielectric and electrooptic spectroscopy. The application of these methods to the study of the dynamic and static properties and structure of AFLC, will be given in the next sections. The data concern mostly one AFLC sample, namely 12OBP1M7 (Hull, UK, also named AS573), with following chemical structure:

$$C_{12}H_{25}O-\!\!\bigcirc\!\!-\!\!\bigcirc\!\!-COO-\!\!\bigcirc\!\!\overset{\displaystyle F}{}-COO-\overset{\displaystyle CH_3}{\underset{\displaystyle |}{C^*H}}-C_6H_{13}$$

and phase transition sequence under cooling is SmC_A^* (78°C), SmC_γ^* (81°C), AF (83°C), FiLC (90°C), SmC^* (92°C), and SmA* (106°C) . We chose this AFLC material for several reasons. First, this sample possesses four basic tilted smectic phases (SmC_A^*, SmC_γ^*, AF, SmC^*) and a high temperature ferrielectric phase FiLC with complicated properties. Second, 12OBP1M7 has been investigated by several research groups and by different techniques, including DSC [35], X-ray [43,44], conoscopy [37–39], dielectric [39,52–56], electrooptic [56] and photon correlation [57] spectroscopy, field-induced macroscopic polarization [35,39,54,55], and apparent tilt angle [35,39,43,44].

V. THE DYNAMIC AND STATIC PROPERTIES OF SmC_A^*

In contrast to other ferrielectric and antiferroelectric phases, the first suggested herringbone structure of SmC_A^* is well defined and its structure is widely accepted. According to this model the molecular tilt directions in any pair of neighboring smectic layers are almost antiparallel to each other, canceling the macroscopic polarization in volume. Therefore, the first attempts to investigate dynamic properties of SmC_A^* (Goldstone mode) by dielectric spectroscopy did not detect any collective modes [58–60], whereas they were observed by photon correlation spectroscopy [61–63]. Later, the collective relaxation modes were discovered using dielectric spectroscopy [56,64–69]. Nevertheless the problem of finding the mechanisms governing the various relaxation processes in the antiferroelectric phase is not straightforward, and a combination of different techniques, such as dielectric, electro-optic and photon correlation spectroscopy, was used to unambiguously establish these phenomena. The mechanism of the collective processes is found by comparing the results of the dielectric and electro-optic spectroscopy with those from theoretical investigations.

In electrooptic spectroscopy, the FLC cell is placed between the crossed polarizers. The signal from the photodiode, placed in the top tube of a polarizing microscope, is fed to a lock-in amplifier to both the fundamental and the second harmonic frequencies of the signal applied across the sample.

Several investigations [64–67] showed that the dielectric spectra in the frequency range from 100 Hz to 1 GHz leads to the existence of two molecular relaxation processes and to at least two collective relaxation processes in the SmC_A^* phase. These are denoted as processes 0, 1, 2, and 3 [56], in increasing order of relaxation frequency. The dielectric spectrum of 12OBP1M7 at 70°C fitted to 4 relaxation processes, for example, is shown in Figure 10. The lower frequency processes (0 and 1) exist only in the two antiferroelectric phases SmC_A^* and AF [56]. The frequency of process 0 is about 2.5 kHz, and that of process 1 lies between 10 and 70 kHz, depending on temperature [67]. Hence it can be concluded only that these processes are specifically related to the common characteristics of antiferroelectric phases; the origin of these processes is still to be explained. Four possible physical mechanisms, some of these already suggested for the collective molecular reorientation in the antiferroelectric phase [64–68], are presented in Figure 11.

1. A deviation from the antiferroelectric order [64] by the azimuthal angle φ such that φ changes in the opposite sense in the adjacent layers or optic like phason [70]. The mode concerning the fluctuation in φ of the type shown in Figure 11(a) in the antiferroelectric phase was called antiferro-

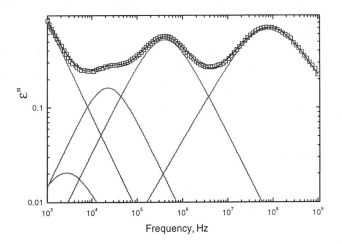

Figure 10. The dielectric losses spectrum of a 50-μm cell in the SmC_A^* phase ($T = 70°C$), fitted to four relaxation processes.

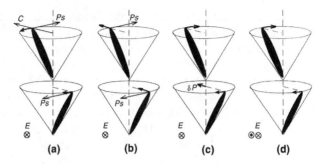

Figure 11. Four physical mechanisms for the molecular rearrangements in the antiferroelectric SmC$_A^*$ phase under the applied voltage. P_s, spontaneous polarization vector; C, C director.

electric Goldstone-like mode by Hiraoka et al [64]. A change of φ in the opposite sense for the two adjacent layers owing to rotations of the directors in the opposite directions is termed the anti-phase motion by Buivydas et al. [66]. The anti-phase motion or the distortion of the antiferroelectric order caused by the antiphase motion is more appropriate because it depicts the mechanism more clearly. The antiferroelectric Goldstone mode should be considered more appropriate for another relaxation process [Fig. 11(c)], the mechanism of which is similar to that of a ferroelectric Goldstone mode.

2. Fluctuations in the tilt angle (θ) or the antiferroelectric soft mode are depicted in Figure 11(b).

3. The helical Goldstone mode owing to antiferroelectric polarization δP is shown in Figures 11(c) and 12. The explanation for the existence of the antiferroelectric polarization is detailed below.

4. A disturbance of antiferroelectric helix owing to the dielectric anisotropy, $\Delta \varepsilon E^2$, is shown in Figure 11(d).

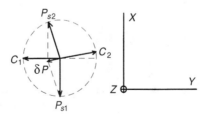

Figure 12. Arrangement of the C directors (C_1, C_2) and local spontaneous polarization (P_{s1}, P_{s2}) of two neighboring smectic layers in the helical antiferroelectric phase with respect to the laboratory frame of reference.

A.   Origin of Process 0

The C-directors in the adjacent smectic layers are not completely anti-parallel to each other even in the antiferroelectric phases, as initially suggested by Buivydas et al. [66]. This gives rise to the antiferroelectric spontaneous polarization δP [Fig. 11(c)] and is found to be almost parallel to the C directors (Fig. 12). We are able to estimate the magnitude of such an antiferroelectric polarization δP. Taking into account that the typical value of the helical pitch ≈ 2000 Å and the smectic layer thickness ≈ 40 Å, a trivial calculation shows that the angle between the spontaneous polarization vectors in the two adjacent smectic layers is 176 to 177° and the magnitude of δP is approximately 3% of the spontaneous polarization P_s. Such an antiferroelectric polarization should spiral along the helical axis as the polarization does in a ferroelectric phase. Hence we can expect the dielectric relaxation process to arise from the distortion of the antiferroelectric helix similar to the cause of a ferroelectric helical Goldstone mode.

B.   Origin of Process 1

The relaxation frequency of process 1 is higher than that of process 0. The mechanism can be either the antiphase motion type [Fig. 11(a)] or the soft mode type [Fig. 11(b)]. Hiraoka et al. [64] suggested the antiferro-electric Goldstone-like mode for this type of motion. We may point out, however, that this is a misleading term because the antiferroelectric Gold-stone mode should be reserved for process 0, because of its similarity to that of a ferroelectric Goldstone mode, as mentioned before. In certain investigations of samples with a direct SmC_A^*–SmA^* transition this process was also assigned to the soft mode [71], mainly because it followed a typical behavior for this mode close to the transition temperature (i.e., relaxation frequency increased with temperature on either side of the transition temperature). Nevertheless, 20°C below this temperature, the relaxation frequency decreases with decreasing temperature. Although a decrease in the relaxation frequency can be explained by an increase in the rotational viscosity with decreasing temperature, the unusual observed dependence of $\Delta\varepsilon$ [56] has no proper explanation (i.e., $\Delta\varepsilon$ increases instead of decreasing close to the transition temperature). Therefore, we rule out the antiferroelectric soft mode as a possible physical mechanism for process 1. The relaxation process 1 is thus tentatively assigned to the distortion of the antiferroelectric order caused by the antiphase motion [66] [Fig. 11(a)]. This mechanism will be confirmed through electrooptic spectroscopic investigations.

Note that among the four proposed mechanisms for the collective relaxation processes, the first three are dielectrically active. The first two are also electro-optically active at the fundamental frequency, and the fourth mechanism is electro-optically active at twice the frequency (second harmonic) of mode shown in Figure 11(c). Therefore, a comparison of the results of dielectric spectroscopy with those of electro-optics can clarify the origin of the two collective relaxation processes in the SmC_A^* phase seen in the dielectric spectra.

Assuming the existence of the antiferroelectric polarization δP, the equation for the director motion in the antiferroelectric phase can be written as follows:

$$\gamma \frac{\partial \varphi}{\partial t} = K \frac{\partial^2 \varphi}{\partial z^2} + \delta P E_0 \cos \omega t \cdot \cos \varphi + \Delta \varepsilon^* E_0^2 \cos^2 \omega t \sin 2\varphi \qquad (5)$$

where $\gamma = \gamma_\varphi \sin^2 \theta$, $K = K_\varphi \sin^2 \theta$, $\Delta \varepsilon^* = \Delta \varepsilon \sin^2 \theta / 8\pi$, γ_φ is rotational viscosity, K_φ is the elastic constant, θ is the molecular tilt angle, φ is the azimuthal angle, and $\Delta \varepsilon$ is dielectric anisotropy. Using the formula $\cos \varphi \sin \varphi = \sin(2\varphi)/2$, Eq. (1) can be written in complex form as

$$\gamma \frac{\partial \varphi}{\partial t} = K \frac{\partial^2 \varphi}{\partial z^2} + \delta P E_0 e^{j\omega t} \cos(\varphi) + \frac{\Delta \varepsilon^*}{2} E_0^2 (e^{j2\omega t} + 1) \sin(2\varphi) \qquad (6)$$

Assuming E_0 to be sufficiently small and applying the standard perturbation technique, let the solution be of the following form:

$$\varphi(z,t) = qz + f(z,t) = qz + \cos(qz) f_{11} e^{j\omega t} + \sin(2qz)(f_{22} e^{j2\omega t} + f_{20}) \qquad (7)$$

where f_{11}, f_{20}, and f_{22} are the coefficients of the Fourier series proportional to E_0 and E_0^2. Substituting Eq. (3) into Eq. (2), we find the coefficients f_{11}, f_{22}, and f_{20}:

$$f_{11} = \frac{\delta P E_0}{Kq^2} \frac{1}{(j\omega \tau_1 + 1)}; \quad f_{22} = \frac{\Delta \varepsilon^* E_0^2}{8Kq^2} \frac{1}{(j\omega \tau_2 + 1)}; \quad f_{20} = \frac{\Delta \varepsilon^* E_0^2}{8Kq^2} \qquad (8)$$

where

$$\tau_1 = \frac{\gamma}{Kq^2} \qquad \tau_2 = \frac{\gamma}{2Kq^2} \qquad (9)$$

Figure 13 shows the dependence of the amplitude and the phase parts of the electro-optic response at the same frequency as the applied signal for a homogeneously aligned cell. The relaxation frequency is 20 to 30 kHz and is almost of the same magnitude as for the dielectric relaxation process 1. Therefore, we can unambiguously assign this relaxation process in the antiferroelectric phase to the distortion of antiferroelectric order by the electric field caused by a change of angle φ arising from the type of motion shown in Figure 11(a). This may also be called the antiphase Goldstone

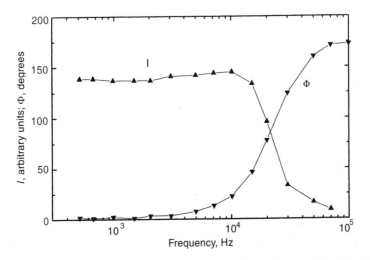

Figure 13. Frequency dependence of intensity (I) and the phase part (Φ) of the linear response for a well homogeneously aligned 20-μm cell at 75°C.

mode. We rule out the assignment owing to the soft mode and the reasons given in the previous section.

Figure 14 presents the dependence of the amplitude and the phase part of the second harmonic electro-optic response at twice the frequency of the signal for an unaligned cell. The relaxation frequency for this process is $\sim 6\,\text{kHz}$, which is two times the frequency of process 0 in the dielectric

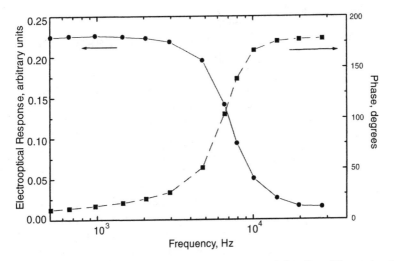

Figure 14. Frequency dependence of the intensity (I) and the phase (Φ) at twice the frequency of the applied signal for an unaligned 20-μm cell at 75°C.

spectra. As mentioned before, the relaxation process 0 exists in the dielectric spectra without bias voltage [66] or only under the bias voltage [64]. The nonlinear electro-optic response appears to be caused by the distortion of the helix owing to the dielectric anisotropy. The relaxation frequency for this response is approximately two times higher than the frequency for process 0 in the dielectric spectra and is in good agreement with theoretical predictions [Eq. (5)]. It may be re-emphasized that the helical distortion is caused by two mechanisms: polar interactions $\delta P.E$ and dielectric anisotropy interactions $\Delta\varepsilon.E^2$. The latter cannot be detected dielectrically but is clearly observed in the absence of the bias voltage in the electro-optic response at the second harmonic frequency of the applied field for process 0.

In summary, the higher frequency relaxation process (Process 1) is the result of distortion of the antiferroelectric order caused by the antiphase motion [Fig. 11(a)]. This may also be called the antiphase antiferroelectric Goldstone mode, which has also been observed in the electro-optic response at the fundamental frequency. The lower frequency relaxation process (process 0) in the dielectric relaxation spectra is the helical distortion mode. Antiferroelectric Goldstone mode is suggested to be the appropriate term, because the mechanism is similar to that for a helical Goldstone mode. The distortion of the helix caused by the dielectric anisotropy gives rise to an electro-optic response at twice the frequency of the applied field for this process. The reasons for having different frequencies for process 0 using different techniques are provided by solving the dynamic equation that governs the motion of the director of an antiferroelectric helix subject to a weak alternating field.

VI. STRUCTURE AND PROPERTIES OF FERRIELECTRIC PHASES

In contrast to the low-temperature antiferroelectric SmC_A^* phase, the molecular structures of other ferrielectric and antiferroelectric phases, such as SmC_γ^* and AF, remain to be determined and explained. As stated, the structure of these phases is model dependent. The situation is even more unclear for new ferrielectric phases, such as FI_L and FI_H [18,19,23,72]; FiLC [38,40], FI1 and FI2 [46]; and spr1, spr2, spr3 [24]. The names of these phases are related to their position in the temperature scale rather than to their molecular structure. Nevertheless, these ferrielectric phases can be divided into two different groups, one of the groups lies in the temperature range below the AF phase (or with $q_T < 1/2$) and the second in the temperature range above the AF phase. The stable existence of ferrielectric phases of the first group has been found experimentally [18,19,23,72], and their molecular structures are determined (at least on the basis of the Ising and ANNNI models). The second group of the ferrielectric phases, which has been observed rather recently in the temperature range between antiferroelectric AF and ferroelectric SmC* phases [24,38,40,73], reveals some unusual properties, and their stability and structure are still under investigation. There does not appear to be any available direct method for the identification of molecular structures of ferrielectric phases. A combination of several methods such as dielectric and electro-optic spectroscopy, conoscopy, field-induced tilt angle, and macroscopic polarization measurements have been used.

The dielectric spectra of the AFLC cells were found to be considerably different for different antiferroelectric and ferrielectric phases, which can be used for identification of different phases. Figure 15 shows the temperature dependence of the dielectric loss spectra of 12OBP1M7 versus temperature without a direct bias voltage for 8-μm and 50-μm cells [39]. A comparison of two plots reveals a remarkable dependence of a part of the spectra on sample thickness. In the temperature range between SmC_γ^* and SmC*, the dielectric spectra for the larger cells look similar to the spectra of the antiferroelectric phase (SmC_A^*). The experimental dielectric spectra are practically independent of the cell thickness for cells thicker than 25 μm. Therefore, in cells with thickness $\leq 8\,\mu$m) some of the ferrielectric subphases are suppressed by the surface interactions.

The dielectric spectra are fitted to Havriliak-Negami equation for n relaxation processes:

$$\varepsilon^*(\omega) = \sum_{i=1}^{n} \frac{\Delta\varepsilon_i}{\left(1 + (j\omega\tau_i)^{\alpha_i}\right)^{\beta_i}} + \varepsilon_\infty \qquad (10)$$

where $\varepsilon^*(\omega)$ is the frequency-dependent complex relative permittivity, ε_∞ is the high-frequency permittivity, i is a variable denoting the number of the relaxation processes up to n, τ_i is the relaxation time of the ith relaxation process, α_i and β_i are the fitting parameters, and $\Delta\varepsilon_i$ is the dielectric relaxation strength (or the static susceptibility) for the ith process. Figures 16 and 17 present the temperature dependence of the dielectric parameters, which were deduced by fitting the dielectric spectra shown in Figure 15(b).

At the high temperature range corresponding to the isotropic phase, two noncollective (molecular) relaxation processes are observed. The higher frequency process is assigned to the molecular relaxation around the long molecular axis and the lower frequency process is that around the short axis. In the SmA* phase, the molecules are parallel to the plane of the electrodes, thus only the higher frequency relaxation process and the soft mode exist. The results presented in Figures 16 and 17 show that the relaxation frequencies of both molecular processes obey the Arrhenius equation; however, the dielectric amplitudes possess complicated behavior. This has the following explanation. The static susceptibility χ of the molecular

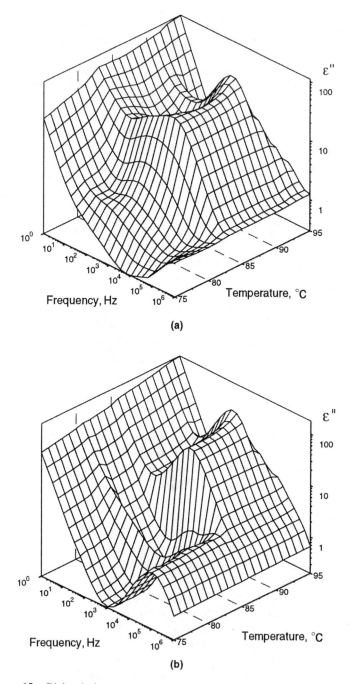

Figure 15. Dielectric loss spectra versus temperature for 12OBP1M7 (AS573), at (a) $d = 8\,\mu m$ and (b) $d = 50\,\mu m$.

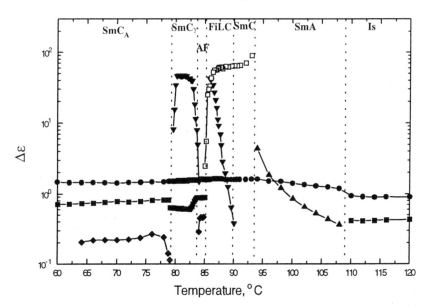

Figure 16. Dependence of the dielectric amplitude ($\Delta\varepsilon$) on temperature for 12OBP1M7; $d = 50\,\mu\text{m}$.

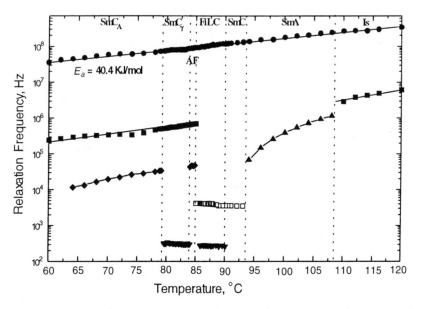

Figure 17. Dependence of the relaxation frequency (f) on temperature for 12OBP1M7; $d = 50\,\mu\text{m}$.

relaxation processes depends on the angle between the electric field and the molecular long axis. For an undisturbed helical structure, the average static susceptibilities around the long $\tilde{\chi}_l$ and short $\tilde{\chi}_s$ molecular axis depend on the tilt angle θ by [39]:

$$\tilde{\chi}_l = \chi_l \left(1 - \frac{\theta^2}{2}\right), \quad \tilde{\chi}_s = \chi_s \frac{\theta^2}{2} \tag{11}$$

where $\chi_l(\chi_s)$ is the maximum static susceptibility of the molecular relaxation process around the long (short) molecular axis found for the planar (homeotropic) molecular orientation. It means that in helical-tilted smectic phases (such as in SmC*, SmC$_\gamma^*$, and SmC$_A^*$), the dielectric amplitude of the molecular relaxation processes depends on the molecular tilt angle θ. For rotation around the long molecular axis, the dielectric amplitude increases as the helix is unwound [39,39a].

In the SmC* phase and at temperatures of several degrees below the SmA*–SmC* phase transition, two relaxation processes are observed: molecular relaxation around the long molecular axis and ferroelectric Goldstone mode. One could also expect the molecular relaxation around the short molecular axis, because some molecules are no longer perpendicular to the electric field owing to a helical SmC* structure. This process, however, is too weak to be detected owing to a superposition of an intense Goldstone mode on to the molecular mode.

In the ferrielectric SmC$_\gamma^*$ phase, there exists only one ferrielectric Goldstone mode. The relaxation frequency of this process agrees with the phenomenologic theory developed by Čepič et al. [74], which describes the dynamic behavior and the dielectric spectra of different SmC$_A^*$ (q_T) subphases. According to this model, the relaxation frequency of the Goldstone mode below the phase transition temperature SmC$^* \rightarrow$ SmC$_\gamma^*$ decreases by approximately one order of magnitude. This behavior was observed experimentally [39,74]. This conclusion can lead to the possible molecular structure that may exist in the ferrielectric SmC$_\gamma^*$ phase. The dielectric parameters of helical Goldstone relaxation process are given as follows:

$$\Delta\varepsilon = \frac{P_s^2}{2\varepsilon_0 K_\varphi q^2} \qquad \tau = \frac{\gamma_\varphi}{K_\varphi q^2} \tag{12}$$

Therefore, the ratio

$$\frac{\Delta\varepsilon}{\tau} = \frac{P_s^2}{2\varepsilon_0 \gamma_\varphi} \tag{13}$$

does not depend on pitch of the helix. In ferrielectric SmC$_\gamma^*$ phase, as suggested by the Ising model, the macroscopic polarization $P_{\text{SmC}_\gamma^*} = P_s/3$; therefore, the ratio $\Delta\varepsilon/\tau$ in ferrielectric phases must be about one order less than in the ferroelectric phase, which agrees with experimental data (Figs. 16 and 17).

As shown in Figures 16 and 17, two antiferroelectric phases SmC_A^* ($< 78°C$) and AF (81–83°C) with similar dielectric properties are observed on both sides of ferrielectric SmC_γ^* phase. The rather low dielectric loss observed in both these phases implies the absence of macroscopic spontaneous polarization. If we consider the molecular structure according to the clock models [27–30, 45], in which the SmC_A^*, SmC_γ^*, and AF phases are characterized by two-, three-, and four-layer superlattices periodicity, respectively, the average spontaneous polarization of all these phases is almost canceled out, and in SmC_γ^* we should expect the dielectric properties similar to SmC_A^* and AF phases.

In contrast to the phases discussed above, the structure and properties of the ferrielectric phase (FiLC), which appears in the temperature range between AF and SmC^* (85–90°C), are even more complicated and have not been explained. Firstly the phase transitions between FiLC and the neighboring AF and SmC^* are not detected by DSC. Secondly, when the temperature is decreased, the second relaxation process gradually appears [39,54,55] in the dielectric spectra at 85 to 90°C [39,53,54]. In addition to the ferroelectric Goldstone process discussed above, a second relaxation process appears about 2°C below the SmC^*–SmA phase transition, which seems to correspond to the ferrielectric Goldstone mode, because its frequency is almost of the same as that of the ferrielectric Goldstone mode in the SmC_γ^* phase (Fig. 17). The coexistence of two Goldstone modes (ferrielectric and ferroelectric) in a similar phase of another AFLC compound have been reported and were explained by the coexistence of two phases at the same temperature induced by the surface interactions [74]. The importance of the surface interactions for the appearance of the ferrielectric was reported [39,75].

The focus of this section is the observation of the coexistence of two relaxation processes in the temperature interval that corresponds (according to DSC data) to the SmC^* phase. According to the temperature-induced Devil's staircase model [8] for temperatures higher than the AF phase, one could reasonably expect the existence of some ferrielectric phase with ½ $< q_T < 1$. In which case, there must be only one corresponding Goldstone mode. Therefore, the coexistence of two Goldstone relaxation processes at the same temperature could be explained by the existence of a mixture of two different subphases (ferrielectric and ferroelectric) at a microscopic level which is also supported by conoscopy [39].

Figure 18 presents the conoscopic images of a 20-μm cell with homeotropic orientation for different temperatures and voltages. In the SmC^* phase, an increase in voltage shifts the center of the image in a direction perpendicular to the applied field and the center continues on

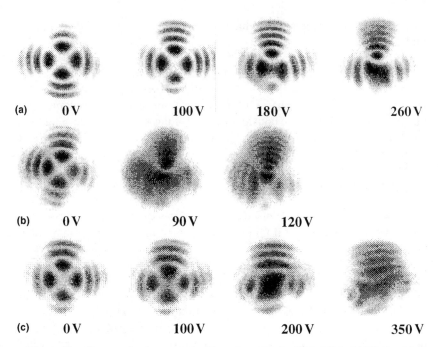

Figure 18. Conoscopic pictures under bias voltage. (a) SmC* at 91°C; (b) FiLC at 86°C; (c) SmC*$_\gamma$ at 81°C; Material 12OBP1M7.

shifting in the normal direction with an increase in voltage. The structure becomes increasingly biaxial and the optical plane continues to be normal to the direction of the applied field. Such behavior is typical of a ferroelectric SmC* phase [6] (see Fig. 8). In the SmC*$_\gamma$ phase, the dependence of the conoscopic image on voltage is similar to that of the SmC* phase, but the optical plane in the SmC*$_\gamma$ phase is parallel to the direction of the field instead of being perpendicular (see Fig. 8). For temperatures between 83 and 85°C and < 78°C, the conoscopic images are typically antiferroelectric [6]. Therefore, we conclude that a basic antiferroelectric phase SmC*$_A$ ($q_T = 0$) exists at low temperatures and an antiferroelectric AF phase ($q_T = 1/2$) at higher temperatures. The unusual behavior was found at 85 and 90°C. The application of a sufficiently small voltage makes the conoscopic image rather blurred [Fig. 18(b)], although a detailed examination of the picture reveals the existence of four centers: one pair as in the ferroelectric phase and the second as in the ferrielectric phase. Such an observation cannot possibly be made for a uniform structure and supports the conclusion about the coexistence of two phases: ferrielectric-like and ferroelectric. A further increase of the applied voltage makes the conoscopic image clearly ferroelectric-like. The identification of this phase is discussed later, but note that in this temperature interval there exist two different Goldstone type relaxation processes with values of the relaxation frequency of the one as in the SmC_γ^* and the second as in the SmC^* phase. Therefore, from dielectric and conoscopic observations, it would appear that the ferrielectric FiLC is actually a mixture (co-existence) of two phases SmC^* and some ferrielectric phase, probably SmC_γ^*, at the level of a finite number of periodic units. At a macroscopic level, the phase has a uniform texture.

The effect of voltage on the dielectric spectra in the SmC^*, SmC_γ^*, and SmC_A^* phases is similar to the results reported in the literature [15–18]. The application of the bias voltage to the FiLC phase considerably changes the dielectric spectra (Fig. 19). An increase in the bias voltage suppresses the low-frequency relaxation mode and causes the amplitude of the high-frequency process to increase initially and then to decrease with an increase in the bias voltage (Fig. 20). This phenomenon could be explained by taking into account the results of the conoscopic investigations presented in Figure 18. If we suggest the coexistence of two phases— SmC_γ^* and SmC^* in this temperature range—then at small electric fields, we have two dielectric relaxation processes and the mixture of two different conoscopic images, which leads to the complicated conoscopic picture already discussed. Application of the electric field causes the transition from FiLC to SmC^*, leading to a suppression in the strength of the ferrielectric mode in dielectric spectra and an increase in the strength in the ferroelectric mode. At these voltages conoscopic pictures become ferroelectric-like. A further increase

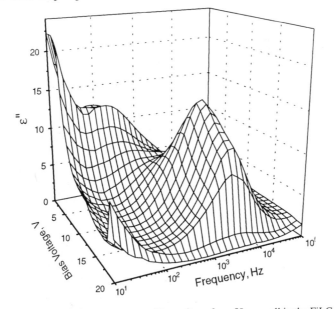

Figure 19. Dielectric losses spectra vs. bias voltage for a 50-μm cell in the FiLC phase at 86°C of 12OBP1M7.

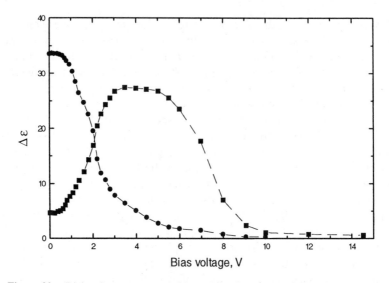

Figure 20. Dielectric parameters vs. bias voltage for a 50-μm cell in the FiLC phase, at 86°C. ●, ferrielectric; ■, ferroelectric; Material 12OBP1M7.

in the electric field will unwind the helix and suppress the ferroelectric Goldstone mode.

The measurements of the apparent tilt angle and macroscopic spontaneous polarization induced by the application of the electric field have been widely used to yield additional information about the structures of antiferroelectric and ferrielectric phases and their transformations with electric field. The dependence of the apparent tilt angle on voltage was found to be similar to the macroscopic spontaneous polarization. Therefore, we will consider only the results of polarization measurements.

Figure 21 presents the temperature dependence of the normalized spontaneous polarization as function of the bias voltage for different phases. At 78°C, which corresponds to the SmC_A^* phase, the macroscopic polarization is extremely low and almost independent of voltage until a threshold value at 27 to 30 V is reached. At this voltage, the antiferroelectric structure changes to the quasi-stable ferrielectric structure with an induced polarization of $1/3P_s$. At the higher voltage (≈ 75 V, not shown in Figure 21), this structure transforms to completely unwound SmC^* phase, and the macroscopic polarization reaches a saturation value of P_s.

The polarization plot corresponding to the SmC_γ^* phase shows typical ferrielectric dependence on voltage. For voltages between 0 and 5 V, polarization obeys an almost linear dependence on voltage, corresponding to a distortion of the helix with voltage, then a saturation value of $P_s/3$ for an unwound ferrielectric structure is reached. Finally, at higher voltage (≈ 60 V), the macroscopic polarization reaches a saturation value of P_s.

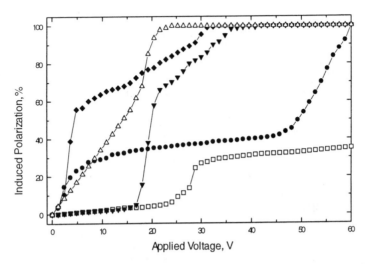

Figure 21. Temperature dependence of the normalized spontaneous polarization as function of bias voltage for different phases. $\triangle$, SmC* at 91°C; $\blacksquare$, FiLC at 86°C; $\blacktriangledown$, AF at 84°C; $\bullet$, SmC$^*_\gamma$ at 81°C; $\square$, SmC$^*_A$ at 78°C. Material 12OBP1M7.

What is really important is the existence of quasi-stable structure with macroscopic polarization of $1/3\,P_s$, which supports the Ising and/or ANNNI models, although this structure does not follow (at least directly) from the clock models. We give the results of the investigations of dielectric amplitude of the molecular relaxation process around the long molecular axis as a function of applied electric field to provide more information about the structure of this ferrielectric phase. In the helical tilted smectics, the average dielectric amplitude for this molecular process is lower than that in the unwound structures owing to the molecular tilt [Eq. (11)].

Figure 22 shows the temperature dependence of the dielectric amplitude of the high-frequency molecular relaxation process as a function of the electric field. It is important to note that in some phases the dielectric amplitude of the molecular relaxation does not follow the induced polarization plot (Fig. 21). For example, in the SmC$^*_A$ phase, the dielectric amplitude reaches a maximal saturation value at 30 to 35 V, at the same voltage that macroscopic polarization reaches the value $P_s/3$. Thus in this structure all the molecules become parallel to electrode plane while the average polarization is only $P_s/3$. A similar situation is observed for the SmC$^*_\gamma$ phase, which is possible for only three-layered ferrielectric structures with two layers tilted in the same direction and third one tilted in the opposite direction. Such a structure agrees with the Ising and ANNNI models and possesses $q_T = 1/3$ (or $q_E = 2/3$) (Fig. 4).

Note that doubts about the validity of the Ising model (and the ANNNI model) have been expressed. These models describe the molecular structure

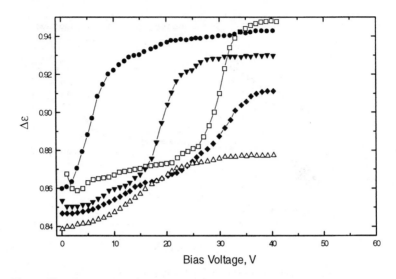

Figure 22. Temperature dependence of the dielectric amplitude of the high-frequency molecular relaxation as function of bias voltage for different phases. $\square$, SmC_A^* at 78°C; $\bullet$, SmC_γ^* at 81°C; $\blacktriangledown$, AF at 84°C; $\blacklozenge$, FiLC at 86°C; $\triangle$, SmC^* at 91°C. Material 12OBP1M7.

in the antiferroelectric and ferrielectric phases [45]. Nevertheless, the experimental results obtained for SmC_A^* (at 78°C) and SmC_γ^* (at 81°C), presented above, can be satisfactorily explained only by the Ising-type models. This means that under a bias voltage, the molecular structure being realized with $q_E = 2/3$ and the induced macroscopic polarization 1/3 of P_s, is that all the molecules in the structure are parallel to the plane of the electrodes.

Figure 22 shows the dielectric amplitude as a function of bias voltage for different temperatures. In particular the dependence of $\Delta\varepsilon$, in SmC* phase at 91°C, is similar to the dependence of polarization on voltage for a helical cell. At low voltages, the effective polarization and the dielectric amplitude are found to gradually increase with an increase in voltage, corresponding to a distortion of the helix. At higher voltages, the normalized polarization and the dielectric amplitude together reach a saturation value for the unwound SmC* structure. We stress that in both the SmC_A^* and the SmC_γ^* phase and in the unwound ferrielectric states, when the induced polarization reaches $P_s/3$ level, the dielectric amplitude of molecular relaxation process also reaches the maximal value.

An application of the DC voltage in the FiLC phase causes the most interesting dependencies of the spontaneous polarization on the applied voltage. The dependence of the effective polarization on bias voltage consists of two almost linear dependencies with different slopes (Fig. 21). For a low-bias voltage (0 to 5 V), the macroscopic polarization grows quickly with voltage and reaches 60% of P_s. This process is accompanied with a decrease in the ferrielectric relaxation intensity (Fig. 20) and a perturbation in the conoscopic image (Fig.18). These observed effects could be explained by a model, that involves a gradual transformation of the ferrielectric phase to the ferroelectric one with increasing bias voltage. When the entire FiLC phase is transformed to SmC^* (≈ 30 V), the conoscopic image again gets sharper and appears ferroelectric-like (Fig. 18b, 120 V). The dielectric spectrum at that stage exhibits only the ferroelectric Goldstone mode (Fig. 20); thus the effective polarization increases gradually with voltage as the helix is strongly being deformed with an increase in the electric field. As mentioned, this state with the effective polarization of 60 to 90% P_s could have been assigned to a step in the field-induced Devil's staircase. In which case the dielectric amplitude of the molecular relaxation around the long axis, according to (Eq. 11), must have a maximal value as in the case of the SmC_γ^* and SmC_A^* phases (Figs. 21 and 22); but this is not observed. At 85 to 90°C, the dielectric amplitude of the molecular mode reaches a maximal value at the same threshold voltage at which the effective polarization reaches a saturation value of P_s, the helix is unwound, and the Goldstone mode is almost suppressed. Therefore, we conclude that the field-induced quasi-stable level corresponds to that of the distorted ferroelectric helix and not to the field-induced Devil's staircase.

Note that the coexistence of different phases can be explained by the surface effects, which has been reported in several publications and presented in Figure 15 as an example. Nevertheless, in the case of the FiLC phase, which is observed for temperatures higher than the AF phase, physical properties were observed in a sufficiently thick cell (≈ 50 μm) and even in a homeotropically aligned cell (≈ 200 μm), where the influence of the surfaces is negligibly small. These properties correspond to a mixture of ferroelectric and ferrielectric phases. Therefore we, believe that the co-existence of two phases in the temperature range between AF and SmC^* (or $1/2 < q_T < 1$) is the volume property of the liquid crystal and it is unaffected by the surface interactions.

VII. ON THE STABILITY OF FERRIELECTRIC PHASES

We can summarize the experimental results as follows:

- The thermodynamically stable antiferroelectric and ferrielectric phases with $q_T \leq 1/2$ (SmC_A^*, SmC_γ^*, and AF) have been found experimentally.
- The herringbone structure of antiferroelectric SmC_A^* phase, in which the molecular tilt directions in neighboring smectic layers are almost opposite to each other, is in agreement with most theoretical models.
- The molecular structure of the SmC_γ^* and AF phases has not been unambiguously established; X-ray diffraction supports the clock

model, whereas dielectric spectroscopy, apparent tilt angle, and macroscopic spontaneous polarization measurements support the Ising and/or ANNNI models.

● The ferrielectric phases, which appear in between the AF and SmC* phases ($1/2 < q_T < 1$), seem to be a mixture of domains of the ferri-electric and ferroelectric phases at the microscopic level, i.e., at the level of a finite number of smectic layers.

The last statement needs further elaboration. According to the Ising model, the q_T parameter of the ferrielectric phase—which exists between AF ($q_T = 1/2$) and SmC* ($q_T = 1$)—must be in the range of 1/2 to 1. The various experimental investigations show that this phase seems to be thermodynamically unstable. According to the numerous experimental investigations, the Ising and/or ANNNI models satisfactorily describe the temperature-induced Devil's staircase for the ferrielectric antiferroelectric phases with $q_T \leq 1/2$, whereas the appearance of higher temperature phases is still under discussion. This disagreement between theory and experi-mental results is owing to the fact that both the Ising and the ANNNI models ignore the electrostatic interaction between the ferroelectric spins. Consider the ferrielectric structure (e.g., SmC$^*_\gamma$) shown in Figure 23(a). In such a structure, the ferroelectric ordering appears as defects in the antiferroelectric structure $AAAA$. All ferroelectric spins are pointed in the same (positive) direction and, therefore, repel each other owing to Coulomb interactions. Such forces stabilize the equidistant arrangement of F-spins ($\uparrow$) in the $AAAA$ matrix, resulting in the SmC$^*_\gamma$ arrangement: ... $\uparrow AA \uparrow AA \uparrow AA$.... In the ferrielectric phases with $q_T > 1/2$ we have a different situation, namely A ordering appears as defects in the $FFFF$ orderings. "For such a structure, what is important for its stable existence of the repulsive forces between A orderings, it is not clear whether A orderings repel one another [8]".

Let us consider, for example, some ferrielectric phase with $q_T = 3/5$. Such a phase possesses the following arrangement of spins: $\uparrow\uparrow A \downarrow A \uparrow\uparrow A \downarrow$ $A \uparrow\uparrow A \downarrow A$..., where $\uparrow$ and $\downarrow$ are the ferroelectric spins directed upward and downward, respectively. The arrangement $\uparrow\uparrow AA \downarrow A$ possesses the same electrostatic energy as $\uparrow A \downarrow\downarrow A$, and both of them are unstable owing to a repulsive interaction between neighboring spins of the same directions ($\uparrow, \uparrow$ or $\downarrow, \downarrow$) and attractive interaction between the next neighboring spins of opposite direction ($\uparrow, \downarrow$), as shown in Figure 23(b). The same situation will be valid for any ferrielectric structure with $q_T = m/n$, with $m \neq 1$. Therefore, applying this restriction to the Ising and ANNNI models, we conclude that the ferrielectric mesophases with $q_T > 1/2$ could not have a stable thermodynamic existence and vice versa; the range of stable ferrielectric mesophases is restricted by the q_T parameter from 0 to 1/2, with $m = 1$.

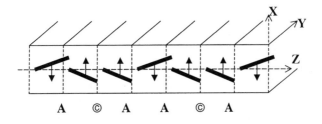

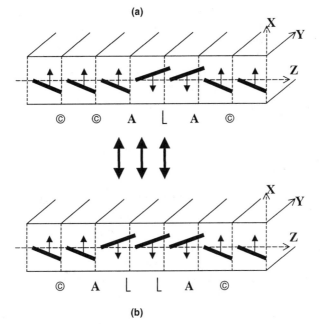

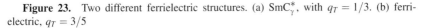

Figure 23. Two different ferrielectric structures. (a) SmC$_\gamma^*$, with $q_T = 1/3$. (b) ferri-electric, $q_T = 3/5$

A similar conclusion follows explicitly from the expanded Landau model [5]. Žekš and Čepič [15] introduced the Lifshitz invariant for chiral systems and obtained 12 possible sequences of the phase transitions between different ferrielectric and antiferroelectric phases. They predicted the existence of incommensurate double-modulated phases between common mesophases. They did not investigate the effect of the direct electric field on these phases, but the coexistence of the two Goldstone relaxation processes (ferrielectric and ferroelectric) follows directly from the nature of these double-modulated incommensurate phases.

From these explanations, the ferrielectric FiLC phase with $q_T > 1/2$ can be considered as an incommensurate double-modulated phase or mixture of SmC$_\gamma^*$ and SmC* domains.

VIII.   THE SUCCESSIVE PHASE TRANSITIONS IN
CHIRAL-TILTED SMECTICS

The unusual, unstable nature of the high-temperature ferrielectric phase FiLC introduces problems in their investigation and determination. This is also the reason why in several publications this phase has been mixed with the so-called SmC_β^* and even with SmC_α^* phases. In the first investigated AFLC material (MHPOBC), several phases were detected by DSC and termed SmC_α^*, SmC_β^* and SmC_γ^*, with a decrease in temperature. The SmC_β^* phase is considered a usual ferroelectric SmC^*. Nevertheless, later investigations show that this assignment is not so straightforward and this phase shows some ferrielectric properties [36]. The careful analysis of experimental data [76] and previous publications on MHPOBC [77–79] shows that SmC_β^* reveals properties that are similar to those of the FiLC. Moreover, such properties have been observed in many other compounds [33,35,43,67,80–84], and this phase has been referred as SmC^*, SmC_β^*, and even SmC_α^*. There are factors that may confuse the precise identification of FiLC phase. This phase appears in the between SmC_γ^* (or AF) and SmA* (or SmC_α^*) and is not detectable by DSC owing to its unstable nature and because the phase transitions between neighboring phases are blurred. The plots of electric field-induced tilt angle and spontaneous polarization in the FiLC phase are somewhat similar to SmC_α^*. Several AFLC compounds, that had been claimed to possess the wide range SmC_α^*, were investigated using dielectric spectroscopy and induced spontaneous polarization [76]. It was found that the SmC_α^* phase exists in only a very narrow range of temperatures ($\approx 1°C$), and the lower temperature region appears to be the FiLC phase. Another complexity in the phase-sequence identification is the strong effect of the surfaces on the appearance of the ferrielectric phases (Fig. 15). The safe cell thickness at which the ferrielectric phases are not affected by the surface was found to be ≥ 50 μm [39]. In summary, the full sequence of the phases in tilted-chiral smectics is shown in Table I [18,24,39,72].

This phase sequence needs additional comments:

- Among these phases, four—SmC_A^*, SmC_γ^*, AF, and SmC^*—are the ground states in the ANNNI model, and their stable existence has been confirmed in numerous investigations.

- The structures of the SmC_A^*, SmC^* phases suggested by different models agree with each other.

- The structures of the SmC_γ^* and AF phases are essentially different, although all the models suggest the three-layer and four-layer periodicity for them.

- Three intermediate states or subphases [24] (spr3–spr1), that appear between those four ground phases are predicted by ANNNI model.

TABLE I
Phases of the Tilted Chiral Smectics

q_T	0	1/5	1/3	?	1/2	?	1	?
	$SmC_A^{*\,a}$	spr3 $Fi_L{}^b$	SmC_γ^*	spr2 $Fi_H{}^b$	AF	spr1 Fi^c, $FILC^d$	SmC^*	SmC_α^*
Suggested names	SmC_A^*		SmC_3^*		SmC_4^*	FeIII	SmC^*	SmC_α^*

Temperature →

a [24].
b [18].
c [72].
d [39].

Nevertheless, only the spr3 (Fi_L) phase was found to be thermo-dynamically stable and ferrielectric-like, and it possesses an electric field–induced state with an apparent tilt angle of $\theta/5$, similar to the $\theta/3$ in SmC_γ^*. Therefore, the possible structure of spr3 phase can be described by $q_T = 1/5$.

- A different situation holds for the spr1 and spr2 phases, which do not show any quasi-stable electric field–induced states and seem to possess no single periodical structure, although their existence has been predicted by the ANNNI model. This inconsistency can be resolved by taking in account the attractive and repulsive interactions between ferroelectric spins. According to these considerations, the q_T of possible stable phase are restricted by the values $1/n$, where n is 1, 2, 3, 5, 7 ... Therefore, only one intermediate phase—spr3—is thermo-dynamically stable.

All the existing models are phenomenologic and do not describe the physical origin of forces, which stabilize ferrielectricity in the tilted chiral smectics. Nevertheless, the essential feature of the most advanced models is a long-range (at least several smectic layers) interaction between the ferroelectric spins and/or local spontaneous polarization of the smectic layers. In this concern, the importance of long-range interactions on the stability or appearance of ferrielectric phases in AFLCs was experimentally confirmed in free-standing films [24,85–90] and in confined geometry [57]. In free-standing AFLC films, the number of ferrielectric phases decreases with the decrease of the thickness of the free-standing film, whereas the antiferroelectric ordering (phase) remains even in a two-layer film. In 12OBP1M7 confined in cylindrical pores, the ferrielectric phases do not appear in 0.02-μm pores, but they exist in bigger (0.2-μm) pores. This fact also supports the importance of long-range interactions for the stability of ferrielectricity in chiral smectics.

IX.   A RECENT EXAMINATION OF THE STRUCTURE OF FERRIELECTRIC PHASES

In summarizing the different approaches to antiferroelectric/ferrielectric liquid crystals, it can be noted that most of these satisfactorily explain the phase sequence changes with temperature and some changes with electric field. The next example shows a good agreement of the ANNNI model with the experimentally found phase transition sequence.

Figure 24 presents a dependence of the normalized induced spontaneous polarization on the applied electric field for AFLC compound 11OBP1M7 [91]: $SmC_A^*-58°C-SmC_\gamma^*-72°C-AF-77.5°C-FiLC-82°C-SmC^*-88°C-SmA^*$. The most interesting results are (1) the existence of field induced state with $P_s/3$ in the temperature range below AF phase and (2) a second state with sufficiently large spontaneous polarization (65%–85% of P_s) in the higher temperature range. On the other hand, the application of electric field in AF phase gives one of the ($P_s/3$ or the high P_s) states depending on the temperature: $P_s/3$ for temperature of 73°C and high P_s state for 77°C.

The experimental curves are in good agreement with the phase diagram on E versus J_1 plane for the normalized temperature $T = 6.0$ [26], presented in Fig. 25. The parameter J_1 is considered to be temperature dependent: $J_1(T) = J_0 + TS$, J_0 accounts for the dipole-dipole interactions. The higher values of J_1 parameter correspond to the higher temperature. Therefore, the X-axis in Fig. 25 shows the possible successive phase transitions with temperature without electric field. The q parameter describing these subphases decreases with temperature increasing from $\frac{1}{2}$ to 0, forming a phase sequence which is similar to "temperature Devil's staircase". The intermediate ferrielectric phases can appear between three ground phases: SmC_A^*, SmC_γ^*, and AF, while the state between AF and SmC*, shown by the dashed line is unstable and it corresponds to FiLC state. This is in good agreement with the experimentally found phase transition sequence for AFLC compound 11OBP1M7, where FiLC phase appears between the AF and SmC*.

The comparison of the experimental results in Fig. 24 with the phase diagram, based on the ANNNI model shows fairly good qualitative agreement between the theory and experiments. The experimental curves in Fig. 24 are denoted by capital letters which correspond to electric field induced phase transitions shown in Fig. 25. For example, at temperature of 77°C (AF phase, curve D) at low electric field, the structure remains antiferroelectric (no induced polarization). A further increase in the electric field induces the high P_s state and finally SmC*. At lower temperature

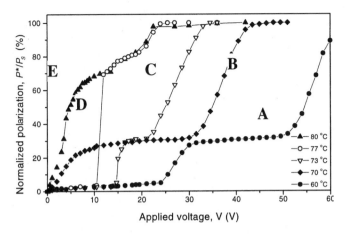

Figure 24. The dependence of normalized induced spontaneous polarization on the applied electric field for the AFLC compound (11OBP1M7) at different temperatures (phases). The capital letters correspond to the electric field induced phase transitions on the diagram shown in Fig. 25.

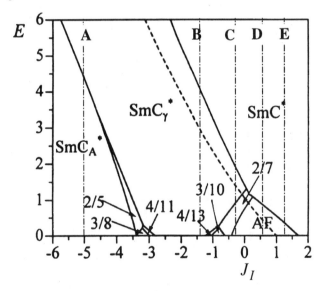

Figure 25. Phase diagram on E versus J_1 plane for the normalized temperature $T = 6.0$. (Redrawn with permission from Ref. [26].)

$T = 73°C$ (AF phase, curve C), the increase in the electric field induces SmC_γ phase ($P = P_s/3$), then the high P_s state, and finally SmC*. Although the ANNNI model is based on the concept of nonphysical spins, the mathematical model explains well both the temperature and the electrically induced phase transition sequences.

312 YU. P. PANARIN AND J. K. VIJ

The major difference between the different models is the disagreement in the structures of ferrielectric phases. Nevertheless, all models suggest a two layer periodicity for SmC_A^*, a three layer periodicity for SmC_γ^*, and a four layer periodicity for the AF. Therefore, the alternative names for these phases can be SmC_2^*, SmC_3^*, SmC_4^*, respectively, where the subscript denotes the structure period. It was shown that spin structures are supported by the polarization and the apparent tilt angle measurements, while the "clock" structure is supported by the X-ray diffraction observation [45]. This contradiction may imply that the real molecular structures are somehow in between these opposite cases. Such structures have recently been suggested by Pikin et al. [30]. In the short/long pitch model, the short pitch structure responsible for the molecular arrangement in the neighboring smectic layers (clock-like structure), is disturbed by the long-pitch mode. This causes the azimuthal angle between the neighboring smectic layers to be different from $\Delta\varphi = 2\pi/n$, where n is the period of the structure. The resulting angle is $\Delta\varphi = 2\pi/n + \delta$. Figure 26 presents the molecular structures in three different phases for the clock model, the spin model, and the short/long pitch model. Last structures are combinations of the two opposite structures. The structure of SmC_γ^* (SmC_3^*) possesses the spontaneous polarization

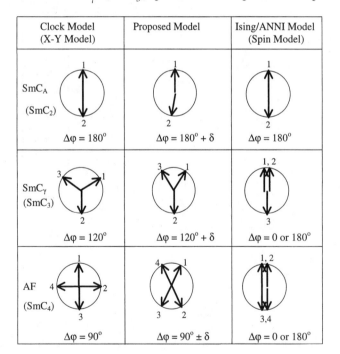

Figure 26. The possible molecular structures in SmC_A^*, SmC_γ^* and AF phases according to the clock model, the spin model, and the short/long pitch model.

$P = (2\cos(60° - \delta) - 1)/3$ while there is no polarization in the four layer structure (AF). It is clear, that for $\delta = 0°$, the structure is pure "clock-like" and for $\delta = 60°$ it is "spin-like". In other words, the proposed structure is a compromise between that given by the two main models.

X. CONCLUSION

We note that although considerable progress in the understanding of the ferrielectric phases in AFLCs has been made, the molecular structure of these phases is still under discussion. Nevertheless, the ANNNI model is the most successful in predicting a variety of the ferrielectric phases, that have been experimentally found (Table 1). The thermodynamic stability of some ferrielectric phases can be assessed by taking into account the attractive and repulsive interactions between the ferroelectric spins. The initial bilayered models are not suitable for describing the structure of even the simplest ferrielectric phases, (e.g., SmC_γ^* and AF) that possess three- and four-layer periodicity. Such periodicity follows from the advanced bilayered models, short-pitch model, and the Ising and ANNNI models, whereas the molecular structure of ferrielectric phases suggested by different models are considerably different from each other. In our opinion, among the theoretical models, only the modified ANNNI model [26] satisfactorily describes the structural changes (or phase transitions) in the ferrielectric phases with electric field, which were observed experimentally.

Despite a significant progress in the understanding of ferrielectricity in tilted chiral smectics, some of the phenomena are still not understood. The most important problems that remain to be solved are the determination of the precise molecular structure in the ferrielectric phases, the successive phase transitions with the electric field, and the physical nature of the short and long-range interactions that stabilize the ferrielectric phases.

Acknowledgments

We thank Professor A. Fukuda for discussions Professor J. W. Goodby and Dr. M. Hird, both of the University of Hull, UK, for a very fruitful collaboration and Ms O. Kalinovskaya for help with the experiments and for preparing the manuscript. The European Community is thanked for financial support.

References

1. L. A. Beresnev, L. M. Blinov, V. A. Baikalov, E. P. Pozhidaev, G. V. Purvanetskas, and A. I. Pavluchenko, *Mol. Cryst. Liq. Cryst.* **89**, 327 (1982).

2. A. D. L. Chandani, T. Hagiwara, Y. Suzuki, Y. Ouchi, H. Takezoe, and A. Fukuda, *Jpn. J. Appl. Phys.* **27**, L729 (1988).

3. A. D. L. Chandani, E. Gorecka, Y. Ouchi, H. Takezoe, and A. Fukuda, *Jpn. J. Appl. Phys.* **28**, L1265 (1989).

4. I. Nishiyama, A. Yoshizawa, M. Fukumasa, and T. Hirai, *Jpn. J. Appl. Phys*, **28**, L2248 (1989).

5. Y. Yamada, N. Yamamoto, K. Mori, K. Nakamura, T. Nagiwara, Y. Suzuki, I. Kawamura, H. Orihara, and Y. Ishibashi, *Jpn. J. Appl. Phys.*, **29**, 1757 (1990); Y. Yamada, N. Yamamoto, M. Yamawaki, I. Kawamura, and Y. Suzuki, *Jpn. Display'92*, 57 (1992).

6. E. Gorecka, A. D. L. Chandani, Y. Ouchi, H. Takezoe, and A. Fukuda., *Jpn. J. Appl. Phys.* **29**, L131 (1990).

7. J. W. Goodby, I. Nishiyama, A. J. Slaney, C. J. Booth, and K. J. Toyne, *Liq. Cryst.* **14**, 37 (1993).

8. A. Fukuda, Y. Takanishi, T. Isozaki, K. Ishikawa, and H. Takezoe., *J. Mater. Chem.* **4**, 997 (1994).

9. A. M. Levelut, C. Germain, P. Keller, L. Liebert, and J. Billard, *J. Phys. Paris* **44**, 623 (1983); A. M. Levelut, R. J. Tarento, F. Hardouin, M. F. Achard, and G. Sigaud, *Phys. Rev. A* **44**, 2180 (1981); F. Hardouin, N. H. Tinh, and A. M. Levelut, *J. Phys. Lett.* **43**, 623 (1982).

10. H. R. Brand and P. E. Cladis, *Mol. Cryst. Liq. Cryst*, **114**, 207 (1984); P. E. Cladis and H. R. Brand, *Liq. Cryst.* **14**, 1327 (1993).

11. S. Galerne and L. Liebert, *Phys. Rev. Lett.*, **64**, 906 (1990); S. Galerne and L. Liebert, *Phys. Rev. Lett.* **66**, 2891 (1991).

12. L. Benguigui and F. Hardouin, *J. Phys. Lett.* **42**, L381 (1981).

13. H. Orihara and Y. Ishibashi, *Jpn. J. Appl. Phys.* **29**, L115 (1990).

14. B. Žekš, R. Blinc, and M. Čepič, *Ferroelectrics* **122**, 221 (1991).

15. B. Žekš and M. Čepič, *Liq. Cryst.* **14**, 445 (1993).

16. V. L. Lorman, A. A. Bulbitch, and P. Toledano, *Phys. Rev. E* **49**, 1367 (1994).

17. P. Bak and R. Bruinsma, *Phys. Rev. Lett.* **49**, 249 (1982).

18. T. Isozaki, T. Fujikawa, H. Takezoe, A. Fukuda, T. Hagiwara, Y. Suzuki, and I. Kawamura, *Jpn. J. Appl. Phys.* **31**, L1435 (1992).

19. T. Isozaki, K. Hiraoka, Y. Takanishi, H. Takezoe, A. Fukuda, Y. Suzuki, and I. Kawamura, *Liq. Cryst.* **12**, 59 (1992).

20. P. Bak and J. von Boehm, *Phys. Rev. B* **21**, 5297 (1980).

21. Y. Yamada and N. Hamaya, *J. Phys. Soc. Jpn.* **52**, 3466 (1983).

22. M. Yamashita and S. Miyazima, *Ferroelectrics* **148**, 1 (1993).

23. T. Isozaki, H. Takezoe, A. Fukuda, Y. Suzuki, and I. Kawamura, *J. Mater. Chem.* **4**, 237 (1994).

24. N. Itoh, M. Kabe, K. Miyachi, Y. Takanishi, K. Ishikawa, H. Takezoe, and A. Fukuda, *J. Mater. Chem.* **7**, 407 (1997).

25. M. Yamashita. *Mol. Cryst. Liq. Cryst.* **263**, 93 (1995); *Ferroelectrics* **181**, 201 (1996).

26. M. Yamashita and S. Tanaka, *Jpn. J. Appl. Phys.* **37**, L528 (1998); S. Tanaka and M. Yamashita, paper presented at ILCC, Strasbourg, 1998.

27. V. L. Lorman, *Mol. Cryst. Liq. Cryst.* **262**, 437 (1995).

28. M. Čepič and B. Žekš, *Mol. Cryst. Liq. Cryst.* **263**, 61 (1995).

29. S. A. Pikin, S. Hiller, and W. Haase, *Mol. Cryst. Liq. Cryst.* **262**, 425 (1995).

30. S. A. Pikin, M. Gorkunov, D. Killian, and W. Haase, *Liq. Cryst.* **26**(8), 1107, 1114 (1999).

31. A. D. L. Chandani, Y. Ouchi, H. Takezoe, A. Fukuda, K. Terashima, K. Furukawa, and A. Kishi, *J. Appl. Phys.* **28**, L1261 (1989).

32. K. Ema, H. Yao, I. Kawamura, T. Chang, and C. W. Garland, *Phys. Rev. E* **47**, 1203 (1993).

33. I. Nishiyama and J. W. Goodby, *J. Mater. Chem.* **2**, 1015 (1992); *J. Mater. Chem.* **3** 149 (1993); *J. Mater. Chem.* **3**, 169 (1993).

34. J. W. Goodby, J. S. Patel, and E. Chin, *J. Mater. Chem.* **2**, 197 (1992).

35. I. Nishiyama and J. W. Goodby, *J. Mater. Chem.* **3**, 161 (1993).

36. K. Hiraoka, Y. Takanishi, K. Skarp, H. Takezoe, and A. Fukuda, *J. Appl. Phys.* **30**, L1819 (1991).

37. Y. P. Panarin, O. E. Kalinovskaya, and J. K. Vij., *Mol. Cryst. Liq. Cryst.* **302**, 93 (1997).

38. L. J. Baylis, H. F. Gleeson, A. J. Seed, P. Styring, M. Hird, and J. W. Goodby, *Proc. 17th Int. Liq. Cryst. Conf.*, ILCC, Strasbourg, 1998, pp. 2–187.

39. Y. P. Panarin, O. E. Kalinovskaya, J. K. Vij, and J. W. Goodby, *Phys. Rev. E* **55**, 4345 (1997).

39a. A. Schönfeld and F. Kremer, *Ber. Bunsenges. Phys. Chem.* **97**, 1237 (1993).

40. Y. Takanishi, K. Hiraoka, V. K. Agrawal, H. Takezoe, A. Fukuda, and M. Matsushita, *J. Appl. Phys.* **30**, 2023 (1991).

41. Y. Takanishi, A. Ikeda, H. Takezoe, and A. Fukuda, *Phys. Rev. E* **51**, 400 (1995).

42. H. F. Gleeson, L. Baylis, W. K. Robinson, J. T. Mills, J. W. Goodby, M. Hird, P. Styring, C. Rosenblatt, and S. Zhang, paper presented at ILCC, Strasbourg, 1998.

43. H. F. Gleeson, L. Baylis, W. K. Robinson, J. T. Mills, J. W. Goodby, A. Seed, M. Hird, P. Styring, C. Rosenblatt, and S. Zhang, *Liq. Cryst.* **26**, 1415 (1999).

44. M. Neundorf, S. Diele, S. Ernst, S. Saito, D. Demus, T. Inukai, and M. Murashiro, *Ferroelectrics* **147**, 95 (1993); M. Neundorf, Y. Takanishi, A. Fukuda, S. Saito, M. Murashiro, T. Inukai, and D. Demus, *J. Mater. Chem.* **5**, 2221 (1995).

45. P. Mach, R. Pindak, A.-M. Levelut, P. Barois, H. T. Nguyen, C. C. Huang, and L. Furenlid, *Phys. Rev. Lett.* **81**, 1015 (1998).

46. V. E. Dmitrienko, *Acta Crystalogr. Sect. A* **42**, 478 (1986).

47. K. H. Kim, Y. Takanishi, K. Ishikawa, H. Takezoe, and A. Fukuda, *Liq. Cryst.* **16**, 185 (1994); K. H. Kim, K. Miyachi, K. Ishikawa, H. Takezoe, and A. Fukuda. *J. Appl. Phys.* **33**, 5850 (1994).

48. K. H. Kim, K. Ishikawa, H. Takezoe, and A. Fukuda, *Phys. Rev. E* **51**, 2166 (1995).

49. K. Miyachi, J. Matsushima, Y. Takanishi, K. Ishikawa, H. Takezoe, and A. Fukuda, *Phys. Rev. E* **52**, R2153 (1995).

50. B. Jin, Z. Ling, Y. Takanishi, K. Ishikawa, H. Takezoe, A. Fukuda, M. Kakimoto, and T. Kitazume, *Phys. Rev. E,* **53**, R4295 (1996).

51. A. Kocot, R. Wrzalik, B. Orgasinska, T. Perova, J. K. Vij, and H. T. Nguyen, *Phys. Rev. E* **59**, 551 (1999).

52. S. Hiller, S. A. Pikin, W. Haase, J. W. Goodby, and I. Nishiyama, *Jpn. J. Appl. Phys.* **33**, L1170 (1994).

53. S. Hiller, S. A. Pikin, W. Haase, J. W. Goodby, and I. Nishiyama, *Jpn. J. Appl. Phys.* **33**, L1096 (1994).

54. Y. P. Panarin, H. Xu, S. T. Mac Lughadha, J. K. Vij, A. J. Seed, M. Hird, and J. W. Goodby, *J. Phys. Condens. Mater.* **7**, L351 (1995).

55. Y. P. Panarin, H. Xu, S. T. Mac Lughadha, J. K. Vij, A. J. Seed, M. Hird, and J. W. Goodby, *Mol. Mat.* **6**, 69 (1996).

56. Y. P. Panarin, O. E. Kalinovskaya, and J. K. Vij, *Mol. Cryst. Liq. Cryst.* **301**, 215 (1997).

57. Y. P. Panarin, C. Rosenblatt, and F. M. Aliev, *Phys. Rev. Lett.* **81**, 2699 (1998).

58. K. Hiraoka, A. Taguchi, Y. Ouchi, H. Takezoe, and A. Fukuda, *J. Appl. Phys.* **29**, L103 (1990).

59. M. Fukui, H. Orihara, A. Suzuki, Y. Ishibashi, Y. Yamada, N. Yamamoto, K. Mori, K. Nakamura, Y. Suzuki, and I. Kawamura, *Jpn. J. Appl. Phys.* **29**, L329 (1990).

60. K. Hiraoka, A. D. L. Chandani, E. Gorecka, Y. Ouchi, H. Takezoe, and A. Fukuda, *J. Appl. Phys.* **29**, L1473 (1990).

61. H. Sun, H. Orihara, and Y. Ishibashi, *J. Phys. Soc. Jpn.* **60**, 4175 (1991).

62. H. Sun, H. Orihara, and Y. Ishibashi, *J. Phys. Soc. Jpn.* **62**, 2066, 2706 (1993); *J. Phys. Soc. Jpn.* **62**, 2706 (1993).

316   YU. P. PANARIN AND J. K. VIJ

63. I. Muševič, R. Blinc, B. Žekš, M. Copič, M. M. Wittebrood, T. Rasing, H. Orihara, and Y. Ishibashi, *Phys. Rev. Lett.* **71**, 1180 (1993).

64. K. Hiraoka, H. Takezoe, and A. Fukuda, *Ferroelectrics* **147** 13 (1993).

65. T. Isozaki, T. Fujikawa, H. Takezoe, A. Fukuda, T. Hagiwara, Y. Suzuki, and I. Kawamura, *Phys. Rev. B* **48**, 13439 (1993).

66. M. Buivydas, F. Gouda, S. T. Lagerwall, and B. Stebler, *Liq. Cryst.* **18**, 879 (1995).

67. S. Merino, M. R. de la Fuente, Y. Gonzalez, M. A. Perez Jubindo, and J. A. Puertolas, *Phys. Rev. E* **54**, 5169 (1996).

68. Y. P. Panarin, O. E. Kalinovskaya, and J. K. Vij, *Appl. Phys. Lett.* **72**, 1667 (1998); Y. P. Panarin, O. E. Kalinovskaya, and J. K. Vij, *Liquid Crystals* **25**, 241 (1998).

69. Y. Kimura, R. Hayakawa, N. Okabe, and Y. Suzuki, *Phys. Rev. E.* **53**, 6080 (1996); *Mol. Cryst. Liq. Cryst.* **304**, 275 (1997).

70. I. Muševič, M. Čepič, B. Zeks, M. Čopič, D. Moro, and G. Heppke, *Phys. Rev. Lett.* **77**, 1769 (1996).

71. H. Orihara, Y. Igasaki, and Y. Ishibashi, *Ferroelectrics* **147**, 67 (1993).

72. T. Isozaki, K. Ishikawa, H. Takezoe, and A. Fukuda, *Ferroelectrics* **147**, 121 (1993).

73. J. Hatano, Y. Hanakai, H. Furue, H. Uehara, S. Saito, and K. Murashiro, *Jpn. J. Appl. Phys.* **33**, 5498 (1994).

74. M. Čepič, G. Heppke, J.-M. Hollidt, D. Lotzsch, and B. Žekš, *Ferroelectrics* **147**, 159 (1993).

75. J.-F. Li, E. A. Shack, Y.-K. Yu, X.-Y. Wang, C. Rosenblatt, E. Neubert, S. S. Keast, and H. F. Gleeson, *Jpn. J. Appl. Phys.* **35**, L1608 (1996).

76. O. E. Kalinovskaya, Y. P. Panarin, and J. K. Vij, unpublished data.

77. J. Hou, J. Schacht, F. Giebelmann, and P. Zugenmaier. *Liq. Cryst.* **22** 409 (1997).

78. T. Sako, Y. Kimura, R. Hayakawa, N. Okabe, and Y. Suzuki, *J. Appl. Phys.* **35**, L114 (1996).

79. J. Lee, A. D. L. Chandani, K. Itoh, Y. Ouchi, H. Takezoe, and A. Fukuda, *Jpn. J. Appl. Phys.* **29**, 1122 (1990).

80. W. K. Robinson, C. Carboni, H. F. Gleeson, M. Hird, P. Styring, and A. J. Seed, *Mol. Cryst. Liq. Cryst.* **263**, 69 (1995).

81. K. Hiraoka, Y. Takanishi, H. Takezoe, A. Fukuda, T. Isozaki, Y. Suzuki, and I. Kawamura, *J. Appl. Phys.* **31**, 3394 (1992).

82. K. Hiraoka, T. Tsumita, Y. Sugiyama, K. Monzen, Y. Uematsu, and Y.-I. Suzuki, *J. Appl. Phys.* **36**, 6847 (1997).

83. K. Hiraoka, Y. Uematsu, H. Takezoe, and A. Fukuda, *J. Appl. Phys.* **35**, 6157 (1996).

84. J. Hatano, M. Sato, K. Iwauchi, T. Tsukamoto, S. Saito, and K. Murashiro, *Ferroelectrics* **147**, 217 (1993).

85. C. Bahr and D. Fliegner, *Phys. Rev. Lett.* **70**, 1842 (1993); *Ferroelectrics* **147**, 1 (1993).

86. C. Bahr, C. J. Booth, D. Fliegner, and J. W. Goodby, *Ferroelectrics* **178**, 229 (1996).

87. C. Bahr, D. Fliegner, C. J. Booth, and J. W. Goodby, *Phys. Rev. E* **51**, 3823 (1996).

88. F. Gouda, A. Dahlgren, S. T. Lagerwall, B. Stebler, J. Bömelburg, and G. Heppke, *Ferroelectrics* **178**, 187 (1996).

89. E. Demikhov, *JETP Lett.* **61**, 977 (1996).

90. D. R. Link, J. E. Maclennan, and N. A. Clark, *Phys. Rev. Lett.* **77**, 2237 (1996).

91. Yu. P. Panarin, O. E. Kalinovskaya, J. K. Vij, D. Parghi, M. Hird, and J. W. Goodby, *Ferroelectrics*, 2000, in press.

ORDER PARAMETER VARIATION IN SMECTIC LIQUID CRYSTALS

STEVE J. ELSTON AND NIGEL J. MOTTRAM*

*Department of Engineering Science, University of Oxford
Parks Road, Oxford, United Kingdom*

CONTENTS

I. INTRODUCTION

In this chapter we discuss some of the external influences on the important order parameters in smectic liquid crystals. These order parameters are defined in Figure 1 and are generally temperature dependent. It is also possible, however, to change them through the application of electric fields and mechanical stresses. In some cases, these stresses occur naturally within certain structures that can form in smectic liquid crystals, and these are of considerable interest. Before considering smectics however, we discuss the more common nematic liquid crystal phase.

Liquid crystal phases form when a substance has a degree of crystalline ordering (orientational and/or positional) while remaining in a liquid state. The simplest of these is the nematic phase, which typically consists of long, thin molecules with a degree of orientational order but no long-range

*Permanent address: Department of Mathematics, University of Strathclyde, Livingstone Tower, 26 Richmond Street, Glasgow G1 1XH, United Kingdom.

Advances in Liquid Crystals: A Special Volume of Advances in Chemical Physics, Volume 113, edited by Jagdish K. Vij. Series Editors I. Prigogine and Stuart A. Rice.
ISBN 0-471-18083-1. © 2000 John Wiley & Sons, Inc.

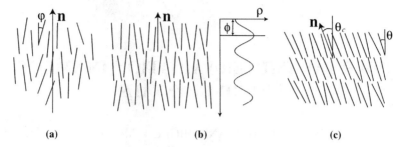

Figure 1. Order parameters in the liquid crystal phases discussed in the text. (a) In the nematic phase, the orientational order parameter is s. $s = <3\cos^2 \varphi - 1 > 2$. (b) In the smectic A phase, the positional order parameter is the complex quantity ψ for which the amplitude ρ depends on the smectic layer density modulation and the angle ϕ depends on the layer position. $\psi = \rho e^{i\phi}$. (c) In the smectic C phase, the molecules tilt at a characteristic angle θ_c to the smectic layer normal (the subscript is normally dropped). $\theta_c = \langle \theta \rangle$.

positional order. This phase forms largely owing to the steric interactions between such molecules, where the bulk energy is minimized by the molecules aligning in roughly the same direction. In fact, a balance is formed between the steric (or mechanical) interactions, which tend toward molecular alignment, and the thermal energy, which tends to randomize this.

The material can be characterized by two key parameters [Fig. 1(a)]. These are the average molecular long axis orientation, given by the unit vector **n**, and a measure of the degree of orientational ordering around this direction, given by the order parameter s, where

$$s = \tfrac{1}{2} \langle 3\cos^2 \varphi - 1 \rangle \tag{1}$$

φ is a measure of the molecular tilt away from **n**. For bulk material, s is then an ergodic variable representing the degree of the orientational ordering. Clearly, if there is no ordering φ is uniformly (spherically) distributed over all angles and consequently $s=0$, in which case the material is in the isotropic phase, whereas for a nematic liquid crystal material $s > 0$.

Now we can write an expansion for the energy density of the nematic liquid crystal phase close to the nematic to isotropic phase transition temperature as [1]

$$f = f_0 + \frac{1}{2}As^2 - \frac{B}{3}s^3 + \frac{C}{4}s^4 + O(s^5) \tag{2}$$

where A is a temperature-dependent coefficient and B and C are largely temperature independent. The minimization of this free energy density at any given temperature then determines the bulk equilibrium value of s. Any external influence on the material (applied electric field, etc.) that changes s away from this equilibrium value will modify the energy density.

The second source of internal energy in a nematic liquid crystal is that owing to elastic distortion away from the equilibrium configuration [Fig. 1(a)]. Any general bulk distortion can be broken into terms associated with splay, twist, and bend deformations. These have the three elastic constants K_{11}, K_{22}, and K_{33} associated with them; and the elastic internal energy density can be expressed as [2]

$$f_d = \tfrac{1}{2}K_{11}(\nabla \cdot \mathbf{n})^2 + \tfrac{1}{2}K_{22}(\mathbf{n} \cdot \nabla \times \mathbf{n})^2 + \tfrac{1}{2}K_{33}(\mathbf{n} \times \nabla \times \mathbf{n})^2 \qquad (3)$$

or in the one elastic constant approximation $K_{11}=K_{22}=K_{33}=K$ as

$$f_d = \tfrac{1}{2}K\{(\nabla \cdot \mathbf{n})^2 + (\nabla \times \mathbf{n})^2\} \qquad (4)$$

the latter of which will be a useful approximation when dealing with smectics. It is also useful to note that the value of K is influenced by the order parameter s. Clearly, when $s=0$, there is no anisotropy and K must be zero. In fact it is found that

$$K \approx K_0 s^2 \qquad (5)$$

Within a nematic material in a typical liquid crystal display device the structures are such that s can be treated as constant. This means that the director profile can be calculated by minimizing the elastic energy given by Eq. (3), together with any external field and boundary conditions imposed. It is this that has allowed the power of nematic continuum theory to be used in the development of liquid crystal display technology.

Close to the phase transition temperature and in regions of high distortion, however, it is important to consider variation in the order parameter s. Two examples of such regions of high distortion are those near to surfaces and defects.

The effect of surfaces has been investigated experimentally and theoretically [3,4]. The authors showed that surface roughness causes a decrease in the nematic order parameter from its bulk value. When the characteristic length scale of the surface roughness is shorter than the nematic–isotropic coherence length, a decrease in the nematic order parameter close to the surface reduces the large curvature energy induced by the boundaries.

In a similar way, the region of high distortion close to a defect such as a disclination line melts to minimize the free energy of the system. Schopohl and Sluckin [5] showed that the nematic order parameter not only decreased near the core of a disclination line but was biaxial there even though the bulk nematic phase was uniaxial. Mottram and Hogan [6] extended this work to consider the effect this region of low order has on the nematic–isotropic phase transition. They found that defect cores were able to seed the transition from nematic to isotropic at a lower temperature than that at which the nematic state lost stability.

Other surface-induced nematic order parameter variation effects were recently considered by Skacej et al. [7]. In this work, the coupling between director deformation and the nematic order parameter leads to a subsurface variation in the average molecular orientation. The consequence of this is a contribution to the surface anchoring condition resulting from the intrinsic anchoring energy owing to a spatial variation of the scalar order parameter close to the surface.

Although the structures in the vicinity of large distortions lead to order parameter changes in nematic liquid crystal materials, it is generally more important to take into account order parameter variations in smectic liquid crystal materials. Often the structures that form in these materials involve order parameter variation as a key part of their formation, and this must, therefore, be included in any modeling undertaken. We now go on to consider such materials.

II. SMECTIC A

The smectic A liquid crystal phase is similar to the nematic phase in that it has orientationally ordered molecules. Now, however, the molecules also form layers; the bulk-layered structure is perpendicular to the orientation of the director. More strictly, the layering is a density wave along the director, with a period approximately equal to a molecular length [Fig. 1(b)]. If the harmonic series of this density wave is considered, then the fundamental term dominates and an order parameter for the smectic A phase can be expressed as

$$\psi = \rho \exp i\phi \qquad (6)$$

where ρ is the amplitude of the density wave and ϕ is the local layer position. The local smectic A density can then be expressed as $\rho_0 + \rho\cos\phi$, where ρ_0 is the average (or nematic) density. Clearly $|\psi| = 0$ in the nematic phase and $|\psi| > 0$ in the smectic A phase, allowing an expansion for the energy density of the bulk material at temperatures close to the nematic to smectic A phase transition temperature to be written as

$$f = f_0 + \tfrac{1}{2}A|\psi|^2 + \tfrac{1}{4}B|\psi|^4 + O(|\psi|^6) \qquad (7)$$

where A is again a temperature-dependent coefficient. Note that only even terms appear in this expansion, because the sign inversion operation on ψ is in fact only a π phase shift in ϕ and, therefore, does not affect the energy density of the material. In particular, we are interested in cases in which the order parameter varies from place to place within a smectic A liquid crystal sample; therefore, gradient terms must be added to Eq. (7) to represent the energy cost of doing this [8]. We can then write

$$f = f_0 + \frac{1}{2}\left\{ A|\psi|^2 + \frac{1}{2}B|\psi|^4 + C_\parallel \left|\frac{\partial\psi}{\partial z}\right|^2 + C_\perp \left|(\nabla_\perp - iq_s\delta\mathbf{n}_\perp)\psi\right|^2 \right\} \quad (8)$$

where $C_\parallel \neq C_\perp$ because of the nematic anisotropy. Here, $\delta\mathbf{n}_\perp$ is the component of the director $\mathbf{n}$, which is not aligned along the z axis; and q_s is the magnitude of the layer wavevector, which is initially along the z axis.

In addition, if distortions occur that lead to variation in the alignment of the director, this energy must be supplemented by that for the director elastic distortion energy, given by Eq. (3) or (4). The resulting combination of energy terms for order parameter variation and elastic distortion is particularly interesting in smectic liquid crystals. It is also analogous to the Landau-Ginzburg functional describing the normal to superconductor transition. It is then also possible to define two important length scales [8]. First, the smectic coherence length

$$\xi = \sqrt{\frac{C}{|A|}} \quad (9)$$

which is the length scale over which variations in ψ are influenced by local perturbations. Second, a penetration length

$$\lambda = \sqrt{\frac{KB}{C|A|}} \quad (10)$$

which is the length scale over which orientational perturbations penetrate. In each case, the subscripts have been dropped from the symbols and we have set $C_\parallel = C_\perp = C$, although strictly there are a number of coherence and penetration lengths to be defined.

Using the analogy between the above equations and the Landau-Ginzburg equation describing the normal-superconductor transition, deGennes [8] showed that two types of behavior of smectic A liquid crystal under imposed deformation could take place, depending on the ratio between λ and ξ, $\kappa = \lambda/\xi$. If $\kappa < 1/\sqrt{2}$, we have type I behavior, in which deformation is resisted until at a critical threshold point a stress-induced transition to the nematic state takes place. If $\kappa > 1/\sqrt{2}$ we have type II behavior, in which at a lower threshold dislocations form.

More recently, interesting experimental results relating to order parameter variation in smectic A liquid crystals were obtained by Cagnon and Durand [9], who applied shear strain to a cell containing homogeneously aligned material in the smectic A phase and observed the transmitted stress. They found that this stress had a periodic component, with the period equal to the smectic layer spacing. This was interpreted as periodic melting of the smectic layer structure close to one surface, and an argument for this based on an algebraic order parameter was presented.

A more complete analysis, based on the equations given above, was presented by Elston and Towler [10]. This ignores any elastic deformation in the system, which is, therefore, governed by Eq. (8). Furthermore, the geometry of the system allows this to be simplified to give

$$ f = f_0 + \frac{1}{2} \left\{ A|\psi|^2 + \frac{B}{2}|\psi|^4 + C\left|\frac{\partial \psi}{\partial z}\right|^2 \right\} \qquad (11) $$

Solving this equation together with suitable boundary conditions (ρ = its equilibrium value at the cell surfaces) leads to interesting behavior. Elston and Towler investigated solutions for a cell thickness of around one order of magnitude greater than the smectic coherence length. They found that in this case externally applied shear leads to a reduction in order at the center of the cell. However, rather than this order monotonically decreasing until melting occurs as the shear is increased (as might be expected intuitively), they observed that at a certain value of shear a limit point is reached. This takes place while the order in the center of the cell remains finite. Beyond this value of shear, no further state exists, so a discontinuous (first-order) transition must take place into the nearest available state. Furthermore, for a range of shears, the energy of the system is greater than the energy of the melt state, so the structure can be considered to be in a super-sheared state.

This was further explored by Mottram et al. [11] who showed that a critical thickness exists at which a change from second- to first-order shear-induced melting takes place. For a cell thickness less than the critical value, the layers continuously melt and reform as the shear increases. In this way, the effective layer tilt remains small, and when the layers melt they do so when the smectic order parameter at the center of the cell is zero [Fig. 2(a)].

For a cell thickness greater than the critical value however, the behavior is significantly different. It is now possible to supershear the layers into a metastable state until a critical value of the shear is reached. With increasing thickness, the critical shear value is linear with respect to the cell thickness, and for large thicknesses, there will, in general, be a large number of metastable states, each associated with a different amount of layer tilt. When the system reaches a critical shear value, the system relaxes into the next

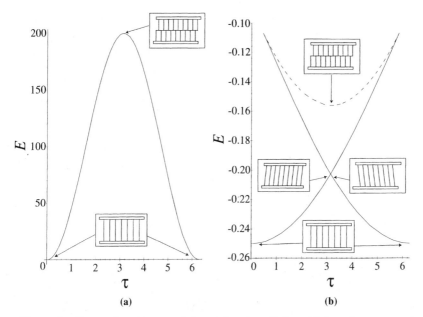

Figure 2. Energy (in arbitrary units) as a function of shear (τ) for a homogeneously aligned smectic A cell. (a) When shearing a smectic A cell with a thickness below a critical value, the smectic layers periodically melt and re-form in a continuous way. The energy as a function of shear, is therefore, continuous, with a maximum when the layers are out of phase. (b) For a cell with a thickness above a critical value, a super-sheared state exists. When the limit point is reached, a first-order transition to the lower energy state must take place.

highest free-energy metastable state, reducing the layer tilt and melting at the center of the cell as it relaxes [Fig. 2(b)].

Kralj and Zumer [12] have considered structures that can form in smectic A materials contained in submicrometer cylindrical cavities. They allowed for variations in both the smectic order parameter ψ and the nematic order parameter s. The energy expression they worked with, therefore, combines the terms in Eqs. (2), (3), and (8), together with surface-anchoring terms. The result is that a number of structures can exist, depending on the values of various important parameters they define. These parameters are the smectic coherence length, with a definition similar to Eq. (9), and a similarly defined nematic coherence length which depends on the ratio between the nematic elasticity and the coefficient A in Eq. (2). Furthermore, a penetration length, similar to that defined in Eq. (10), and a surface extrapolation length, which depends on the ratio between the elasticity and the surface-anchoring energy, are defined. Concentrating on the effect of the last of these, two particular cases were considered. First, strong homeotropic anchoring, which leads to a smectic-escaped-radial structure, which the smectic layers form concentric cylinders in most of the cavity. But near the center, the smectic material melts into the nematic phase, and in this region the director reorients to be parallel to the axis of the cylinder [Fig. 3(a)]. Second, in the regime of weak-anchoring, chevron (conical) or bookshelf smectic layer structures form. In the former, there is melting of the smectic order at the surfaces and in the center of the cylinder [Fig. 3(b)], whereas in the latter, melting only takes place at the surface of the cavity [Fig. 3(c)]. The predicted structures tie in well with experimental observations of smectic A materials in microcapillaries.

A naturally occurring spatial variation in the smectic A order parameter takes place in the structure of the twist grain boundary (TGB) phase. This phase occurs in some highly chiral smectic A liquid crystals and is characterized by an array of equally spaced TGBs. Discovered by Goodby et al. [13] in 1988, it leads to characteristic X-ray diffraction patterns (Fig. 4). Typically, the phase occurs between the chiral nematic and smectic A phases of some materials and can be understood in terms of the smectic layer melting discussed above. Generally, if a chiral nematic forms a smectic A phase, then the chirality must be suppressed. This is because the chirality in a nematic liquid crystal is manifest as a twist structure, with the twist axis perpendicular to the nematic director. Twist is not an allowed deformation in a smectic material, however, so the twist must disappear. Nevertheless, close to a nematic to smectic A phase transition, the chiral forces can be sufficiently strong to locally melt the smectic ordering and allow the chiral nematic phase to reform. In these regions, an array of parallel screw dislocations form, which allows a plane of twist to occur perpendicular to the smectic layering. The TGB smectics are, therefore, of type II, with the dislocation phase existing between the chiral nematic and smectic A phases.

Mechanically induced instabilities in smectic A liquid crystals have been considered that can also lead to the formation of periodic structures [14]. Experimentally, this appears in a homeotropically aligned smectic A cell as an undulation in the smectic layers when a dilative stress is applied across the structure. It comes about because of the balance between layer distortion and dilative stresses across the smectic layers.

We see that in each case the structures that can form in a smectic A liquid crystal cell under the influence of some externally applied constraints depend on the balance between the forces involved. Generally, one of these forces originates from the energy cost of the variation in the smectic A order parameter, and others from the external conditions.

III. SMECTIC C

Assuming that it is formed from a material initially in the smectic A phase the characteristic order parameter describing the transition into the smectic

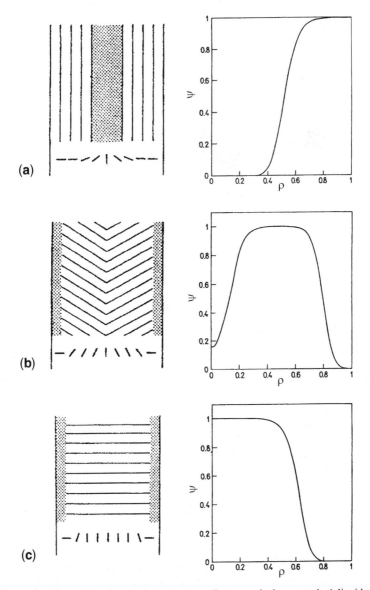

Figure 3.   Three of the characteristic structures that can exist in a smectic A liquid crystal confined in a microcapillary. (a) Melting takes place in the center of the structure. (b) Melting takes place at both the center and surface. (c) Melting takes place at the surface only. ψ, the magnitude of the smectic A order parameter; ρ, normalized radius.

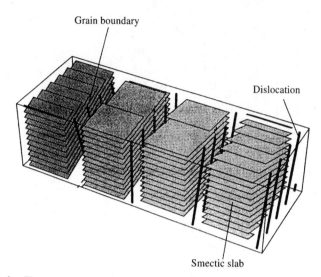

Figure 4.   The structure of the smectic A TGB phase. Chirality in the material leads to a periodic melting of the smectic layers, where screw dislocations form. This results in a structure that allows the smectic layers to rotate around a helical axis.

C phase is the molecular tilt angle θ [Fig. 1(c)]. This allows an expansion for the energy density of the material close to the smectic A–smectic C phase transition temperature to be written as

$$f = f_0 + \frac{1}{2}a\theta^2 + \frac{1}{4}b\theta^4 + \frac{1}{6}c\theta^6 + O(\theta^8) \qquad (12)$$

where the temperature dependence of the leading coefficient can be explicitly included by writing

$$a = \alpha(T - T_{AC}) \qquad (13)$$

where T_{AC} is the transition temperature.

Some of the earliest experimental work on smectic C materials involved studies of the influence of external disturbances on this order parameter. In particular, Garoff and Meyer [15] studied the influence of an applied electric field on the chiral version of smectic C liquid crystals. At temperatures close to the smectic A to smectic C phase transition point, applied electric fields parallel to the smectic layers coupled to the molecular dipole and induced a change in the tilt angle θ. In particular, at temperatures just above the phase transition (i.e., in the smectic A phase), an applied field induced a tilt angle roughly proportional to its magnitude. This is referred to as electroclinic behavior. Its origin can be easily understood if it is assumed that the spontaneous polarization in the chiral smectic C phase is proportional to the smectic tilt angle. A term of the form $-P_0\theta E$ can then be added to Eq. (12) to include the influence of an applied electric field. Now if only the leading terms are taken, minimization of the energy results in

$$\theta = \frac{P_0 E}{\alpha(T - T_{AC}^*)} \qquad (14)$$

where the smectic A–smectic C phase transition temperature is now T_{AC}^*, because it has been modified by the polarization interactions. Lee and Patel [16] showed that close to the phase transition temperature, the electroclinic effect becomes nonlinear, and it is necessary to include the higher order terms from Eq. (12) to explain this. We will return to the electroclinic effect when we consider the influence of elasticity.

A further area of experimentation that shows the influence of external disturbances on the order parameter of smectic C liquid crystals is in freely suspended films. In this case, the smectic liquid crystal is supported as a bubble across a hole, rather than in a conventional liquid crystal cell. Here the surface tension effects that stabilize the bubble can change the order parameter (tilt angle θ) at the surfaces. Generally, this is represented by including a separate energy expansion for the tilt angle of the surface smectic layers, where owing to the broken symmetry (free surface), the parameters can be significantly different from those in the bulk of the material [17]. Furthermore, it is important to consider how the surface layers and bulk material (within the film) couple together through elastic terms such as those in Eq. (4). If the tilt takes place in one plane the energy density of the bulk of the film can be expressed as

$$f = f_0 + \frac{1}{2}\alpha(T - T_{AC})\theta^2 + \frac{1}{4}b\theta^4 + \frac{1}{6}c\theta^6 + \frac{1}{2}K\left(\frac{d\theta}{dz}\right)^2 \qquad (15)$$

which together with an expression for the surface layers allows the director profiles within films to be calculated. Strictly, of course, this expression becomes quantized in the layer spacing for very thin films, and an example structure for a film with 25 smectic layers is shown (as a function of temperature) in Figure 5.

There has been considerable interest in smectic C free films owing not only to their intrinsic scientific interest, but also because characterizing them allows the parameters in Eq. (15) to be determined. Heinekamp et al. [17] undertook such studies for the common chiral smectic C liquid crystal

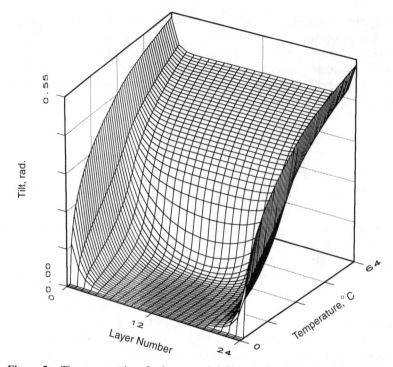

Figure 5. The structure in a freely suspended film of smectic C material consisting of just 25 smectic layers. At temperatures above the bulk smectic A–smectic C phase transition point, a surface transition occurs that penetrates into the bulk by a few layers. Close to the bulk phase transition temperature, this penetration extends farther into the film; and as the bulk smectic A–smectic C phase transition is crossed, tilt occurs throughout the film's thickness. Note that the temperature is measured *down* from the surface transition temperature.

material DOBAMBC. They used an ellipsometric technique to determine the depolarization of light transmitted through a film at an incident angle of 45°. The existence of the spontaneous polarization in the chiral material allows the smectic tilt direction to be selected and the optical effects observed then allows information about the average tilt in the film to be determined. Comparison of this with that predicted as outlined above allowed determination of the parameters in Eq. (15) for this material.

Recently, Bahr et al. [18,19] used this method to study smectic C films and other more complex materials. For example, they studied materials that show an unexpected tilt inversion for thin films (fewer than around 20 smectic layers), which is not present in thicker films. This is still not fully understood, although its appears to be the result of the influence that the very thin films have on the conformational structure and dipole orientation of the molecular species involved [20].

We [21] have used this method in an attempt to characterize both pure materials and the chiral smectic C mixtures SCE8 and SCE13. Although the results for pure materials were consistent with the results other workers obtained, the results for SCE8 and SCE13 neither were self-consistent nor compared well with results presented by others. We suggested two possible explanations for this. First, it is possible that selective evaporation of the species making up the materials is changing the nature of the mixtures SCE8 and SCE13. A second possibility is that components of the mixtures separate into either layers or small islands of pure material. Although these effects may make the use of free-standing film experiments inappropriate for the determination of the physical constants of chiral smectic C mixtures, in themselves they appear worthy of study. Together with the observations noted above, they indicate that the physics of thin films of smectic liquid crystal materials is particularly interesting.

Closely related to these free film studies are studies of smectic C materials in homogeneously aligned cells. Such structures have recently been investigated using optical-guided-mode techniques to determine the director profile in the cells. This has produced a number of interesting, although not self-consistent, results. Mengnan et al. [22] used guided-mode techniques to study the effect of the BDH material mix783 on lecithin alignment. In this work boundary layers of around $0.5\,\mu$ thickness were observed, although the temperature dependence was not discussed. Studies by Ruan et al. [23] for the Merck material SCE13R on lecithin indicated no boundary layer at all. Yang et al. [24] studied the chiral version of the same material and found boundary layers of around 20 nm thickness when no surface alignment was used. Our own studies [25] of this material on lecithin alignment showed a boundary layer thickness that decreased monotonically as the temperature was decreased below the smectic A–smectic C phase transition temperature. This is consistent with the predictions from Eq. (15) for a homeotropic boundary condition that leads to a boundary-layer-like profile with a boundary layer thickness given by

$$z_{bl} = \sqrt{\frac{2k}{\alpha(T - T_{AC})}} = \frac{1}{\theta_e}\sqrt{\frac{2k}{b}} \qquad (16)$$

where θ_e is the equilibrium smectic tilt angle in the smectic C phase. This can also be used to determine the characteristic value of K/α, which is found to be $60 \times 10^{-15}\,\mathrm{m^2K}$. This is consistent with values found using the electroclinic effect (discussed below) and is not inconsistent with the value for the material SCE8: $\approx 120 \times 10^{-15}\,\mathrm{m^2K}$ [26]. It is, however, not consistent with results on the same material by Mazzulla et al. [27], who indicate a boundary layer that increases in thickness with decreasing temperature. The apparent contradictions between different results for homeotropically aligned smectic C structures may be owing to the problems of the layer structure, which are not entirely clear.

In some situations, it is also important to consider the influence of elasticity on the electroclinic effect. For temperatures significantly above the phase transition, analyzing the effect using Eq. (12) is sufficient to explain most observations. Close to the phase transition temperature, however, it is important to extend the analysis to the form of Eq. (15) and include an elastic term. We have shown that this is particularly important for small applied voltages, for which the contribution of the elastic deformation energy can be significant [28]. The difference can easily be seen if we consider the behavior predicted by Eq. (12), together with an electric field term, at the phase transition temperature, where $T = T_{AC}$. In this case we see that for small θ we have

$$\theta \sqrt[3]{\frac{E}{b}} \qquad (17)$$

In an experiment, however, it can be seen that θ does not show this form. The addition of the elastic term limits the growth of the tilt, leading to a much better agreement with the experiment. A model with the addition of elastic deformation energy, therefore, provides a more complete description close to the phase transition. This can be seen by the improved comparison with experimental results shown in Figure 6.

IV. SMECTIC C LAYER STRUCTURE

In the above discussion we focussed on a smectic layer structure that was simple and unimportant (such as in free films) or we largely ignored it (in homeotropic cells and the electroclinic effect). There are, however, many cases in which the smectic layer structure is highly important and can itself influence the order parameter.

A key example of this that has been considered recently is the smectic A Frederiks transition. Traditionally, it was assumed that it is not possible to have a Frederiks transition in a smectic A liquid crystal, because there is no mode of director reorientation or distortion that preserves the smectic layer structure [29]. Thus when an electric field is applied to a smectic A liquid crystal with a positive dielectric anisotropy, an irreversible (plastic) transition is observed above a threshold voltage [30]. This is not the case, however, if the material is in the smectic A phase close to the smectic A to smectic C phase transition temperature. In this case, director tilt can occur in the smectic A phase, accompanied by smectic layer shrinkage. The

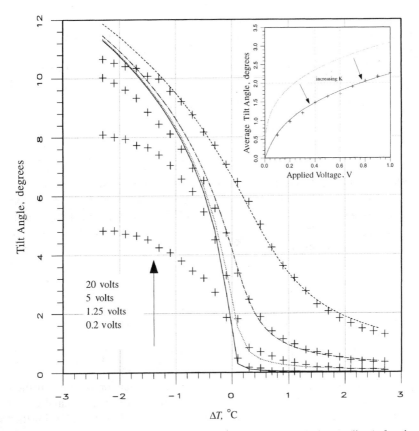

Figure 6. Comparison between data (individual points) and theory (lines) for the electroclinic effect in a chiral smectic C material. *Inset*, The low voltage behavior at the smectic A–smectic C phase transition temperature. At this temperature, it is clear that the influence of elasticity (increasing K) on the structure that forms in the electroclinic effect is significant.

occurrence of this allows retention of the layer packing formed in the smectic A phase together with molecular tilt relative to the layer normal, provided the smectic layers tilt as well [31]. The energy density can then be expressed as

$$f = f_0 + \frac{1}{2}\alpha(T - T_{AC})\theta^2 + \frac{1}{4}b\theta^4 + \frac{1}{2}K\left(\frac{d\chi}{dz}\right)^2 - \frac{1}{2}\varepsilon_0\Delta\varepsilon E^2 \chi^2 \quad (18)$$

where χ is the tilt of the director relative to the sample boundaries and $\Delta\varepsilon$ is the (positive) dielectric anisotropy. Of course, to solve this it is necessary to relate the director tilt relative to the smectic layer normal (θ) to the director tilt in the cell (χ). This is done through the approximation $\delta \approx c\theta$, where δ is the layer tilt angle relative the cell surface normal and c is a constant just smaller than unity. The layer tilt and smectic tilt angle then add together to give $\chi \approx (1 + c)\theta$. Taking this together with Eq. (18) leads to the following Euler-Lagrange equation governing the system

$$\frac{\alpha(T - T_{AC})}{(1 + c)^2}\chi + \frac{b}{(1 + c)^4}\chi^3 + K\frac{d^2\chi}{dz^2} - \varepsilon_0\Delta\varepsilon E^2 \chi = 0 \qquad (19)$$

When a field is applied, the lowest order mode of distortion must maintain the total layer displacement across the cell and will, therefore, be an odd mode. As a consequence, we can then look for solutions to Eq. (19) of the form $\chi = \chi_0 \sin(2\pi z/d)$, where d is the liquid crystal cell thickness. This allows the determination of threshold voltage for the smectic A Frederiks transition (SAFT) as

$$V_{SAFT} = \frac{1}{\sqrt{\varepsilon_0\Delta\varepsilon}}\left\{4\pi^2 K + \alpha(T - T_{AC})\frac{d^2}{(1 + c)^2}\right\}^{\frac{1}{2}} \qquad (20)$$

Elston [31] showed that experimental data for a positive dielectric anisotropy material at a temperature close to the smectic A to smectic C phase transition were consistent with this. The behavior of the threshold voltage squared was linear with reduced temperature over a range on the order of a degree above the phase transition temperature (Fig. 7). Furthermore, it was shown that the behaviour of the reorientation above the threshold voltage was consistent with the above by the use of numerical solutions to equation (19).

Closely related to this is the much earlier work by Ribotta and Durand [14], who discussed how a tilt angle can be induced in a smectic A phase close to a smectic C phase transition through layer compression forces. In this case, the induction of a smectic tilt angle allows the layers to reduce thickness in response to an externally imposed compressive force.

The most commonly observed layer structure effect in smectic C liquid crystals is the formation of the so-called chevron structure [32]. This generally forms in thin cells of homogeneously aligned smectic liquid crystal when the smectic A–smectic C phase transition temperature is crossed. A number of people have given consideration to this structure, and a common feature of models of the structure variation in the smectic tilt order parameter across the thickness of the cell. For example, we [33] have recently been investigating the stability of the chevron structure as it forms

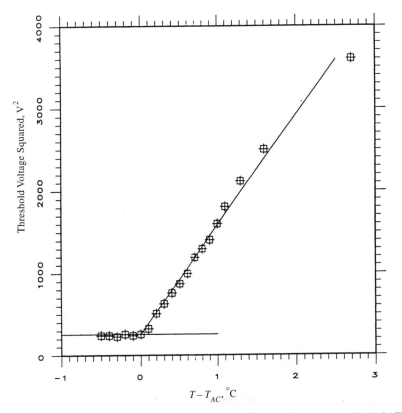

Figure 7. Threshold voltage squared as a function of reduced temperature in the SAFT. Data are shown as points, together with two straight-line fits with the break point at $T - T_{AC} = 0$. The sloping line is consistent with Eq. (20).

from the smectic A phase. To do this we used an energy expression similar to that of Eq. (18), but without the field-driven term. We added a constraint to control the total layer displacement over the thickness of the liquid crystal cell so that the layers may move. This results in an Euler-Lagrange equation

$$(T - T_{AC})\theta + b\theta^3 + (1 - c)^2 K \frac{d^2\theta}{dz^2} + \lambda = 0 \qquad (21)$$

where λ is a multiplier to control the above constraint. Solving this with various values of the coefficients and controlling the constraint to give a particular total layer displacement across a cell allows the stability of the chevron structure to be analyzed. When the total layer displacement is fixed at zero, it is found that above a critical temperature the only solution to this equation leads to a bookshelf structure, as expected in the smectic A phase.

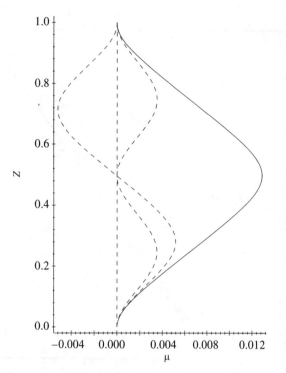

Figure 8. The structures that can form in a smectic C liquid crystal cell under the constraint of zero total smectic layer displacement across the cell. Below a critical temperature, only the single chevron structure (*solid line*) is stable; above this temperature, the bookshelf structure is stable.

As the temperature is decreased into the smectic C phase, however, various other solutions can occur: the chevron structure (with a single kink in the center of the cell) and multiple chevron structures (with a number of kinks across the thickness of the cell) (Fig. 8). In these cases, it is always the single kinked chevron structure that is stable [34]. If the total layer displacement is increased, this chevron structure can evolve into a tilted layer structure, basically through motion of the chevron cusp toward one cell surface until it merges with the surface [Fig. 9(a)]. It can be seen that this process always results in a decrease in energy [Fig. 9(b)]; therefore, it might be expected that the chevron structure is unstable to such a process and should naturally evolve into a tilted layer structure. If a surface positional anchoring energy of the magnitude measured by Cagnon and Durand [9] is added to the system, however, this is sufficient to prevent such motion. In fact, it is seen that only a small positional anchoring energy is needed to pin the chevron structure once it has formed [33].

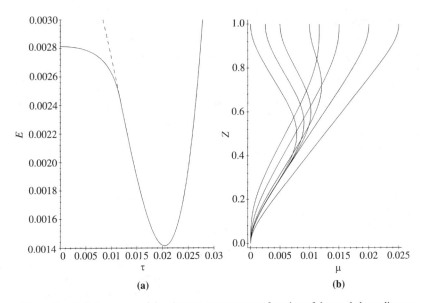

Figure 9. (a) The energy of the chevron structure as a function of the total shear distance. (b) The evolution of the chevron structure into a tilted layer structure as a smectic C liquid crystal cell is sheared or as the layering slips along one surface. Although the tilted layer structure is clearly of lower energy, we expect the chevron structure to be stabilized by the presence of layer positional anchoring energy as measured by Cagnon and Durand [9].

Although this discussion allows the stability of the chevron structure to be predicted, it does not consider in any detail the structure around the chevron cusp. Clearly, however, the structure in the region of the cusp is important for understanding the details of the chevron formation and to allow the switching process in chiral smectic C liquid crystals to be understood. Consideration of this structure has been given by a number of authors, and various detailed models have been suggested.

Clark and Rieker [32] put forward the original theoretical model of the chevron, which explained it in terms of a kink, or discontinuous change in the smectic layering. Although this model has a discontinuity in the layer tilt δ, the **n**-director structure is continuous at the chevron interface.

Since then a number of other models of the chevron interface have been put forward. Generally, a feature that these models have in common is that the discontinuity in δ is avoided by allowing one or more of the parameters of the system—such as the layer tilt δ, the cone angle θ, the azimuthal angle around the smectic cone ϕ, or the layer thickness—to vary smoothly within the cell. Nakagawa [35] proposed a continuous model based on the minimization of an energy consisting of terms from layer dilation, layer bending, and variations of ϕ.

Limat and Prost [36] described the chevron structure in terms of the layer tilt angle, which was assumed to be continuous across the cell. By writing the free energy in terms of δ, they were able to investigate the second-order transition between the smectic A bookshelf structure and the smectic C chevron structure.

By assuming that the layer tilt, δ was coupled to θ and then allowing θ to vary through the cell de Meyere et al. [37] were able to remove the layer tilt discontinuity while preserving the layer thickness. Their model, which balances energy contributions from layer curvature and variations in θ, found solutions for the chevron structure. The assumption of constant layer thickness and layer curvature leads to the necessary condition that θ is zero at the chevron interface (i.e., the liquid crystal is in the smectic A phase).

Limat [38] later extended Nakagawa's [35] model to included the possibility that the layer tilt angle does not equal the smectic cone angle. Vaupotic and co-workers [39,40] removed the constant layer thickness condition and found solutions for which θ is nonzero but small at the chevron interface, where there is a small region of layer dilation.

Mottram et al. [41] recently proposed a model for the chevron interface that is based on the requirement for continuity in the smectic C biaxial order parameter across the chevron tip. This continuity condition forces the liquid crystal to adopt a uniaxial phase at the chevron tip (Fig. 10). The continuity of biaxial ordering at the chevron interface has an important consequence for chiral smectic C materials. In the bulk of a chiral smectic C liquid crystal, there exists a spontaneous polarization owing to a permanent molecular dipole; however, the rotational symmetry of the uniaxial state implies that there is zero polarization at the chevron interface. This result may be extremely important when considering ferroelectric devices that switch through the coupling between the spontaneous polarization and an applied electric field.

Of interest to display engineers is the control of the chevron tilt direction by suitable treatment of the surfaces. If the surface alignment is truly homogeneous, with no pre-tilt of the director, then the two possible chevron tilt directions are degenerate. In this case, different parts of the cell contain regions in which the chevron structure tilts each way; and at the interfaces between tilt directions, zigzag defects are seen. This is not desirable, so in practice the surfaces are treated to introduce a pre-tilt direction to break this degeneracy. This pre-tilt is typically a few degrees, and means that chevron structures tilted in the two directions have different energies.

This has been investigated theoretically by Ulrich and Elston [42], who used an energy expression similar to that proposed by Nakagawa [35]. This includes terms for smectic layer distortion, reorientation of the axis parallel to the smectic C tilt angle but in the plane of the smectic layers, and

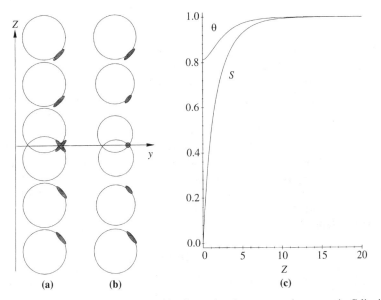

Figure 10. (a) The structure that could exist at the chevron cusp in a smectic C liquid crystal cell. (b) Clearly, if the biaxiality is not to be discontinuous at the chevron cusp point it must melt in some way. (c) The calculated behavior of the smectic tilt angle (θ) and biaxial order parameter (S) when the condition of biaxiality continuity is imposed at the chevron cusp.

variations in the cone angle away from the equilibrium value. Thus the energy can be expressed as

$$f = f_0 + \frac{K}{2}\left\{(\nabla \cdot \mathbf{a})^2 + \theta^2\left[(\nabla \cdot \mathbf{c})^2 + (\nabla \times \mathbf{c})^2\right]\right\} + \frac{b}{4}(\theta_e^2 - \theta^2)^2 \quad (22)$$

where the layer normal is given by the vector $\mathbf{a}$, the so-called c-director is given by the vector $\mathbf{c}$. Using this equation, they showed that most of the changes in the smectic tilt angle required to match the surface pre-tilt condition take place very close to the surface. In addition, they showed that there are a range of pre-tilt angles over which each of the possible tilt directions are stable. Thus by tuning the pre-tilt values, the chevron tilt direction can be controlled and the zigzag defects (theoretically) avoided.

V. CONCLUSION

In this chapter we have outlined some of the issues of order parameter variation in smectic liquid crystals. This has concentrated on areas in which external perturbations can modify a particular order parameter from its bulk equilibrium value. We have seen that externally applied stress can lead to melting of the smectic A order parameter. Similarly, we have seen that the stresses in structures that can form in smectic C liquid crystals (such as the chevron) can lead to changes in the smectic tilt angle order parameter.

There is still much work to be done in the field. For example, there has been little consideration of structures in more than one dimension in smectic materials. Some work on the zigzag structures seen in smectic C liquid crystal cells has been undertaken by Mukai and Nakagawa [43], but it is incomplete. It will be particularly interesting to develop modeling of smectic structures in two and three dimensions that can then be used to understand not only the defects that form in these materials but also the important switching processes in chiral smectic C materials. These involve the formation of domains and domain wall structures, which are complex systems to understand.

References

1. P. G. deGennes and J. Prost, *The Physics of Liquid Crystals*, Oxford University Press, Oxford, UK, 1993.

2. F. C. Frank, *Discuss. Faraday Soc.* **25**, 19 (1958).

3. R. Barberi and G. Durand, *Phys. Rev. A* **41**, 2207 (1990).

4. G. Barbero and G. Durand, *J. Phys. II* **1**, 651 (1991).

5. N. Schopohl and T. J. Sluckin, *Phys. Rev. Lett.* **59**, 2582 (1987).

6. N. J. Mottram and S. J. Hogan, *Phil. Trans. Roy. Soc.* **355**, 2045 (1997).

7. G. Skacej, A. L. Alexe-Ionescu, G. Barbero, and S. Zumer, *Phys. Rev. E* **57**, 1780 (1998).

8. P. G. deGennes, *Solid State Commun.* **10**, 753 (1972).

9. M. Cagnon and G. Durand, *Phys. Rev. Lett.* **70**, 2742 (1993).

10. S. J. Elston and M. J. Towler, *Phys. Rev. E* **57**, 6706 (1998).

11. N. J. Mottram, T. J. Sluckin, S. J. Elston, and M. J. Towler, unpublished data.

12. S. Kralj and S. Zumer, *Phys. Rev. E* **54**, 1610 (1996).

13. J. Goodby, M. A. Waugh, S. M. Stein, E. Chin, R. Pindak, and J. S. Patel, *Nature* **337**, 449 (1988).

14. R. Ribotta and G. Durand, *L. J. Phys.* **38**, 179 (1977).

15. S. Garoff and R. B. Meyer, *Phys. Rev. Lett.* **38**, 848 (1977).

16. S.-D. Lee and J. S. Patel, *Appl. Phys. Lett.* **54**, 1653 (1989).

17. S Heinekamp, R. A. Pelcovits, E. Fontes, E. Y. Chen, R. Pindak, and R. B. Meyer, *Phys. Rev. Lett.* **52**, 1017 (1984).

18. C. Bahr and D. Fliegner, *Ferroelectrics* **147**, 1 (1993).

19. C. Bahr, C. J. Booth, D. Fliegner, and J. W. Goodby, *Phys. Rev. E* **52** R4612 (1995).

20. D. Schlauf, C. Bahr, and C. C. Huang, *Phys. Rev. E* **55**, R4885 (1997).

21. N. U. Islam, S. J. Elston and M. J. Towler, *Ferroelectrics* **214**, 9 (1998).

22. G. Mengnan, J. R. Sambles, and F. Yang, *Liq. Cryst.* **13**, 637 (1993).

23. L. Ruan, J. R. Sambles, and M. J. Towler, *Liq. Cryst.* **18**, 81 (1995).

24. F. Yang, G. W. Bradberry, and J. R. Sambles, *Phys. Rev. E* **50**, 2834 (1994).

25. N. U. Islam and S. J. Elston, *Liq. Cryst.* **26**, 709, (1999).

26. M. Osipov, J. R. Sambles, and F. Yang, *Liq. Cryst.* **21**, 727 (1996).

27. A. Mazzulla and J. R. Sambles, *Liq. Cryst.* **22**, 727 (1997).

28. S. M. Beldon and S. J. Elston, *Liq. Cryst.* **26**, 143, (1999).

29. I. M. Blinov and V. Chigrinov, *Electroptic Effects in Liquid Crystal Materials*, (Springer-Verlag, Berlin, 1994).

30. D. A. Dunmur and T. W. Walton, *Mol. Cryst. Liq. Cryst. Lett.* **2**, 7 (1985).

31. S. J. Elston, *Phys. Rev. E* **58**, R1215 (1998).

32. N. A. Clark and T. P. Rieker, *Phys. Rev. A* **37**, 1053 (1988).

33. N. U. Islam, N. J. Mottram, and S. J. Elston, *Liq. Cryst.* **26**, 1059 (1999).

34. N. J. Mottram and S. J. Elston, *Eur. Phys. J. B.* **12**, 277–284 (1999).

35. M. Nakagawa, *Displays* **11**, 67 (1990).

36. L. Limat and J. Prost, *Liq. Cryst.* **13**, 101 (1993).

37. A. de Meyere, H. Pauwels, and E. de Ley, *Liq. Cryst.* **14**, 1269 (1993).

38. L. Limat, *J. Phys. II Paris* **5**, 803 (1995).

39. N. Vaupotic, M. Copic, and T. J. Sluckin, *Phys. Rev. E* **57**, 5651 (1998).

40. N. Vaupotic, S. Kralj, M. Copic and T. J. Sluckin, *Phys. Rev. E* **54**, 3783 (1996).

41. N. J. Mottram, N. U. Islam, and S. J. Elston, *Phys. Rev. E* **60**, 613, (1999).

42. D. C. Ulrich and S. J. Elston, *Appl. Phys. Lett.* **68**, 185 (1996).

43. S. Mukai and M. Nakagawa, *Jpn. J. Appl. Phys.* **33**, 6255 (1994).

STRUCTURE AND ORIENTATION OF MOLECULES IN DISCOTIC LIQUID CRYSTALS USING INFRARED SPECTROSCOPY

T. S. PEROVA AND J. K. VIJ

Department of Electronic and Electrical Engineering, Trinity College, University of Dublin, Ireland

A. KOCOT

Institute of Physics, University of Silesia, Katowice, Poland

CONTENTS

Advances in Liquid Crystals: A Special Volume of Advances in Chemical Physics, Volume 113, edited by Jadish K. Vij. Series Editors I. Prigogine and Stuart A. Rice.
ISBN 0-471-18083-1. © 2000 John Wiley & Sons, Inc.

I. INTRODUCTION

A. Discotics and Their Potential for Applications

Chandrasekhar et al. [1] discovered the discotic mesophase in 1977 while investigating the optical and X-ray properties of hexa-substituted esters of benzene. The discovery of the discotic mesophase took place because molecular disks, as uniaxial basic building blocks with long chains attached to it, were predicted to exhibit liquid crystalline mesophase. This was successfully realized [1,2].

A tremendous growth of interest during the last decade has strongly been motivated by the potential of technological application [3–5]. The successful commercialization will undoubtedly continue to come, not in applications where they compete directly with well-established inorganic materials but where they provide unique properties and processing advantages. This activity seems to have increased because of the ability of discotics for self-organization, the ease of alignment, and the large potential of their applications especially in the area of photoconductivity. The face-to-face assembly of the disk-like molecules promotes the overlap of π orbitals and thus enhances the transfer rate, perhaps to the point of formation of energy bands as in solids. The continued success of organic photoconductors arises not because they are necessarily better than selenium or amorphous silicon but because large areas can be coated more cheaply. Another advantage that organics have over selenium is that they are environmentally friendlier. A related area of activity is that of light-emitting diodes, in which a thin film of a guest (e.g., 10% by weight of tristilbene in a columnar triphenylene-based discotic) can lead to a reduced threshold voltage for light emission in the visible region. This characteristic is connected to the strongly anisotropic nature of discotics, which may serve not only as hole-conducting but also as electron-conducting materials

Discotics can induce alignments of other organic materials, and their thin films have the potential of applications that extend from electro-optics to holography. They are predicted to be prime candidates for negative retardation films for use in the front of twisted nematic devices to improve their viewing angles [6]. Retardation films can be prepared through the self-assembly of discotic liquid crystals (DLCs) in either nematic or columnar phases (which both have negative birefringences). The self-assembled array must, however, be aligned appropriately relative to the surfaces of the device and be not allowed to crystallize. This is possible, because crystallization can be prevented by self-organization followed by polymerization.

DLCs are composed of flat or nearly flat cores with semiflexible tails. In discotic mesophases, the disks form molecular cores, similar to the rod-shaped molecules that form the basic building block of the calamitic phases. The cores have six or eight (or sometimes four) long-chain substituents, commonly with ester or ether linkage groups. Available experimental evidence indicates that the presence of these side chains is crucial to the formation of discotic liquid crystals. The molecular packing in discotic liquid crystalline phase is, however, different from that of the calamitic phases. These molecular cores form long-range ordered columns. The centers of the cores form a rectangular or hexagonal lattice in the plane normal to the column axis. The intercolumnar distance is 20 to 40 Å, depending the lateral chain length, whereas the stacking distance within the column is less than 4.5 Å. Therefore, interactions among the neighboring molecules within the same column should be much stronger than inter-actions among molecules of neighboring columns. Aromatic cores may be normal or inclined at a certain angle to the column's axis. The inclination depends on the lengths of the tails and the strength of the polar groups. Such inclination translates the centers of the cores, and consequently the lattice becomes pseudo-hexagonal.

It has been shown that polymers with discotic mesogens [7], either as part of the main chain or as side groups attached to the polymer backbone via a flexible spacer, form liquid crystalline polymers. These polymers are also of particular interest because they exhibit new properties. It is possible for instance to orient these polymers macroscopically either by magnetic fields or by mechanical forces. Cooling down to below the glass transition temperature results in highly ordered anisotropic glasses. Discotic liquid crystals can be doped to form charge transfer complexes [8]. In fact, it is even possible to induce liquid crystalline behavior by doping amorphous discotic polymers with electron acceptors [9].

It is now widely accepted that chiral molecules of a suitable form and shape exhibit ferroelectricity in liquid crystals. Some 4 years after the discovery of discotics, Prost [10] predicted that tilted columnar phases of chiral disk-like molecules can be ferroelectric. Ferroelectricity in discotics was first successfully realized by Bock and Helfrich [11] in 1992. These discotics form a tilted columnar phase. The chirality was introduced in the chain close to its connection to the core. The nearby bonds with electric dipole moments can give rise to a nonzero time-averaged molecular dipole moment. The total dipole moment is normal to both the axis of the column and the tilt direction. Boden et al. [12] and independently Praefcke and Holbey [13] recently realized much stronger chirality in the disk itself by the α-halogenation of triphenylene-based discotics. This converted the disk from being planar to a propeller-like geometry. Surface-induced chirality in a self-assembled monolayer of a discotic liquid crystal has recently been demonstrated [14]. Thus the modern tendency in the ferroelectric phenomenon of liquid crystals (including discodic liquid crystals) is to move from chiral molecules to chiral phases [15].

These discoveries have not so far been fully exploited. Although it is unlikely that ferroelectric discotics in future will be used in display-type devices, nevertheless these will have the potential for applications, that require self-organization, order, and mobility. Work on a ferroelectric discotic-dibenzopyrene is also briefly reviewed later in this chapter.

B. Mesophases in Discotics

The various mesophases of discotics were beautifully reviewed by Chandrasekhar [2]. The mesophases of known structures are classified into columnar and nematics. The columnar phases can have different types of lattices. These are mainly hexagonal and rectangular. The hexagonal phase can be ordered, in which there is a regularity in the stacking of the triphenylene cores in each column or it can be disordered, in which case the column is liquid like. The disks in the nematic phase have an orientational order but no long-range positional order. A twisted nematic (or cholesteric) phase with the helical axis normal to the director was also been found [16].

C. Scope of the Chapter

In this chapter we review the recent works on the various techniques of alignment. This is important, because applications not only require an ordered structure relative to the surface of the substrate but also require that the ordered arrangement be stable without crystallization occuring over longer periods of time and over repeated cycling of temperature and mechanical stress. We also review and discuss the use of Fourier transform infrared (FTIR) spectroscopy for finding the type of alignment that is achieved under certain conditions of the experiment, whether it is homogeneous or heterogeneous, and the extent of the long-range order that is realized in the mesophase. We find that FTIR polarizing spectroscopy used in conjunction with polarizing optical microscopy is a powerful technique for finding the extent and type of alignment and the order parameter. The substrates on which the alignment is achieved have been varied, and the surface treatments have been altered. This is to find the reasons for a particular alignment or to explain a transition from one type of alignment to another that may occur under certain experimental conditions. Scanning electron microscopy and white light interferometry have been used to find the extent of flatness of the surface of the substrate.

We have also investigated a number of compounds whose structural formulas are similar to those of the discotics but that are not known to exhibit a mesophase. This is carried out to demonstrate the changes that occur in IR spectra for a similar system that exhibits discotic mesophases. Furthermore, certain compounds that are used as dopant for forming charge transfer complexes have also been studied. These compounds are 2,4,7-trinitrofluorenone (TNF) and (2,4,7-trinitro-9-fluorenylidene)-malonic bishexadecylester (TNF16MB). The IR spectra of these compounds are given in Appendix A.

II. THEORETICAL BACKGROUND

A. Order Parameter

The orientational order in a mesomorphic phase can be described by an orientational distribution function. The infrared absorption for a single molecule is proportional to $(\vec{E} \bullet \bar{p}_i)^2$ where $\vec{E}$ is the electric field of IR radiation and p_i is the transition dipole moment vector with respect to the normal co-ordinate q_i of the ith mode.

$$p_i = \frac{du_i}{dq_i} \qquad (1)$$

where p_i corresponds to the ith vibration of a molecule. The main advantage of IR spectroscopy lies in its possibility of studying the orientational distribution of dipoles, for those parts of the molecule in which the molecular vibrations are localized. This contrasts with other techniques in which the molecular orientation as a whole is investigated.

The orientational order parameter S is given by

$$S = \tfrac{1}{2}\langle 3\cos^2\theta - 1\rangle \tag{2}$$

where θ is an angle between the optical axis (director) of the sample and the orientation position of an individual molecule (i.e., between $\vec{n}$ and $\vec{L}$, $\vec{n}$ denotes the director) (Fig. 1). The dichroic ratio R_i for an unpolarized beam can be defined as follows:

$$R_i = \frac{I_D}{I_I} \tag{3}$$

Here I_D is the integrated absorbance of the band in the discotic phase, and I_I is the integrated absorbance in the isotropic phase. For the case of the isolated and well-separated Lorentzian bands, these can be taken as the peak intensities of the bands. Using Neff et al. [17] calculations for the unpolarized radiation, we can define I_D as follows

$$I_D = \cos^2\beta\langle\sin^2\theta\rangle + \tfrac{1}{2}\sin^2\beta\langle 1 + \cos^2\theta\rangle \tag{4}$$

where θ is the angle between the normal to the plane of aromatic core and the direction of an incident IR beam, β is the angle between the normal to

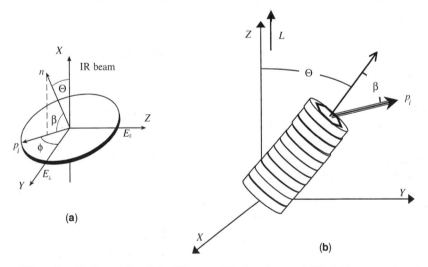

Figure 1. The interaction of the IR beam with a discotic material for (a) homeotropic and (b) planar orientation of molecules.

the plane of the core and direction of the transition dipole moment vector p_i (see Fig. 1). The angle θ is different for different molecules, and its average value determines the order parameter for the central parts of molecules. Taking into account that in the isotropic phase

$$\langle \sin^2 \theta \rangle_I = \tfrac{2}{3} \tag{5}$$

$$\langle \cos^2 \theta \rangle_I = \tfrac{1}{3} \tag{6}$$

I_I for unpolarized radiation will be given as

$$I_I = \cos^2 \beta \langle \sin^2 \theta \rangle_I + \tfrac{1}{2} \sin^2 \beta \langle 1 + \cos^2 \theta \rangle_I \tag{7}$$

Substituting Eqs. (5) and (6) into Eq. (7) we obtain

$$I_I = \tfrac{2}{3} \langle \cos^2 \beta + \sin^2 \beta \rangle = \tfrac{2}{3} \tag{8}$$

From Eq. (2), S becomes

$$S = 1 - \tfrac{3}{2} \langle \sin^2 \theta \rangle \tag{9}$$

or

$$\langle \sin^2 \theta \rangle = \tfrac{2}{3}(1 - S) \tag{10}$$

Substituting Eqs. (4), (8) and (10) into Eq. (3) we obtain

$$R_i = \frac{2(1 - S) + 3S \sin^2 \beta}{2} \tag{11}$$

Assuming $\beta = 90°$ or $\beta = 0$ we obtain

$$S = 2(R_i - 1) \qquad (\text{for } \beta = 90°) \tag{12}$$

for vibrations with the transition dipole moment p_i lying in the plane of the molecular core and

$$S = 1 - R_i \qquad (\text{for } \beta = 0°) \tag{13}$$

for vibrations with p_i normal to the molecular core.

B. Intensity Distribution Functions

Two main orientation of discotic liquid crystals are shown in Fig. 2. These are planar (or edge-on) alignment and homeotropic (or side-on) alignment (discussed in more detail below). It is accepted that the orientation of the disk is considered with respect to the column's axis.

To obtain orientational information about the selected molecular group located in the core, we need to know all spatial components of the absorbance. In the uniaxial columnar phase, the components of the absorbance A_X, A_Y, and A_Z can be expressed using the formulas obtained for a classical SmA phase [18], provided the dispersion parameter is ignored,

$$A_Y, A_Z = \frac{A}{6}\{(2(1-S) + 3S\sin^2\beta)\}$$

$$A_X = \frac{A}{3}\{(2S + 1 - 3S\sin^2\beta)\} \qquad (14)$$

where: $A = A_X + A_Y + A_Z$ is the trace of absorbance, and β is the angle between the transition dipole moment and the axis of the column.

For the case of the planar-homogeneous alignment, the orientational distribution of the parts of the molecules can be found from the profile of the absorbance measured as a function of the angle of polarization [19]. The absorbance profile can be fitted using the following formula [20,21]:

$$A(\phi) = -\log\{10^{-A_\parallel} + (10^{-A_\perp} - 10^{-A_\parallel})\sin^2(\omega - \omega_0)\} \qquad (15)$$

where $A_\parallel$ is the maximum absorbance, $A_\perp$ is maximum absorbance, ω is a polarizer angle, and ω_0 is the angle for which the maximal absorbance is obtained. These can be related to the components along the orthogonal coordinate system, i.e. A_X, A_Y, A_Z.

The most repeated (and mainly used for practical applications) alignment for discotic liquid crystals however, is side-on alignment with the core lies

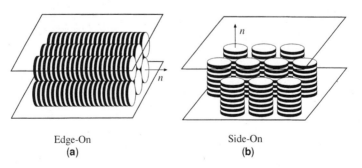

Edge-On   Side-On
(a)   (b)

Figure 2. Different types of alignment of discotic liquid crystals.

parallel to the plane of windows. In such a case, owing to the axial symmetry of the discotic phase, both absorbance components in the window plane are the same, therefore, the polarization angular dependence of the absorbance is independent of the angle of polarization. The third component of the absorbance can be obtained by tilting the sample out of the normal angle of incidence (oblique transmission technique) [Fig. 3(b)] [22,23].

Although the angular range of the sample rotation is limited ($-60°$ to $60°$), we can reasonably reproduce the angular dependence of the absorbance and extrapolate the values of the absorbance to the tilt angle of $90°$. The fitting procedure has been carried out using the general formula derived for the polarized IR absorption [22,23]:

$$A(\alpha) = A_Z + (A_X - A_Z)\sin^2\alpha \qquad (16)$$

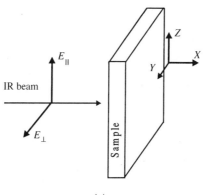

(a)

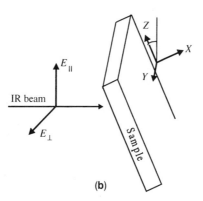

(b)

Figure 3. Polarized infrared transmission technique at (a) normal and (b) oblique incidence of light.

where α is the angle between the electric field vector and the window plane within the medium, A_Z is the absorbance component in the plane, and A_X is normal to the window plane. For normal incidence $\alpha = 0°$, $A(\alpha = 0) = A_Z$, and for $\alpha = 90°$, $A(\alpha = 90°) = A_X$.

These studies provide another possibility of finding the order parameter, simply using polarized IR measurements at normal and oblique incidence of light only in the discotic phase. This is quite important, because a multipath transition through the discotic–isotropic phase transition temperature i.e., passing through the transition several times, influences the alignment of the liquid crystalline cell. Using the aforementioned technique, the order parameter for the homogeneous planar alignment can be calculated from the dichroic ratio obtained for perpendicular-to-the-core bands from polarized measurements. To get a quantitative measure of the order parameter for the side-on orientation, one can use the following approach. For axial symmetry $A = A_X + 2A_Z$; and in the temperature range of the mesophase, A is assumed to be constant. This is true only if A is considered to be the integrated intensity. If we consider A as peak intensity, the broadening of the band with the increase in temperature should be taken into account. Once the trace of the absorbance is obtained for a particular band, we can reproduce the order parameter S using only the A_Z component of the absorbance, which is in the plane of the window.

III. EXPERIMENTAL

A. Chemicals

The discotics discussed in this review with structural formulaes, phase sequences, and references to the synthesis procedures are given in appropriate sections and in Appendix A. For our study, we used a number of triphenylene derivatives: hexa-trioxytriphenylene (H3T), hexa-pentyloxytriphenylene (H5T), hexa-heptyloxytriphenylene (H7T), hexa-n-decanoyloxytriphenylene (H10OT), hexa-(hexylthio)-triphenylene (H6TT), 1-nitro-2,3,6,7,10,11-hexaheptyloxytriphenylene (H7T-NO$_2$) [24–28]; 2,3,7,8,12,13-hexa-n-decanoyloxytruxene (HA10TX) [29]; 1,2,5,6,8,9,12,13-octakis–$((S)$-2-(heptyloxy)-propanoyloxy)-dibenzo-[e,$_L$]-pyrene (D8m*10) [30]; hexa-decanoyloxy-rufigallol (RHC10) [31]; a number of anthracene derivatives (A4n10, A6n10, A8n7) [32], pyramioic V6n10 [32,33], 2,4,7-trinitrofluorenone (TNF) [24], and (2,4,7-trinitro-9-fluorenylidene)-malonic bishexadecylester (TNF16MB) [24].

B. Alignment of Discotic Liquid Crystals

We review the various alignment techniques suitable for discotic liquid crystals. The alignment for discotic liquid crystals is defined below. The different possible methods for finding the alignment are obtained. For discotic liquid crystals, the accepted nomenclature for the orientation of the phase is considered with reference to the columnar axis (or column's director). In planar alignment, the director of the columns lies in the window plane; and in homeotropic alignments, the director of the column is oriented perpendicular to the window plane (Fig. 2). The director in discotic material is normal to the molecular plane; this implies that the plane of the molecules is, on average, parallel to the windows for homeotropic alignment (or sometimes called a side-on alignment). For the planar alignment, the plane of the molecules is, on average, normal to the windows. Hence it is also called edge-on alignment. Noted that the homeotropic alignment is always uniform (or homogeneous) in nature, because only one direction normal to the substrate or to the plane of the windows is possible. This is not the case, however, for the planar alignment, because there are many possible orientations for the director in which it can lie in the plane of the window (Fig. 4). This means that the planar alignment can be homogeneous planar, if the director orientation is uniform (in one direction) and heterogeneous planar if the director orientation is randomly distributed in the plane of the window.

To introduce a uniform (or homogeneous) planar alignment different treatment of the LC cell (or window's surface) is necessary. For rodlike molecules, different surface treatments are widely used and have been reviewed [34–37]. The various surface treatments include rubbing of the

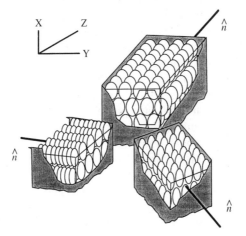

Figure 4. The domain distribution in a sample formed by slow cooling of the discotic mesogen in a magnetic field. The disks represent the liquid crystalline molecule and $\hat{n}$'s are the directors of the various domains, which lie in the YZ plane, whereas the magnetic field is parallel to X. [Reproduced with permission from *J. Physique* **42**, 1303 (1981)]. Ref. [52].

polymer films, SiO coating at oblique evaporation, shearing method, and application of electric or magnetic fields. Using these techniques, a reasonably good alignment of the nematic and smectic LC can be achieved. The alignment techniques for DLCs are not as developed. A number of different techniques, including the use of various surface treatments, were investigated first by Vauchier et al. [38] with respect to some triphenylene derivatives. We discuss the relevant findings below.

1. Homeotropic (Side-On) Alignment

Vauchier et al. [38] showed that one way to obtain the homeotropic alignment for discogens on the glass surface is to use flat molecules with six polar side functions. These gave a good anchoraging to glass. These surfactants are hexphenol; mellitic acid; 2,3,6,7,10,11-hexa-hydroxytriphenylene (HHXT); and rufigallol. The effect of these surfactants on the alignment of different triphenylene-derivative DLCs has been discussed [38]. The glass slides are cleaned with acetone and treated in boiling water for 30 min. After immersion for 10 min in appropriate solution at 20°C, the slides are rinsed with the solvent, drained, and dried for 1 h at 110°C in nitrogen. The coating solutions are hexaphenol 1% and mellitic acid 5% by weight in pure water; HHXT 0.25%, and a saturated solution (0.1%) of rufigallol in ethanol. With these surfactants, it was possible to obtain a homeotropic alignment for some of the triphenylene derivatives and planar for others [38].

The possibility of achieving a homeotropic alignment using other surfactants and thin layers of polymers has also been given. In particular, Dubois et al. [39] showed that a surfactant such as hexahydroxybenzene spin-coated on ITO-coated glass plates produced homeotropic alignments of nonyloxybenzoate of hexahydroxytriphenylene (NBHT).

Phillips et al. [40] obtained a homeotropic alignment for three triphenylene derivatives: triphenylene hexa(heptoxybenzoate) H7OBT and two new triphenylene derivatives DB126 and DB119 (with benzoate series in alkyl chain). The nematic discotic phase of all three materials could be aligned successfully using a thin layer of polyimide (PI) or poly(vinyl alcohol) (PVA). Conoscopic observation confirmed that the sample was homeotropically aligned (i.e., director perpendicular to the plane of the cell).

Warmerdam et al. [41] found that using mellitic acid and HHXT the quality of alignment of truxene samples was not good. These authors used a thin PI film to ensure uniform alignment of some truxene derivatives such as 2,3,7,8,12,13-hexa-(decanoyloxy)truxene and 2,3,7,8,12,13-hexa(tetradeca-noyloxy)truxene.

Kruk et al. [42] used nylon 6,6 (1% solution in methanol) for obtaining a homeotropic alignment for H5T. They used ZLI 2650 (0.5% solution) for homeotropic alignment of the discotic polymer (triphenylene derivative).

Mourey et al. [43] showed that the homeotropic alignment of hexa-*n*-dodecanoyloxytruxene (HDT) is quite easily obtained by taking the sample between two clean or ITO-coated glass plates. To get a planar alignment of HDT, the authors used glass plates with an oblique coating of silicon monoxide (SiO). Such a treatment yielded only homeotropic alignment for the case of hexa-*n*-dodecanoyloxy truxene (C_{12}HATX) [44]. The aromatic cores of this molecule appeared to have strong affinity for the SiO surface. Raghunathan et al. [44], therefore, treated the obliquely SiO-coated plates with octadecyl triethoxy silane, thus obtaining an aliphatic surface that can be expected to have the undulations of the underlying SiO coating. They found that this combined treatment resulted in planar alignment of C_{12}HATX.

2. *Planar (Edge-On) Alignment*

The phenomenon of the surfaces, especially the crystalline surface, being able to control the alignment of nematic liquid crystals has been known for over 80 years [45]. The orientation of calamitic mesophases by crystalline surfaces is also well known [46]. Vauchier et al. [38] reported a planar orientation of molecules in the columnar discotic phase of some triphenylene derivatives using a perfect cleavage (001) surface of apophyllite, a lamellar tetragonal silicate, and a perfect cleavage of muscovite (a monoclinic mica). The authors also showed that another possibility for obtaining planar alignment of discogens was to use ITO-coated glass slices or glass slices coated with SiO at oblique evaporation. In particular, the planar alignment for some triphenylene derivatives was achieved using this technique (the angle between the beam and the slide plane was 20°) [38]. Mourey et al. [43] reported a planar alignment of hexa-heptyloxybenzoate and some truxene derivatives on a SiO-evaporated surface.

The attempt to use oblique SiO evaporation for obtaining planar alignment did not always work [44]. Raghunathan et al. [44] showed that to get a proper planar alignment for a C_{12}HATX discotic liquid crystal a combination of SiO evaporation and of special polymer layer on the top of SiO must be used.

Some other surface treatments have been reviewed for alignment of DLC [47]. Recently we reported IR investigations of some derivatives of triphenylene and truxene enclosed in between different crystalline and polycrystalline windows [48–50]. A strong influence of the substrate's structure on the bulk alignment of these discotic materials was demonstrated.

Other possibilities for achieving the uniform alignment in discotic liquid crystals are the alignments in strong magnetic fields [51–53], using shear flow [54,55], strand suspension [56], and orientation of the switchable discotic LC in an electric field [29,57]. Some of these techniques, however, require specialized equipment, a special cell for the shearing technique [54,55], or a strong magnetic field [51–53]. Also, in the latter method for obtaining a heterogeneous planar alignment, the cell (or the magnet) needs to be rotated.

It is well known that rodlike liquid crystalline molecules have a positive diamagnetic anisotropy ($\Delta\chi$), so that a strong magnetic field aligns the molecules parallel to the field. In contrast, for disk like materials, $\Delta\chi$ is negative; hence we would expect that all configurations with the director in a plane perpendicular to the magnetic field are equally favored (See Fig. 4). To circumvent this problem Levelut et al. [51] and Goldfarb et al. [52] suggested the use of a rotating magnetic field to obtain a homogeneous planar alignment. At a magnetic field of 2 T, only planar alignement was possible [51,52]. Recently Ikeda et al. [53] showed that a strong magnetic field ($\sim$5 T) applied along the cell can give rise to homeotropic alignment.

C. Sample Preparation

In our investigations we have used the following solutions for aligning the discotics [42,58]: (1) nylon-6,6 (1% solution in methanol) and (2) poly(amic acid) solution (ZLI 2650) (E. Merck, Darmstadt, Germany). A uniform coating of the alignment layer was obtained by placing four drops of the nylon 6,6 solution on two windows (suitable for IR measurements), followed by rotation at the rate of 300 rpm. An identical procedure was adopted for other preparations by using two drops of the ZLI 2650 solution. The windows coated with surface-aligning agents thus obtained were first dried by keeping them at room temperature for $\sim$1 h, after which they were then placed in an oven at 343 K for a further 24 h. The cell with two windows was assembled. The cell thickness was estimated from the measurements of the interference fringes observed in the IR spectra.

The sample cells were prepared using $\sim$1 to 4 mg of sample in the isotropic phase; a sample spacing of 3 to 15 μm was achieved using a Mylar spacer. The windows with the sample inside were sealed at the outer rims using an epoxy adhesive (Araldite) to prevent leakage of the liquid crystalline materials from the cell when the sample is heated to the isotropic phase. Although sample flow is possible in the IR cell during measurements due to inadequate filling and sealing, reproducibility in the results on the same cell during heating and cooling showed that no such leakage took place.

D. IR Spectroscopy

1. Conventional IR Measurements

The IR spectra of these spin-coated windows were recorded over a temperature range of 300 to 470 K. The IR absorbance bands for the surface-aligning agents were found to be of negligible intensity, and the dependence of their intensity, on temperature was equally negligible. Hence subtraction of the nylon 6,6 and ZLI 2650 spectra from that of the liquid crystalline material was regarded as being unnecessary for the wave number range of interest. The strongest band for nylon 6,6, centered at ~ 1640 cm^{-1}, was separated from the band of the C-C aromatic stretching vibration (centered at 1619 cm^{-1}) using the computer program BANDFIT (BioRad), described below. The nylon band at 1640 cm^{-1} was easily eliminated using a stepit procedure in BANDFIT. Temperature control of the cell was achieved within ± 0.2 K. The temperature increment adopted for the measurements was 1 K/20 min, near the phase transition temperatures.

A Digilab FTS60A FTIR spectrometer with a liquid-nitrogen-cooled MCT detector was used to record the spectra. A total of 64 scans were co-added to increase the signal:noise ratio, and a resolution in the region of $0.25 \div 2$ cm^{-1} in the computation of the absorbance spectra in the wave number range 450 to 4000 cm^{-1} was achieved.

The spectra of the windows recorded at various temperatures were substracted from the sample spectra. Spectra were smoothed, and baseline corrections made. BANDFIT was used to find the peak position, to separate out the complex peaks into their constituent bands, to fit them to the sum of the Lorentzian and Gaussian functions, and to calculate the integrated areas under the peaks. Well-separated bands in the IR spectra were fitted to a Lorentzian function to determine the frequency and amplitude of maximum absorbance. It was occasionally necessary to fit a complicated band structure in the wave number range 700 to 900 cm^{-1} to four subbands, to fully characterize the perpendicular bands for H5T. The curvefitting was carried out using the standard Grams-Research program with a Voight function for the shape of the bands.

IR measurements were performed in IR transmission mode at normal and oblique incidence of light (Fig. 3). At least two methods have been proposed for characterizing an oriented film. These involve tilting the oriented film in the incident beam to obtain polarized spectra as a function of tilt angle. This method was introduced in early 1970s by Schmidt [22] and Koenig et al. [23] for investigating thin films of polymers. Since that time the technique has been used extensively for the 3-D study of the molecules, orientation of molecules in Langmuir-Blodgett (L-B) films [59,60] and molecular crystals [61]. This method was not, however, applied for liquid crystalline thin films owing to technical difficulties. In particular, the need to use bulky heating stages prevented the beam being incident on the sample at relatively small tilt angles. This difficulty was however succesfully overcome in our work by using a special hot stage [19,62].

Another method for the orientational study of the liquid crystalline thin films was succesfully developed for investigating the discotic liquid crystals [63]. In this method, the conformation was characterized in each phase and changes were followed as a function of temperature, allowing association of structural changes with the transition observed by other techniques. We will discuss these techniques in some detail in Section 3.

2.   Polarized Oblique IR Transmission Measurements

A special sample holder was designed for oblique transmission investigations [19,62]. The arrangement allows the sample to be rotated around the Y axis and in the polarization plane around the X axis (Fig. 5). The sample cell and the holder contained within the heating stage can be rotated between the angles $-60°$ and $+60°$ around the Y axis.

Measurements were carried out for angles of polarizations between $0°$ and $180°$ through steps $10°$. Oblique transmission measurements were performed at polarizer positions $0°$ and $90°$ only. These measurements were conducted under the following conditions:

2.1.   Planar Alignment

1. The column's axis was oriented along the Z axis of the laboratory frame (the Z axis coincides with the polarizer position at $0°$; Fig. 4a). The measurements were carried out for the sample rotated around the Y axis, and angle α' is varied from $-60°$ to $+60°$ with $10°$ steps for the polarizer positions for $0°$ and $90°$.

2. The column's axis coincides with the Y axis of laboratory frame [Fig. 5(b)]. Measurements for the sample tilt were made with the same conditions as above.

2.2.   Homeotropic Alignment.   For homeotropic alignment (the axis of the columns is oriented normal to the window plane), the oblique measurements

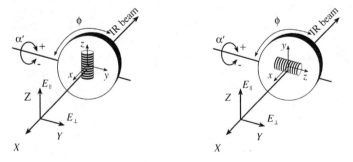

Figure 5.   The oblique transmission method. X, Y, and Z, represent the axes of the laboratory frame; x, y, and z, correspond to the molecular axes.

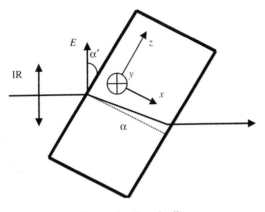

Figure 6. Sample tilt.

were performed for $0°$ and $90°$ angles of polarization at accidentally chosen sample position.

The orientation of the liquid crystal with respect to the window plane can easily be controlled through comparing of the in-plane (or out-of plane) core band intensity in the mesophase compound to the intensity in the isotropic phase. To compare the intensity of bands from different tilt angles, we must consider that the actual optical path length varies with the angle of the sample tilt (Fig. 6) and the orientation of transition dipole moments changes with respect to the electric vector of the IR beam.

Because light is refracted when going through media of different refractive indices, the angle α the electric vector makes with the Y axis in the film is related to the angle of the sample tilt, α', through Snell's law [22]

$$n_1 \sin\alpha' = n_{2X} \sin\alpha'' = n_{3X} \sin\alpha \qquad (17)$$

where n_1 is the refractive index of the first medium, n_{2z} is the refractive index of the second medium, α'' is the angle of refraction, n_{3X} is the refractive index of the third medium, and α is the angle of refraction in this medium. If we regard the first medium as the air, second medium as the cell window, and the third medium as the liquid crystalline layer itself, we can ignore the second medium for analysis of the results, so that

$$n_1 \sin\alpha' = n_{3X} \sin\alpha \qquad (18)$$

Because of sample tilt, the thickness of the layer for the measured absorbance is different from the sample thickness. By definition, the absorbance is proportional to the thickness through which the IR beam travels.

The true absorbance A_t of the film is related to the measured absorbance A_m by the following expression

$$A_t = A_m \cos \alpha \qquad (19)$$

where α can be found from Eq. (18) if n_{3X} and α' are known. α' is read out from the scale connected to the cell holder, and n_{3X} is estimated in the first instance from the refractive index of the sample in the isotropic phase.

3. Anisotropic Sample/KBr Mull Comparison Method

An alternative IR technique for obtaining an experimental information on the molecular orientation on the external surface was suggested by Yang et al. [63]. They suggested measuring the molecular orientation on the external surfaces at each temperature by comparing the integral intensity of each isolated band of the anisotropic sample to that of the mull. The integrated band intensity $I(T)$ can be expressed for each temperature as

$$I(T) = NkE^2 p^2 [\langle a_Y(T)^2 + a_Z(T)^2 \rangle] \qquad (20)$$

where k is a coefficient that depends on universal constants and absorption geometry, N is the number of molecules exposed by the incident beam, E is the incident electric field strength, p is the change in the dipole moment with respect to the normal coordinate, $a_i(T)$ is the directional cosines of p measured with respect to the surface normal defined as the X direction, and the angle brackets refer to averages taken over all absorbing molecules in the beam. At normal incidence, only the Y and Z components can be observed. Direct observation of the X component requires the surface normal to be inclined to the beam propagation direction as shown in Section 2. Because the squared directional cosines must sum to unity for any contributor to the averages (and thus for the averages), Eq. (20) may be written as

$$I(T) = NkE^2 p^2 [1 - \langle a_X(T)^2 \rangle] \qquad (21)$$

and if $NkE^2 p^2$ can be determined, $\langle a_X(T)^2 \rangle$ can be measured by measuring $I(T)$. Such a direct approach is not experimentally convenient, as the coefficients $NkE^2 p^2$ are difficult to measure directly; so in the work that follows, a ratio method is adopted to remove them from the analysis.

Let the degree of molecular orientation on the surface be represented by a function f, defined as

$$f = \langle a_Y(T)^2 \rangle + \langle a_Z(T)^2 \rangle \qquad (22)$$

For a completely isotropic system, each orientation to the beam axis is equally probable, the squared directional cosine angle averages are each 1/3, and f is then 2/3. Any value of f that departs from 2/3 would then provide a measure of molecular orientation, which can be obtained by comparing the band intensity for the cast film with that for the mull. Let $I_A(T)$ represent the measured integrated band intensity for the anisotropic film and $I_B(T)$ that for the mull. Then

$$\frac{I_A(T)}{I_B(T)} = \left(\frac{N_A}{N_B}\right)\left(\frac{f}{(2/3)}\right) \tag{23}$$

where N_A and N_B are the numbers of molecules in the beam, respectively, in the film and mull. At temperatures above discotic-isotopic phase transition, T_i, the samples are all in the isotropic state, and the intensity ratio measures the number ratio N_A/N_B. Experimentally, T_i lies well above the discotic–isotropic phase transition and was selected to measure the ratio $N_A/N_B = I_A(T_i)/I_B(T_i)$. The function f at different temperatures is then

$$f = \left(\frac{2}{3}\right)\left[\frac{I_B(T_i)}{I_A(T_i)}\right]\left[\frac{I_A(T)}{I_B(T)}\right] \tag{24}$$

The dichroic ratio R_X can be defined as

$$R_X = \frac{\langle a_X(T)^2 \rangle}{\langle a_Y(T)^2 \rangle} \tag{25}$$

Because the Y and Z axes are equivalent, $\langle a_Y(T)^2 \rangle$ and $\langle a_Z(T)^2 \rangle$ are the same, and the dichroic ratio is

$$R_X = \frac{2\langle a_X(T)^2 \rangle}{[1 - \langle a_X(T)^2 \rangle]} \tag{26}$$

In terms of the function f, R_X becomes

$$R_X = \frac{2(1-f)}{f} \tag{27}$$

When the average transition moment is inclined to the surface normal, the orientational distribution function S is

$$S = \left[\frac{(R_X - 1)}{(R_X + 2)}\right] \tag{28}$$

S may also be written as

$$S = \frac{1}{2}\langle 3\cos^2\theta - 1 \rangle \tag{29}$$

where θ is the angle between the transition moment direction and the surface normal. In terms of f the order parameter S may be written as

$$S = \frac{(2 - 3f)}{2} \tag{30}$$

or, if N_A and N_B are the same, simply as $1 - I_A/I_B$. Experimentally, however, it is more convenient to use Eq. (24) to measure f and Eq. (30) to calculate S.

E. Investigations of Surface Structure

The scanning electron microscope (JEOL JSM-35 model) with its high magnification, resolution, and depth-of-field capabilities was used to examine the surface of windows with and without the nylon spacer. To investigate the surfaces further, a white light interferometer was used. White light interferometry is a technique in which the optical path difference between a reference beam and the sample beam generates an interference pattern consisting of bright and dark fringes. From the interference patterns, the depth profiles of the surface can be obtained. In this study a ZYGO New View 100 3D Image Surface structure analyzer was used. Three $0.18 \times 0.14 \, \text{mm}^2$ areas are scanned on each window. From each region on the window the peak to valley distance is calculated, from which an average peak to valley distance for the window is determined.

IV. RESULTS OF FTIR INVESTIGATIONS AND DISCUSSION

In this section we mainly review our results of IR investigations performed over the last 10 years; however, the IR results obtained for DLCs in the literature will also be briefly reviewed. Hsu and co-workers [63–67] performed IR investigations on the model DLCs, or the homologous series of benzenehexa-n-alkanoates (Fig. 7).

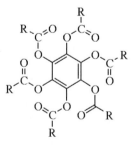

Figure 7. Benzenehexa-n-alkanoates. $R=C_5H_{11}$, benzene-hexa-hexanoate, (BH6); $R=C_6H_{13}$, benzene-hexa-heptanoate (BH7); $R=C_7H_{15}$, benzene-hexa-octanoate (BH8); $R=C_8H_{17}$, benzene-hexa-nanonoate (BH9).

A. Simple Discotics (Benzenhexa-*n*-Alkanoates)

The series of benzenehexa-*n*-alkanoates $(BH_n; n = 6-9)$ were investigated by Hsu and co-workers [63–67] using IR and Raman techniques. Several transitions associated with these macromolecules were determined using thermal studies and optical microscopy [1]. Primarily on the basis of X-ray diffraction studies [68], the structures associated with some of the phases have been proposed. The most interesting finding is that the homologue to $n = 6$ does not exhibit an intermediate liquid crystalline phase between the crystalline and isotropic phases. The homologous of $n = 7-9$, do possess an intermediate state. The proposed structure for the intermediate state contains a lamellar order with hexagonal symmetry in two dimensions and a liquid-like disorder in the third [68]. In fact, this intermediate phase is referred to as the discotic or the columnar phase, in which the disks are stacked on top of each other, forming columns. The spacing along the column axis is irregular. There are some uncertainties associated with the interpretation of the phase transitions found for the $n = 6$ molecule.

The IR spectra as a function of temperature obtained for different compounds and spectra in different spectral ranges are shown in Figures 8–12. As seen from the figures, the complicated IR spectra obtained for these DLCs can change significantly as a function of temperature, particularly in the vicinity of the phase transitions. The assignment of the bands is shown in Table I. The orientation of the core and the chain packing

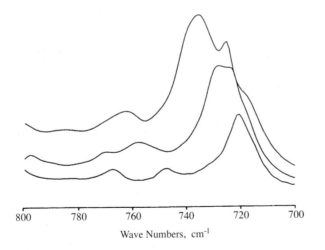

Figure 8. IR spectra of the CH_2 rocking region of benzenehexa-*n*-alcanoates obtained at room temperature. Band resolution, 2 cm^{-1}; 1000 scans. *Top*, BH6; *middle*, BH7; *bottom*, BH8. [Reproduced with permission from *Macromolecules*, **19**, 616 (1986)]. Ref. [64].

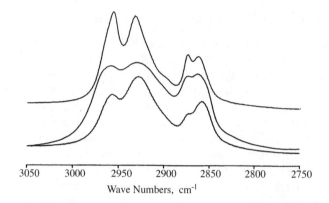

Figure 9. IR spectra of the CH stretching region of BH*n* at room temperature. Band resolution, 2 cm^{-1}. *Top*, BH6; *middle* BH7; *bottom*, BH8. [Reproduced with permission from *Macromolecules*, **19**, 616 (1986)]. Ref. [64].

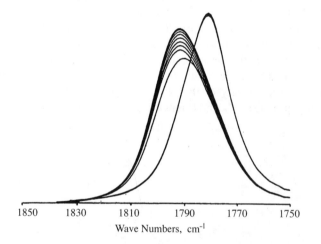

Figure 10. IR spectra of the C=O stretching region of BH6 at various temperatures. Band resolution, 2 cm^{-1}. [Reproduced with permission from *Macromolecules*, **19**, 616 (1986)]. Ref. [64].

can be extracted from the analysis of the vibrational bands that belong to the core and to the chain.

1. Alkyl Chain Vibrations

As seen in Figure 8, the CH$_2$ rocking vibrations of the three molecules differ considerably. The strong peak observed in the 700 cm^{-1} region is found at 733, 730, and 721 cm^{-1} for the BH6, BH7, and BH8 molecules,

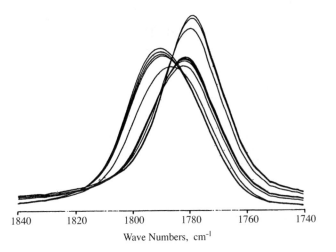

Figure 11. IR spectra of the C=O stretching region of BH8 at various temperatures. Band resolution, 2 cm^{-1}. [Reproduced with permission from *Macromolecules*, **19**, 616 (1986)]. Ref. [64].

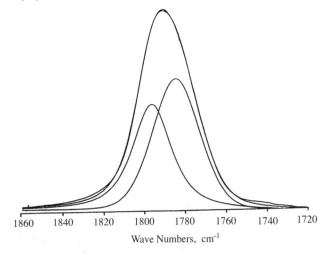

Figure 12. Results of a two-component fit to the data. [Reproduced with permission from *Macromolecules*, **19**, 616 (1986)]. Ref. [64].

respectively. Numerous studies have shown that the CH_2 rocking vibration in the 700 cm^{-1} region is particularly sensitive to the chain packing in polymethylens [69,70]. The normal spectroscopic features observed include a doublet at approximately 720 and 730 cm^{-1} arising from the crystal field splitting [71–73]. None of the molecules studied shows evidence of

TABLE I
Dichroism of IR Bands for Disk-like Molecules and Polymers

BH8-6,6-d_2 (Cr-353 k-D-355-k-I)		PBHA-18 (Cr-329-D-340-I)	
cm^{-1}	Dichroism	cm^{-1}	Dichroism
1778	Perpendicular	1780	Perpendicular
1760	Parallel		
1483	Perpendicular		
1439	Parallel		
1433	Parallel		
		1416	Perpendicular
1415	Parallel		
		1379	Parallel
1375	Parallel		
1321	Parallel		
1287	Perpendicular		
		1233	Parallel
		1220	Parallel
1220	Parallel		
		1202	Parallel
1188	Perpendicular		
1173	Parallel		

Reproduced with permission of *Macromolecules* **22**, 2611 (1988), Ref. [63].

intermolecular interactions, making it unlikely that "branches" pack in a well-defined unit cell. The CH_2 rocking vibration has also been found to be characteristic of the localized conformation of polymethylene sequences [74]. This rocking vibration is usually found near 720 cm^{-1} only when a minimum of four consecutive trans sequences exist [69,74]. Therefore, the authors drew the conclusion that the polymethylene sequences in BH6 or BH7 do not exist in the fully trans conformation. Only BH8 (and probably BH9) has a fully extended branch.

This conclusion is consistent with observations in the relatively simple IR-active CH-stretching region shown in Figure 9. The methyl stretching vibrations are exhibited at ~2873 cm^{-1} (methyl symmetric stretching vibration) and at 2956 and 2967 cm^{-1} (methyl asymmetric stretching vibration) [75]. If the polymethelene chains are indeed fully trans in configuration, two bands near 2920 and 2855 cm^{-1} are expected. They are assigned to the CH_2 asymmetric and symmetric stretching vibrations. This is observed for the BH8 molecules only. The two bands for the BH6 and BH7 molecules are found at approximately 2933 and 2862 cm^{-1}. Essentially all of the characteristic bands assigned to fully trans polymethylene sequences, including the CH_2 bending vibrations (not shown), are consistent with the interpretation made about the conformations associated with the three molecules [64,68]. It is interesting to note, however, that the characteristic bands for all the three molecules are located at identical positions at high temperatures when the order of the polymethylene sequences in BH8 is destroyed. But it is emphasized that it is difficult to assess quantitatively the degree of disorder using these spectroscopic methods.

Even though the polymethylene bands do not exhibit any differences between the discotic and isotropic phases, it was found that the vibrational bands associated with the ester group are extremely sensitive indicators of the phase transitions observed by thermal measurements [63,64,66,67]. For example, the changes observed for the C=O stretching vibrations in the 1700 cm^{-1} region are shown in Figures 10 and 11. It is clear that the frequencies stay relatively constant at 1781 cm^{-1} for BH6 in the crystalline state and then abruptly shift to 1792 cm^{-1} at melting. For the BH8 molecule, however, the band observed at 1780 cm^{-1} in the crystalline phase gradually shifts to 1781 cm^{-1} in the discotic phase and then shifts to 1790 cm^{-1} in the isotropic phase.

In the band deconvolution procedure, two bands, 1781 and 1793 cm^{-1}, gave the best fit for the series of data obtained as a function of temperature. The frequencies obtained from the fitting procedure stay relatively constant (Fig. 13). These results give additional confidence that only two components

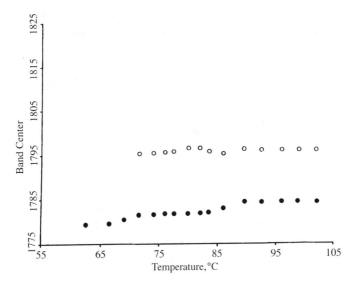

Figure 13. The frequency of the two fitted components as a function of temperature. [Reproduced with permission from *Macromolecules*, **19**, 616 (1986)]. Ref. [64].

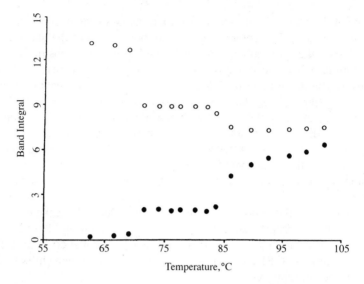

Figure 14. The relative intensity of two fitted components for BH8 as a function of temperature. $\bigcirc$, 1781 cm^{-1}; $\bullet$, 1795 cm^{-1}. [Reproduced with permission from *Macromolecules*, **19**, 616 (1986)]. Ref. [64].

are associated with the C=O stretching vibration of these disk-like molecules.

The gradual shift of the band position as a function of temperature cannot be correlated with the structural changes of the molecules. Therefore, instead of measuring the frequency of the overall band as a function of temperature, the integrated intensities obtained from the fitting procedure of the two components were plotted as a function of temperature (Fig. 14). As seen in the figure, the relative intensities change abruptly at temperatures associated with the crystalline–discotic and discotic–isotropic phases. For the BH6 molecule, only one transition was found, at the crystalline–isotropic transition.

Based on the analysis of the C=O stretching vibration, Kardan et al. [64] suggested that two possible conformations exist for the ester groups of the disk-like molecules. One exists primarily in the crystalline state, and the other exists primarily in the isotropic and discotic phases. The relative population can vary significantly as a function of temperature. Conformational calculations have been carried out for the disk-like molecules [76]. Several conformations can exist with the C=O bond at different angles with respect to the benzene ring. Because of the sterically crowded nature of these molecules, these conformations are sensitive to changes in nonbonded intermolecular interactions. The most favorable packing suggested by the theoretical analysis is that the disk-like molecules form pairs with the C=O bonds pointing at an angle in the opposite direction. It is extremely unlikely that the angle between the C=O bond and the plane containing the benzene ring can decrease. Therefore, Kardan et al. [64] suggested that the angle may actually increase, changing the distance between the disks. No other experimental evidence shows that this occurs; however the authors carried out a normal vibrational analysis of a disk-like molecule with the ester grouping at various geometries. They found that the calculated results suggest that the C=O stretching vibration varies in frequency corresponding to changes in this angle. Nevertheless, the magnitude of this change cannot be reproduced accurately.

2. Core Vibrations

In the spectra obtained at room temperature, the bands assignable to aromatic ring stretching or bending vibrations are the candidates for assessing the order associated with the aromatic cores. Most of these, however, lie in the regions of complex hydrocarbon side-chain absorptions and presumably disappear at higher temperatures into broadened composite bands. But there is one weak IR band centered around 1615 cm^{-1} that is observed in all of the BHn compounds and that is in an uncluttered region of the spectrum; it can be correlated with discotic phase changes [2].

The 1615 cm^{-1} band is not easily detected in the crystalline phase for any of the samples studied in this section, because it is not IR active for the benzene core in the crystalline state [65,77]. Its integrated intensity increases slightly at the solid–solid transitions. At the crystal–discotic liquid crystal phase transition, its intensity increases by a factor of 2 compared to the crystalline state. The intensity increases again by factor of almost 2 at the discotic–isotropic phase transition. In any given phase, the band intensity remains roughly constant throughout the temperature range of the existence of the phase. The integral intensity of this band was estimated from a diagonally drawn baseline, which extended 25 cm^{-1} on either side of the band center. No internal calibration of the band intensity was possible, because observed bands were conformationally or temperature sensitive. The intensity of the 1615 cm^{-1} band measured in this way is shown as a function of temperature for the mixture of BH8 and BH7 in Figure 15. We choose this example from the work of Yang et al. [65], because the range of the discotic phase is wider and the differences in the spectral behavior for the various phases is more obvious. The temperature ranges of BH7 and BH8 in the discotic phase are only 6° and 2°, respectively. For some binary mixtures, the temperature range for a discotic phase can be up to 15° or even higher. It should be emphasized that these mixtures are homogeneous [78,79].

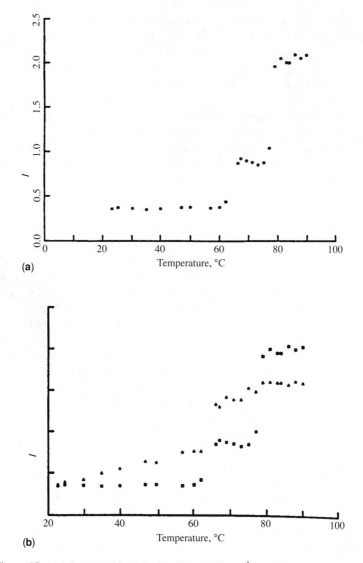

Figure 15. (a), Integrated intensity (I) of the 1615cm^{-1} in a 1:1 molar mixture of BH7 and BH8-6,6-d$_2$ as a function of temperature. (b), Integrated intensity of 1615 cm^{-1} (■) and the ratio of tg/u bands (▲) as a function of temperature for 1:1 molar mixture of BH7 and BH-6,6-d$_2$. [Reproduced with permission from *J. Chem. Phys.* **89**, 5950 (1988)]. Ref. [65].

As pointed out, the intensity of the 1615 cm^{-1} band is substantially lower in the discotic phase than in the isotropic phase and even lower in the high temperature stable crystalline phase than in the discotic. If it is accepted that the intensity of the 1615 cm^{-1} band is correlated with core–core order, then Figure 14 suggests that considerable order does exist between the rigid cores above the crystalline–discotic phase transition. This contrasts strongly with the substantial disorder found for the flexible alkane chains even in the high temperature crystalline phase [66].

At the discotic–isotropic phase transition, the core is almost completely disordered, which correlates with the distribution of the rigid core packing. In the isotropic phase, the spectra are almost identical to those of samples in benzene solution at room temperature. However, there is some evidence that the discotics retain some order well in the isotropic phase [80,81].

In conclusion, the 1615 cm^{-1} band can be used to assess the degree of order associated with the mesogens in the disk-like thermotropic liquid crystals. The discotic phase transition behavior has been postulated to depend on changes in mesogen order. The experiments described above support this view.

The IR studies of the site-specific deuteriated hexahydroxybenzene hexa(n-octanoate-6,6-d_2) (BH8-6,6-d_2) and poly(benzene hexa-n-alkanoate) (PBHA-18) discotic liquid crystals were carried out by Yang et al. [67] to find the spectroscopic evidence of the side-chain disordering processes in these compounds. For this analysis, the regions of CD$_2$ stretching vibration of 2000 to 2300 cm^{-1} and CD$_2$ rocking vibration in 500 to 700 cm^{-1} were considered. The information about the chain disorders can be obtained using the ratio of the band intensity A_{650}:A_{622} which gives the population ratio of the disordered to the ordered sequences tg/tt. This is because, the vibration at 622 cm^{-1} can be assigned to the rocking vibration with tt chain conformation flanking the CD$_2$ group. Because polymethylene chains are disordered in the isotropic phase, a band was observed at 650 cm^{-1}, which is characteristic of the tg sequences [67]. Hence the deuteriated molecules have provided valuable spectroscopic features for the use in the structural characterization of these disk-like molecules. There is substantial disruption at 32°C crystal–crystal transition involving the entire polymethelene chain. These studies show that the side-chain disordering is associated with the crystal–discotic and the discotic–isotropic transitions and that the side chains in the discotic phases are more highly ordered (fewer gauche bonds) than those in the isotropic state.

3.   Analysis of the Molecular Orientation on the External Surfaces

Significant differences were found when IR spectra obtained of the film cast on a KBr plate were compared to the spectra of samples mulled in KBr [63].

These spectroscopic differences are associated in part with the molecular orientation on the external surfaces. As the temperature is varied, the average orientation and conformation change, which produces changes in intensities that reflect alterations in molecular interactions, including those with the external surfaces. Figure 16 displays the C=O stretching vibration for a disk-like molecule and a polymer in KBr mull as a cast film as a function of temperature. Changes in both frequency and intensity are different for the different cases. The same effect is observed for the CH and CD stretching regions [63]. The relative intensities of these bands differ appreciably for the disk-like molecules, depending on the sampling method used. The anisotropic cast film should display orientational, interactional, and conformational effects, whereas orientational effects in these spectra are expected to vanish in the isotropic KBr mulls. When the spectra of the film and mull at each temperature are compared, the orientational effects may be isolated, because conformational and interactional effects are expected to be common. The method for extracting the information about the alignment of molecules at the external surfaces was described earlier. We will restrict this discussion to only the core vibrations. The details are given in Yang et al. [63].

The goals of the investigation were to demonstrate that the oriented film/ KBr mull technique could be used to assess molecular orientation on external surfaces, to elucidate the structures of disk-like liquid-crystalline molecules in each phase, and to identify structural changes at phase transitions. The method also promises the polarization characteristics of IR active vibrations.

The carbonyl stretching vibration was used for the structural characterization because it is highly localized. The intensities for the $C^*=O^*$ (C^* and O^* are the carbon and the oxygen of the ester group) stretching vibration observed in BH8-6,6-d_2 and PBHA-18 are shown in Figure 17. By using Eqs. (24), (27) and (30), f, R_X, and S can be calculated; and these parameters are shown in Figures 18–20. As expected, the orientational function for the disk-like molecules decrease with increase in temperature. As shown in the figures, the first decrease in the f or R_X value occurs at the 32°C solid–solid phase transformation. The solid structure changes from an ordered crystalline phase to a less ordered crystalline phase in this phase transition [63]. A decrease in the molecular orientation at elevated temperatures is consistent with a loss of molecular order. The orientation of the disk-like molecules is maintained at temperatures below the crystalline–discotic transition, although a decrease occurs near the 32°C solid–solid transition. The orientational distribution function decreases with increasing temperature. In Figure 20, the orientational function changes abruptly at the transition temperatures: ordered crystal–disordered crystal phase (phase II–

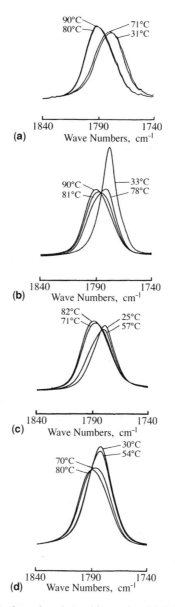

Figure 16.   IR spectra in the carbonyl stretching region. (a) Oriented film of BH8-6, 6-d$_2$. (b) Unoriented KBr mull of BH8-6,6-d$_2$. (c) Oriented film of PBHA. (d) Unoriented KBr mull of PBHA-18. [Reproduced with permission from *Macromolecules*, **22**, 2611 (1989)]. Ref. [63].

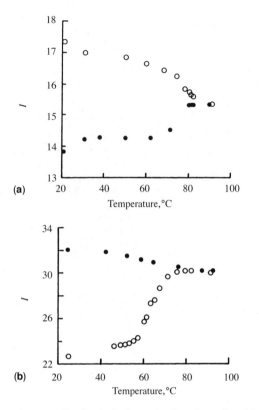

Figure 17. Intensity normalization in the isotropic phase for carbonyl band. (a) BH8-6,6-d_2. (b) PBHA-18. [Reproduced with permission from *Macromolecules*, **22**, 2611 (1989)]. Ref. [63].

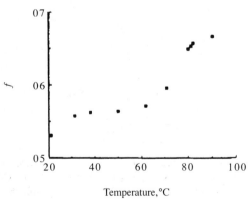

Figure 18. Measurable absorption ratio, I_A/I_B, as a function of temperature for BH-6,6-d_2. [Reproduced with permission from *Macromolecules*, **22**, 2611 (1989)]. Ref. [63].

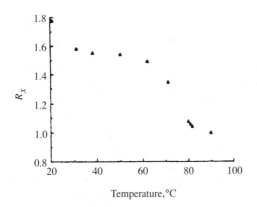

Figure 19. Dichroic ratio R_x as a function of temperature for BH8-6,6-d_2. [Reproduced with permission from *Macromolecules*, **22**, 2611 (1989)]. Ref. [63].

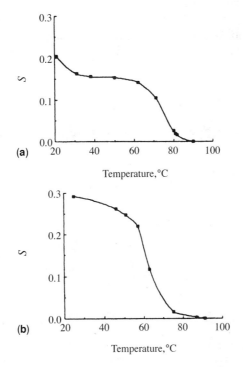

Figure 20. Order parameter S for BH8-6-6-d_2. [Reproduced with permission from *Macromolecules*, **22**, 2611 (1989)]. Ref. [63].

phase I) transition and at the discotic–isotropic transition. The tilt angle between the disks and the column axis may change at the disordered crystal–discotic transition in BH8-6,6-d_2 [82].

This analysis together with the analysis performed for the alkyl chain vibrations enabled Yang et al. [63] to draw conclusions about the orientation of individual chemical groups. All spectroscopic changes they observed indicate that at low temperatures the plane containing the zigzag chain is nearly parallel to the benzene ring. These results are consistent with the rotational isomeric model for the C-C*O* bond in ester molecules proposed by Abe [83]. In that model, the lowest energy state is the gauche conformer. Figure 21 indicates that the trans conformation will cause the zigzag carbon chain's plane to be approximately perpendicular to the central core. Only g or g' conformations can yield the dichroism observed for the CH$_2$, CH$_3$, and CD$_2$ stretching vibrations [63] and be consistent with the Abe [83] model. Thus the gauche conformer of the C-C*O* bond has the least energy state, in which the zigzag carbon chain lies in a plane nearly parallel to that of the benzene ring.

Thus the method suggested by Yang et al. [63] allows the orientation of molecules in cast film to be obtained as a function of temperature without tilting the sample. Discotic cores in the crystal phase of the cast film orient predominantly parallel to the external surfaces. The measured orientation

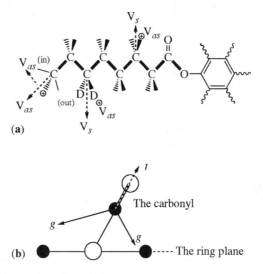

Figure 21. Orientation of the C*O*–C bond with respect to the core. ●, Carbon; ○, Oxygen. V$_s$ and V$_{as}$ show the orientation of the transition dipole moment for symmetric and asymmetric stretching vibrations respectively. [Reproduced with permission from *Macromolecules*, **22**, 2611 (1989)]. Ref. [63].

function decreases with the increase in temperature and can be used to follow changes in molecular orientation with temperature through various phase changes and characterizing some of the accompanying structural changes. Orientation of the flexible hydrocarbon chains in the crystal phase of the oriented film is nearly parallel to the disk-like cores, indicating that the gauche conformation of the C-O-C*O*-C bond is more stable than the trans one. Significant disordering of the flexible side chain occurs at the solid–solid phase transition, and the chain ends disorient much more readily than do groups near the aromatic core. Polarization characteristics of the observed bands are also reported on the basis of this analysis.

Nevertheless, as will be shown, the polarized IR study combined with a sample tilt geometry is useful for the study of ordering and the orientation of molecules in the discotic phase.

B.   Triphenylene Derivatives

1.   H3T, H5T, H7T, and H10OT

In this section, we review the results of IR investigations for the following triphenylene-based materials: (H3T), (H5T), (H7T), (H10OT), (H6TT), (H7T-NO$_2$) and the discotic liquid crystal polymer (DLCP). The DLCs were mainly synthesized by Ringsdorf's group in Mainz and Helfrich's group in Berlin. H7T-NO$_2$ was synthesized at Ivanovo State University (Russia), and H6TT was synthesized at the Centre for Liquid Crystal Research at Bangalore (India). The structural formulas and the phase sequences obtained using polarized microscopy and DSC are shown in Figure 22.

Before discussing the results, it is useful to briefly present the results of Destrade et al.'s [82] study of the polymorphism of these compounds (Tables II and III). We begin from the simple derivatives such as H3T, H5T, H7T, and H10OT and then describe H6TT and H7T-NO$_2$. As seen from Figure 22 and Table II H3T does not show a discotic phase. However, H5T and H7T exhibit the hexagonal ordered (D$_{ho}$) mesophase, and H10OT exhibits rectangular disordered (D$_{rd}$) mesophase. D$_{ho}$ is a discotic hexagonal ordered phase, because it exhibits a helical order and a three-column superlattice. Consequently, the distances between the cores in a column are approximately the same. A different situation, however, is observed for a discotic hexagonal (or rectangular) disordered phase (D$_{hd}$), in which the molecules form a hexagonal array of columns; both helical and superlattice orders are lost. The distance between the aromatic cores is the same for H5T and H7T ($d = 3.27$ Å) and is different for H10OT ($d_1 = 3.6$ Å; $d_2 = 4.5$ Å) (Fig. 23) [84].

1.1.   Results and Discussion. The spectra of H3T, H5T, H7T, and H10OT are shown in Figure 24. The frequency for the maximum intensity for some

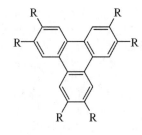

(a)

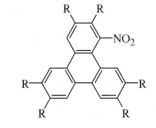

(b)

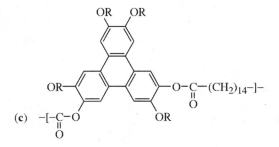

(c)

Figure 22. Structures of triphenylene-based liquid crystals. (a) H3T, $R = OC_3H_7$ (Cr-445-I); H5T, $R = OC_5H_{11}$ (Cr-342-D_{ho}-395-I); H7T, $R = OC_7H_{15}$ (Cr-341.6-D_{ho}-366-I); H10OT, $R = OCOC_9H_{19}$ (Cr-348-D_{rect}-398.5-I); H6TT, $R = SC_6H_{13}$ (Cr-335-H-343-D_{hd}-366-I). (b) H7T-NO_2, $R = OC_7H_{15}$ (G-268-D_h-407-I). (c) The structure of the main chain DLCP. $R = CH_2$-CH_2- CD_2- CH_2- CH_2- CH_2- CH_3.

vibrational bands for the studied triphenylene derivatives are summarized in Table IV. The assignment for the bands are made on the basis of the published data [77, 85–90].

Differences in the band positions and intensity are observed between H3T and H5T for CH_2 asymmetric stretching and CH_2 symmetric stretching vibrations near 2800 to 3000 cm^{-1} [58], owing to a smaller number of C atoms in the alkyl tails in H3T. This diference is smaller when the length of alkyl chain is substantially increased (compare the spectra for H7T and H10OT). The C–O–C symmetric bands in the range 1000 to 1100 cm^{-1} also show some differences in the absorbance. For CH out-of-plane vibration

TABLE II

Transition Temperatures of the Hexa-n-Alkoxy-Triphenylenes (HnT)[a]

n		Cr		D	I
1	•	317	–		•
2	•	247	–		•
3	•	177	–		•
4	•	88.6	•	145.6	•
5	•	69	•	122	•
6	•	68	•	97	•
7	•	68.6	•	93	•
8	•	66.8	•	85.6	•
9	•	57	•	77.6	•
10	•	58	•	69	•

Data from Ref. [82]. [Reproduced with permission from *Mol. Cryst. Liq. Cryst.* **106**, 121 (1984)].
[a]$R = OC_nH_{2n+1}$.

TABLE III

Transition Temperatures of the Hexa-n-Alcanoyloxy-Triphenylenes[a]

n		Cr		D_0		D_1		D_2	I
1	•	296	–		–		–		–
3	•	230	–		–		–		•
4	•	193	–		–		–		•
5	•	146	–		–		–		•
6	•	108	–		•	120	–		•
7	•	64	–		•	130	–		•
8	•	62	–		•	125	–		•
9	•	75	–		•	125.5	–		•
10	•	67	•	56	•	108	•	121.5	•
11	•	80	•	93	•	111	•	122.3	•
12	•	83	•	81	•	99.2	•	118	•
13	•	86.5	–		•	96	•	111	•

[a]$R = OCOC_nH_{2n+1}$.
Data from Ref. [82]. [Reproduced with permission from *Mol. Cryst. Liq. Cryst.* **106**, 121 (1984)].

near 837 cm^{-1}, the value of absorbance is higher for H3T than for H5T. During the heating cycle, significant changes in the shape of the C–O–C symmetrical bands in the range 1000 to 1100 cm^{-1} are observed (Fig. 25). The most significant changes in the shape of these bands occurrs near the phase transition temperatures; however, changes in the magnitude of the absorbance for the discotic material with temperature are small for the CH$_2$ and CH$_3$ stretching vibrations, which suggests that the tail is highly disordered even in the discotic phase. Comparing the spectra of the discotic

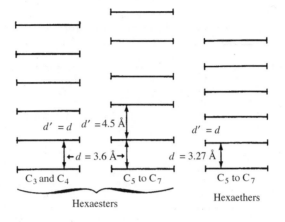

Figure 23. Molecular stacking for hexaesters and hexaethers. [Reproduced with permission from *J. Physique*, **43**, 355 (1982)]. Ref. [84].

phase to that of the isotropic phase, we find that the order type in H5T, H7T, and H10OT is side-on. The order type was deduced from the observation that absorption is higher in the discotic than in the isotropic phase for vibrations parallel to the core and the reverse is true for the vibrations with transition dipole moments oriented perpendicular to the core. It should be mentioned that these DLC cells were prepared using nylon 6,6 coated Si (or ZnSe) windows. The alignment can be different for differently treated substrates, as shown by our investigations [48–50]. These results will be discussed below.

To investigate the type and quality of the alignment achieved we used two different methods: the Neff's method and the polarized measurements combined with a sample tilt geometry. The order parameter S versus temperature for the C–C aromatic stretching vibration at ~ 1617 cm^{-1} is shown in Figure 26 for H5T. We have found $S \cong 0.85$ for this vibration using Neff's method. The value of S is so large that we conclude that the cores of the discotic material are almost perfectly aligned. The H5T material seems to be aligned by self-organization of the cores of the molecules. S is large because the central triphenylene disks may be almost normal to the column's axis [91]. A deuterium NMR spectroscopy study of hexa-hexyloxy-triphenylene (a DLC) provided the value of the order parameter $S = 0.95$ for the core [92]. Previous IR measurements of liquid crystals have shown that the band near 1600 cm^{-1} and the CH aromatic out-of-plane vibrations near 800 to 850 cm^{-1} are the most sensitive indicators of ordering in liquid crystals, liquid crystal polymers, and discotic liquid crystals [18, 26, 74, 93]. The plot for the order parameter for CH aromatic

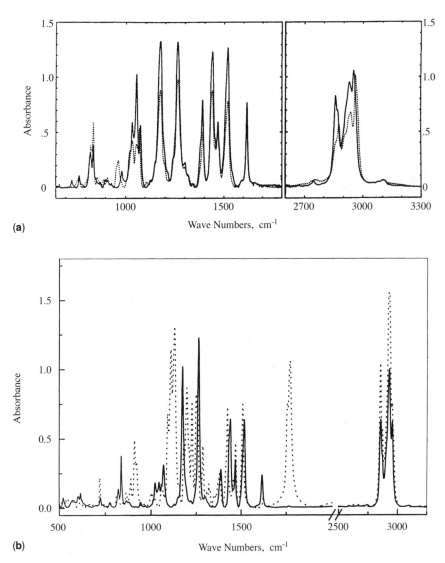

Figure 24. (a) The IR spectra of H3T and H5T. (b) The IR spectra of H7T (*solid line*) and H10OT (*dotted line*).

out-of-plane vibration as a function of temperature is shown in Figure 26, and its maximum value in the D_{ho} phase is $S \cong 0.8$. The maximum value of S for C-O-C asymmetric stretching vibration near 1175 cm^{-1} is $S \cong 0.38$ (see Fig. 27). Although the C-O-C bonds are directly connected to the central cores, the directions of the dipole moments for these vibrations

TABLE IV
IR peaks position for different triphenylene derivatives (in discotic phase)

Band assignment with Respect to the Core	Orientation with Respect to the Core	Position, cm^{-1}				
		H5T	H7T	H6TT	H10OT	H7T-NO$_2$
CH$_2$ symmetric stretching	⊥	2861	2857	2857	2855	2859
asymmetric stretching		2934	2927	2927	2926	2929
CH$_3$ symmetric stretching	⊥	2871	2872	2873	2872	2873
asymmetric stretching		2959	2956	2959	2958	2958
C=O stretching	⊥	—	—	—	1767	—
N=O asymmetric stretching	⊥	—	—	—	—	1532
C–C aromatic stretching	∥	1506	1508		1508	1512
		1517	1516	1550		1530
		1589		1590	1588	
		1617	1616		1620	1613
CH$_2$ scissoring mode	∥	1468	1468	1465	1465	1468
CH$_3$ asymmetric bending	∥	1457	1457	1452	1457	1458
CH$_3$ symmetric bending	∥	1434	1432		1419	1432
		1390	1383	1380	1376	1387
N=O symmetric stretching	∥	—	—	—	—	1365
		1263	1263		1251	1267
C–H aromatic in-plane deformation	∥	1124		1120	1108	1117
C–N stretching vibration	∥	—	—	—	—	1082
C–C alkyl stretching	∥	893	913	902	910	907
NO bending	⊥	—	—	—	—	845
C–H aromatic out-of-plane deformation	⊥	866	870	866	885	868
		800–866	800–860	800–860	800–860	800–860
NO$_2$ wagging	∥	—	—	—	—	750
C–H rocking mode	⊥	729	723	725	720	724

are not exactly parallel to the plane of the core and vary with the conformation.

A quite opposite situation has been found for CH$_2$ and CH$_3$ stretching vibrations near 2900 cm^{-1}. The order parameter for the entire band at 2780 to 3050 cm^{-1} has been calculated and is found to be extremely small: $S \cong 0.015$. Calculations for each of the CH$_2$ and CH$_3$ asymmetric vibrations found S to be close zero, within experimental error. This follows from the hypothesis that CH$_2$ and CH$_3$ stretching vibrations of the alkyl tails have many possible directions for the transition dipole moment vectors, owing to flexibility of the tails and the presence of different conformations. All the presented plots for the order parameter as a function of temperature show that S decreases with an increase in temperature; $S \rightarrow 0$ at temperatures of 5 K before the phase transition to the isotropic phase. The analysis of CH$_2$

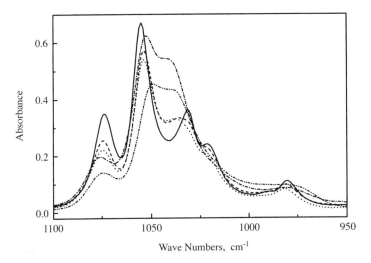

Figure 25. The IR spectra of the C−O−C vibration of H5T at different temperatures. - -, 300K (C); - - · ·, 338K (C); - - - -, 340K (D$_{ho}$), - · ·, 390K (D$_{ho}$); · · · ·, 400K (I). [Reproduced with permission from *Liq. Cryst.*, **14**, 807 (1993)]. Ref. [58].

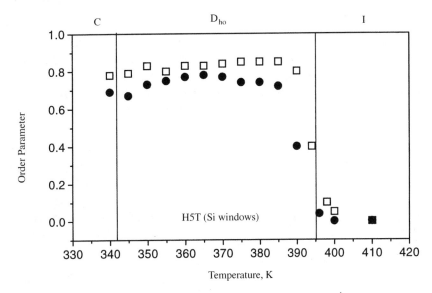

Figure 26. The order parameter for aromatic in-plane (1617 cm^{-1}) and out-of-plane (837 cm^{-1}) vibrations calculated using Eqs. (12) and (13).

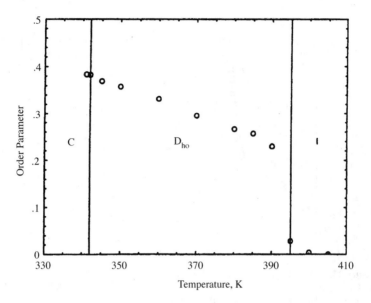

Figure 27. The order parameter for the C-O-C stretching asymmetric vibration near 1174 cm^{-1}. [Reproduced with permission from *Liq. Cryst.*, **14**, 807 (1993)]. Ref. [58].

rocking vibration are summarized for the triphenylene derivatives at the end of this section.

1.2 Frequency Changes. The frequency changes for a few vibrational bands obtained for H5T are shown in Figures 28 and 29. The strong discontinuity in the frequency (1 to 4 cm^{-1}) occurs some 2 to 3 K below the Cr-D$_{ho}$ phase transition for all the bands that were investigated. It is known that the wave number is related to the force constant K and the effective mass by

$$\bar{v} \sim \sqrt{\left(\frac{\langle k \rangle}{m_r}\right)} \qquad (31)$$

where $\langle k \rangle$ is an average value of the force constant over the molecular interactions of a vibration, and m_r is the reduced mass of a vibration. Because repulsive forces become much larger than the attractive ones at short interatomic distances, a significant increase in the wave number can be expected, as $\langle k \rangle$ increases from the crystalline to the discotic phase. Such an increase has been observed for most of the investigated bands, except the CH out-of-plane vibrations near 837 cm^{-1}. For the CH out-of-plane

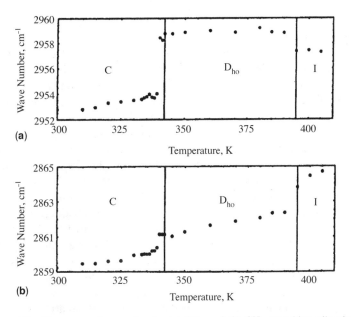

Figure 28. Frequency changes for the (a) CH_3 and (b) CH_2 stretching vibrations near 2900 cm^{-1}. [Reproduced with permission from *Liq. Cryst.*, **14**, 807 (1993)]. Ref. [58].

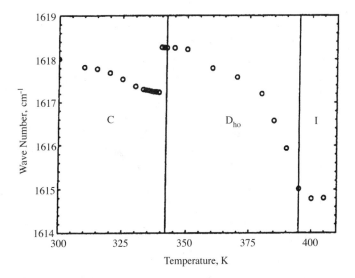

Figure 29. Frequency changes for the C–C aromatic stretching vibration near 1617 cm^{-1}. [Reproduced with permission from *Liq. Cryst.*, **14**, 807 (1993)]. Ref. [58].

vibration, the frequency decreased by 1 to 2 cm^{-1}, at 3 K below the Cr–D$_{ho}$ transition temperature.

At the D$_{ho}$–I phase transition, the frequency decreased abruptly for the CH$_3$ asymmetric stretching vibration near 2958 cm^{-1}, the C–C aromatic stretching vibration near 1617 cm^{-1}, the CH bending mode near 1389 cm^{-1} and the C–O–C symmetric stretching vibrations. The increase in frequency for the D$_{ho}$–I phase transition occurred for the CH aromatic out-of-plane vibration near 836 cm^{-1} and the CH$_2$ symmetric stretching vibration near 2863 cm^{-1}. Different changes in frequency for these two bands appeared during the (Cr–D$_{ho}$) phase transition (Fig. 28 and 29). The frequency increased for the CH$_2$ symmetric stretching vibration and decreased for CH aromatic out-of-plane vibration. Nevertheless, in all cases we have observed large changes in frequency versus temperature some few degrees below the phase transition temperature. This means that the short-range order in the discotic H5T material changes at 2 to 3 K below the phase transition temperatures. The CH bending modes are sensitives to the alkyl chain conformation and to chain packing [74]. A frequency change by about 2 cm^{-1} for the CH bending mode near 1390 cm^{-1} suggests that during the Cr–D$_{ho}$ phase transition changes occur in the conformation and in the packing of chains.

The results of the temperature dependence of the peak intensity for different vibrational bands of H7T and H10OT are shown in Figures 30 and 31. We can see from these plots that H7T also shows quite good alignment

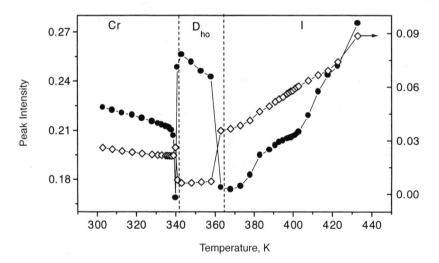

Figure 30. The dependency of the peak intensity of the 1600 cm^{-1} band on temperature for H7T. ●, 1616 cm^{-1}; ◇, 866 cm^{-1}.

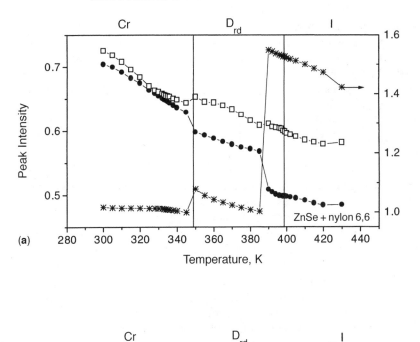

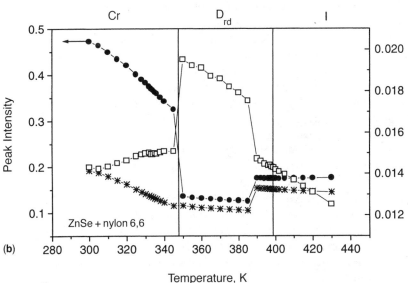

Figure 31. The dependence of the peak intensity on temperature of H10OT contained between nylon-coated ZnSe windows for few vibrational bands. (a) ●, 1508 cm^{-1}; □, 1620 (×10) cm^{-1}; *, 1767 cm^{-1}. (b) ●, 911 cm^{-1}; *, 889 cm^{-1}; □, 1588 cm^{-1}.

and a large-order parameter; however, the mesophase range is much narrower in this case than for H5T. As seen in Figure 30, there is some evidence for the existence of another phase in a temperature range 380 to 400 K. We assume that this can arise from the existence of the second mesophase (probably nematic) for this discotic material. As was shown by Veerman and Frenkel [91], the appearance of different mesophases for discotic molecules depends on the ratio of the thickness L to the diameter D of the disk. For molecules with a longer alkyl chain the ratio $L:D$ becomes smaller. For $L:D = 0.1$, the nematic and columnar phases can be observed. It quite possible, however, that for this particular compound there is some local ordering at the beginning of isotropic phase that may also cause some differences in temperatures dependencies.

The results of temperature dependencies observed for H10OT show that the alignment of these compounds is not as good as shown by H5T and H7T, despite quite noticeable observed features around the phase transition temperatures. The highest order parameter has been obtained for the C=O vibrations. We conclude that the core is tilted with respect to the windows plane, whereas C=O groups in the discotic phase are nearly perpendicular to the window plane. As shown earlier, the C=O group is usually oriented at $57°$ with respect to the core plane. The results obtained for the disordered mesophases of other compound leads to the conclusion that, owing to a difference in the distance between the different cores inside the column, the intermolecular interactions can easily be broken for large distances with increase in temperature (as a result of the fluctuations). This means that pretransitional phenomena to the isotropic phase can be started in the discotic phase and is discontinuous, as opposed to taking place slowly over a range of temperatures. A detailed analysis of the alkyl chain vibrations for the similar triphenylene derivative (hexa-nanoyloxytriphenylene) was performed by Rey-Lafon et al. [94].

It is important to mention that the results discussed above were obtained during the heating cycle. Figure 32 show the results obtained for H7T during the cooling cycle. We find that the differences observed for the phase transition from the isotropic to discotic phase is not as large as for the discotic-crystalline state. In the case of the edge-on alignment, however, the differences for heating and cooling cycles are even much larger. We will discuss this in more detail for the H7T-NO$_2$ and H6TT compounds.

As stated, we have always observed the side-on alignment during the cooling cycle, whereas during the heating cycle, the behavior of the DLC depends strongly on the substrate and its structure, topology, and surface treatment. Polarized microscopy performed for H5T and H7T for the heating and cooling cycles show this clearly. As was shown by Wang et al. [95], this can be explained by the memory effect existing for the discotic materials.

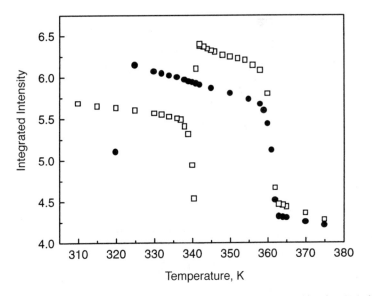

Figure 32. The temperature behavior for H7T under cooling (●) and heating (□) shown for C–C aromatic stretching vibration at 1600 cm^{-1}. Reproduced with permission from *J. Molec. Struct.* **271**, 511–512 (1999). Ref. [97].

Using molecular modeling, they predicted the existence of two different D_{ho} phases for H5T. These phases appear to be the result of different temperature treatments.

We review the results of H5T and H7T obtained using polarized oblique infrared transmission spectroscopy [19,62,96,97]. The technique of oblique incidence allows us to define more precisely the orientation of the different parts of molecules. This leads to better assignment for the bands. This can also be used to find the order parameter in the discotic phase without heating the sample to the isotropic phase. The later is a necessary condition for finding S using Neff's method [97]. Figure 33 shows the results obtained for H5T sandwiched between nylon 6,6-coated Si windows when the sample tilt is varied from $-60°$ to $+60°$. As seen from the Figure, the sample tilt strongly influences the intensity of some of the vibrational bands.

We performed a set of experiments to show that the transition dipole moment p_i for the C–C aromatic band (1617 cm^{-1}) is randomly distributed along the disks (Fig. 34). We measured the absorbance as a function of the angle of the polarization of the IR beam and found that the absorbance is independent of this angle. We used this fact to find n_{3X} (Eq. (18)). We aligned the polarizer with the Z axis (polarization angle $=90°$); the true absorbance must be independent of the sample tilt angle as the sample is being rotated along the Z axis. We then aligned the polarizer with the Y axis

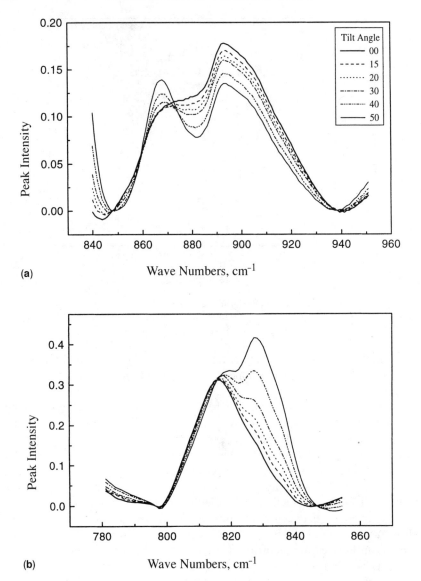

Figure 33. FTIR spectra of the discotic phase of H5T at different tilt angles. –, 0°; - - -, 15°; · · · ·, 20°; -·-·-, 30°; -· ·-· ·-, 40°; —, 50° [Reproduced with permission from *Supramol. Sci.*, **4**, 529 (1987)]. Ref. [62].

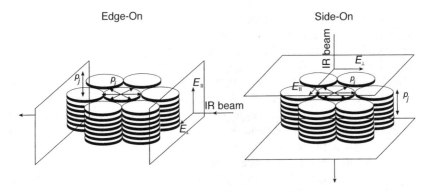

Figure 34. The orientation of the transition dipole moments for the in-plane and out-of-plane vibrations for the edge-on (a) and side-on (b) alignment of H5T in its columnar phase. [Reproduced with permission from *Europhys. Lett.*, **44**, 198 (1998)]. Ref. [49].

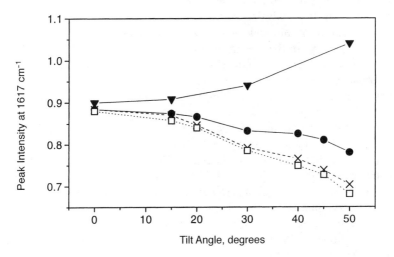

Figure 35. The dependence of the peak intensity of C–C aromatic vibration on sample tilt angle. ▼, experimental curve for the polarizer position at 0°; ● experimental curve for the polarizer position at 90°; **X**, experimental curve for the polarizer position at 90° corrected on the basis of n_{3X} estimated as given in the text; □, experimental curve for the polarizer position at 90° corrected on the basis of $n_{3X} = 1.5$ (for the isotropic phase). [Reproduced with permission from *Supramol. Sci.*, **4**, 529 (1997)]. Ref. [62].

for zero sample tilt (polarization angle $= 0°$) and measured the absorbance as a function of the sample tilt angle. The measured absorbances for the two polarizations are shown in Figure 35. The corrections for n_{3X} and the variation of the sample thickness are applied by slightly varying n_{3X} about its estimated value; with these corrections, the true absorbance along the X

axis is found to be independent of the sample tilt angle, whereas the true absorbance along the Y axis decreases with the increase of the angle (Fig. 40). This can be easily explained because p_i is no longer coincident with the E of the IR beam and makes an angle α with it.

The results for the dependence of the perpendicular bands (i.e., normal to the core) are also interesting. Figure 36 shows the dependence of the peak intensity on the sample tilt angle for polarizer positions of 0° and 90°. We find that the 828 cm^{-1} band does not show up when the sample is not tilted for both polarizations and when the sample is tilted at an angle of polarization 90°. For the 0° angle of polarization, this band starts to appear when the sample is tilted slightly and shows a big increase in the intensity when the tilt angle (without correction) approaches 50°. This result shows that vibrational band with a frequency maximum of 828 cm^{-1} (C–H aromatic out-of-plane vibration) has the transition dipole moment oriented exactly perpendicular to the plane of windows for the columnar mesophase of H5T. From this analysis, we understand why the band does not appear in the spectrum of columnar phase, whereas it does appear (with a slightly higher frequency) in the spectra of the crystalline and the isotropic phases with normal incidence of light (Fig. 37). This result confirms the side-on alignment of the sample.

The dependence of the true absorbance on the sample tilt angle for some of the aforementioned bands are shown in Figure 38, which demonstrates a strong decrease in the absorbance (with p_i oriented parallel to the window plane), a strong increase in the absorbance (p_i oriented perpendicular to the windows plane) and a small change in intensity for some bands (p_i randomly oriented along the columns). The C–C aromatic stretching vibration (1617 cm^{-1}) belongs to the parallel bands, because it has a strong preferable orientation in the plane of windows. From this study, we can concluded that C–C vibration is indeed parallel to the window, because this band has shown a dramatic decrease in absorbance when the cell is tilted with respect to the IR beam. The band with the frequency maximum at 828 cm^{-1} (C–H out-of plane deformation) belongs to the second type. At the same time, the intensity of the 816 cm^{-1} band changes only slightly as a consequence of the variation in the thickness of the cell (nearly in the same manner as in the isotropic phase). Interestingly, the 868 and 893 cm^{-1} bands exhibit opposite dependencies on the sample tilt angle (Fig. 38). Although the intensity of the 868 cm^{-1} band increases with an increasing sample tilt angle, its absorbance decreases. Usually, the low-frequency vibrational bands in the region of 800 to 900 cm^{-1} are assigned to the C–H aromatic out-of-plane deformations. The bands in the region 840 to 900 cm^{-1}, however, could alternatively be assigned to the C–C alkyl stretching vibrations [86,90]. This implies that the projection of the transition dipole

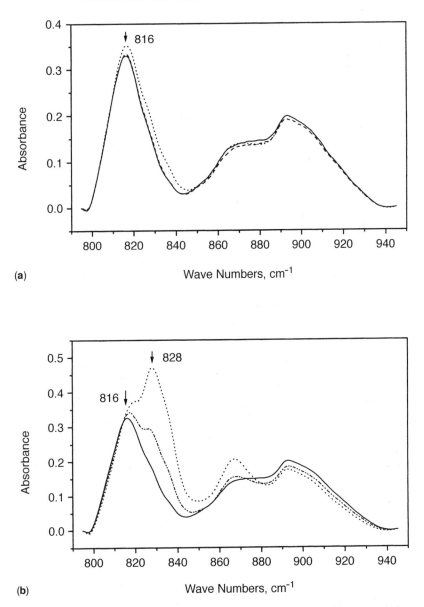

Figure 36. Polarized FTIR spectra of H5T at different tilt angles. (a) Polarizer position at 90°. —, 0°; ---, 15°; ·····, 45°. (b) Polarizer position at 0°. —, 0°; - · - · -, 30°; ······, 50°. [Reproduced with permission from *Supramol. Sci.*, **4**, 529 (1997)]. Ref. [62].

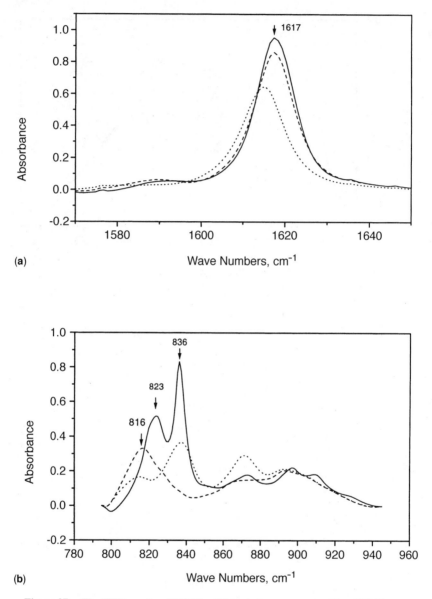

Figure 37. The FTIR spectra of H5T for different phases. —, crystalline (300 K); ----, columnar (360 K); · · · ·, isotropic (400 K). [Reproduced with permission from *Supramol. Sci.*, **4**, 529 (1997)]. Ref. [62].

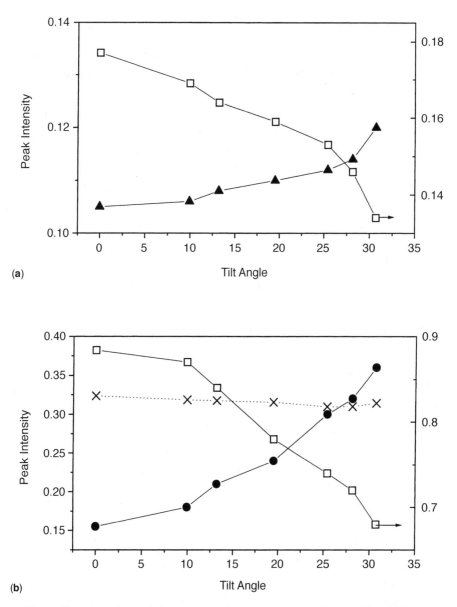

Figure 38. Dependence of the peak intensity on the corrected tilt angle for different vibrational bands. (a) ▲, 868 cm^{-1}; □, 893 cm^{-1}. (b) ●, 828 cm^{-1}; **X**, 816 cm^{-1}; □, 1617 cm^{-1}. [Reproduced with permission from *Supramol. Sci.*, **4**, 529 (1997)]. Ref. [62].

moment for this vibration on the $Y-Z$ plane is much greater than in a plane normal to it. From these investigations, we make a precise assignment of the 893 cm^{-1} band to be the C–C in-plane alkyl stretching vibration, because its intensity exhibits an opposite dependence on the tilt angle to that of the perpendicular bands of 828 and 868 cm^{-1}.

To evaluate the order parameter, the oblique IR measurements were performed with a smaller step of sample rotation angle ($2°$) to increase the accuracy of the results and the fitting procedure. Figure 39 shows the angular dependence of the corrected absorbance as a function of α (between the electric field vector of incident light and the window plane) for side-on alignment of the H5T. We observed an increase in the intensity of the perpendicular band $830\,\text{cm}^{-1}$ ($\beta = 0°$ because the transition dipole is parallel to the column's axis) and a decrease in the intensity of the parallel band at 1617 cm ($\beta = 90°$ because the transition dipole is perpendicular to the column's axis) as predicted by Eqs. (2) and (14). The curves show the fit to Eq. (16) of the experimental data. At the low temperature the order parameters are reasonably high and are in an excellent agreement [Fig. 40(a)]; however, for the heating cycle, the parameter S for the C–C stretching band seems to be strongly affected, showing a rapid increase in the tail disorder, whereas S for the C–H out-of-plane band remains nearly constant.

Figure 40(b) shows the temperature dependence of S for H7T obtained for two vibrational bands belonging to the core: in plane C–C stretching vibration at the 1617 cm^{-1} and C–H out-of-plane vibration at the 836 cm^{-1}. The results for order parameter are very much the same, except that the S values are slightly lower than those for H5T.

2. The Influence of the Substrate's Structure on the Alignment of DLC

From our investigations, we found that the alignment of the DLCs strongly depends on the substrate surface [48–50,96]. To investigate this effect we used a number of substrates, including amorphous, crystalline, and polycrystalline materials transparent in the IR region and using different surface treatments. Results for these investigations are presented here.

From an extensive work carried out on nematics and smectics [34,37], certain factors (the type of the orienting layer, the mechanical rubbing or its grooving) are known to control the surface alignment. Microscopically, the following contributions to the anchoring energy are considered. These correspond to the topology, steric, polar, and dispersive (caused by van der Waals) interactions [34,37]. Except the topology, the other anchoring energies are primarily governed by the intermolecular interactions between the substrate and the liquid crystalline molecules. The mechanisms for the surface-induced bulk alignment are based on the short-range surface-molecule and then molecule-molecule interactions [98] and the minimiza-

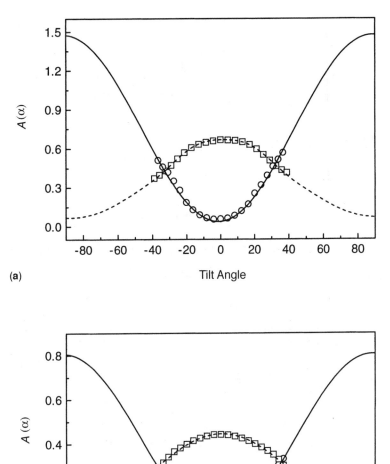

(a)

(b)

Figure 39. (a) Peak intensity dependence vs. the tilt angle for H5T at 360 K □, C–C aromatic in-plane vibration (1617 cm^{-1}); ○, C–H aromatic out-of-plane vibration (830 cm^{-1}). (b) Peak intensity dependence vs. the tilt angle for H7T at 342 K. □, C–C aromatic in-plane vibration (1617 cm^{-1}); ○, C–H aromatic out-of-plane vibration (836 cm^{-1}). [Reproduced with permission from *J. Molec. Struct.* **271**, 511–512 (1999)]. Ref. [97].

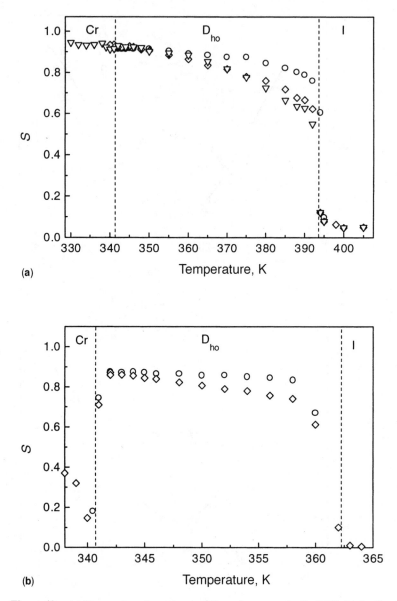

Figure 40. (a) Temperature dependence of the order parameter for H5T. $\square$, heating the
C–C aromatic in-plane vibration (1617 cm^{-1}); ∇, cooling the C–C aromatic in-plane
vibration (1617 cm^{-1}), $\bigcirc$, heating the C–H aromatic out-of-plane vibration (830 cm^{-1}). (b)
Temperature dependence of the order parameter for H7T. $\diamondsuit$, C–C aromatic in-plane vibration
(1617 cm^{-1}); $\bigcirc$, heating the C–H aromatic out-of-plane vibration (836 cm^{-1}). [Reproduced
with permission from *J. Molec. Struct.* **271**, 511–512 (1999)]. Ref. [97].

tion of the long-range interactions between the surface and the bulk [99]. The observations show that the condition of the minimum surface energy is fulfilled if the director (axis of the column) has a planar orientation (i.e., edge-on alignment); and this type of alignment for the discotic LC is normally preferred for untreated substrates. Whether this alignment has a homogeneous or heterogeneous character depends on whether a substrate has one or more than one easy directions of orientations.

Orientation of the discotic LC and its stability of alignment in the discotic phase has been investigated for four substrates: Si, CaF$_2$, ZnSe, and ZnS first for the clean but untreated substrates and then coated with a polymer [49,50,96]. The substrates of Si, ZnS, ZnSe, and CaF$_2$ were selected to provide various types of interactions between the different surfaces and the discotic material and because they are transparent to IR radiation. Because the alignment of rodlike molecules forming the nematic phase [35–37] depends on whether the substrate has one or more than one easy axis of orientations, it is imperative to examine the dependence of any alignment of the discotic in the presence or absence of the easy axis of orientation of a substrate. For the crystalline-cut (110) of ZnSe substrate, one easy direction of orientation is found using X-ray diffraction [100–102]. The ZnS substrate shows only one easy axis of orientation [101]. As already pointed out, Si is analogous to glass, because a native oxide layer may exist on the surface of Si similar to that on glass. For glass substrates, however, the planar-heterogeneous orientation of triphenylene derivatives is observed using different techniques, X-rays, NMR, etc. [51,52,103], therefore, it is natural to accept a similar alignment of discotic on Si as on a glass substrate. The characteristics of CaF$_2$ windows are rather similar to those of Si because there are three easy directions of orientation instead of a number of easy directions in Si.

For the analysis of an alignment, two bands are selected: the aromatic C–C stretching vibrations in the range 1500 to 1600 cm^{-1} and the C–H out-of-plane vibrations close to 700 to 900 cm^{-1}. As mentioned, these bands are the most sensitive indicators of ordering in the discotic phase, because their intensities are directly correlated with the core–core order. Figure 34 shows the distributions of the transition dipole moments for the in-plane and out-of-plane vibrations with respect to the plane of the window for both the edge-on and the side-on alignment of the discotic material. For these cases of alignment, the absorbance as a function of the polarization angle can be predicted for both normal and oblique angles of incidence. For the side-on alignment, the absorbance for the in-plane vibration should be independent of the angle of polarization, because the transition dipole moments for these type of vibrations are uniformly distributed in the plane of the window. The intensity of the out-of-plane vibrations (with the transition dipole moment perpendicular to the window plane) should be extremely low and would probably have the same projections on the Y and Z axes. A strong angular dependence, however, is expected for the absorbance as a function of the oblique angle of incidence due to a difference between the projections of the transition moments parallel and perpendicular to the electric vectors of the field.

Furthermore for the edge-on alignment, one can expect a large polarization dependence of the absorbance for both types of vibrations.

2.1. Uncoated Substrates. Figure 41 shows an edge-on orientation of the DLC in the range of 340 to 397 K for the untreated substrates. Measurements were made with a reasonably slow temperature scan of 0.5 K/min. When the sample is kept much longer (for a few hours) in the discotic phase, the range of temperature in the discotic phase for the edge-on orientation is significantly reduced for Si and CaF$_2$ substrates and the phase transition to side-on orientation is observed (Fig. 42), whereas the edge-on alignment for the ZnSe and ZnS remains almost unaltered.

Large differences in the dependence of the absorbance on the angle of polarization are observed for the two sets of substrates (Fig. 43). For Si and

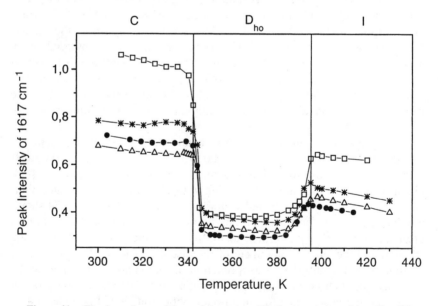

Figure 41. The dependence of the peak intensity of C–C aromatic stretching vibration on temperature for HPT contained in between different untreated substrates. The cell thicknesses are Si ($\triangle$); 12 μm; CaF$_2$ (*), 12 μm; ZnS ($\square$), 17 μm; and ZnSe ($\bullet$), 10 μm. [Reproduced with permission from *Europhys. Lett.*, **44**, 198 (1998)]. Ref. [49].

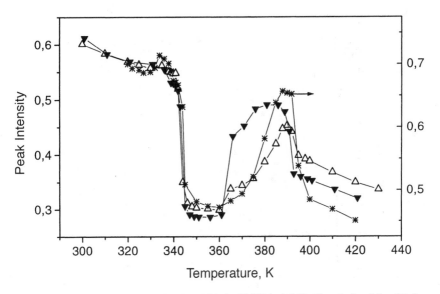

Figure 42. As in Figure 46 after 1 to 1.5 h for Si ($\triangle$) and CaF$_2$ (*) and after 1.5 to 2 h for Si ($\triangledown$). [Reproduced with permission from *Europhys. Lett.*, **44**, 198 (1998)]. Ref. [49].

CaF$_2$ cells, no dichroism is observed; whereas a pronounced dichroism is found in the Y-Z plane of a ZnS cell. Results of the polarization measurements for cells with ZnSe windows are found to be similar to those of the ZnS windows. Note here that for the 866 cm^{-1} band, the peak intensities for different angles of polarization were taken directly from the spectra, whereas peak intensities for the 836 cm^{-1} band were obtained after fitting an observed complex band in the 780 to 870 cm^{-1} region. An example of the band fitting procedure for the angles of polarization of 0° and 90° is shown in Figure 44. The frequencies of a number of bands used in the analysis are identified from measurements in the crystalline and isotropic phases. The frequency maxima for these bands are 814, 824, 836, and 866 cm^{-1}.

We obtained a different set of results for the oblique IR incidence for the two classes of substrates Si and CaF$_2$ on one side and ZnSe and ZnS on the other. For H5T sandwiched between ZnSe and ZnS windows, a strong dependence of the band intensities on the sample tilt angle is observed; the results for Si and CaF$_2$ windows, on the contrary, show a smaller dependence of the absorbance on the angle of polarization. Results in Figure 45, demonstrate the 1517 cm^{-1} band for cells with ZnS and Si windows. These results are found to be consistent with a type of the substrate. From the above data, we conclude that for ZnSe and ZnS windows the orientation of the director of the columns is planar-homogeneous, whereas for the Si and CaF$_2$ windows the orientation of $\hat{n}$ is planar-heterogeneous (Fig. 46). These

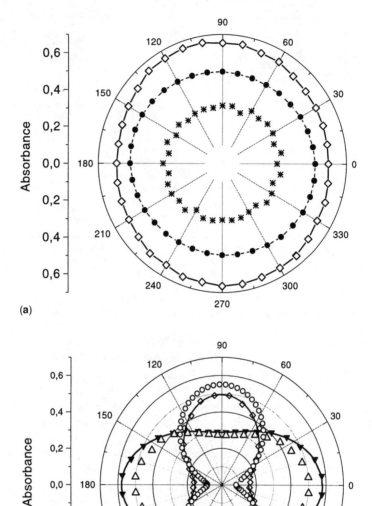

Figure 43. The intensity distribution for in-plane and out-of-plane vibrations in the window plane (Y-Z plane). (a) For Si and CaF_2 cells. ●, 1517 cm^{-1} (Si); ◇, 1517 cm^{-1} (CaF_2); *, 866 cm^{-1}; (Si). (b) For ZnS cell. *Solid lines*, calculated functions using Eq. (15). ▼, 1517 cm^{-1}; ◇, 866 cm^{-1}; △, 1617 cm^{-1}; ○, 836 cm^{-1} (calculated). [Reproduced with permission from *J. Mater. Sci. Eng. C*, **283**, 8–9 (1999)]. Ref. [96].

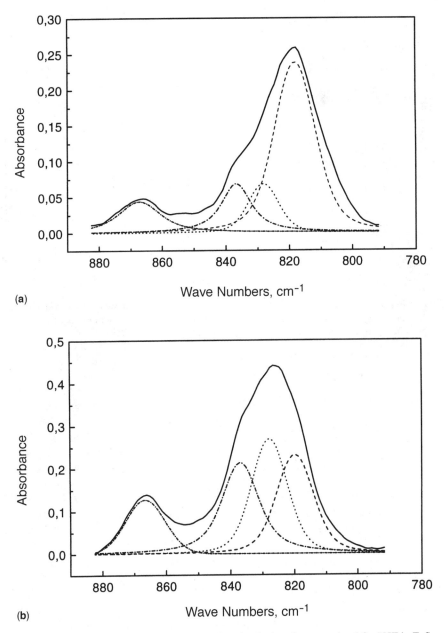

(a)

(b)

Figure 44. Example of the fitting procedure for the low-frequency band for H5T in ZnS cell for polariser position at (a) 0° and (b) 90°. —, experiment; - - - - -, 867 cm^{-1}; - - - - -, 828 cm^{-1}; · · · ·, 836 cm^{-1}; - - - -, 818 cm^{-1}. [Reproduced with permission from *J. Mater. Sci. Eng. C*, **283**, 8–9 (1999)]. Ref. [96].

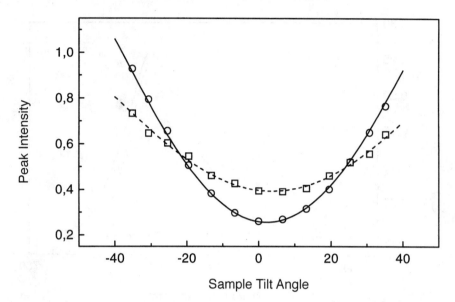

Figure 45. Results of oblique IR measurements at polariser position 0°. The intensity distribution in *Z-X* plane (perpendicular to the window plane) for heterogeneous edge-on alignment (Si cell) and homogeneous edge-on alignment (ZnS cell). *Curves*, calculated functions using Eq. (16) to fit the experiment. ○, Si (1517 cm^{-1}); □, ZnS (1517 cm^{-1}). [Reproduced with permission from *J. Mater. Sci. Eng. C*, **283**, 8–9 (1999)]. Ref. [96].

observations are consistent with the surface symmetry [100, 102], homogenous for ZnSe and ZnS windows and heterogeneous for Si and CaF$_2$ windows.

To check the influence of the symmetry of the surface on the alignment of liquid crystals, we studied the alignment obtained of H5T on the different cuts of ZnSe polycrystalline material [50]. In particular, the curve, □, in Figure 47(a) shows results of IR measurements for H5T contained between conventionally optical polished pure ZnSe windows with a cut parallel to the surface of deposition, ZnSe (II)$_1$ (the same windows were used for results of Figure 41). Curve, ●, shows results for H5T contained between [ZnSe (⊥)] cut perpendicular to the surface of deposition. These figures show that the alignment is planar in both cases. However, the results of absorbance as a function of angle of polarization carried out at temperature of 350 K for both cases are different. A noticeable dichroism is observed for H5T contained between ZnSe (II)$_1$ windows whereas no dichroism is found for ZnSe (⊥) [see Figure 47(b)]. This implies that the alignment in the latter case is heterogeneous-planar and shows the dependence of the alignment of the discotic on the substrate's easy axis of orientation and or of topology. For the nylon coated ZnSe(⊥) windows [see curve, *, in Figure 47(a)], the

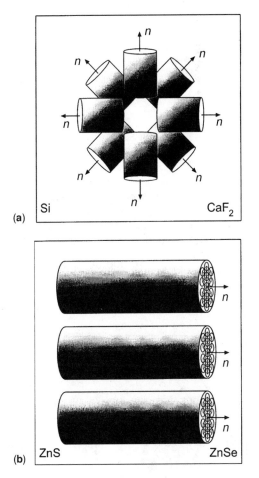

Figure 46. (a) Heterogeneous and (b) homogeneous planar alignment of the columnar phase of discotic liquid crystals.

alignment achieved is homeotropic. On the contrary, for $ZnSe(II)_1$ the alignment is still homogeneous-planar, shown in Figure 49, curve □. In the latter case, the effect of the easy axis of orientation and/or of the topolgy of the surface on the alignment is found to be much greater than the interactions between the polymer and molecules of the discotics. The latter tend to make the disks lie parallel to the substrate.

2.2. Nylon Spin-Coated Substrates. Figure 48 shows temperature dependence of the intensity of the C–C aromatic stretching vibrations for H5T cells assembled using nylon spin-coated substrates (Si, ZnS, CaF_2, and ZnSe) as windows of cells. The temperature dependence of the absorbance

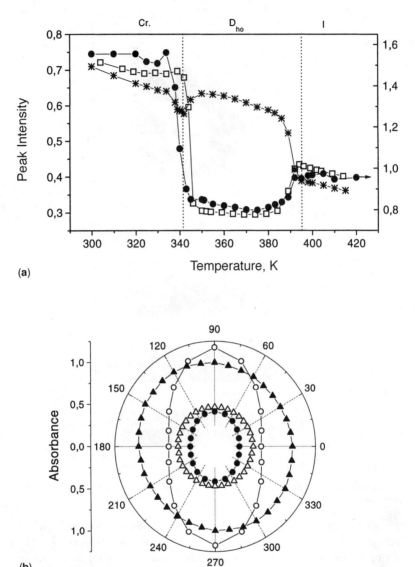

Figure 47. (a) Dependence of the peak intensity of the C–C vibrations on temperature for H5T contained between ZnSe windows differently oriented with respect to the surface of deposition ZnSe (II)₁ (●) and ZnSe (⊥), (□). *, nylon-coated ZnSe(⊥) windows. (b) Intensity distribution for in-plane vibrations in the window plane. ○, 1517 cm⁻¹ (ZnSe (II)₁); ●, 1617 cm⁻¹ (ZnSe (II)₁); △, 1617 cm⁻¹ (ZnSe (⊥) ×1.2); ▲, 1517 cm⁻¹ (ZnSe (⊥) ×1.2). [Reproduced with permission from *Molec. Mater.* **11**, 267 (1999)]. Ref. [50].

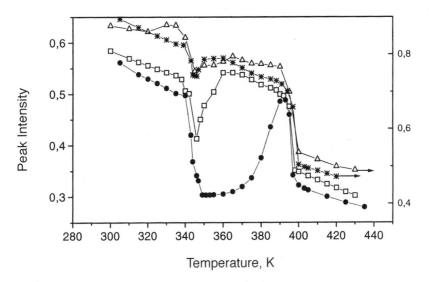

Figure 48. Peak intensity of the C–C aromatic stretching vibrations vs. temperature for H5T contained between nylon-coated windows. Cell thickness are Si, $10\,\mu m$ ($\triangle$); ZnSe, $8\,\mu m$ ($\bullet$); ZnS, $9\,\mu m$ ($\square$); CaF_2, $10\,\mu m$ (*).

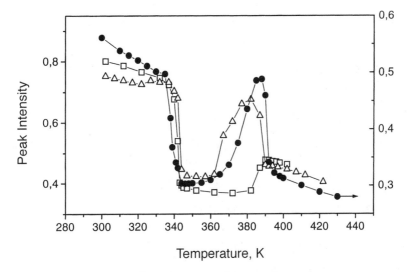

Figure 49. Peak intensity of the $1617\ cm^{-1}$ band vs. temperature for H5T in ZnSe windows without time treatment ($\square$); with a 1.5 h treatment at 360 K; ($\triangle$) and without any waiting period for the thinner cell ($8\,\mu m$) ($\bullet$). [Reproduced with permission from *Europhys. Lett.*, **44**, 198 (1998)]. Ref. [49].

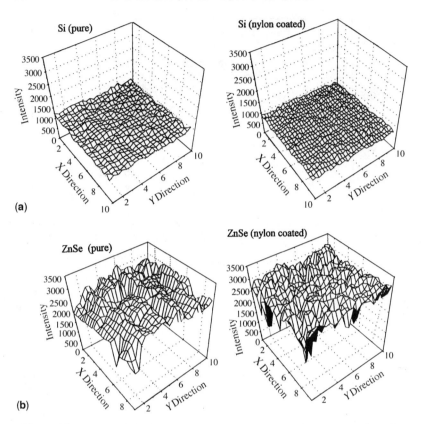

Figure 50. Results of the SEM (a, b, c) and white light interferometry (d) for different substrates.

for the various bands varies by substrate. For Si and CaF$_2$ windows, the alignment in discotic phase is side-on (intensity of parallel bands in discotic phase greater than in the isotropic phase), whereas for ZnSe and ZnS windows the orientational transition from the edge-on to side-on alignment is observed with temperature [49,96]. Figure 49 shows that the stability of the edge-on orientation is improved with the increase in the thickness of the LC sample, and the transition occurs at a higher temperature for a thicker sample than for a thinner sample.

The topology, however, can play an important role. The easier reorientation of columns for these substrates leads to a conclusion that the topological effect is small for these substrates. Investigations on the roughness of the surface of different substrates show that the surface of these windows is much smoother compared to the surfaces of other substrates because these substrates are flat within ±1 μm. Our surface investigations

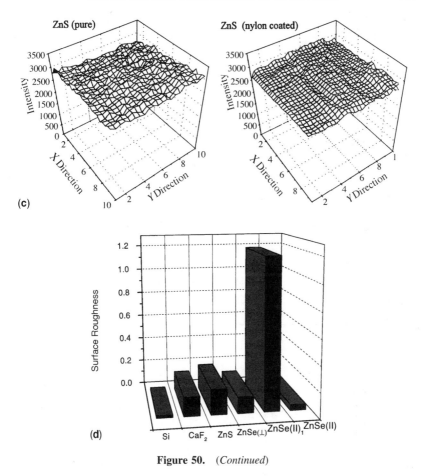

Figure 50. (*Continued*)

show that the roughness of ZnSe windows is much higher than for other substrates. Moreover some regularity of this roughness (groove-like structure) can be observed using electron micrography and white light interferometry (Fig. 50). This result is also supported by the X-ray diffraction of polycrystalline ZnSe material [101,102]. An investigation of the polycrystalline ZnSe using the diffraction method of inverse pole figures shows that on the surface of its deposition, the degree of texturing is increased [102]. The ZnSe windows were sliced parallel to the surface of the deposition.

The IR transmission measurements for oblique and normal incidences of the IR beam have revealed an edge-on alignment for nylon spin-coated ZnSe cell at 355 K and side-on alignment at 384 K. Figures 51 shows the dependence of the absorbance for the edge-on alignment for bands at 1617 and 836 cm^{-1} on the angle of polarization. The figure also shows a strong

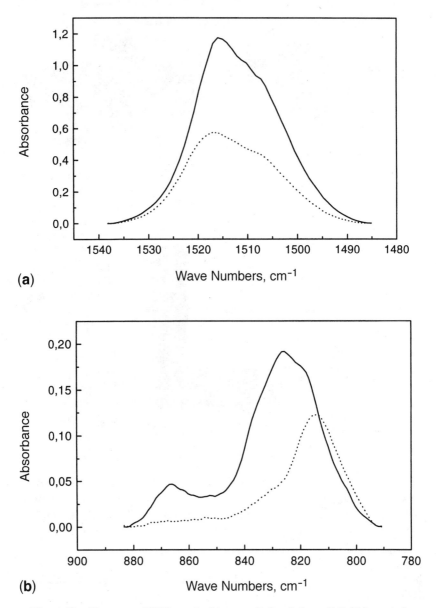

(a)

(b)

Figure 51.   IR spectra of H5T contained between ZnSe windows. Cell thickness is 8 μm.
(a) For in-plane vibrations at polarizer positions 0° (· · · ·) and 90° (——). (b) Out-of plane
vibrations at polariser positions 0° and 90° (- - - -). [Reproduced with permission from *J.
Mater. Sci. Eng. C*, **283**, 8–9 (1999)]. Ref. [96].

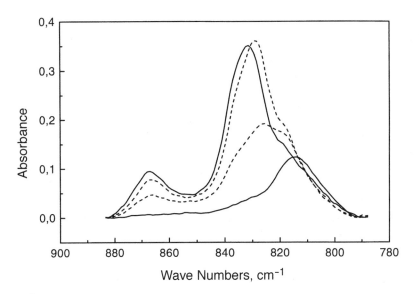

Figure 52. IR spectra of H5T in ZnSe cell (with nylon-coated windows) for edge-on (at 355 K) and side-on (at 384 K) alignment for normal and oblique at 60° incidence at polarizer position 0°. - - -, 355 K; — 384 K. [Reproduced with permission from *J. Mater. Sci. Eng. C*, **283**, 8–9 (1999)]. Ref. [96].

IR dichroism, implying that an easy direction of orientation in ZnSe plays an important role for even nylon spin-coated substrate. Polarized measurements at 384 K show no dependence of the peak intensity on the angle of polarization, confirming a side-on alignment.

The results of the oblique incidence of IR, the measurements for the edge-on and side-on alignment for ZnSe cell are shown in Figure 52, which demonstrates the changes in spectra for the sample tilt angles of 0° and 60° for the side-on and the edge-on alignments of H5T molecules. Figure 53 shows the dependence of peak intensity versus tilt angle. The dependencies of the absorbance on the sample tilt angle are different for these two cases.

The results of the polarized and oblique IR measurements comply with the characteristics of the various substrates. For ZnSe and ZnS, a large dichroic ratio for the vibrational bands is observed owing to the existence of only one easy axis of orientation, whereas a low dichroic ratio for the aromatic in-plane and out-of-plane vibrations is found for the untreated Si and CaF$_2$ windows (Fig. 43). The orientation of the director in the plane of these windows for the latter two substrates is uniformly distributed (a heterogeneous planar alignment). The weak surface anchoring energy for Si and CaF$_2$ is expected, and this arises from the heterogeneity of alignment on these substrates.

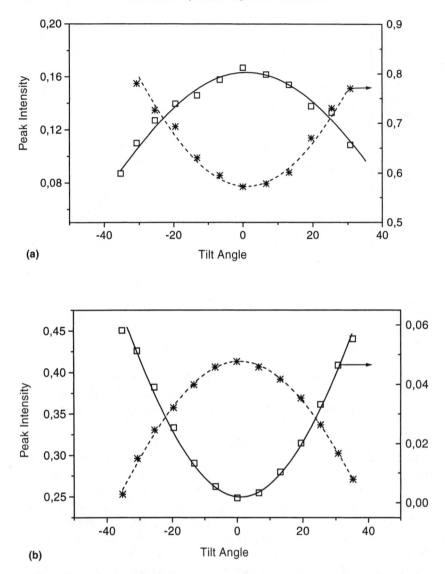

Figure 53. The intensity distribution for in-plane and out-of-plane vibrations in the Z-X plane for H5T in a ZnSe cell (a) Edge-on alignment at 355 K. □, 818 cm^{-1}; *, 1517 cm^{-1}. (b) Side-on alignment at 384 K. □, 867 cm^{-1}; *, 1616 cm^{-1}. [Reproduced with permission from *J. Mater. Sci. Eng. C*, **283**, 8–9 (1999)]. Ref. [96].

When considering the characteristics of the substrates, we can explain the results in terms of the anchoring phase transition. First, we note that the transition is somewhat similar to the "local Frederiks transition" observed in nematics [104–107]. A comparison of the experimental results of discotics with nematics shows that the nature of the transition in discotics is different than in nematics. The theory of Dubois-Violette and de Gennes [104] for nematics to discotics is adopted. The anchoring energy per centimeter squared is given by

$$F = W^s(\theta, \varphi) + \int_0^r \left[\frac{1}{2} U(z, \theta, \varphi) + \frac{1}{2} K \left(\frac{d\theta}{dz} \right)^2 \right] dz$$

where $W^s(\theta, \varphi)$ is the surface anchoring energy owing to the short-range forces. Because the edge-on alignment is observed on the untreated substrates, it is safe to conclude that $W^s(\theta, \varphi)$ as a result of the topology of the substrates causes this alignment. The long-range van der Waals energy $U(z, \theta, \varphi)$ integrated over the radius r also promotes the edge-on alignment, because the stability of this type of alignment is observed to increase with the sample thickness. Here θ is the angle that the director of the column makes with the normal to the substrate, φ is the azimuthal angle with respect to an easy axis of orientation, and z is measured along the normal to the substrate. The term involving the elastic constant K is the distortion energy and restores the order. For Si and CaF_2, however, the stability of alignment is quite weak owing to a degenerate anchoring of the discotic with these substrates; the fluctuations with an increase of temperature drive the transition from edge-on to side-on alignment (Fig. 42). In the side-on (homeotropic) alignment the fluctuations give rise to an additional effective anchoring energy [106] (fluctuations in θ of the director in the side-on are more degenerate than in the edge-on alignment). The phase transition appears to be of the second order.

For nylon-coated substrates, the short-range interaction $W^s(\theta, \varphi)$ is different. This arises mostly from the chemical bonds between the polymer and the discotic material. The benzene rings of the polymer seemingly lie flat on the substrate, and these promote alignment of the core of discotics parallel to the surfaces. A side-on or homeotropic alignment is, therefore, observed, except when the topology of ZnSe is predominant. The van der Waals term $\int_d^r \frac{1}{2} U(z, \theta, \varphi) dz$ promotes the edge-on alignment as before, but its contribution becomes smaller because the integral now covers a shorter distance between d and r; here d is the thickness of the nylon film. Thus a competition between the two terms occurs: the short-range term favors the side-on and the long-range favors the edge-on orientation. Depending on the relative strength of the two contributions, the anchoring transition can be observed. Thus for ZnSe and ZnS substrates, a transition from edge-on to the side-on orientation is observed (Fig. 48). Note, however, that the topology of ZnSe (groove-like structure) increases the stability of the edge-on orientation for this substrate, so the anchoring transition is observed at a higher temperature. For Si and CaF_2 coated with nylon, the edge-on orientation is not seen, because the contribution of the short-range interactions is $> \int_d^r \frac{1}{2} U(z, \theta, \varphi) dz$ over the entire range of the discotic phase. Figure 49 also shows that an increase in the liquid crystalline sample thickness shifts the transition toward the higher temperature. This is easily explained because the contribution of the van der Waals term increases with the increase in the sample thickness owing to r as the upper limit in the integral.

3. H6TT

H6TT is one of the few known discotic mesogen systems with both short-range intracolumnar stacking in a conventional discotic hexagonal liquid crystalline phase (D_{hd}) and a long-range intracolumnar order in the helical columnar phase H. The helical columnar phase of this discotic material is suitable for a fast transport of photogenerated charge carriers [3]. This material was studied using different techniques, including X-ray [108,109], NMR [110,111], fluorescence [112–114] and FTIR spectroscopy [26,115].

Lee et al. [26] based their study on the analysis of changes in the frequency maxima for different vibrations (alkyl chain CH_3 and core C−C stretch) and drew conclusions about the structural change at the phase transitions. We obtained [115] some supplementary information about the orientation of molecules in different phases and the orientational transition of H6TT contained between two CaF_2 windows in the D_{hd} phase.

The synthesis of H6TT has been detailed [27,116]. The structure and phase sequences of H6TT, was first studied by Gramsbergen et al. [117] [Fig. 22(a)]. A more detailed X-ray diffraction study revealed the following features [108,109] (Fig. 54).

1. The crystalline (Cr) phase has a monoclinic structure. The molecules are ordered into columns, within which there is a repeat unit of two molecules. The aliphatic tails are not confined to the plane of the core but tilt up and down; two of the six tails contain isolated gauche bonds, resulting in considerable displacement of the end of the tail from the plane of the core. Substantial thermal motion is seen at the end of the tails at room temperature.

2. The second phase is a three-dimensional crystalline phase with the helical columnar order. Therefore, Lee et al. [26] named the helically ordered phase the H phase, although this phase sometimes [102,110,118] is referred to as D_{ho} [101,109,117]. In this phase, molecules are ordered into

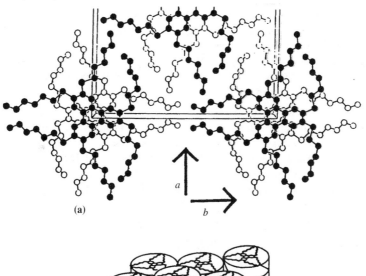

(a)

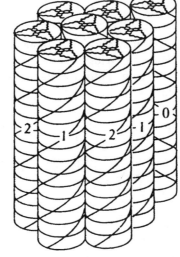

(b)

Figure 54. (a) Crystal structure of H6TT. Shown in the view of half of the unit cell, seen from the c-axis. ●, atoms in the upper plane; ○, atoms in the lower plane. (b) The structure of the H phase. Column 0 has the opposite helicity of columns 1 and 2 and is displaced by $z = p/2$ from those columns, where p is the intracolumn molecular spacing. [Reproduced with permission from *Phys. Rev.*, **A 37**, 1329 (1988)]. Ref. [108].

columns in a hexagonal array. Within each column, adjacent discs are rotated by ~45.5°, resulting in long helical order with a pitch P equal to 7.92 molecular spacings (i.e., incommensurate with the positional order along the columns). Similar short-range helical order has been observed in hexa-n-alkoxy derivatives of triphenylene and other columnar materials by

Levelut et al. [118–120]. Furthermore, a three-column superlattice results from both the relative phases and signs of helices in different columns and also from the translations of molecules along the columns. In the modeling of this phase to fit the X-ray, results it was assumed that the aliphatic tails were fully extended (with all trans bonds) but that the tails were tilted $\pm 3°$ to 5°above and below the plane of the cores to form a "propellor" structure.

3. In the liquid crystalline D_{hd} phase, the molecules still form a hexagonal array of columns; however, the positional order within the columns is liquid-like, and both helical and superlattice order is lost. Little is known about the quantitative extent of tail disorder in the H6TT D_{hd} phase. A linear decrease in the intercolumnar lattice constant with increasing temperature is consistent with a thermally activated increase in the density of gauche bonds [26].

4. Although the isotropic liquid I phase was not studied in detail, it can be inferred from many other studies of discotic compounds that there is considerable nematic-like short-range order. It is reasonable to assume that there is a high level of tail disorder in the I phase.

The complete spectra in the region 500 to 3000 cm^{-1} for all phases of H6TT are shown in Figure 55. With the exception of a few new peaks at higher temperatures, the principal effect of increasing the temperature is to decrease the peak amplitude and broaden individual lines, so that closely

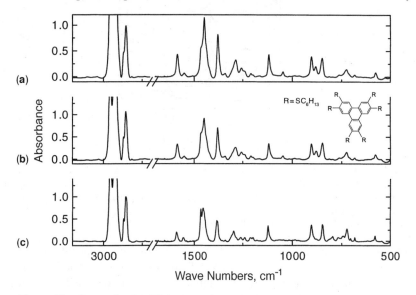

Figure 55. IR spectra of H6TT in the (a) isotropic (380 K) (b) discotic (362 K), and (c) crystalline (320 K) phases.

separated peaks become unresolved at higher temperatures. Several peaks from these spectra can be easily identified (Table IV).

We will first review the results obtained by Lee et al. [26], whose main attention was given to the analysis of changes in the peak position with temperature. The liquid crystal for this study was sandwiched in between two pure KBr substrates with a diameter of 25 mm and thickness of 2 mm. The cell preparation was slightly different from that described here. The information about sample preparation can be found in Lee et al. [26].

Figure 56 shows the evolution of three selected regions in the IR spectra with increasing temperature: CH bending and scissoring modes, benzene stretching modes, and CH stretching modes. Clearly, there are both gradual and sudden amplitude and frequency changes with temperature. Figure 57(a) shows the weighted average of the CH_3 symmetric bending absorption frequencies as a function of temperature. Although the effect is small, it is clear that the slope of frequency vs. temperature changes at around 50°C and 70°C. While the change at 70°C corresponds to the $H \rightarrow D_{hd}$ transition, the change at 50°C is actually about 12°C lower that the nominal melting point of 62°C. A similar change of slope is seen in Figure 57(b), in which the frequency of a benzene ring stretching mode is plotted against temperature. Again, the slope changes at both 50°C and 70°C. Figure 57(c) shows the temperature dependence of the frequency of the asymmetric stretching mode of the CH_3 group at the end of the alkyl tails. In this case, dramatic changes are seen at 50°C, 60°C, and 70°C. The symmetric stretching mode of CH_3 at around 2875 cm^{-1} (not shown) reveals the same behavior as its asymmetric counterpart. It is generally understood that CH_3 stretching mode frequencies are increased by the introduction of gauche bonds [121,122]. Thus a plausible source for the peak movement shown in Figure 57(c) is an increase in the statistical distribution of gauche bonds in the alkyl tails.

Analysis of these peak shifts is facilitated by a detailed comparison with X-ray diffraction results. In the Cr phase, for which the crystal structure is known, four of the alkyl tails are fully extended, although they are somewhat tilted and distorted from an ideal conformation consisting of all trans (antiperiplanar) bonds, while two of the tails contain a single gauche bond [26,109]. The root-mean-square amplitudes of carbon-atom displacements range from < 0.25 Å near the core of the molecule, to almost 0.50 Å near the ends of the tails. Models used in least-squares fits of the H phase X-ray diffraction data assumed that the tails were fully extended [108,109], and this assumption is further strengthened by the observation that the intercolumnar distance in the H phase is temperature independent. Both the presence of diffuse off-axis scattering and the calculated Debye-Waller factors, however, indicated that a certain amount of tail disorder or at least tail vibrational motion is present. (Note that the incommensurate structure of

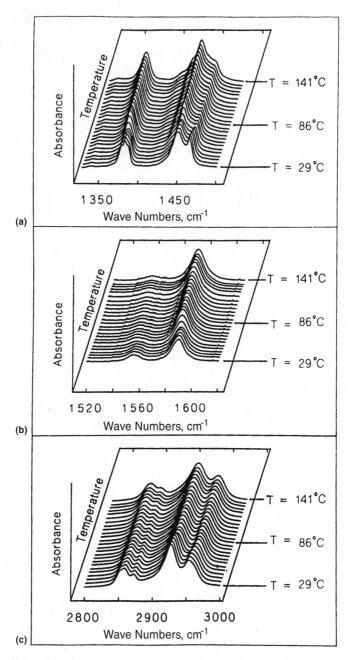

Figure 56. FTIR absorption profiles of (a) CH bending modes, (b) benzene stretching modes, and (c) CH stretching modes as a function of temperature. Data were taken upon heating at the following temperatures: 29, 33, 39, 43, 48, 53, 63, 68, 72, 81, 86, 91, 96, 100, 105, 109, 114, 118, 123, 127, 132, 136, and 141°C. [Reproduced with permission from *Molec. Cryst. Liq. Cryst.*, **198**, 273 (1991)]. Ref. [26].

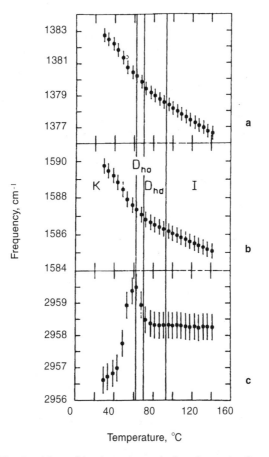

Figure 57. Fitted positions of the absorption peaks from Lorentzian fits to the data shown in Figure 55(a). (a) Methyl symmetric bending mode. (b) Benzene stretching mode. (c) Methyl group asymmetric CH stretching mode. When more than one Lorentzian fit was needed to give a reasonable fit, the weighted averaged of the positions were plotted. Error bars indicate estimated errors from the fits. [Reproduced with permission from *Molec. Cryst. Liq. Cryst.*, **198**, 273 (1991)]. Ref. [26].

the H phase implies that no two molecules find themselves in exactly identical environments). Diffuse scattering is also seen in the diffraction from the D_{hd} phase [109]; in addition, the intercolumnar distance decreases rapidly and linearly with increasing temperature. This indicates an increase in the level of tail disorder, and almost certainly an increase in the statistical density of gauche bonds, leading to tails which are, on the average, shorter.

Thus from X-ray data alone, one would anticipate that the Cr→H transition involves straightening out the pair of gauche bonds in the Cr phase and an increase in the extent of tail vibrational motion, while the $H \rightarrow D_{hd}$ transition involves a rapid increase in the population of gauche bonds. From this viewpoint, the rapid evolution of the CH_3 stretching modes around 60°C may imply a large increase in thermal motion of the tails just below the transition temperature, followed by the two tails with gauche bonds straightening out when the H phase develops. Note that no dramatic change is seen at either the $H \rightarrow D_{hd}$ or $D_{hd} \rightarrow I$ transition.

While the changes in the CH_3 stretch as a function of temperature are easily observable, there is hardly any change in the frequency of the CH_2 stretching bands. The CH_2 stretching frequencies have been shown to be sensitive only to nearby gauche bonds [75,123]; it thus appears that gauche bonds are only present, if at all, near the ends of the tail. By contrast, the CH_3 bending modes and the benzene stretching modes show clear anomalies at the $Cr \rightarrow H$ and $H \rightarrow D_{hd}$ transitions. This is to be expected for the benzene breathing modes, because both of these transitions involve changes in intracolumnar ordering, and hence in core-core interactions. The CH_3 bending modes are sensitive to tail packing and conformation [74] and thus also show clear anomalies at these transitions.

Lee et al. [26] noted that no absorbance bands, including those not detailed above, show any anomaly at the $D_{hd} \rightarrow I$ transition. From the results in the CH stretching region, they concluded that there is no sudden increase in the number of gauche bonds in going from the H to the D_{hd} phase or from the D_{hd} to the I phase. The results from the benzene stretching region also suggest that the changes in the core environment in going from the D_{hd} to the I phase must be subtle. From the plots in Figure 57, we note that although the nominal melting point of H6TT is 62°C the changes can be observed in the slopes of the plots at about 50°C. Because all the frequencies shown in Figure 57 are affected by tail vibrations, this suggests that the alkyl tails actually gain considerable thermal motion at about 10°C below the nominal melting point.

We now review the results of IR investigations of H6TT sandwiched between varyingly treated substrates (ZnSe, CaF_2, and MgF_2) [115]. These results give additional information about the orientation of molecules in discotic phases on different substrates. As was shown, the behavior of peak intensity for different vibrational bands depends on the type and the quality of alignment. So we focus our attention mainly on amplitude changes, although the frequencies changes were also analyzed and compared with the results obtained by Lee et al. [26]. The analysis shows that the peak frequeny behaviors are quite similar.

Figures 58 and 59 show the temperature dependence of the intensity of some of the chosen vibrational bands which are C–C aromatic stretching vibrations at ~ 1590 and 1550 cm^{-1}, CH bending modes near 1380 cm^{-1},

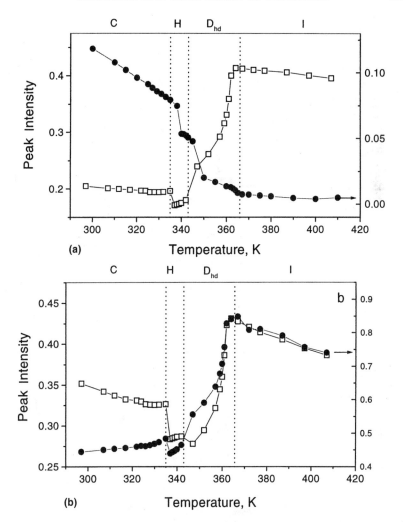

Figure 58. (a) The dependence of the peak intensity of aromatic in plane at 1590 cm^{-1} (□) and aromatic out-of-plane at 794 cm^{-1} (●) vibrations on temperature for HHTT contained in between pure ZnSe windows. The thickness of the cell was ~11 μm. (b) Same as part **a** for alkyl chain vibrations at 1120 cm^{-1} (□) and 1380 cm^{-1} (●). (Ref. [115]).

C–H out-of-plane deformation at 700 to 800 cm^{-1}, CH$_2$ and CH$_3$ symmetric and asymmetric stretching vibrations in the range of 2820 to 3000 cm^{-1}.

For our case of the alignment of H6TT the phase transitions can be easily identified as seen from Figures 58 and 59. A dramatic changes in the amplitude was observed at C→H$_{ho}$ and H$_{ho}$→D$_{hd}$ phase transitions. For the D$_{hd}$→I phase transition, a different behavior for different cells was found.

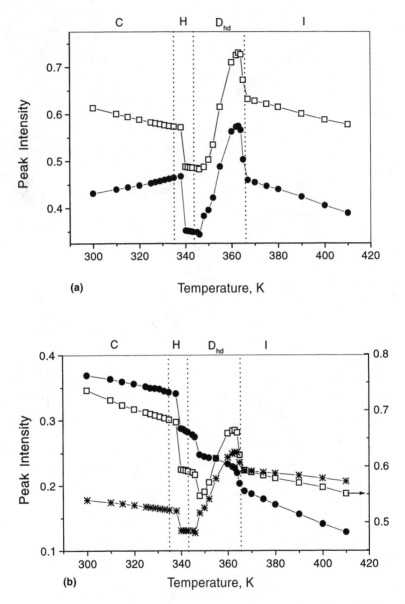

Figure 59. (a) The dependence of the peak intensity of aromatic in-plane at 1450 cm^{-1} ($\square$) and aromatic out-of-plane at 1380 cm^{-1} ($\bullet$) vibrations on temperature for HHTT contained in between CaF$_2$ windows coated by GeO at oblique evaporation. (b) Same as part (a) for alkyl chain vibrations at 1120 cm^{-1} ($\square$); 1590 cm^{-1} (*), and 2858 cm^{-1} ($\bullet$). (Ref. [115]).

From the figures, we can see continuous transition of the discotic in a ZnSe cell and quite a dramatic phase transition for the discotic contained in a CaF_2 cell. This difference will be discussed below. Note here that the $D_{hd} \rightarrow I$ phase transition would also be noticeable on the diagram of the frequency peak versus temperature (not shown). At the same time, the amplitude and frequency maxima are quite stable in the D_{ho} phase. This shows that there are no large changes in the intracolumnar order in between those phases except that intracolumnar distance changes owing to the conformational changes of alkyl chains (as shown above). It follows from the X-ray studies [108,109], that the intercolumnar distance dramatically decreases linearly with increase in temperature. This affects the intermolecular interactions and causes a change in the intensity of the bands, as seen for a majority of vibrational bands studied.

Figures 58 and 59 show that the intensity of the in-plane vibrations continuously increases in the D_{hd} phase. This behavior, however, is different for H6TT contained between ZnSe and CaF_2 windows. We can see that the intensity of the in-plane vibrations (C−C aromatic stretching vibration near 1590 cm^{-1}) in the D_{hd} phase (I_{hd}) changes with respect to the intensity of this band in the isotropic phase (I_I) from $I_{hd} < I_i$ to the $I_{hd} > I_i$. So it is obvious that for H6TT contained between CaF_2 windows, we observe the orientational transition from the edge-on to the side-on orientation. The same effect was observed for H5T, discussed earlier.

To find why such a large difference is observed for CaF_2 and ZnSe cells, we carry out the polarized IR transmission measurements at the oblique and normal incidence of light at 343 K (in helical H phase). Figure 60 shows the results of the polarized measurements for both cells, together with the calculated intensity distribution function using Eq. (15). From the figure, one can see quite a large difference between the two cases. For the CaF_2 windows, the dichroism is small, whereas a large dichroism is observed for the ZnSe windows. The same effect was observed for hexapentyloxytriphenylene sandwiched between ZnSe and CaF_2 windows. The easy orientation direction of the surface of the windows is presumably responsible for these effects. For ZnSe windows, there is a preferable direction of orientation on the surface of these cuts [102]. This causes the homogeneous-planar orientation in the H phase for H6TT contained between ZnSe windows. For CaF_2, there are at least three easy axes of orientation on the surface of these cuts. These cause heterogeneous-planar orientation of H6TT. The latter orientation is not stable, and by changing the temperature it can be converted to the side-on alignment. The side-on alignment is more homogeneous, thus again a long-range order appears in the D_{hd} phase. As a result, we can find more dramatic changes in between the $D_{hd} \rightarrow I$ phase, both in the amplitude and in the frequencies of vibrational bands.

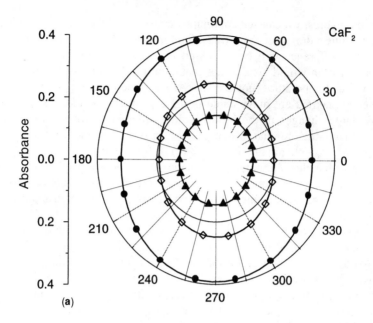

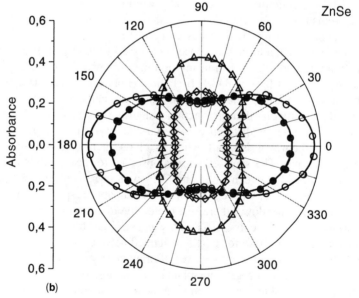

Figure 60. The intensity distribution for in-plane and out-of-plane vibrations in the window plane ($Y - Z$ plane). (a) For the $CaF_2 + GeO$ cell. $\diamond$, 1120 cm^{-1}; $\bullet$ 1380 cm^{-1}; $\blacktriangle$ 1590 cm^{-1}. (b) For the ZnS cell. $\bigcirc$, 848 cm^{-1}; $\diamond$, 1590 cm^{-1}; $\triangle$ 1120 cm^{-1}; $\bullet$ 794 ($\times 6$) cm^{-1}. *Solid lines*, the calculated functions using Eq. (15). (Ref. [115]).

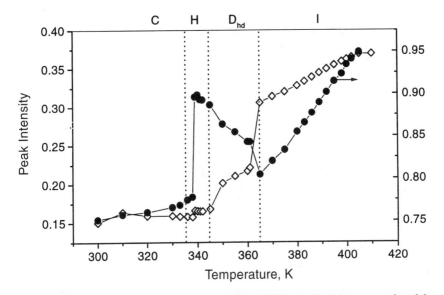

Figure 61. Peak intensity vs. temperature for H6TT contained between nylon 6,6-coated MgF$_2$ windows. Measurements carried out during the heating cycle. $\diamond$, 1590 cm^{-1}; $\bullet$, 2858 cm^{-1}.

Moreover, we note that despite the similarity in the behavior of H6TT with other triphenylene derivatives, there are some differences in the orientation of this compound to the various substrates. The difference in H6TT alignment on pure and coated (by nylon 6,6 or GeO) substrates (Figs. 58, 59, and 61) is not as pronounced that for other discotic liquid crystals. We never observed a side-on alignment during a heating cycle, even on the coated surfaces (Fig. 61).

In conclusion, we mention some interesting results found for H6TT during the cooling and the heating cycles. As discussed, the transition temperatures and the orientational behavior during heating and cooling cycles using IR spectroscopy is different (Fig. 32). The results for the cooling cycle for H6TT show the sudden changes in orientation for the D$_{hd}$ → H phase transition (Fig. 62). The phase transition in between the two mesophases is much more pronounced during the cooling than during the heating cycle (Figs. 58 and 59). We note that a side-on alignment during the cooling cycle usually turns out to be the result. When the temperature reaches the transition point to the H phase, however, a sudden change in the intensities of the in-plane and the out-of-plane vibrations of the core were observed. The structure probably becomes 3-D and lattice-like, and some microdomains have changed their orientation with respect to the window plane. This is seen from the polarized microscopy performed for these cells

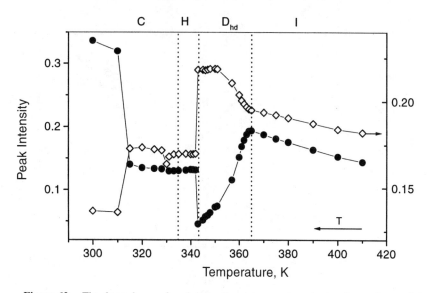

Figure 62. The dependence of peak intensity on temperature (under the cooling cycle) for H6TT contained between untreated ZnSe windows for in-plane (1590 cm^{-1}; $\diamond$) and out-of-plane (847 cm^{-1}; $\bullet$) vibrations.

[115]. At the $D_{hd} \rightarrow H$ phase transition, the picture from a typical side-on alignment (mainly gray and black) becomes more mosaic-like, with a wider distribution of black-gray-white colors. We assume that these sudden changes are the result of changes in the distance between neighboring cores. Quantum chemical calculations performed recently for two triphenylene cores show that the total energy decreases sharply below a certain value when disks are pushed face to face (Fig. 63) [124]. The distance at the minimum energy between the two molecules is presumably closely related to the minimum packing distance achievable in the D_{ho} and H columnar phases. Etchegoin [124] noticed that the rotation of one molecule with respect to another minimizes the total energy in the interaction, which is presumably related to the microscopic mechanism that brings forth the helical columnar H phase when the temperature is lowered from D_{hd} to H phase.

It has been postulated that the H6TT in H phase consists of microdomains [113]. In each microdomain, the columns are hexagonally packed, and the orientation of the column's axis changes from one microdomain to another. This assertion is consistent with the results of polarized microscopy and IR spectroscopy presented here.

4.   H7T-NO$_2$

The triphenylene derivatives discussed so far have a symmetrical structure with respect to the normal of the core. Recently, however, a class of new

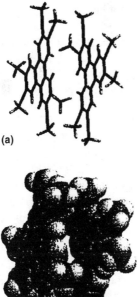

(a)

(b)

Figure 63.   Two interacting H1T molecules in stick (a) and (b) space-fill representations. The structure of the individual molecules was obtained from the AM1 geometry optimization given in Table 1 [124] and are not reoptimized. The distance between the centers of the molecules is 4 Å, and they are allowed to rotate to minimize the total energy. The overlap of p_z orbitals results in a rotation by 60° of one molecule with respect to the other. The band level diagram of the two interacting molecules for this separation distance is given in Ref. [124]. [Reproduced with permission from *Phy. Rev. E*, **56**, 538 (1997)].

disymmetrical triphenylene derivatives was synthesized with a potential for applications. In particular, it was shown that for triphenylenes with a longer alkyl chain, the asymmetry might appear owing to the differences in molecular packing, which, in turn, could introduce a chirality [14].

The mixed alkoxy and alkylthio or seleno-substituted triphenylene can provide another new and a large group of disk-like liquid crystal materials with interesting properties [125,126]. In particular, such mixed derivatives with a single sulfur or selenium bridge between core and the chain [27,127], acting for instance, as internal electron withdrawing unit [13], should be materials from which one might expect the formation of ordered, wider range and more stable columnar mesophases like those of the classical triphenylene hexaesters [13,128]. Moreover Praefcke et al. [125] and Boden et al. [126] assumed that a much stronger chirality than for chiral discogens would presumably appear if a helical twist could be imparted to the central aromatic core.

To meet this objective, three sevenfold-substituted triphenylene derivatives were recently synthesized [12, 125, 126], with the M group replacing the aromatic H atom in position C-1 (M=NO₂, F, Cl, Br) or with one differently structured alkyl chain in position C-11 [129].

The results of the preliminary IR investigation for the orientational behavior of one of these discotics (H7T-NO₂) on different substrates is presented here. The results of the structural analysis for this compound have been given [125,126]. These lead to the conclusion that H7T-NO₂ shows the helical structure with the crowding core (Fig. 64). This may affect the IR spectra because now the triphenylene core is not flat anymore as it was for symmetrical triphenylenes such as H5T and H7T [124–126]. The spectra for H7T-NO₂ and H7T are shown in Figure 65 in discotic phase at 350 K and in isotropic phase at 420 K for both compounds. We cannot compare the spectra for crystalline phases, because the phase transition for H7T-NO₂ is 268 K, which is out of the temperature limit used in our experimental setup. As seen from these figures, the position of some bands is slightly influenced by the presence of one NO₂ group in a core. Some new bands have appeared, which belong to the stretching and bending vibrations of N=O and C–N bonds. The biggest effect one can see is on the band in the region 1520 cm^{-1}. In particular, a new strong band with a frequency maximum at ~1532 cm^{-1} is developed at the higher frequency side of the band of C-C aromatic stretching vibration (at 1506 cm^{-1}). This new band belongs to N=O asymmetric stretching vibration [77]. The N=O symmetric stretching vibration appears at 1365 cm^{-1} as a shoulder on the main band at 1380

Figure 64. Ball-and-stick and space-filling models of the hypothetical 1-nitro-triphenylene showing the helical deformation and coiled shape of their tetracyclic aromatic cores caused by the substituents at C-1. As shown, the energetically favored orientation of the NO₂-group is at about 82° with respect to the aromatic core. The carbons and hydrogens of the bay region under consideration are numbered: C-1 and C-12 on the left and 12-H on the right. For simplification, the conformational calculations were carried out for the methoxy-homologues, although they have not yet been synthesized. [Reproduced with permission from *Liq. Cryst.*, **22**, 113 (1987)]. Ref. [125].

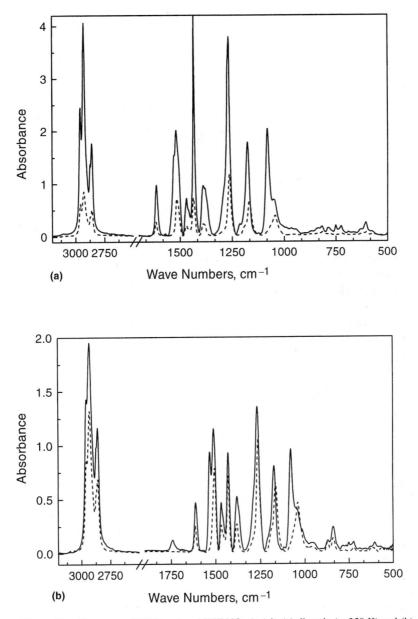

Figure 65. IR spectra of H7T (- - -) and H7T-NO$_2$ (——) in (a) discotic (at 350 K) and (b) isotropic (at 420 K) phases.

cm^{-1} seen in H5T, which belongs to the alkyl chain (CH_3 symmetric bending). We can also see the strong band at a frequency of 1100 cm^{-1}. In fact, a weak band at the same frequency was also observed in the spectra of H5T. Nevertheless, we propose that for H7T-NO_2 the strong band at ~ 1100 cm^{-1} belongs to the C–N stretching vibration [77]. The out-of-plane vibrations in the region of 760 to 900 cm^{-1} are also slightly altered by NO_2 group. We can also see the NO bending mode at a frequency of 845 cm^{-1}. This band is hidden, however, by the more intense band of the C-H out-of plane deformations at 832 cm^{-1} and is resolved only with the sample tilt (Fig. 66a). The spectral region of 700 to 800 cm^{-1} is shown in Figure 67 for H7T-NO_2 together with the other triphenylene derivatives. These figures show a band with the peak frequency at ~750 cm^{-1}, which appears only for H7T-NO_2. Another reason to show this particular spectra region is to compare the results for CH_2 rocking vibration at ~ 730 cm^{-1}. As seen from Figure 67; only H10OT has an alkyl chain in all trans configuration, because only in this case did we observe a band at 720 cm^{-1}. More detailed analysis of these vibrations and the effect of the core bending will be discussed elsewhere. Here, we focus on the effect of the structure of the substrate on the alignment of this compound.

We made oblique IR transmission measurements and varied the temperature of the sample. Figures 68 and 69 show that the intensity of the in-plane and the out-of-plane vibrations versus temperature being different when the material is sandwiched between pure ZnSe windows than when between nylon 6,6-coated ZnSe windows. The behavior is similar to that for H5T. The results for H7T-NO_2 sandwiched between pure ZnSe windows is opposite to that of H5T and H7T. During the heating cycle, we note that the alignment of molecules is continuously converted from the side-on to the edge-on (observed just before the phase transition at 380 K). HTT-NO_2 contained between nylon 6,6-coated ZnSe windows shows typical side-on alignment. This behavior coincides with that obtained for H5T and H7T. The order parameter is different for different bands, which shows that some parts of molecules are not exactly parallel or perpendicular to the plane of the window but are inclined at an angle. As has been shown [125,126] using molecular structural calculations, the core of this molecule is bent. The NO_2 group is oriented at 82° with respect to the core. We can prove this assertion using the results of the N=O stretching vibrations as a function of temperature and sample tilt. The results of oblique transmission measurements are shown in Figures 66 and 70, from which we note that the band at 1532 cm^{-1} behaves in an opposite manner to that for the C–C aromatic stretching vibration at 1506 and 1620 cm^{-1} and is similar to that of 833 cm^{-1} band (Fig. 70). Figures 66(a) and 70(b) show a dramatic change in the region of the out-of plane deformation and no bending (see spectral region

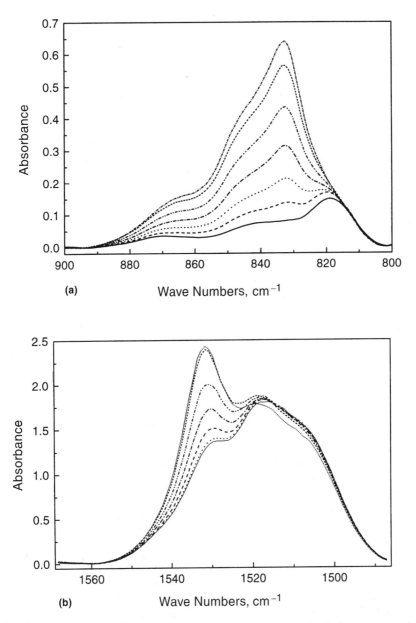

Figure 66. IR spectra of H7T-NO$_2$ in mesophase at the sample tilt for two spectral ranges. (a) Combination of aromatic C–H out-of-plane (836 cm^{-1}) and NO bending (845 cm^{-1}) modes. —, 0°; - - -, 10°; ····, 20°; ----, 30°; -·-··-··, 40°; ---, 50°; -·-·, 60°. (b) Combination of the aromatic C–C stretch (at ∼1507 cm^{-1}) and asymmetric N=O stretch (at 1533 cm^{-1}) vibrations. —, 0°; - - -, 10°; - - -, 20°; -·-·, 30°; -·-··-··, 40°; ····, 50°; ----, 60°.

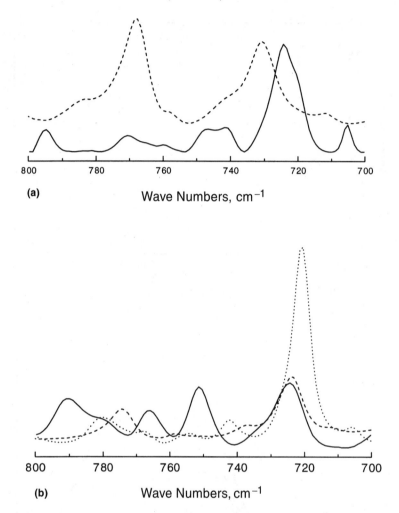

Figure 67. IR spectra in the CH_2 rocking vibration (at ~ 730 cm^{-1}) for triphenylene derivatives. (a) - - - HST; —, HSTT. (b) - - - - 10100T; —, H7T-NO$_2$; - - -, H7T.

780 to 900 cm^{-1}). We can see that the intensity of band at 836 cm^{-1} is strongly increased. This leads us to conclude that 836 cm^{-1} band is perpendicular to the plane of the window (and that of the core). At the same time, the C–C aromatic stretching vibration (oriented in the plane of the core) does not show a dramatic change, as has been observed for H5T and H7T, for example (Figs. 30 and 31). This is probably due to the fact that the part of the core is bent and hence we will have some component of C–C vibrations inclined at certain angle to the plane of the window. This is also

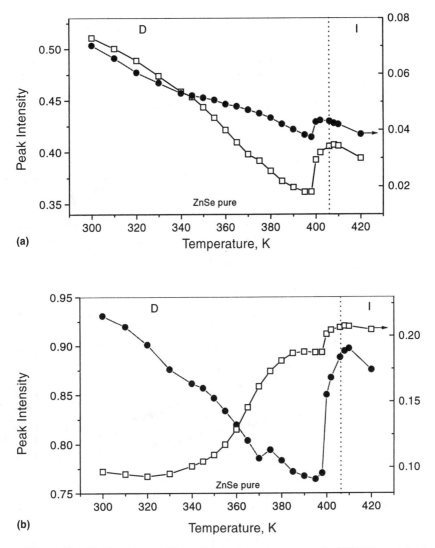

Figure 68. The dependence of the peak intensity vs. temperature for H7T-NO$_2$ contained between untreated ZnSe windows for different vibrational bands. (a) □, 1615 cm^{-1}; ●, 724 cm^{-1}. (b) ●, 1082 cm^{-1}; □, 839 cm^{-1}.

the reason that the bands at 1532 and 845 cm^{-1} largely depend on the tilt of the sample. These results are consistent with the structural analysis of this compound [125,126]. Using these results, we can make a proper assignment of the strong band observed at 750 cm^{-1}. In the literature, this band is assigned to the out-of-plane vibrations by some and to the wagging motion

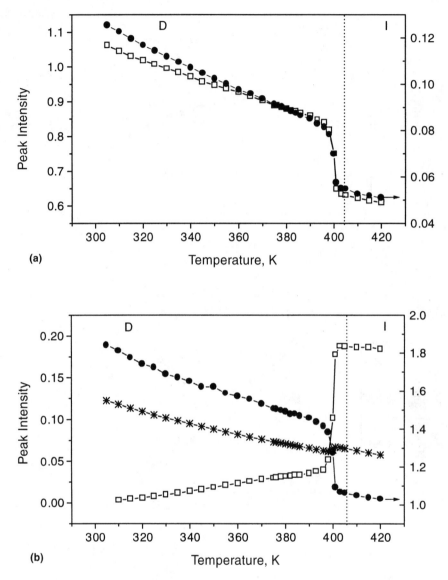

Figure 69. The same as in Figure 68 for H7T-NO$_2$ contained between nylon 6,6 coated ZnSe windows. (a) □, 1651 cm^{-1}; ●, 751 cm^{-1}. (b) □, 839 cm^{-1}; *, 724 cm^{-1}; ●, 1082 cm^{-1}.

Figure 70. The dependence of the peak intensities vs. corrected tilt angle at polarizer position 0° for the same H7T-NO$_2$ cell for which temperature dependence was shown in Figure 69. (a) □, 1615 cm^{-1}; ◆, 1469 cm^{-1}; *, 1380 cm^{-1}. (b) *, 1533 cm^{-1}; □, 1522 cm^{-1}; ●, 836 cm^{-1}. (c) *, 751 cm^{-1}; ●, 724 cm^{-1}; ▽, 689 cm^{-1}; □, 605 cm^{-1}.

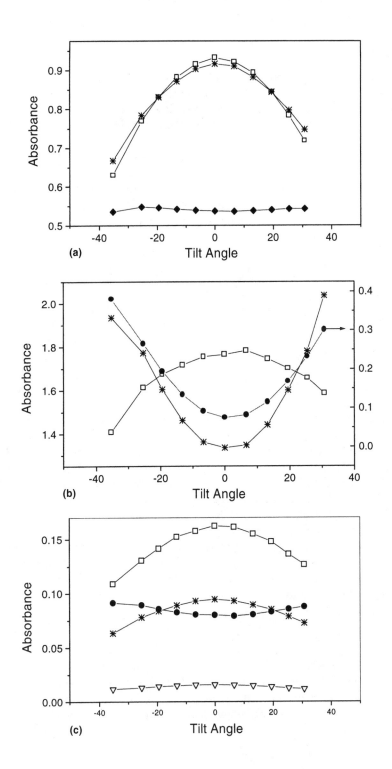

(a)

(b)

(c)

of the nitro group by others. The behavior of the peak intensity of this band on temperature and on tilt of the sample led us to conclude that the band at ~ 750 cm^{-1} belongs to the in-plane wagging motion, because the transition dipole moment for this vibration is oriented parallel to the core (Figs. 69 and 70).

We performed a fitting of the intensity distribution function in X–Z plane using Eq. (16) (Fig. 71). Then we calculated an order parameter for different vibrational bands, using the method described earlier. The results are shown in Figure 71, which reveals that the order parameter for the 836 cm^{-1} band in particular is quite large for the sample; thus the sample was well aligned.

C. Truxene Derivatives

In this section we review the IR investigations [48,130] of three truxene-based discotic liquid crystals. Those discotics are 2,3,7,8,12,13-hexa-tetradecanoyloxy-truxene (HA14TX), 2,3,7,8,12,13-hexakis-decanoyloxy-truxene (HA10TX), and 2,3,7,8,12,13-hexa(4-undecyloxybenzoyloxy) (HBTX). The molecular structures of these compounds are shown in Figure 72. The phase sequences for these compounds were originally reported by Destrade et al. [131] and listed in Table V. The most intriguing feature of these compounds is fact that a re-entrant isotropic phase is seen in mixtures of HA14TX with $\sim 13\%$ HBTX [131,132].

An X-ray diffraction study of the suspended strands of HA14TX [108] indicated that the high temperature phase should in fact, be classified as D_{hd} (hexagonal disordered) rather than as D_{ho} (hexagonal ordered). These X-ray measurements also showed no sign of the reported D_{ho}–D_{rd} transition at 112°. The intensities of the peaks for low-temperature rectangular symmetry were reduced from those of the hexagonal symmetry by a factor of at least 10^4. The distortion of the 2-D lattice normal to the columns, breaks the hexagonal symmetry, by $< 0.5\%$.

The IR measurements for HA14TX and HBTX were carried out using an IBM IR/97 FTIR instrument in the transmission mode [130]. KBr windows were used for these measurements. The infrared measurements for HA10TX contained between nylon 6,6-coated Si and ZnSe windows were carried out using a Bio-Rad 60A spectrometer.

1. HA14TX

Figure 73 shows the FTIR absorption spectra of HA14TX at 29°C and 169°C. The two spectra are similar; with increasing temperature, peaks are broadened and their intensity reduced. Hence the nearby peaks become unresolved. The band assignment for the studied truxene derivatives are listed in Table VI.

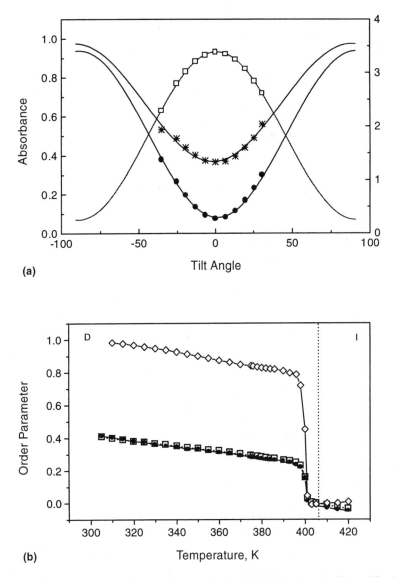

Figure 71. (a) Fitting to the experimental data for H7T–NO$_2$ shown in Figure 70 using Eq. (16). ●, 836 cm^{-1}; □, 1615 cm^{-1}; *, 1533 cm^{-1}. (b) The order parameter for some vibrational bands shown in Figure 69. □, 1615 cm^{-1}; ●, 1082 cm^{-1}; ◇; 836 cm^{-1}.

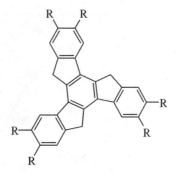

Figure 72.   Structures of truxene derivatives. HA14TX, R = OCOC$_{13}$H$_{27}$ (Cr-334-N$_d$-357-D$_{rd}$-385-D$_{ho}$-514-I);   HA10TX,   R = OCOC$_9$H$_{19}$ (Cr-341-N$_d$-358-D$_t$-411-D$_{ho}$-553-I); HBTX, R = OCO—⟨◯⟩— OC$_{11}$H$_{23}$ (Cr.-363-D$_{rd}$-410-N$_d$-452-D$_{rd}$-557-N$_d$-570-I).

TABLE V
Transition Temperatures (°C) of the Hexa-*n*-Alcanoyloxy-Truxenes

HA10TX	Cr	68	N$_D$	85	D$_{rd}$	138	D$_{ho}$	280	I		
HA14TX	Cr	61	N$_D$	84	D$_{rd}$	112	D$_{ho}$	241	I		
HBTX	Cr	90	D$_{rd}$	137	N$_D$	179	D$_{rd}$	284	N$_D$	297	I

Data from [131].

Figure 74 shows the evolution of selected individual absorption peaks with increasing temperature. In Figure 74(a), the main peaks near 2855 and 2923 cm^{-1} are the result of the symmetric and asymmetric stretching modes of CH$_2$, respectively. The peak shoulders near 2870 and 2955 cm^{-1} are the result of the symmetric and asymmetric stretching modes of CH$_3$, respectively. Figure 74(b) shows the carbonyl C=O stretching bands. Kardan et al. [64] observed that the carbonyl stretching bands of a benzene hexa-*n*-alkanoate discotic liquid crystal is actually made up of two different bands owing to the two possible different orientations of the C=O bond relative to the neighboring C–C bond. This peak was fitted to the three Lorentzian bands [130]. Both bands in Figure 74(c) are assigned to the benzene stretching modes. The band near 1380 cm^{-1} in Figure 74(d) is most likely owing to the symmetric bending mode of CH$_3$, while the bands at 1468 and 1477 cm^{-1} are most probably owing to the asymmetrical bending of CH$_3$ and the scissoring mode of CH$_2$, respectively, although it should be noted that the latter two appear to be centered about 10 cm^{-1} higher than the position that is normally observed for *n*-alkanes [69,70].

Figure 75 shows the frequencies of various modes extracted from least squares fitting to the Lorentzian profiles. Figure 75(a) shows the weighted average of the methylene asymmetric CH stretching frequencies as a

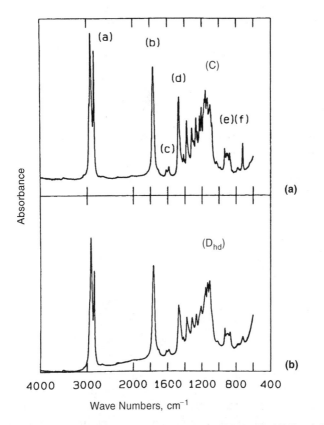

Figure 73. Complete FTIR absorption spectra of HA14TX at (a) 29°C and (b) 169°C. The spectrum for HA14TX was truncated in the 400 to 600 cm^{-1} region owing to increasing noise levels near 400 cm^{-1}. a, methyl and methylene CH stretches; b, carbonyl C=O stretches; c, benzene ring breathing modes; d, methyl and methylene CH bends and scissoring; e, out-of-plane deformation of aromatic hydrogen; f, methylene rocking mode. Other bands have no definite assignments. Note the compressed scale from 4000 to 2000 cm^{-1}. [Reproduced with permission from *Liq. Cryst.*, **8**, 839 (1990)]. Ref. [130].

function of temperature. Note the large jump in the absorption frequency at about 45°C and the change in the slope at about 65°C. From 65°C onward, the absorption frequency appears to increase slowly with increasing temperature, with a slight leveling off at about 160°C. Figure 75(b) shows the carbonyl C=O stretching frequency as a function of temperature. A jump in the frequency is seen at about 62°C, and the slope changes at about 150°C. The frequency of a benzene breathing mode is shown in Figure 75(c). Changes in slope are seen approximately at 45°C and 62°C. Finally, the CH$_2$ scissoring bending mode frequency is shown in Figure 75(d). Changes in slope can be seen at ~45°C, 62°C, 110°C, and 150°C.

TABLE VI
Bands Assignment of Truxene Derivatives

Assignment	Position, cm^{-1}		
	HA14TX	HA10TX	HABT
CH$_3$ asymmetric stretching vibration	2955	2956	2960
CH$_2$ asymmetric stretching vibration	2923	2926	2925
CH$_3$ symmetric stretching vibration	2870	2872	2875
CH$_2$ symmetric stretching vibration	2855	2855	2855
C=O stretching vibration	1760	1768	1735
C–C in-plane aromatic stretching vibration	1600	1613	1600
C–C aromatic in-plane deformation		1586	1510
CH$_2$ scissoring vibration	1477	1476	
CH$_3$ asymmetric bending	1468	1468	1480
CH$_3$ symmetric bending	1380	1377	1380
O=C–O stretching mode	1260	1268	1250
C–O–C stretching mode	1160	1157	
C–H aromatic in-plane deformation			
C–H out-of-plane bending	875	873	
CH aromatic out-of-plane vibration		840	840; 760
CH$_2$ rocking	720	721	720

Asher and Levin [123] and Snyder et al. [75] showed that the CH stretching frequencies are primarily sensitive to the chain conformation and that their absorption frequencies increase in the presence of nearby gauche bonds [121,122]. Therefore, the sudden increases in absorption frequency seen in Figure 75(a) can be interpreted as sudden increases in the statistical distribution of gauche bonds on the alkyl tails of the HA14TX molecule. The presence of substantial tail disorder is consistent with the large-angle diffuse scattering seen in the X-ray powder patterns [130]. It is interesting to note that the large increase in the absorption frequency occurs at 45°C, which is 20°C lower than the melting transition temperature. Thus the tail disorder increases abruptly and significantly well below the Cr–N$_D$ transition, an effect that was seen in compounds such as benzene hexa-n-alkanoates [64] and hexahydroxybenzene hexanoate. This premelting effect was observed by two independent measurements of two different HA14TX

Figure 74. HA14TX FTIR absorption profiles of (a) CH stretching modes, (b) C=O stretching modes, (c) benzene stretching modes, and (d) CH bending or scissoring modes at different temperatures. Data were taken upon heating to the following temperatures: 29, 33, 39, 43, 53, 58, 63, 68, 72, 77, 81, 86, 91, 100, 105, 109, 114, 118, 123, 127, 132, 136, 141, 146, 151, 155, 160, 164, 173, 178, 182, and 187°C. The plots have arbitrary and different vertical scales. [Reproduced with permission from *Liq. Cryst.*, **8**, 839 (1990)]. Ref. [130].

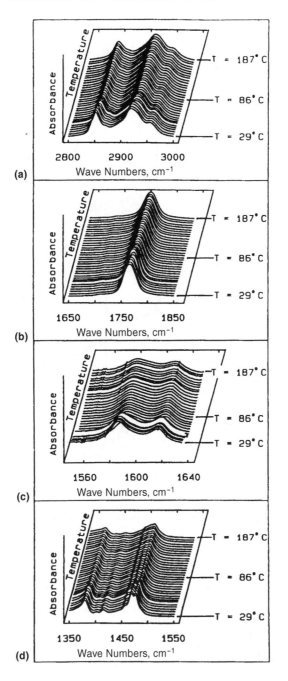

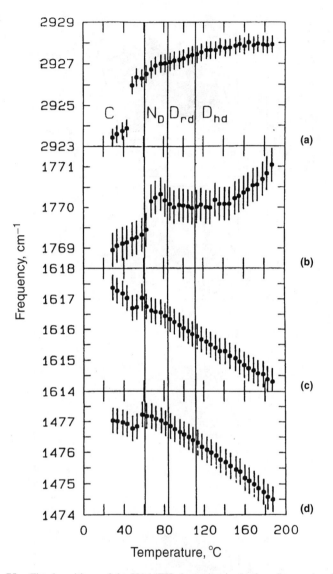

Figure 75. Fitted positions of the HA14TX absorption peaks from Lorentzian fits of (a) methylene asymmetric CH stretching mode, (b) carbonyl C=O stretching mode, (c) benzine stretching mode, and (d) methylene scissoring mode. In cases for which more than one Lorentzian was needed to give a reasonable fit, the weighted average of the positions are plotted. Data were taken by heating the sample from ~29°C to ~187°C in 5°C increments. Error bars indicate estimated errors from the fits. Vertical lines show the phase boundaries from previous studies. [Reproduced with permission from *Liq. Cryst.*, **8**, 839 (1990)]. Ref. [130].

sample cells. It could either result from a true thermodynamic transition between two solid phases, in which the tails collectively become more disordered at a well-defined transition temperature, or merely a rapid but continuous evolution of tail disorder that acts as a precursor to the actual melting. The X-ray data [130] are consistent with a solid–solid phase transition; however, a nonequilibrium double-melting transition, as suggested by Warmerdam et al. [133], cannot be ruled out.

On heating the sample above the melting temperature, the CH stretching frequency undergoes a small gradual increase and plateaus at $> 160°C$. Thus any increases in the statistical distribution of gauche bonds in the alkyl tails upon heating after the melting temperature must be quite small. The leveling out of the CH stretching frequency at the melting temperature is similar to that seen by Cameron et al. [134] in 1,2-dipalmitoyl-*sn*-glycero-3-phosphocoline.

The C=O stretching band is best modeled by a sum of three Lorentzians. The weighted average position of these peaks is plotted against temperature in Figure 75(b). The absorption frequency clearly jumps at the melting temperature of 65°C. There is also a change in slope at around 150°C. Kardan et al. [64] also observed such an increase in the C=O stretching frequency in the melting transition of benzene hexa-*n*-alkanoates. They suggested that this could be a result of the molecular cores moving apart upon melting. There do not appear to be any anomalies in the data at around 45°C. Because the carbonyl bond is close to the truxene core, this again suggests that the observed premelting transition involves only the alkyl tails.

Figure 75(c) shows the temperature dependence of a benzene stretching frequency. No significant features are seen beyond 45°C. The temperature dependence of the CH_2 scissoring mode is plotted in Figure 75d. Anomalies are seen at about 45°C and 65°C. In addition, there appear to be small changes in slope at about 110°C and 150°C. Optical microscopy measurements [42,133] indicate a $D_r \rightarrow D_h$ transition at $\sim 112°C$, although no transition at 112°C was observed in the course of single-strand X-ray studies [108]. The present IR absorption data suggest that there is, in fact, some transition in the vicinity of 122°C, in which the alkyl tails undergo a subtle change in the environment, without significant changes to the core or to the number of gauche bonds. X-ray diffraction intensities are expected to be fairly insensitive to small changes in tail conformation, because the tail–tail correlation peak is quite broad. IR absorption spectra of CH bending modes have been shown to be sensitive to the alkyl chain conformation and the chain packing [103]. Any sliding or rotating of the alkyl chains with respect to the neighboring chains might show up in the IR spectra but not in the X-ray data. Therefore, the results show that the transition at 112°C from

D_{rd} to D_{hd} is most likely. This primarily involves the motions of the alkyl tails rather than any motions of the cores.

2. *HA10TX*

Here we review the results of our IR investigation of the HA10TX, which is similar to HA14TX, except that the alkyl chain is shorter for the former. The complete spectra for some phases (Cr, D_{rd}, and D_{ho}) of this compound are shown in Figure 76.

The analysis of the IR data of truxene HA10TX also shows a strong influence of the substrate's structure on the alignment of this discotic as was found for H5T. Note that this discotic material exhibits a larger number of phases that are spread over a wider range of temperatures than does H5T. Our experimental setup did not allow us to cover the temperature range up to the isotropic phase transition. The isotropic phase starts at 510 K; however, our measurements are made up to a maximum temperature of 480 K. Thus the dichroic ratio could not be determined using Neff's method. The dependence of the peak intensity on temperature, however, can give a complete picture of the phase transitions and the orientational behavior in the discotic phase, as shown in Figure 77 for the C–C aromatic in-plane and C–H out-of-plane vibrations of HA10TX contained in two different cells (ZnSe and Si). As seen, the dependence of the peak intensity on temperature for HA10TX in ZnSe windows is exactly opposite to that in Si windows. The

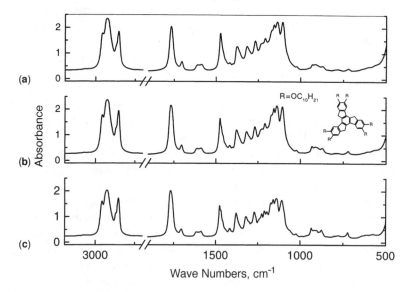

Figure 76.   IR spectra of HA10TX in (a) discotic D_{ho} (420 K) (b) discotic D_{rd} (360 K), and (c) crystalline (310 K) phases.

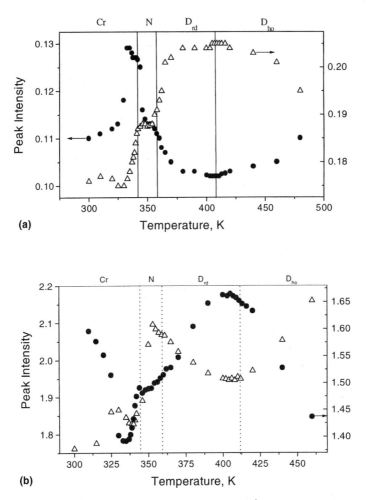

Figure 77.   The peak intensity of (a) parallel (1617 cm^{-1}) and (b) perpendicular (2928 cm^{-1}) bands for HA10TX contained in nylon 6,6-coated ZnSe (●) and Si (△) cells vs. temperature.

frequency of the band maximum, however, shows similar temperature dependencies for both types of windows (Fig. 78). It is interesting to note that the frequency maximum for C–C aromatic stretching vibrations differs by ∼2 cm^{-1} in the entire temperature range for different cells. At the same time, the peak frequencies of the other vibrational bands are similar [Fig. 78(b)].

Note that the dependence of the frequency maximum on temperature for HA10TX is similar to those for HA14TX; however, the dependence of peak

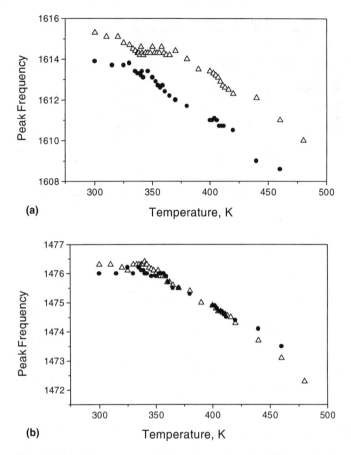

Figure 78. The peak frequency against temperature for two different vibrational bands for the same cells shown in Figure 77.

intensity on temperature observed by Lee et al. [130] is not as sensitive to phase transitions as that obtained in our work (Figs. 77). This is probably because our method of sample preparation allows one to obtain a better alignment. Figure 77 shows that even the D_{rd}–D_{ho} transition is significantly noticeable. Moreover, the existence of the two crystalline phases (with the phase transition at about 57°C) was established from our IR measurements for HA10TX, as shown in HA14TX by Lee et al. [130].

3. HBTX

Figure 79 shows the complete FTIR absorption scans of HBTX at 29°C and 169°C. As seen from Figures 73 and 79, the relative intensities of the benzene stretching modes shown by HBTX are much higher than those of

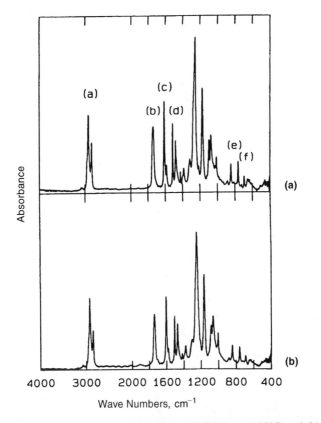

Figure 79. Complete FTIR absorption spectra of HBTX at (a) 29°C, and (b) 169°C. *a*, Methyl and methylene CH stretches; *b*, carbonyl C=O stretches; *c*, benzene ring breathing modes; *d*, methyl and methylene CH bends and scissoring; *e*, out-of-plane deformation of aromatic hydrogen, and *f*, methylene rocking mode. Other bands have no definite assignments. Note the compressed scale from 4000 to 2000 cm^{-1}. [Reproduced with permission from *Liq. Cryst.*, **8**, 839 (1990)]. Ref. [130].

HA14TX [130], because the additional benzoyl groups are attached to the alkyl tails near the core. The two HBTX spectra taken at different temperatures are qualitatively similar. As in HA14TX, the amplitudes decrease and peak broadens so that nearby peaks become unresolved with increasing temperature.

Figure 80 shows the variation of the absorption peaks for different regions with increasing temperature. Detailed band assignments for the different regions shown are identical to those of HA14TX and HA10TX (Table VI). The only difference is that the carbonyl C=O stretching bands appear near 1738 cm^{-1} and the benzene band appears near 1606 cm^{-1}.

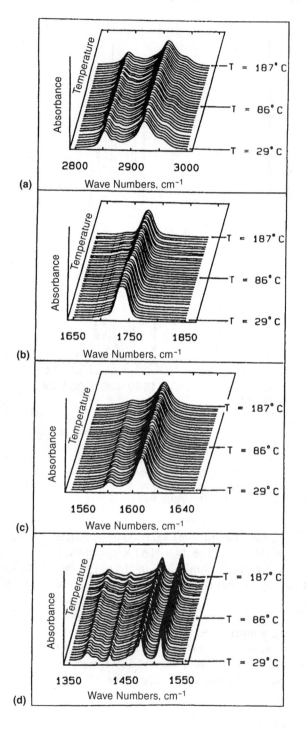

Figure 81(a) shows the temperature dependence of the CH_2 asymmetric stretching frequency. There is clearly a small break in the position and change in slope at $\sim 85°C$. There may also be a small change in slope at $\sim 135°C$. Figure 81(b) shows the plot of the weighted average of the carbonyl C=O stretching frequency as a function of temperature. There is also a sharp increase in the absorption frequency at $\sim 85°C$. The absorption frequency of a benzene stretching mode is plotted against temperature in Figure 81(c). An anomaly is again seen at around 85°C and a small change in slope at $\sim 135°C$. The CH_2 scissoring and the CH_3 asymmetric bending modes are unresolved in the HBTX spectra. The weighted average of their peak positions is plotted against temperature in Figure 81(d).

The temperature dependence of the CH stretching bands clearly suggests that even in the crystalline phase, the statistical distribution of gauche bonds in the alkyl tails increases significantly upon heating. The rate of increase decreases in the D_{rd} phase and becomes small or zero near the $D_{rd} \rightarrow N$ transition. This is considerably different from the evolution in HA14TX, in which two distinct transitions are observed. As is the case for HA14TX, the 1.4 $Å^{-1}$ diffuse peak observed in powder X-ray measurements of HBTX suggests that the alkyl tails of HBTX are significantly disordered even in the crystalline phase. This notion is supported by the IR data.

The comparison of the CH stretching bands of HA14TX and HBTX as a whole, shows that the HBTX vibrational bands are located at a higher frequency. Studies [75,123] have shown that the CH stretching modes are quite insensitive to the environment except for the existence of nearby gauche bonds. Although comparisons of spectra from different molecules must be treated with caution, Lee et al. [130] concluded that the tail of HBTX are more disordered than that of HA14TX.

The weight average of the carbonyl C=O stretching frequency is plotted against temperature in Figure 81(b). The scatter in the data points indicates the uncertainty caused by the existence of the multiple or shallow minima in the least-square fitting procedure. The only clearly identifiable feature is the large increase in absorption frequency near the melting temperature of 85°C. Lee et al. [130] noted that this band is best fitted by a sum of three Lorentzians. As in the case of HA14TX, the jump observed at 85°C was interpreted as a sudden increase in the core–core distance upon melting.

Figure 80. HBTX FTIR profiles of (a) CH stretching modes, (b) C=O stretching modes, (c) benzene ring stretching modes, and (d) CH bending or scissoring modes at different temperatures. Data were taken upon heating to the same temperatures as in Figure 74. The plots have arbitrary and different vertical scales. [Reproduced with permission from *Liq. Cryst.*, **8**, 839 (1990)]. Ref. [130].

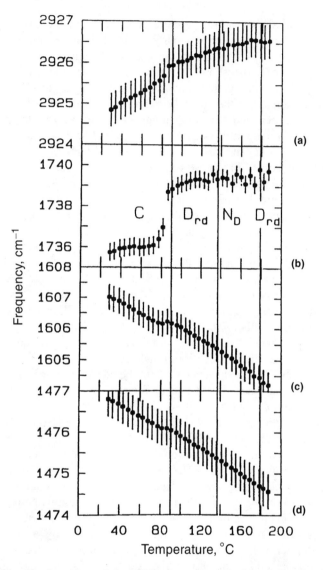

Figure 81. Fitted positions of the HBTX absorption peaks from Lorentzian fits for (a) methylene asymmetric CH stretching mode; (b) carbonyl C=O stretching mode, (c) benzene stretching mode, and (d) methylene scissoring mode. In cases in which more than one Lorentzian was needed to give a reasonable fit, the weighted average of the positions is plotted. Data were taken by heating the sample from ~29°C to ~187°C in increments. Error bars indicate estimated errors from the fits. Vertical lines show the phase boundaries from previous studies. [Reproduced with permission from *Liq. Cryst.*, **8**, 839 (1990)]. Ref. [130].

In addition to an anomaly at the melting point near 85°, the plot of a benzene stretching mode frequency versus temperature, Figure 81(c), shows a small, but detectable change in slope at $\sim 135°C$, corresponding to the $D_{rd} \rightarrow N$ transition. Because this transition involves changes in the core–core interactions, and the benzoyl groups are attached near the core, it is not surprising that an anomaly in the benzene stretching frequency at this temperature can be seen. The CH_2 scissoring and the CH_3 asymmetric bending bands are unresolved in the HBTX spectra. The weighted average frequency of these two bands is plotted against temperature in Figure 81(d), which clearly shows an anomaly at about 85°C and, as in HA14TX, a decrease in the absorption frequency with increasing temperature.

D. Discotic Polymer

This section reviews the FTIR absorption study [42] of a main-chain DLCP, for which NMR spectroscopic studies [103] were also made. The structural formula for the main-chain liquid crystalline polymer is shown in Figure 22c. The triphenylene disk, with ester groups attached on either side, forms a part of the main chain of the polymer, with the alkyl spacer inserted between the two adjacent disks. The average degree of polymerization was calculated to be 19. Polarizing microscopy and X-ray studies show that the discotic liquid crystalline phase is hexagonally ordered (D_{ho}), and this phase is exhibited over a wide range of temperatures.

In the D_{ho} phase [92], the centers of the cores form a hexagonal lattice in the plane normal to the column's axis. Several experiments have confirmed both the long-range order of the 2-D lattice and the regularity in the stacking of the triphenylene cores in each column.

The spectrum of the discotic liquid crystalline polymer in the glass phase at 300 K is shown in Figure 82. The bands assignment is listed in Table VII. The band wave numbers are somewhat approximate and correspond to the molecular vibrations in the glassy state at 300 K. The dichroism ratio for the ith band R_i, is defined using Eq. (3).

For analysis of the results we first turn our attention to the vibrational band centered at $\sim 1619 \, cm^{-1}$ and identify this vibration with the C–C in-plane stretching vibration of the benzene ring. For the unsubstituted benzene ring, particularly in the crystalline phase, this vibration is forbidden in the IR region owing to symmetry considerations. When the hydrogen atoms of the ring are substituted by other molecular groups, however, this band is of measurable intensity in the liquid crystalline and isotropic phases [18,65]. The triphenylene ring consists of three benzene rings: Two hydrogen atoms on each of the rings are replaced by either OR or ester groups [Figure 22(c)]. The symmetry of the disk is thus reduced, and consequently the band at $1619 \, cm^{-1}$ is found to be quite intense. For the H5T discotic liquid crystal,

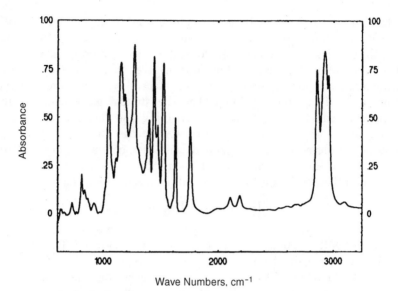

Figure 82. The IR spectrum of the DLCP in the glassy state at 300 K. [Reproduced with permission from *Supramol. Sci.*, **2**, 51 (1995)]. Ref. [42].

TABLE VII
Bands Assignment of the DLCP

Assignment	Wave number, cm^{-1}
CH$_3$ asymmetric stretching vibration	2960
CH$_2$ asymmetric stretching vibration	2925
CH$_3$ symmetric stretching vibration	2875
CH$_2$ symmetric stretching vibration	2855
CD$_2$ stretching vibration	2195 and 2100
C=O stretching vibration	1753
C−C in-plane aromatic stretching vibration	1619
C−C aromatic in-plane deformation	1517
CH$_2$ scissoring vibration	1467
C−C aromatic in-plane deformation	1434
C−O stretching vibration	1394
CH bending mode	1379
Core−O stretching vibration	1266
C−H aromatic in-plane deformation	1182
C−O−O asymmetric stretching vibration	1150
C−O−O symmetric stretching vibration	1050
CH aromatic out-of-plane vibration	917
CH aromatic out-of-plane vibration	866

the C–C in-plane aromatic stretching vibration appeared at 1617 cm^{-1} and was found to be as strong as the C–O–C stretching vibration. For the polymeric liquid crystal, the absorbance centered at 1619 cm^{-1} was found to decrease with increase in temperature for the C–C in-plane stretching vibrations. This implies that with an increase in temperature the cores start tilting away from the window of the cell. The absorbance for the C–C in-plane vibrations reduces from 0.6 to 0.3 over a temperature change of 120°C, indicating a significant tilting of the disks and an increase in disorder in the arrangement of the disks with temperature, because the absorbance is proportional to the length of the orthogonal projection of the transition dipole moment vector p_i on the surface of the cell. Therefore, the alignment of the discotic polymer in the D$_{ho}$ phase is the side-on type.

For the vibrations that have p_i parallel to the plane of the core (e.g., the C–C in plane aromatic stretching vibration near 1619 cm^{-1}) and p_i perpendicular to the plane of the core (such as CH aromatic out-of-plane vibrations) the order parameters were calculated using Eqs. (12) and (13), respectively.

The figures given below show the order parameter and the frequency changes for several bands of the sample aligned using nylon 6,6. The plot of the order parameter S versus temperature for the C–C aromatic in-plane stretching vibration near 1619 cm^{-1} is shown in Figure 83. The maximum value of $S=0.65$ in the D$_{ho}$ phase was found to be almost twice the value of the order parameter for the same band when the sample was aligned using ZLI 2650. This demonstrates that alignment of the cores achieved when using nylon 6,6 is better than with ZLI 2650. For the same vibration, the maximum value of the order parameter for the very similar (i.e., with the same core) nonpolymeric discotic liquid crystal H5T was $S=0.85$, and the sample was self-aligned [58]. For the discotic liquid crystalline polymer cell prepared without using any surface alignment procedures, the absorbance heights did not vary with temperature, thus showing that the sample was randomly aligned.

The plot for the frequency, vs. temperature, of the C–C aromatic in-plane stretching vibration near 1619 cm^{-1} is given in Figure 84. No discontinuities in the frequencies for this vibration near the D$_{ho}$-I phase transition temperatures were found. In contrast, for the non-polymeric H5T material, the frequency of the C–C in-plane aromatic stretching vibration decreased by 4 cm^{-1} in going from the Cr–D$_{ho}$ to the D$_{ho}$–I phase transition temperatures. The frequency of this vibrational band decreases almost linearly with an increase in temperature, and a total decrement in frequency of 4.5 cm^{-1} was obtained.

Two perpendicular CH aromatic out-of-plane vibrations have been found. One of these appeared near 917 cm^{-1}, and the other was near 866 cm^{-1}

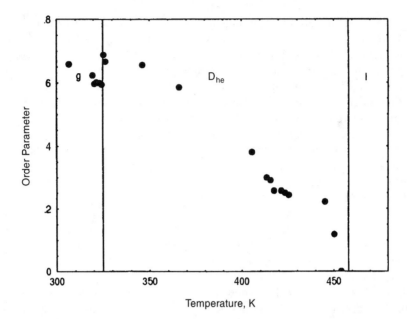

Figure 83. Order parameter S as a function of temperature for the C–C in-plane aromatic stretching vibration near 1619 cm^{-1}. [Reproduced with permission from *Supramol. Sci.*, **2**, 51 (1995)]. Ref. [42].

(Fig. 85). It was possible to separate these out by fitting the observed bands in the wave number range from 770 to 950 cm^{-1} with a total of eight peaks, by using the BANDFIT program. Plots showing changes of the order parameter vs. temperature for the vibrations near 917 cm^{-1} and near 866 cm^{-1} were measured (Fig. 86). The values of the order parameter after the Cr–D$_{ho}$ phase transition were approximately the same as for the C–C in-plane aromatic stretching vibration near 1619 cm^{-1}.

Strong frequency changes with temperature for the band near 917 cm^{-1} were observed [Fig. 87(a)]. The frequency decreased in the D$_{ho}$ phase by ~ 10 cm^{-1}. The observed change of frequency in the discotic phase for the vibration near 866 cm^{-1} is ~ 4 cm^{-1} [Fig. 87(b)]. At 425 K (i.e., 28 K below the transition to the isotropic phase) a sharp decrease in the frequency by 3.5 cm^{-1} was observed; after the transition to the isotropic phase had occurred, the frequency increased again. The possibility of either a phase transition near 425 K, or a sudden rearrangement of the structure, was futher investigated. The DLCP material under similar conditions was carefully studied using polarizing optical microscopy, but no new phase was found. These results, therefore, support those obtained from NMR spectroscopic

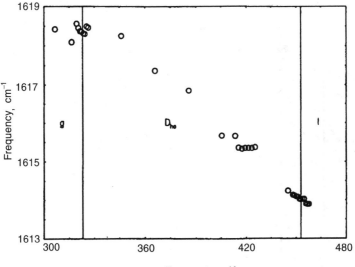

Figure 84. Frequency change with temperature for the C–C in-plane aromatic stretching vibration near 1619 cm^{-1}. [Reproduced with permission from *Supramol. Sci.*, **2**, 51 (1995)]. Ref. [42].

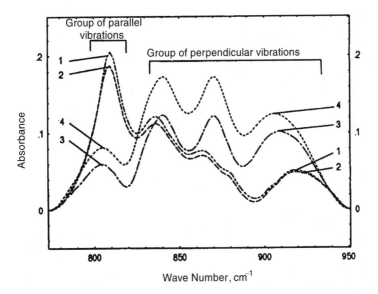

Figure 85. The IR spectra of a group of vibrations in the range from 770 to 950 cm^{-1} at different temperatures. - - -, 310 K (g); ·····, 330 K (D$_{ho}$); -·-·-·, 450 (D$_{ho}$); - - - -, 469 K (1). [Reproduced with permission from *Supramol. Sci.*, **2**, 51 (1995)]. Ref. [42].

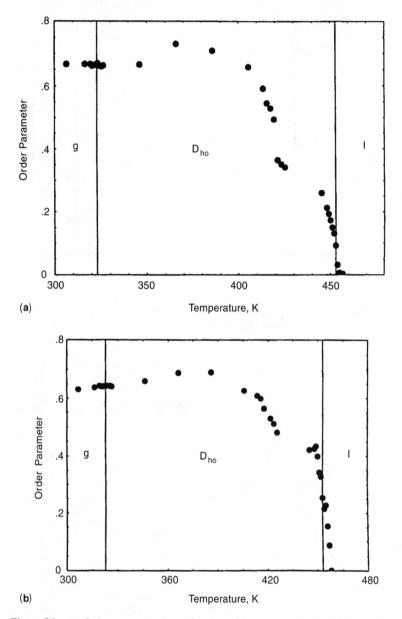

Figure 86. (a) Order parameter S as a function of temperature for the CH aromatic out-of-plane vibration near 917 cm^{-1}. (b) Order parameter S as a function of temperature for the CH aromatic out-of-plane vibration near 866 cm^{-1}. [Reproduced with permission from *Supramol. Sci.*, **2**, 51 (1995)]. Ref. [42].

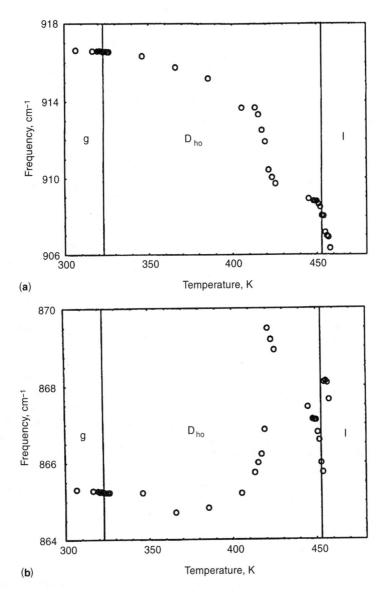

Figure 87. (a) Frequency changes with temperature for the CH aromatic out-of-plane stretching vibration near 917 cm^{-1}. (b) Frequency changes with temperature for the CH aromatic out-of-plane stretching vibration near 866 cm^{-1}. [Reproduced with permission from *Supramol. Sci.*, **2**, 51 (1995)]. Ref. [42].

studies [103], suggesting a biphasic behavior of the polymer, which is observed over a broad temperature interval. The order parameter for this biphasic behavior is found to be lower than for the D_{ho} phase.

Alignment by using nylon-6,6 provided large values of the order parameter (compared with those obtained when using poly(amic acid) solution) for the C=O stretching vibration centered at ~ 1753 cm^{-1}. A strong decrease in the absorbance with temperature was found for this band. The plot of S vs. temperature for the C=O vibration is given in Figure 88a, which shows a maximum value of $S = 0.45$. A strong frequency increase (9 cm^{-1}) with temperature for the C=O vibration can be seen in Figure 88b.

Interesting results were obtained when the DLCP was aligned using the poly(amic acid) solution ZLI 2650. The following alignment procedure was used for one of the samples. The top window of the cell was sheared a few times relative to the bottom when the sample in the cell was in its isotropic phase. This was carried out in an attempt to stretch (i.e., flatten) the main polymeric chains and to make the alkyl tails parallel to the windows of the cell. Some evidence for this process lies in the observation of strong changes in the absorbance with temperature for the CH$_2$ and CH$_3$ bands, which are observed during the first heating cycle from the discotic to the isotropic phase following the alignment. On cooling the sample to the glassy state and then "heating it up once again," these vibrations appear to be highly disordered during the second heating cycle from the discotic to the isotropic phase. The changes in absorbance for some bands (e.g., C–C in-plane aromatic stretching vibration) remained the same during the first and the second heating cycles. Many CH$_2$ groups are present in both the main polymeric chain and the alkyl tails [Fig. 22(c)]. The CH$_3$ groups exist as terminal groups in the alkyl tails attached to the aromatic core, and as the terminal groups of the main polymeric chains.

When the sample was heated to the isotropic phase during the first heating cycle, the absorbance of the C=O band decreased with increasing temperature (Fig. 89). Next, the sample was cooled extremely slowly to the glassy state, and the spectra were recorded. In this case, a decrease in the absorbance with decreasing temperature for the C=O stretching vibration near 1753 cm^{-1} was found (in contrast to that observed during the heating cycle shown in Figure 89). The decrease occurred in the first instance because the transition dipole moment vectors of the C=O vibrations during the first heating cycle possibly become disoriented while reaching the isotropic phase, and then during the cooling cycle these vectors tend to become almost normal to the windows of the cell. These explanations follow from an observed reduction in the absorbance of the band, with an increase in temperature during the first heating cycle and a subsequent continued reduction in the absorbance with a decrease in temperature during the first

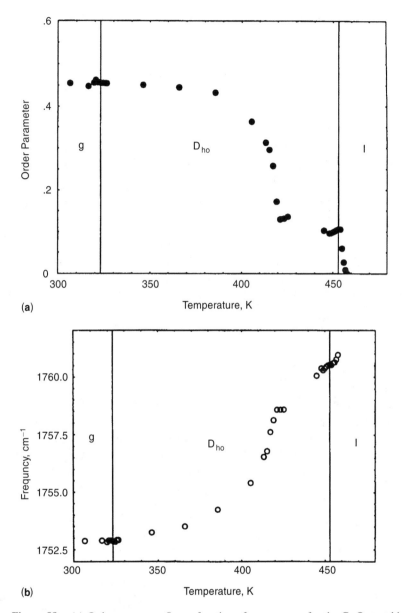

Figure 88. (a) Order parameter S as a function of temperature for the C=O stretching vibration near 1753 cm^{-1}. (b) Frequency changes with temperature for the C=O stretching vibration near 1753 cm^{-1}. [Reproduced with permission from *Supramol. Sci.*, **2**, 51 (1995)]. Ref. [42].

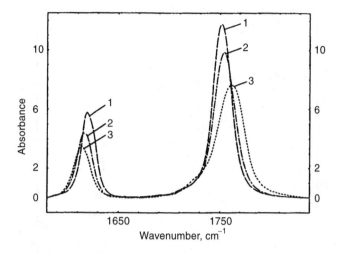

Figure 89. Changes in the absorbance values of the C=O stretching vibration near 1753 cm^{-1} and for the C–C in-plane aromatic stretching vibration near 1619 cm^{-1} when heating the sample up to the isotropic phase for the first time after preparation. The sample was aligned with a poly(amic acid) solution. - - -, 300 K (g); - · - ·, 420 K (D_{ho}); - - - -, 462 K (I). [Reproduced with permission from *Supramol. Sci.*, **2**, 51 (1995)]. Ref. [42].

cooling cycle. We can, therefore, safely conclude that reorientation of the C=O bonds of the ester group during temperature cycling does occur, but without significantly affecting the cores, which tilt to only a limited extent (as discussed previously). The surface alignment makes the C=O bonds become parallel to the disks, and these then have a tendency to reorientate and become approximately normal to the cores during the temperature cycling. The measurements made during the second heating cycle gave similar results to those found for the first cooling cycle. This confirmed our findings of the behavior of the C=O band and showed that the sample did not leak from the cell during the first heating cycle. The preferred orientation of the C=O bond is, therefore, found to be out-of-plane and lies along the column's axis. This is in agreement with observations of a low-molar-mass, triphenylene-based discotic liquid crystal [52].

Figures 89 and 90 show that the absorbance heights of the C–C in-plane aromatic stretching vibrations near 1619 cm^{-1} are approximately the same in the glassy state during both the heating and the cooling cycles. The value for this vibration is also the same for the isotropic phase. The absorbance decreased with increasing temperature as expected; hence the orientations of the cores change only slightly with temperature, but overall they continue to lie parallel to the windows of the cell in the discotic phase, thus retaining

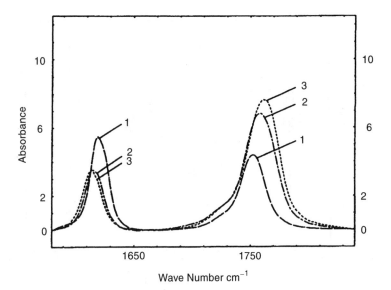

Figure 90. Changes in the absorbance values of the C=O stretching vibration near 1753 cm^{-1} and for the C–C in-plane aromatic stretching vibration near 1619 cm^{-1} when cooling the sample down to the glassy state. The sample was aligned with a poly(amic acid) solution. ---, 300 K (g); ····, 420 K (D_{ho}); - - -, 462 K (I). [Reproduced with permission from *Supramol. Sci.*, **2**, 51 (1995)]. Ref. [42].

the side-on alignment. The order parameter for the C–C in-plane aromatic stretching vibration appeared be $\sim 10\%$ greater in the glassy state during the cooling cycle when compared to that observed during the first heating cycle. However, both values of the order parameter were lower than those found for the same band when the sample was aligned using nylon-6,6.

The order parameter for the group of symmetric stretching vibrations of CH$_2$ and CH$_3$ near 2855 and 2875 cm^{-1} is presented in Figure 91. Separation of these vibrations using the BANDFIT program and calculations of the order parameter for the individually separated bands should have provided reliable results, but the value was found to be small. The measured value of $S = 0.06$ for this group when the sample is aligned using nylon-6,6 appears to be particularly small. The alkyl tails are highly disordered [58], which arises from a tilt in the disks with temperature, as indicated by measurements of the C–C in-plane aromatic stretching vibration. As the disks tilt and become somewhat disordered, the alkyl chains attached to these disks become highly disordered. These results support those found by Goldfarb et al. [52] who reported that the alkyl chains are highly disordered, in agreement with the short effective radius determined by X-ray studies, of

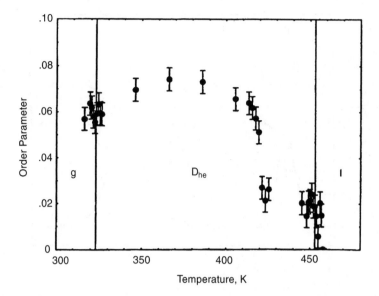

Figure 91. Order parameter S as a function of temperature for the group of CH_2 and CH_3 symmetric stretching vibrations near 2855 and 2875 cm^{-1}. The sample was aligned with a nylon-6,6 solution in methanol.

the discotic columns in certain discotic mesophases. These results, however, contradict those found by Yang et al. [65] for model discotic compounds.

E. Switchable DLC

Prost [10] proposed that the main prerequisites for the appearance of ferroelectricity in columnar mesophase are the chirality of the molecules and the occurrence of a tilt of the individual molecules with respect to the axis of the column. Experimentally, this suggestion was first realized by Bock and Helfrich [11] who described bistable ferroelectric switching for some dibenzopyrene derivatives in 1992. Subsequently, columnar ferroelectrics were synthesized by Scherowsky and Chen [135] and Heppke et al. [136]. It was found that some dibenzopyrene derivatives [11,138] and their mixtures [136] exhibit different phases in varying external electric fields. The low field phase (electric fields < 10 V/μm) shows a smaller spontaneous polarization and smaller reorientation angles of the optical axis than the high field phase. Although tentative structures of these phases were described [30], a detailed analysis of segmental arrangements has not yet been given.

Three possible mechanisms of switching in columnar ferroelectrics were suggested in [11,30,135]. In one case, the column rotates as a whole around the axis of the column. In another case, the tilt reverses through an untilted state with no molecular rotation around the column axis. According to a third mechanism, the molecular tilt direction, but not the molecules themselves, rotate around the column axis. At present, it is unknown which mechanism dominates during ferroelectric switching.

Bearing in mind that IR spectroscopy is a powerful tool for studying the segmental orientations in organic materials (and in particular for liquid crystals), two attempts to analyze the properties of two different electric field phases in swichable discotic were made [57,138].

The experimental setup and structural formula of the DLC material 1,2,5,6,8,9,12,13-octakis-((S)-2-(heptyloxy)-propanoyloxy)-dibenzo-[e,l]-pyrene (D8m*10) are shown in Figures 92 and 93. The clearing temperature of this material is approximately 388 K. No phase transition above 248 K could be detected using DSC.

In our work [57], the chiral material D8m*10 was sandwiched between two indium tin oxide (ITO) coated ZnSe windows (resistivity, 2 kΩ/cm) to make the surface conductive. The cell thickness was fixed at ~8 μm using mylar spacers. The cell was filled by using the shearing method. The texture of the mesophase between crossed polarizers, when the sample was cooled from its isotropic phase, consisted of a characteristic flower-like domains with some gray to black areas. The sample was found to achieve a uniform edge-on alignment after subjecting the cell to an alternating electric field of

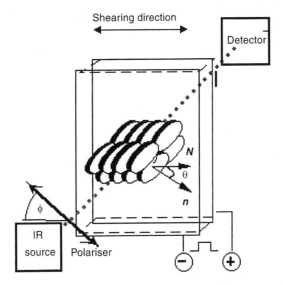

Figure 92.   Experimental setup. Polarization rotation angle φ, core normal n and column axis N are indicated. [Reproduced with permission from authors of Ref. [138].]

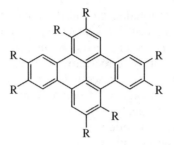

Figure 93.   Structure of dibenzopyrene. R = OCOC$_{10}$H$_{21}$ (D-388-I).

20 V/µm at 10 Hz in the columnar phase at 373 K for 2 to 3 h. This edge-on alignment remained stable after the field was switched off.

The spectra of D8m*10 and HA10TX at 340 K without applied field are shown in Figures 94 and 95. The later is shown for comparison, because the compound has a structural arrangement similar to D8m*10 but does not exhibit ferroelectricity and can ideally be used for the band assignments. The tentative assignments of some IR-bands are given in Table VIII. The most intense bands are found in the wave number range 2700 to 3000 cm^{-1} (alkyl chain vibrations), 1700 to 1800 cm^{-1} (C=O stretch vibration), and 950 to 1200 cm^{-1} (C–O–C stretch vibrations).

In the spectrum of D8m*10, all bands corresponding to benzene ring vibrations and core-oxygen vibrations are of low absorbance. This behavior is different from the other discotic liquid crystals discussed above. One of the small peaks in the frequency range of benzene ring vibrations (at 1540 cm^{-1}) was nicely isolated from the others and, therefore, could be precisely integrated.

The intensity of the bands assigned to the benzene ring vibrations (in the range 1400 to 1650 cm^{-1}) and to the core-oxygen vibration (approximately 1250 cm^{-1}) are temperature dependent, increasing when heating to the isotropic phase to a value far higher than in the columnar phase. A similar increase was found for the C–O–C vibrations. The dependence of the magnitude of absorbance on temperature were repeatable. The temperature dependence of the peak intensity and the frequencies of the maximum absorbance for the isolated benzene ring vibration (at 1540 cm^{-1}) and the carbonyl stretch vibration (at 1777 cm^{-1}) are shown in Figures 96 and 97. Figure 96(a) shows that the alignment achieved is edge-on, because the C–C aromatic stretching band at 1407 cm^{-1} has a lower intensity in the discotic phase than in the isotropic phase. This is also confirmed by polarized microscopy. The Figure also shows that three other bands (1455, 1540, and 1777 cm^{-1}) have an opposite behavior and hence are oriented perpendicular to the core. Moreover, such an orientation allows us to conclude that the

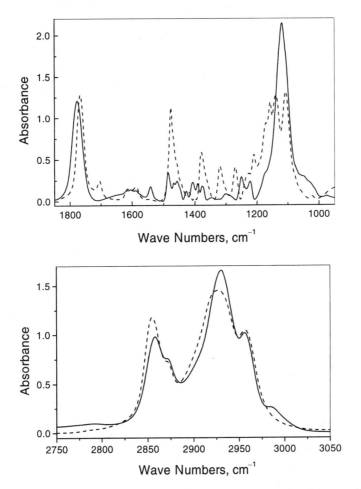

Figure 94. IR spectra of D8m*10 (*solid line*) and HA10TX (*dotted line*) at 343 K. [Reproduced with permission from *Molec. Cryst. Liq. Cryst.*, **263**, 293 (1995)]. Ref. [57].

alkyl chain is mainly in gouche configuration. We can also see that the dependence of the peak intensity on temperature for the 1540 and 1777 cm^{-1} bands is similar: near the phase transition to the isotropic liquid, the absorbance decreases abruptly and then, on further heating, increases. We find that the frequency of the benzene ring vibration band shifts slowly to lower frequencies with increasing temperature (Fig. 97), whereas the frequency maximum of the C=O stretching vibration does not change between 300 and 365 K. When heating into the isotropic liquid, the latter increases rather sharply by 5 cm^{-1} and then again remains constant at

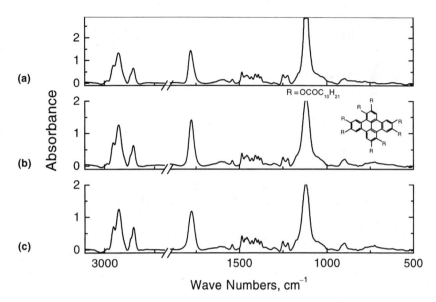

Figure 95. IR spectra of D8m*10 at (a) the isotropic phase at 395 K, (b) in the discotic phase at 363 K, and (c), the discotic phase at 310 K.

TABLE VIII
Band assignment, dichroic ratios, and apparent tilt angles for different phases of the dibenzopyrene derivative D8m*10

Band Position, cm^{-1}	Assignment	No Field, R	Low-Field Phase R	Low-Field Phase θ, degrees	High-Field Phase R	High-Field Phase θ, degrees
2930	$\nu_{as}(CH_2)$	1.07	1.09	4	1.09	9
2859	$\nu_s(CH_2)$	1.09	1.12	3	1.12	5
1777	$\nu(C=O)$	2.61	3.20	13	5.98	25
1540	$\nu(C-C)_{ar}$	1.00	0.63	—	0.23	—
1484	$\nu(C-C)_{ar}$	0.79	0.57	32	0.31	38
1407	$\nu(C-C)_{ar}$	0.57	0.61	18	0.35	30
1248	Core-oxygen stretching	0.38	0.33	14	0.16	25

[Reproduced with permission from authors of Ref. [138].]

higher temperatures (Fig. 97). The temperature dependence of the frequency of the C=O vibration, namely the jump to a higher frequency at the clearing point is easily explained by a lower interaction between the carbonyl and aromatic electrons in the isotropic liquid. Such frequency shifts of aromatic ester C=O vibrations at the clearing temperature are known from smectic liquid crystals [18].

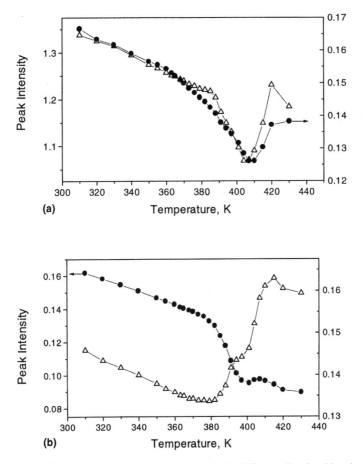

Figure 96. Peak intensity changes on temperature for different vibrational bands. (a) $\triangle$, 1777 cm^{-1}; $\bullet$, 1540, cm^{-1}. (b) $\bullet$, 1455 cm^{-1}; $\triangle$, 1407 cm^{-1}. [Reproduced with permission from *Molec. Cryst. Liq. Cryst.*, **263**, 293 (1995)]. Ref. [57].

Now we discuss the intensity and frequency of the IR bands as a function of the electric field. We find that the DC bias influences only the C=O bond. The intensity of the isolated benzene ring vibration remains constant when the field is increased, and all further less-isolated bands seem not to be influenced by the applied field, as far as can be detected. The peak intensity of the C=O (carbonyl group) as a function of the increasing electric field is shown in Figure 98. The most prominent feature is the sharp decrease of the peak intensity when the field is increased above about 10 V/μm. The same field strength was also found by Bock and Helfrich [11] for the phase transition inferred from a discontinuous change of the tilt angle, the

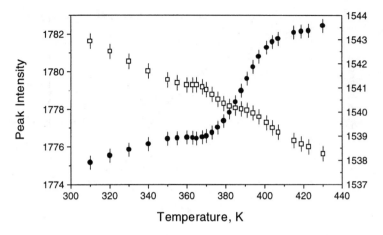

Figure 97. Frequency changes for the aromatic C–C and C=O stretch vibrations vs. temperature. ●, 1777, cm^{-1}; □, 1540 cm^{-1}.

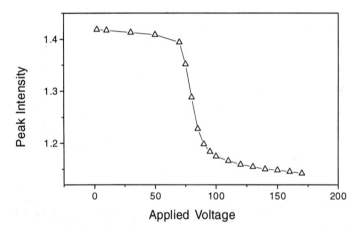

Figure 98. Dependence of the peak intensity of the C=O band on D.C. bias for D8m*10 at 371 K. [Reproduced with permission from *Molec. Cryst. Liq. Cryst.*, **263**, 293 (1995)]. Ref. [57].

spontaneous polarization, and the birefringence color. The IR absorption decreases at this field strength to about 80% of the low field value. Above and below 10 V/μm, the C=O absorption intensity is only weakly field dependent, decreasing continuously with increasing field strength. This weak but detectable continuous decrease of the C=O band intensity with increasing field strength within each of the two columnar phases supports suggestions by Rey-Lafon et al. [94] and of Samulski and Toriumi [139] regarding the flexibility of the alkyl chains in discotic liquid crystals. The ester group can change its relative orientation with respect to the ring without bringing about a change in the overall molecular orientation.

When the electric field is subsequently increased and decreased, we observe around 10 V/μm a hysteresis of the peak intensity versus the field strength. The field dependence of the C=O band intensity is similar at different temperatures, except that the width of the hysteresis decreases with increasing temperature (Fig. 99). This behavior is different from the behavior of the IR bands observed by Reynolds et al. [140] for some piezoelectric copolymers under applied electric field. This is probably because DLCs show more reversable behavior in electric field.

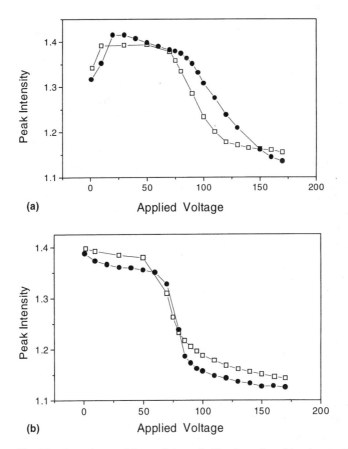

Figure 99. The dependence of the peak intensity for the carbonyl band on applied D.C. bias with increasing (□) and decreasing (●) voltages at (a) 363 K and (b) 376 K.

More specific information about the orientation of different moieties under application of static external electric field can be extracted if the IR absorbance is measured with polarized light as a function of the angle of polarization [93]. Its combination with time-resolved FTIR spectroscopy allows one to follow the molecular motion in the course of the ferroelectric switching [141,142]. Polarized and time-resolved FTIR measurements of D8m*10 have been performed [138].

To establish the segmental arrangements in the low and the high field phases and to elucidate the mechanism of the molecular reorientation, FTIR spectroscopy was employed to study the orientation of the molecular moieties in columnar ferroelectric liquid crystals under the influence of external static and alternating electric fields [138]. From this study, some additional information about low and high-field phases was obtained. We will review the main results.

The sample was oriented by shearing the liquid crystalline melt (at 110°C) between CaF$_2$ windows coated with an ITO layer. An additional SiO layer protected the ITO electrodes from short-circuits during the shearing. Polyethylenterephtalate spacers with a thickness of 2 μm were used to maintain a well-defined spacing between the electrodes.

The FTIR spectrometer (FTS-6000, BioRad) together with an IR microscope (UMA-500) made the measurements possible on the identical spot in the visible and in the infrared range. IR spectra were measured as a function of the polarizer rotation angle ω from 0° to 170° in steps of 10°. For each band under study, the positions of maximal and minimal (ω_{max} and ω_{min}) absorbance were calculated by the center gravity method [144]. The values ω_{max} of the different bands determine the average orientations of the corresponding transition moment component perpendicular to the IR-propagation. Therefore, the shifts of ω_{max} or ω_{min} that appear in response to the alternating external electric field reflect the angular excursions of the different segmental reorientations [139]. This shift delivers the apparent angle $\theta = \left(\omega_{max}^{(+E)} - \omega_{max}^{(-E)}\right)/2$ of the different molecular segments. The dichroic ratio R was calculated as $R = A(\omega_{max})/A(\omega_{min})$ for selected bands assigned to the carbonyl group and the alkyl tails. The A(ω_{min}) and A(ω_{max}) values are the heights of the absorption peaks at ω_{min} and ω_{max}, respectively.

Figure 100 shows that the polar plot for the absorbance function vs. angle of polarization for three different vibrational bands in the absence of DC bias. We can see a quite noticeable dichroism for C–C and C=O stretching vibrations, which allows us to conclude that the alignment achieved was more or less homogeneous planar (edge-on). As was shown for the H5T sample, the alignment on CaF$_2$ windows with GeO coating was hetero-geneous planar; however, the uniform alignment in the case of D8m*10 was

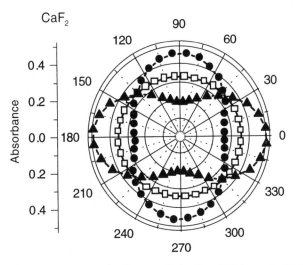

Figure 100. Dependence of absorption on polarization of IR beam for 2930 cm^{-1} ($\square$; CH$_2$), 1777 cm^{-1}($\triangle$; CO), and 1407 cm^{-1} ($\bullet$; core) bands for fresh sheared dibenzopyrene at $E=0$ V/μm. Absorption of the 1407 cm^{-1} band is multiplied by 6. Direction 0° corresponds to the shearing direction. [Reproduction with permission from authors of Ref. [138].

achieved by the application of the external electric field, which probably dominates the effect of the substrate structure.

The low field phase for D8m*10 is formed at a field strength $E \leq 10$ V/μm. Bistable ferroelectric switching was observed in this phase when the polarity of the external field was reversed. The different molecular orientations for positive and negative electric fields are demonstrated by the polar plots of the IR absorbance vs. polarizer angle A(ω) (Fig. 101). For the different bands, the polarizer angles for which maximum absorbance is obtained (ω_{min} or ω_{max}) are no longer zero. Furthermore ω_{max} or ω_{min} values thus obtained do not coincide with each other, neither for the positive nor for the negative voltage. The apparent angle (half of the field induced change in ω_{min} or ω_{max}) is maximal for the 1484 cm^{-1} core band ($\theta = 32°$), whereas it is minimal for the 2930 cm^{-1} band (CH$_2$ tails). The nonzero values for θ_{2924} and θ_{1484} provide proof that neither the average alkyl tail nor the core plane is perpendicular to the shearing direction anymore. The difference in these two θ values shows that the alkyl chains and the plane of the core are inclined to each other [142]. This result supports an earlier conjecture concerning tail deflection [11].

The apparent θ angles and the dichroic ratios for selected bands obtained from the polar plots are summarized in Table VIII (columns 3 and 4). From the table, one can observe that the θ values for the bands assigned to the core

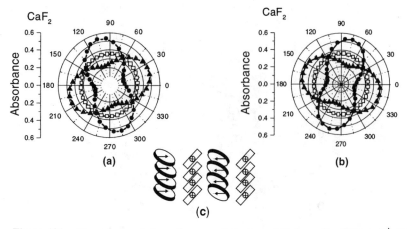

Figure 101. Dependence of absorption on polarization of IR beam for 2930 cm^{-1} ($\square$, CH$_2$), 1777 cm^{-1}($\triangle$, CO), and 1407 cm^{-1} ($\bullet$, core) bands for dibenzopyrene in the low field phase at (a) $E = +9$ V/μm and (b) $E = -9$ V/μm. (c) Proposed molecular arrangement. [Reproduction with permission from authors of Ref. [138].]

(1484, 1407, 1248, and 1116 cm^{-1}) differ from each other. As shown by Jakli et al. [144], the difference in apparent angles results from the non-cylindrical symmetry of the orientation distribution function of IR transition moments around the core axis. For of D8m*10, the molecular axis coincides with the normal to the plane of the core $\boldsymbol{n}$. Thus the rotation of the polar plots for core bands with a reversal of the field reflects a change in the position of core normal $\boldsymbol{n}$ and the biased rotation of the core around $\boldsymbol{n}$ (Fig. 101).

The low value of dichroism of the core band gives the evidence for large randomization of the cores in the low field phase. According to the model proposed by Bock and Helfrich [30], this can be related to a different inclination (with respect to electrodes plane) of the cores in the neighboring columns [Fig. 101(c)].

The transition to the high field phase is observed in an external electric field $E \geq 10$ V/μm. Polar plots of the IR absorbance vs. polarizer angle are shown in Figure 102 for positive and negative external electric fields. The apparent angles θ and dichroic ratios R for selected bands are given in Table VIII. One can conclude that the transition to the high field phase leads to an increase of the θ values for all molecular segments. This is in accordance with the earlier electro-optical observation of higher tilt angles in the high field phase [11,30]. Similar to the low field phase, the differences in the θ values for bands assigned to the alkyl (2930 and the 2859 cm^{-1}) and to the core reflect the deflection of alkyl chains with respect to the core plane. The noncoincidence of the θ values for the different core bands is again owing to the biased rotation of the core around its axis $\boldsymbol{n}$.

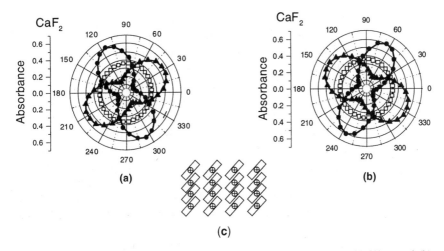

Figure 102. Same Figure 105 for the high field phase at (a) $E = +20$ V/μm and (b) $E = -20$ V/μm. (c) Proposed molecular arrangements. [Reproduction with permission from authors of Ref. [138].]

The dichroic ratios R for the different bands are considerably higher for the high field phase than for the low field phase, providing evidence that the core axis n is less inclined with respect to the electrode surface. The minimal dichroic ratio is observed for the 1777 and 1248 cm^{-1} bands. The orientational order parameter S of the core axis n can be estimated (in first approximation that orientation distribution function of n vectors has cylinder symmetry around n axis) from

$$S = \frac{R-1}{R+2} \cdot \frac{1}{0.5(3\cos^2\beta - 1)} \tag{35}$$

where β is the angle formed by core axis n and transition moment of the absorption band. Using $R = 0.16$ for 1248 cm^{-1} and assuming $\beta = 90°$, $S = 0.78$ as the minimum value of the order parameter. This high S value obtained for high field phase is not consistent with the model proposed earlier [30]. According to this model, the molecular core in high field phase is inclined to the plane of the electrodes as in the low field phase [Fig. 101(c)]. The phase transition is brought about only by a 180° rotation of the tilt direction of some of the columns [30]. That rotation should not affect the dichroic ratio. The strong increase of R values and the high value of S in the high field phase suggest that all molecular disks are nearly perpendicular to the electrodes [Fig. 102(c)]. As shown by Bock and Helfrich [30], the transition between molecular arrangements leads to the tripling in polarization that was observed experimentally. It should be mentioned that the model depicted in Fig. 102(c) was also considered among the possible models, but it was ruled out owing to observed faster switching in low field phase. The switching then turned out to be faster in the high but not the low field phase [137].

The segmental response in the columnar FLC with a change of the electric field polarity was investigated using time-resolved FTIR spectroscopy by Shilov et al. [138]. Fast and slow reorientation processes were observed for all molecular segments. The fast process is probably owing to a reorientation of the director with respect to the plane of the electrodes (electroclinic response). The slower process is the result of the reorientation of the director along a cone (ferroelectric response). The results are in agreement with optical measurements. The lower absolute values of tilt angles for the carbonyl band is related to the hindered rotation of the carbonyl groups.

Shilov et al. [138] monitored the evolution of the IR bands in course of the electric field–induced reorientation (using time-resolved technique), and noted that the reorientation process is divided into three steps: a fast initial response followed by a slowing down of the reorientation, which is followed by an acceleration of the reorientation. For the high field phase, the fast process after reversal of the electric field can be assigned to the reorientation of the core normal n lying in a plane parallel to electrodes by a few degrees (electroclinic-like response). In contrast, the succeeding reorientation is described by a rotation of n on a cone (ferroelectric-like response).

SUMMARY

We reviewed our work on the alignment, the order parameter, and the structure of a range of DLCs using FTIR spectroscopy. Some of the literature concerning this work was also referenced. For the analysis of these problems, a number of substrates: crystalline ($ZnSe$, MgF_2, CaF_2, BaF_2), polycrystalline ($ZnSe$, ZnS) and amorphous (Si), were initially used without a surface treatment and then coated by a polymer (nylon 6,6, PVA, polyimide), ITO, SiO, or GeO at oblique evaporation. Discotic liquid crystals with different discotic phase structures have also been used. These exhibit phases such as helical (H) hexagonal-ordered (D_{ho}), hexagonal-disordered (D_{hd}), rectangular-disordered (D_{rd}), tilted (D_t) and obliquely ($D_{oblique}$) ordered.

It was established that three different factors influence the alignment of discotic LCs (which are similar to those for the calamitic liquid crystals) on the surfaces of the pure substrate: a topology of the surface, a symmetry of the surfaces (whether or not there is an easy axis of orientation), and the structure of the liquid crystal itself. To fully investigate the problem, the discotic liquid crystal H5T was extensively used, although other DLC materials with different substrates were also studied. H5T was selected because the D_{ho} phase possesses the best alignment of the other known discotic liquid crystals.

It was shown that the alignment of H5T on the surface of a pure substrate is planar (edge-on). The stability of this alignment strongly depends on the type of the alignment: homogeneous-planar or heterogeneous-planar. The alignment will be homogeneous for the substrates, which have a groove-like structure or a preferable orientation of the easy axis. The homogeneous-planar alignment of H5T and of H6TT was found for the surfaces of ZnS and ZnSe and heterogeneous-planar for all other investigated surfaces. It was also shown that the stability of alignment depends on the thickness of the LC film in the range studied (3 to 17 μm). Orientational transition from the edge-on to the side-on alignment was observed for the cells with a hetero-geneous planar alignment and for thinner cells, during the heating cycle.

The alignment of DLCs under the influence of the external fields (electric or magnetic) depends on the direction of these fields and on the conditions of their application.

The alignment of H5T (and of many other discotic materials) on the polymer-coated surfaces is mainly side-on, except if the surface roughness is quite large (~ 1 μm). In the latter case, the alignment observed is edge-on. This alignment however, is not stable, and the orientational transition from edge-on to side-on is observed with an increase in temperature. This transition is not instantaneous and sharp but takes a definite duration of time, depending on the temperature. It also depends on the history of the LC cell, whether it was heated to its isotropic phase and then cooled to its discotic phase.

Note that for the liquid crystals with highly ordered phases (H and D_{ho}), the edge-on alignment is quite stable, moreover the phase transition to the next phase (D_{hd} or isotropic) is sharp (closer to a first-order transition). For all other discotic phases, the transition between the phases is discontinuous. These findings are in agreement with the recent quantum chemical structural calculations that show that for H and D_{ho} phases the packing of molecules is different (and involves a lower distance in between the neighboring molecules and consequently a lower energy). For all other structures, the distances between the molecules are large (and not even uniform as for the disordered phases). This will probably cause an increase in the heterogeneity due to an increase in the molecular fluctuations with an increase in temperature. The temperature dependence of the alignment depends on the structure of the mesophase.

It was found that the method of the sample filling strongly influences the alignment of discotic liquid crystals. The side-on alignment on the polymer-coated substrate is much more uniform using the shearing method than the capillary method of cell filling.

Three methods were suggested for determining the order parameter of discotic liquid crystals using IR spectroscopy: a comparison of the intensity of the characteristic vibrational bands in the discotic and the isotropic phases, a comparison of the intensity of the characteristic bands for the oriented film in discotic phase with the intensity of these bands in KBr mull (at the same temperature), and a combination of the polarized and sample tilt IR transmission measurements in a discotic phase. The latter method, in our view, has the greatest advantage because it allows the measurements to be restricted in the phase of interest and does not require a heating of the sample to its isotropic phase.

The method of the sample tilt that was developed for an investigation of the DLCs allows one to obtain a more precise assignment of some of the vibrational bands, the assignment of which cannot otherwise be easily determined.

Using the temperature dependence of the infrared spectrum of the ferroelectric discotic liquid crystal D8m*10, we can extract valuable information about the orientation of different parts of molecules in the discotic phase and their behavior with the electric field. The field-induced phase transition is clearly shown. The field-induced phase transition between the two ferroelectric phases of this compound does not affect the quadrupolar order of the aromatic cores, whereas the order of one of the chiral dipolar chain heads increases considerably when passing from the low field to the high field phase. This suggests that the transition implies only 180° rotations of columnar tilt directions, and that the conformation of the dipolar chiral groups is affected considerably by these rotations.

FTIR spectroscopy with polarized light allows one to elucidate the static arrangements in ferroelectric switchable columnar liquid crystals. It was confirmed that in the low field phase, the normal of the molecular disks is inclined to the plane of the electrodes. It was established that in the high field phase this inclination of the core normals vanishes. In both phases, the alkyl tails are deflected with respect to the core planes.

Time-resolved FTIR spectroscopy makes it possible to follow the reorientation dynamics under the influence of an alternating external electric field with a time resolution of 5 μs. It was detected that in both the high and the low field phase the reorientation from the first to the second state passes three regions: a fast initial response, a slowing down of the reorientation rate, and finally a second acceleration. It was shown for the fast reorientation in the high field phase that immediately after reversal of the external electric field the core normal n reorients in a plane parallel to the electrodes by a few degrees (electroclinic-like response). The final reorientation to the second surface stabilized state is characterized by a rotation of n on a cone around the column axis (ferroelectric-like response). A similar switching behavior is found in the low field phase.

Acknowledgments

We most sincerely thank H. Ringsdorf (Mainz Universität, Germany) and H. Bock (CNRS Pessac, France) for collaborations and discussions and for giving us most of the samples for these investigations. S. Kumar of the Centre of Liquid Crystal Research, Bangalore, India, and A. Akopova, Liquid Crystal Research Laboratory, Ivanovo University, Russia, are thanked for giving us some of the triphenylene derivatives. F. Kremer, G. Kruk, L. Blinov, S. Shilov, and G. Anan'eva are thanked for useful discussions. Sergey Tsvetkov is thanked for the IR measurements and technical assistance during manuscript preparation. The work was supported by two EU grants SCI*0291 and HCM CHRX-CT930353 (and PECO CIPD-CT94-0616) by the Science and the Physics Committees, respectively. Enterprise Ireland is also thanked for their award of small grants.

APPENDIX A

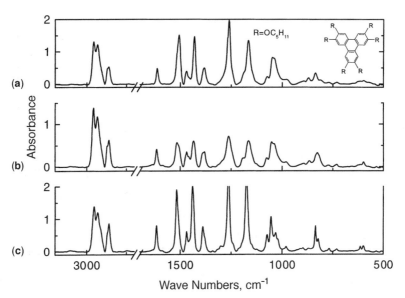

Figure A1.   IR spectra of H5T in the (a) isotropic (384 K), (b) discotic (360 K), and (c) crystalline (320 K) phases.

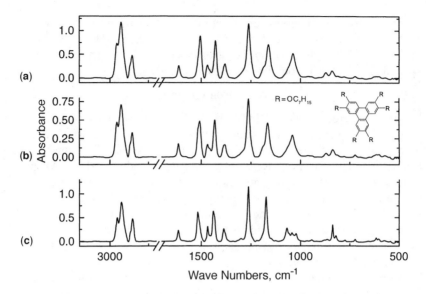

Figure A2. IR spectra of H7T in the (a) isotropic (420 K), (b) discotic (360 K), and (c) crystalline (320 K) phases.

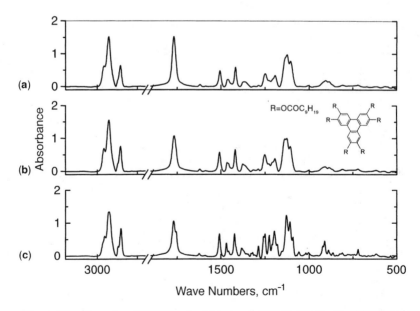

Figure A3. IR spectra of H10OT in the (a) isotropic (420 K), (b) discotic (365 K), and (c) crystalline (320 K) phases.

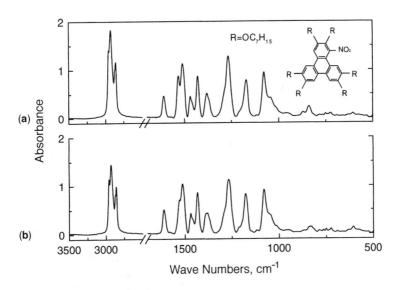

Figure A4.   IR spectra of H7T-NO$_2$ in the (a) isotropic (420 K) and (b) discotic (350 K) phases.

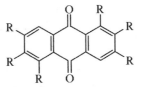

Figure A5.   Structure of hexa-decanoyloxy-rufigallol.   R=OCOC$_9$H$_{19}$ (Cr-382-D$_{rect}$-401-I). Ref. [31].

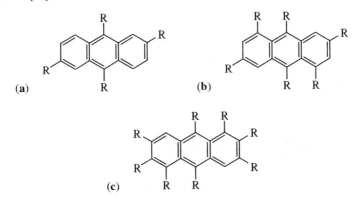

Figure A6.   Structures of anthracene-based DLCs. (a) R=OCOC$_9$H$_{19}$ (Cr-394-I) (b) R= OCOC$_9$H$_{19}$ (Cr-396-D$_t$-425-I) (c) R= OCOC$_6$H$_{13}$ (Cr-373-D$_t$-420-I). Ref. [32].

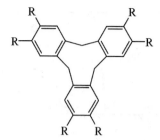

Figure A7.   Structure of hexa-*O*-decanoylglucopyranose. R=OCOC$_9$H$_{19}$ (Cr-306-K-D$_h$-419-K-I).

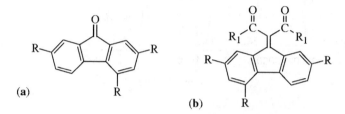

Figure A8.   Structure of trinitrofluorenone derivatives. (a) R=NO$_2$ (TNF). (b) R=NO$_2$ R1=OC$_{16}$H$_{33}$ (TNF16MB).

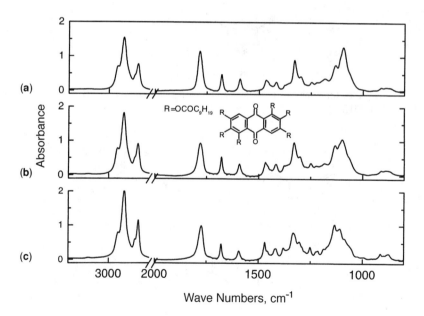

Figure A9.   IR spectra of RHC10 in the (a) isotropic (410 K), (b) discotic (378 K), and (c) crystalline (330 K) phases.

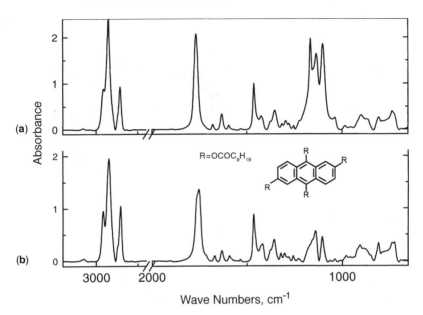

Figure A10. IR spectra of the anthracene-based discotic molecule A4n10 in the (a) isotropic (405 K) and (b) crystalline (350 K) phases.

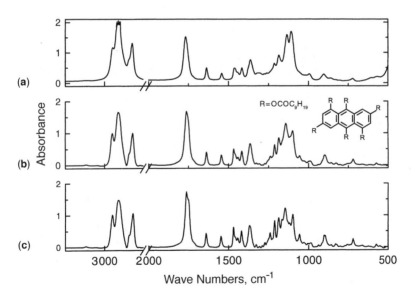

Figure A11. IR spectra of the anthracene-based DLC A6n10 in the (a) isotropic phase (395 K), (b) discotic phase (363 K), and (c) crystalline phase (310 K).

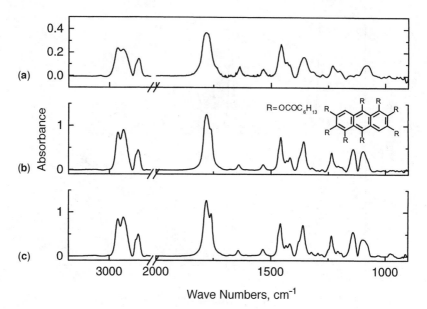

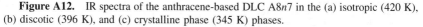

Figure A12. IR spectra of the anthracene-based DLC A8n7 in the (a) isotropic (420 K), (b) discotic (396 K), and (c) crystalline phase (345 K) phases.

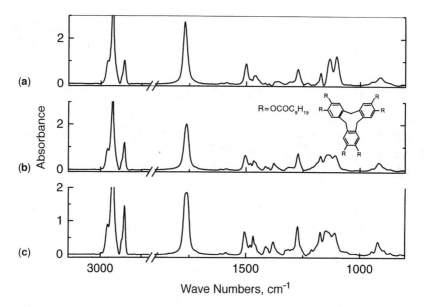

Figure A13. IR spectra of hexa-*O*-decanoylglucopyranose in the (a) isotropic (420 K), (b) discotic (396 K), and (c) crystalline (345 K) phases.

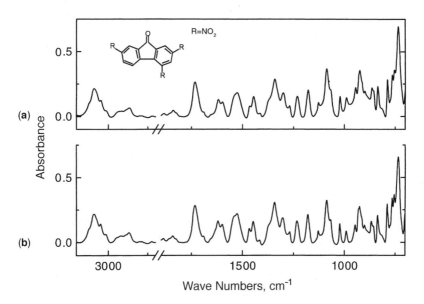

Figure A14. IR spectra of TNF in the (a) crystalline (320 K) and (b) isotropic (400 K) phases.

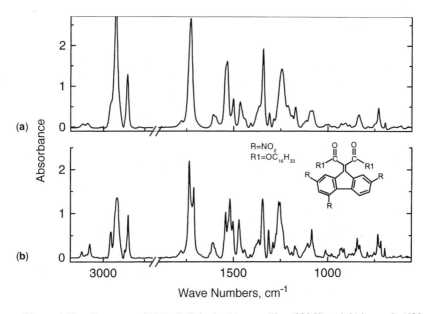

Figure A15. IR spectra of TNF16MB in the (a) crystalline (320 K) and (b) isotropic (400 K) phases.

References

1. S. Chandrasekhar, B. K. Sadashiva, and K. A. Suresh, *Pramana* **9**, 471 (1977).

2. S. Chandrasekhar, *Liquid Crystals*, 2nd eds., Cambridge University Press, Cambridge, UK, 1992.

3. D. Adam, P. Schuhmacher, J. Simmerer, L. Häussling, K. Siemensmeyer, K. H. Etzbach, H. Ringsdorf, and D. Haarer, *Nature* **371**, 141 (1994).

4. S. Jin, T. H. Tiefel, R. Wolfe, R. C. Sherwood, and J. J. Mottine, *Science* **255**, 446 (1992); N. C Maliszewskyj, P. A. Heiney, J. Y. Josefowicz, J. P. McCauley Jr., and A. B. Smith III, *Science* **264**, 77 (1994).

5. N. Boden, R. Bissell, J. Clements, and B. Movaghar, *Curr. Sci.* **71**, 599 (1996); N. Boden, R. J. Bushby, J. Clements, B. Movaghar, K. J. Donovan, and T. Kreouzis, *Phys. Rev. B* **52**, 13274 (1995).

6. J. W. Goodby, personal communication.

7. W. Kreuder and H. Ringsdorf, *Makromol. Chem. Rapid. Commun.* **4**, 807 (1983); W. Kreuder, H. Ringsdorf, and T. Tschirner, *Makromol. Chem. Rapid. Commun.*, **6**, 367 (1985).

8. H. Ringsdorf, R. Wüstfeld, E. Zerta, M. Ebert, and J. H. Wendorff, *Angew. Chem. Int. Edn.* **28**, 914 (1989).

9. H. Ringsdorf and R. Wüstfeld, *Phi. Trans. R. Soc. London* **A330**, 95 (1990).

10. J. Prost, in N. Boccara, ed., Symmetries and Broken Symmetries, *Proc. of the Colloque Pierre Curie*, IDSET, Paris, 1981, p. 159.

11. H. Bock and W. Helfrich, *Liq. Cryst.* **12**, 703 (1992).

12. N. Boden, R. J. Bushby, and A. N. Cammidge, *Liq. Cryst.* **18**, 673 (1995).

13. K. Praefcke and J. D. Holbrey, *J. Inclusion Phen. Molec. Recogn. Chem.* **24**, 19 (1996).

14. F. Charra and J. Cousty, *Phys. Rev. Lett.* **80** 1685 (1998).

15. H.-G. Kuball, *Liq. Cryst. Today* **9**, 1 (1999).

16. J. D. Lister, R. J. Birgeneau, M. Kaplan, C. R. Safinya, and J. Als-Nielsen, in T. Risle, ed., *Ordering in Strongly Fluctuating Condensed Matter Physics*, Plenum Press, New York, p. 357 (1980).

17. V. D. Neff, L. W. Gulrich, and G. H. Brown, *Mole. Cryst.* **1**, 225 (1966).

18. E. Hild, A. Kocot, J. K. Vij, and R. Zentel, *Liq. Cryst.* **16**, 783 (1994).

19. T. Perova, A. Kocot, and J. K. Vij, *Mol. Cryst. Liq. Cryst.* **301**, 111 (1997).

20. W. G. Jang, C. S. Park, J. E. Maclennan, K. H. Kim, N. A. Clark, *Ferroelectrics* **180**, 213 (1996).

21. A. Kocot, R. Wrzalik, and J. K. Vij, *Liq. Cryst.* **21**, 147 (1996).

22. P. G. Schmidt, *J. Polym. Sci. Part A* **1**, 1271 (1963).

23. J. L. Koenig, S. W. Cornell, and D. E. Witenhafer, *J. Polym. Sci. Part A* **2**, 301 (1967).

24. H. Bengs, O. Karthaus, H. Ringsdorf, C. Baehr, M. Ebert, and J. H. Wendorff, *Liq. Cryst.* **10**, 161 (1991).

25. N. Boden, R. C. Borner, R. J. Bushby, A. N. Cammidge, and M. V. Jesudason, *Liq. Cryst.* **15**, 851 (1993).

26. W. K. Lee, P. A. Heiney, J. P. McCayley, and A. B. Smith III, *Molec. Cryst. Liq. Cryst.* **198**, 273 (1991).

27. B. Kohne, W. Poules, and K. Praefcke, *Chem. Ztg.* **108**, 113 (1984).

28. N. Boden, R. J. Bushby, and A. N. Cammidge, *Molec. Cryst. Liq. Cryst.* **260**, 307 (1995).

29. C. Destrade, J. Malthete, T. H. Nguyen, and H. Gasparoux, *Phys. Lett. A* **78**, 82 (1980).

30. H. Bock and W. Helfrich, *Liq. Cryst.* **18**, 387 (1995).

31. C. Carfagna, A. Roviello, and A. Sirigu, *Molec. Cryst. Liq. Cryst.* **122**, 151 (1985).

32. H. Bock, *Elektrooptische effekte mit kolumnaren flussigkristallen,* Ph.D. Thesis, Freie Universität, Berlin, 1994.

33. R. G. Zimmermann, G. B. Jameson, R. G. Weiss, and G. Demailly, *Molec. Cryst. Liq. Cryst. Lett.* **1**, 183 (1985).

34. J. Cognard, *Molec. Cryst. Liq. Cryst. Suppl. Series* **A5**, 1 (1982).

35. B. Jerome, *Rep. Prog. Phys.* **54**, 391 (1991).

36. L. M. Blinov and V. G. Chigrinov, *Electrooptic Effects in Liquid Crystal Materials*, Spring-Verlag, New York, 1994.

37. A. A. Sonin, *The Surface Physics of Liquid Crystals*, Gordon & Breach, Amsterdam, 1995.

38. C. Vauchier, A. Zann, P. Le. Barny, J. C. Dubois, and J. Billard, *Molec. Cryst. Liq. Cryst.* **66**, 103 (1981).

39. J. C. Dubois, M. Hareng, S. LeBerre, and J. N. Perbet, *Appl. Phys. Lett.* **38**, 11 (1981).

40. T. J. Phillips, J. C. Jones, and D. G. McDonnell, *Liq. Cryst.* **15**, 203 (1993).

41. T. Warmerdam, D. Frenkel, and R. J. J. Zijlstra, *J. Phys. France* **48**, 319 (1987).

42. G. Kruk, J. K. Vij, O. Karthaus, and H. Ringsdorf, *Supramol. Sci.* **2**, 51 (1995).

43. B. Mourey, J. N. Perbet, M. Hareng, and S. Le Berre, *Molec. Cryst. Liq. Cryst.* **84**, 193 (1982).

44. V. A. Raghunathan, N. V. Madhusudana, and S. Chandrasekhar, *Molec. Cryst. Liq. Cryst.* **148**, 77 (1987).

45. F. Grandjean, *Bull. Mineral.* **39**, 164 (1916).

46. J. Billard, *Bull. Mineral.* **103**, 444 (1980).

47. O. Albrecht, W. Cumming, W. Kreuder, A. Laschensky, and H. Ringsdorf, *Colloid Polym. Sci.* **264**, 659 (1986).

48. T. S. Perova and J. K. Vij, *Adv. Mater.* **7**, 919 (1995).

49. T. S. Perova, J. K. Vij, and A. Kocot, *Europhys. Lett.* **44**, 198 (1998).

50. S. E. Tsvetkov, T. S. Perova, J. K. Vij, D. Simpson, S. Kumar, and F. Vladimirov, *Molec. Mater.* **11**, 267 (1999).

51. A. M. Levelut, F. Hardouin, H. Gasparoux, C. Destrade, and N. H. Tinh, *J. Phys.* **42**, 147 (1981).

52. D. Goldfarb, Z. Luz, and H. Zimmermann, *J. Phys. Paris* **42**, 1303 (1981).

53. S. Ikeda, Y. Takanishi, K. Ishikawa, and H. Takezoe, *Molec. Cryst. Liq. Cryst.* **329**, 1201 (1999).

54. J. T. Mang, S. Kumar, and B. Hammouda, *Europhys. Lett.* **28**, 489 (1994).

55. B. Hammouda, J. T. Mang, and S. Kumar, *Phys. Rev. E* **51**, 6282 (1995).

56. L. J. Chiang, C. R. Safinya, N. A. Clark, K. S. Liang, and A. N. Bloch, *J. Chem. Soc. Chem. Commun.* 695 (1985).

57. T. S. Perova, J. K. Vij, and H. Bock, *Molec. Cryst. Liq. Cryst.* **263**, 293 (1995).

58. G. Kruk, A. Kocot, R. Wrzalik, J. K. Vij, O. Karthaus, and H. Ringsdorf, *Liq. Cryst.* **14**, 807 (1993).

59. P.-A. Chollet, *Thin Solid Films* **52**, 343 (1978).

60. H. Nakahara and K. Fukuda, J. *Colloid Interface Sci.* **69**, 24 (1979).

61. F. Kaneko, O. Shirai, H. Miyamoto, M. Kobayashi, and M. Suzuki, *J. Phys. Chem.* **96**, 10554, (1992); and *J. Phys. Chem.* **98**, 2185 (1994).

62. T. S. Perova, A. Kocot, and J. K. Vij, *Supramol. Sci.* **4**, 529 (1997).

63. X. Yang, S. A. Nitzsche, S. L. Hsu, D. Collard, R. Thakur, C. P. Lillya, and H. D. Stidman, *Macromolecules* **22**, 2611 (1989).

64. M. Kardan, B. B. Reinhold, S. L. Hsu, R. Thakur, and C. P. Lillya, *Macromolecules* **19**, 616 (1986).

65. X. Yang, D. A. Waldman, S. L. Hsu, S. A. Nitzsche, R. Thakur, D. M. Collard, C. P. Lillya, and H. D. Stidman, *J. Chem. Phys.* **89**, 5950 (1988).

66. M. Kardan, A. Kaito, S. L. Hsu, R. Thakur, and C. P. Lillya, *J. Phys. Chem.* **91**, 1809 (1987).

67. X. Yang, M. Kardan, S. L. Hsu, D. Collard, R. B. Heath and C. P. Lillya, *J. Phys. Chem.* **92**, 196 (1988).

68. S. Chandrasekhar, B. K. Sadashiva, K. A. Suresh, N. V. Madhusudana, S. Kumar, R. Shashidar, and G. Venkatesh, *J. Phys. Paris* **40**, C3 (1979).

69. J. H. Schachtshneider and R. J. Snyder, *Spectrochim. Acta* **19**, 117 (1963).

70. R. J. Snyder and J. H. Schachtshneider, *Spectrochim. Acta* **19**, 85 (1963).

71. R. S. Stein, and G. B. B. M. Sutherland, *J. Chem. Phys.* **22**, 1993 (1954).

72. M. Tasumi and S. Krimm, *J. Chem. Phys.* **46**, 755 (1967).

73. M. Tasumi and T. J. Shimanouchi, *J. Chem. Phys.* **43**, 1245 (1965).

74. R. G. Snyder, *J. Chem. Phys.*, **47**, 1316 (1967).

75. R. G. Snyder, S. L. Hsu, and S. Krimm, *Spectrochim. Acta Part A* **34A**, 395 (1978).

76. M. Pesquaer, M. Cotrait, P. Monson, and V. Volpilbac, *J. Phys. Paris* **41**, 1039 (1980).

77. G. Varsányi, *Vibrational Spectra of Benzene Derivatives*, Academic Press, New York, 1969.

78. R. E. Goozner and M. M. Labes, *Mol. Cryst. Lett.* **56**, 75 (1979).

79. S. Chandrasekhar, in G. H. Brown, ed., *Advances in Liquid Crystals* Vol. 5, Academic Press, 1982, p. 47.

80. M. Sorai and H. Suga, *Molec. Cryst. Liq. Cryst.* **73**, 47 (1981).

81. M. Sorai, H. Yoshioka, and H. Suga, *Molec. Cryst. Liq. Cryst.* **84**, 39 (1982).

82. C. Destrade, P. Foucher, H. Gasparoux, N. H. Tinh, A. M. Levelut, and J. Malthete, *Mol. Cryst. Liq. Cryst.* **106**, 121 (1984).

83. A. Abe, *J. Am. Chem. Soc.* **106**, 14 (1984).

84. M. Cotrait, P. Marsau, M. Pesquer, and V. Volpilhac, *J. Phys. Paris* **43**, 355 (1982).

85. E. I. Kats and M. I. Monastyrsky, *J. Phys. Paris* **45,** 709 (1984).

86. N. Kirov and P. Simova, *Vibrational Spectroscopy of Liquid Crystals*, Bulgarian Academy of Sciences, Sofia, 1984.

87. L. J. Bellamy, *The Infrared Spectra of Complex Molecules*, Chapman & Hall, Cambridge, 1975.

88. R. M. Sirverstein, G. C. Bassler, and T. C. Morill, *Spectrometric Identification of Organic Compounds*, John Wiley & Sons, New York, 1981.

89. F. Scheinmann, *An Introduction to Spectroscopic Methods for the Identification of Organic Compounds*, Vol. 1, Pergamon Press, New York, 1974.

90. D. Lin-Vien, N. B. Colthup, W. G. Fateley, and J. G. Graselli, *Infrared and Raman Characteristic Frequencies of Organic Molecules*, Academic Press, London, 1991.

91. J. A. C. Veerman and D. Frenkel, *Phys. Rev. A* **45**, 5632 (1992).

92. S. Chandrasekhar and G. S. Ranganath, *Rep. Prog. Phys.* **53**, 59 (1990).

93. A. Kocot, G. Kruk, R. Wrzalik, and J. K. Vij, *Liq. Cryst.* **12**, 1005 (1992).

94. M. Rey-Lafon, C. Destrade, and A. T. Hemida, *Molec. Cryst. Liq. Cryst.* **137**, 381 (1986).

95. T. Wang, D. Yan, E. Zhou, O. Karthaus, and H. Ringsdorf, *Polymer* **39**, 4509 (1998).

96. B. Orgasińska, T. S. Perova, K. Merkel, A. Kocot, and J. K. Vij, *J. Material Sci. Engin. C* **293**, 8–9 (1999).

97. B. Orgasińska, A. Kocot, K. Merkel, R. Wrzalik, J. Ziolo, T. Perova, and J. K. Vij, *J. Molec. Struct.* 511–512, 271 (1999).

98. J. M. Geary, J. W. Goodby, A. M. Kmetz, and J. S. Patel, *J. Appl. Phys.* **62**, 4100 (1987).

99. D. W. Berreman, *Phys. Rev. Lett.* **28**, 1683 (1972).

100. G. V. Anan'eva, personal communication.

101. T. Koyama, T. Yodo, and K. Yamashita, *J. Cryst. Growth.* **94**, 1 (1989).

102. G. D. Anan'eva, A. A. Dunaev, and T. I. Merkulyaeva, *Russ. J. High Purity Compounds* **4**, 114 (1995).

103. W. Kranig, C. Boeffel, and H. W. Spiess, *Macromolecules* **23**, 4061 (1990).

104. E. Dubois-Violette and P. G. de Gennes, *J. Phys. (Paris) Lett.* **36**, L255 (1975).

105. E. I. Kats, *Zh. Eksper. Theoret. Fiz.* **70**, 1394 (1976).

106. G. Ryschenkov and M. Kleman, *J. Chem. Phys.*, **64**, 404 (1976); and *J. Chem. Phys.* **64**, 413 (1976).

107. L. M. Blinov and A. A. Sonin, *Mol. Cryst. Liq. Cryst.* **179**, 13 (1990).

108. E. Fontes, P. A. Heiney, M. Ohba, J. N. Haseltine, and A. B. Smith III, *Phys. Rev.* **A37**, 1329 (1988).

109. P. A. Heiney, E. Fontes, W. H. de Jeu, A. Riera, P. Carroll, and A. B. Smith III, *J. Phys. Paris*, **50**, 461 (1989).

110. R. J. Dong, *Nuclear Magnetic Resonance of Liquid crystals*, 2nd ed., Springer-Verlag, New York, 1997.

111. X. Shen, R. Y. Dong, N. Boden, R. J. Bushby, P. S. Martin, and A. Wood, *J. Chem. Phys.* **108**, 4324 (1998).

112. D. Markovitsi, I. Lecuyer, P. Lianos, and J. Malthete, *J. Chem. Soc. Far. Trans.* **87**, 1785 (1991).

113. S. Marquet, D. Markovitsi, P. Millie, H. Sigal, and S. Kumar, *J. Phys. Chem.* **B102**, 4697 (1998).

114. D. Naunsgaard, M. Larsen, N. Harrit, J. Frederiksen, R. Wilbrandt, and H. Stapelfeldt, *J. Chem. Soc. Far. Trans.* **93**, 1893 (1997).

115. T. S. Perova, S. Tsvetkov, J. K. Vij, and S. Kumar, *Molec. Cryst. Liq. Cryst.* (2000), in press.

116. Ger. Offen. DE 3.346.980 (C1. C07C149/36) 04 July 1985, K. Praefcke, B. Kohne, W. Poules, and E. Poetsch, (to Merck G.m.b.H.).

117. E. F. Gramsbergen, H. J. Hoving, W. H. de Jeu, K. Praefcke, and B. Kohne, *Liq. Cryst.* **1**, 397 (1986).

118. A. M. Levelut, *J. Phys. Paris* **40**, L-81 (1979).

119. A. M. Levelut, P. Oswald, A. Ghanem, and J. Malthete, *J. Phys. Paris* **45**, 745 (1984).

120. A. M. Levelut, J. Malthete, and A. Collet, *J. Physique, Paris* **47**, 351 (1986).

121. D. G. Cameron, H. L. Casal, H. H. Mantsch, Y. Boulanger, and I. C. P. Smith, *Biophys. J.,* **35**, 1 (1981).

122. J. Umemura, D. G. Cameron, and H. H. Mantsch, *Biochem. Biophys. Acta* **602**, 32 (1980).

123. I. M. Asher and I. W. Levin, *Biochem. Biophys. Acta* **468**, 63 (1977).

124. P. Etchegoin, *Phys. Rev. E* **56**, 538 (1997).

125. K. Praefcke, A. Eckert, and D. Blunk, *Liq. Cryst.* **22**, 113 (1997).

126. N. Boden, R. J. Bushby, A. N. Cammidge, S. Duckworth, and G. Headdock, *J. Mater. Chem.* **7**, 601 (1997).

127. K. Praefcke, W. Poules, B. Scheuble, R. Poupko, and Z. Luz, *Z. Naturforsch.* **39b**, 950 (1984).

128. C. Destrade, M. C. Mondon, and J. Malthete, *J. Phys.* **40**, C3-17 (1979).

129. B. Glusen, W. Heitz, A. Kettner, and J. H. Wendorff, *Liq. Cryst.* **20**, 627 (1996).

130. W. K. Lee, P. A. Heiney, M. Ohba, J. N. Haseltine, and A. B. Smith III, *Liq. Cryst.* **8**, 839 (1990).

131. C. Destrade, P. Fouchier, J. Malthete, and N. H. Tinh, *Phys. Lett.* **A88**, 187 (1982).

132. W. K. Lee, B. A. Winter, E. Fontes, P. A. Heiney, M. Ohba, J. N. Haseltine, and A. B. Smith III, *Liq. Cryst.* **3**, 87 (1989).

133. T. W. Warmerdam, R. J. M. Nolte, W. Drenth, J. C. van Miltenburg, D. Frenkel, and R. J. J. Zijlstra, *Liq. Cryst.* **2**, 1087 (1988).

134. D. G. Cameron, H. L. Casal, and H. H. Mantsch, *Biochemistry* **19**, 3665 (1980).

135. G. Scherowsky and X. H. Chen, *Liq. Cryst.* **17**, 803 (1994).

136. G. Heppke, D. Lötzsch, M. Müller, and H. Sawade, paper presented at the 6th Int. Conf. on Ferroelectric Liq. Cryst., Brest, 1997.

137. G. Heppke, D. Krüerke, M. Müller, and H. Bock, *Ferroelectrics*, **179**, 203 (1996).

138. S. V. Shilov, M. Müller, D. Krüerke, H. Skupin, G. Heppke, and F. Kremer, *Phys. Rev. E,* (to be published).

139. E. T. Samulski and H. Toriumi, *J. Chem. Phys.* **79**, 5194 (1983).

140. N. M. Reynolds, K. J. Kim, C. Chang, and S. L. Hsu, *Macromolecules* **22**, 1092 (1989).

141. F. Hide, N. A. Clark, K. Nito, A. Yasuda, and D. M. Walba., *Phys. Rev. Lett.* **75**, 2344 (1995).

142. S. V. Shilov, H. Skupin, F. Kremer, T. Wittig, and R. Zentel., *Phys. Rev. Lett.* **79**, 1686 (1997).

143. D. G. Cameron, J. K. Kauppinen, D. L. Moffatt, and H. H. Mantsch, *Appl. Spectrosc.* **36**, 245 (1982).

144. A. Jákli, M. Müller, D. Krüerke, and G. Heppke, *Liq. Cryst.* **24**, 467 (1998).

ROTATIONAL DIFFUSION AND DIELECTRIC RELAXATION IN NEMATIC LIQUID CRYSTALS

WILLIAM T. COFFEY

Department of Electronic and Electrical Engineering, Trinity College, Dublin, Ireland

YURI P. KALMYKOV*

Institute of Radio Engineering and Electronics of the Russian Academy of Sciences, Moscow Region, Russia

CONTENTS

*Present address: Centre d'Études Fondamentales, Université de Perpignan, 52 Avenue de Villeneuve, 66860 Perpignan Cedex, France

Advances in Liquid Crystals: A Special Volume of Advances in Chemical Physics, Volume 113, edited by Jagdish K. Vij. Series Editors I. Prigogine and Stuart A. Rice.
ISBN 0-471-18083-1.   © 2000 John Wiley & Sons, Inc.

I. INTRODUCTION

Dielectric relaxation processes in liquid crystalline materials are determined by structural properties, intermolecular interactions, and reorientations of the molecules. The nematic liquid crystal (LC) mesophase has the simplest dielectric behavior and the advantage that qualitative results obtained for nematics also provide, to some extent, a basis for understanding the molecular dynamics of other mesophases, such as ferroelectric or antiferroelectric smectic. The complex dielectric permittivity of a nematic liquid crystal has different dispersion regions and qualitative behavior in the parallel ($\mathbf{E} \| \mathbf{n}$) and perpendicular ($\mathbf{E} \perp \mathbf{n}$) alignments ($\mathbf{E}$ is the measuring AC field and $\mathbf{n}$ is the director of the nematic). The transverse component of the dielectric susceptibility tensor always exhibits the Debye dispersion (owing to reorientation of the dipolar molecules), which usually occurs for liquid systems in the microwave region. The longitudinal component may also exhibit an additional dispersion at radio frequencies (the low frequency region). The origin of these dispersion regions is, in general, the result of hindered reorientations of the molecules about the long and short molecular axes in the strong orientational forces of the nematic phase. If there are additional dipolar groups of the molecules with their own internal degrees of rotation, the corresponding dispersion regions will again be similar to those in the isotropic liquid phase.

Dielectric relaxation in nematic liquid crystals is usually interpreted in the context of a model of noninertial rotational Brownian motion of a particle in a mean field potential field U [1,2], although the mean field approximation has a restricted area of applicability, because it ignores local order effects. Despite this drawback, the model is easily visualized and allows us to carry out quantitative evaluations of the dielectric parameters of nematics. The usual approach to the problem is to solve the Fokker-Planck equation for the distribution function W of the orientation of a unit vector $\mathbf{u}$, fixed in the molecule [3]:

$$2\tau_D \frac{\partial}{\partial t} W = \nabla_\Omega^2 W + \frac{1}{kT} \mathrm{div}(W \mathrm{grad} U) \qquad (1)$$

where ∇_Ω^2 is the Laplace operator in angular variables, k is the Boltzmann constant, T is the temperature, $\tau_D = (2D_\perp)^{-1}$ is the orientational relaxation time in the isotropic phase, $D_\perp$ is the rotational diffusion coefficient with respect to the axis that is perpendicular to the axis of symmetry of the molecule. Eq. (1) can be formally solved by the separation of variables method, which reduces the solution to a Sturm-Liouville problem [4]. An alternative method is to expand the distribution function W in spherical harmonics [5]. The problem is then reduced to the solution of infinite-dimensional systems of linear differential-recurrence equations for the statistical moments (averaged spherical harmonics), which may be written in the matrix form:

$$\frac{d}{dt}\mathbf{X}(t) = \mathbf{A}\mathbf{X}(t) \tag{2}$$

Various methods for the solution of Eqs. (1) and (2) have been discussed [3, 4,6,7].

The basic model for the calculation of the components of the complex dielectric permittivity tensor and dielectric relaxation times of nematics is the rotational diffusion model of the symmetric top molecules in the self-consistent mean field uniaxial potential of Maier and Saupe [8, 28–30]

$$\frac{U}{kT} = -\sigma\cos^2\vartheta \tag{3}$$

where ϑ is the angle between the axis of symmetry of the molecule and the Z axis of the laboratory coordinate system. The calculation of the complex dielectric susceptibility and associated relaxation times for a potential of the form of Eq. (3) has already been considered [3,4,6,7], and a clear physical understanding of dielectric relaxation in nematics has been achieved. Nevertheless, only numerical and approximate solutions have generally been obtained. These solutions are either so complicated or so limited in applicability that their use in practice is difficult. In recent studies, however, we have obtained both exact and rather simple approximate solutions when the dipole moment of the molecule μ is directed along the axis of symmetry of the molecule [9]. Here the general case, in which the vector μ is directed at an angle β to the molecular axis of symmetry, is considered, allowing us to substantially extend the scope of the applicability of the model. To accomplish this, we propose using the basic concepts of a recently developed method for the evaluation of the parameters governing the linear response of systems having dynamics described by the Fokker-Planck equation [10]. This method is based on an exact solution of the recurrence relations for the spectra of the statistical moments using either ordinary or matrix continued fractions. Moreover, the approximate solutions are derived by using the concepts of the integral relaxation (correlation) time and the effective eigenvalue, allowing us to generalize the theory to an arbitrary axially symmetric mean field potential. The simple approximate formulas so obtained relate the longitudinal and transverse components of the complex dielectric susceptibility and corresponding relaxation times to the order parameter S and are valid for any strength of the nematic potential. They are presented in a form suitable for comparison with dielectric relaxation measurements and can easily be tested with experimentally available data. We note that from a mathematical point of view the calculation of the dielectric response of polar molecules in a mean field potential closely resembles the problem of magnetic relaxation of single domain ferromagnetic particles [11,12].

II. DIELECTRIC SPECTRA AND CORRELATION FUNCTIONS

The complex dielectric permittivity tensor $\varepsilon_{ij}(\omega) = \varepsilon'_{ij}(\omega) - i\varepsilon''_{ij}(\omega)$ of a nematic is diagonal and has only two independent components, one perpendicular $(\varepsilon_\perp(\omega) = \varepsilon_{XX}(\omega) = \varepsilon_{YY}(\omega))$ and the other parallel $(\varepsilon_\|(\omega) = \varepsilon_{ZZ}(\omega))$ to the director vector $\boldsymbol{n}$ in the laboratory coordinate system XYZ, where the Z axis coincides with the director.

The components of the complex dielectric permittivity tensor are determined by the relations [13]:

$$\varepsilon_\gamma(\omega) - \varepsilon_{\gamma\infty} = \frac{4\pi\mu^2 N_0 R_\gamma(\omega)}{kT}[\Phi_\gamma(0) - i\omega\tilde{\Phi}_\gamma(\omega)], (\gamma = \|, \perp) \qquad (4)$$

where

$$\tilde{\Phi}_\gamma(\omega) = \int_0^\infty \Phi_\gamma(t)e^{-i\omega t}dt, (\gamma = \|, \perp) \qquad (5)$$

$$\Phi_\gamma(t) = \langle M_\gamma(0)M_\gamma(t)\rangle_0 \qquad (6)$$

are the macroscopic autocorrelation functions of the parallel and perpendicular components of the dipole moment per unit volume $\mathbf{M}$, $\varepsilon_{\gamma\infty}$ are the high frequency limits of the components of the complex dielectric permittivity tensor, $R_\gamma(\omega)$ is the frequency-dependent factor of the internal field. The dipole moment of the ensemble of N interacting dipolar molecules is defined as

$$\mathbf{M}(t) = \sum_{i=1}^N \boldsymbol{\mu}^i(t) \qquad (7)$$

where $\boldsymbol{\mu}^i(t)$ is the dipole moment vector of the ith molecule. Hence

$$\Phi_\gamma(t) = \left\langle \sum_{i,j}^N \mu_\gamma^i(0)\mu_\gamma^j(t) \right\rangle_0 \qquad (8)$$

The internal field factor $R_\gamma(\omega)$ for an ellipsoidal cavity surrounded by an infinite dielectric continuum with the same complex dielectric permittivity tensor $\varepsilon_{ij}(\omega)$ is given by [13]:

$$R_\gamma(\omega) = \frac{\varepsilon_\gamma(\omega)}{\varepsilon_\gamma(\omega) - \sigma_\gamma(\omega)[\varepsilon_\gamma(\omega) - \varepsilon_{\gamma\infty}]} \tag{9}$$

where $\sigma_\gamma(\omega)$ are the components of the depolarization tensor, which are defined as

$$\sigma_\gamma(\omega) = \frac{a_X a_Y a_Z}{2[\varepsilon_{XX}(\omega)\varepsilon_{YY}(\omega)\varepsilon_{ZZ}(\omega)]^{1/2}} \int_0^\infty \frac{ds}{[s + a_\gamma^2/\varepsilon_\gamma(\omega)]D(s)} \tag{10}$$

with

$$D(s) = \left[s + \frac{a_X^2}{\varepsilon_{XX}(\omega)}\right]\left[s + \frac{a_Y^2}{\varepsilon_{YY}(\omega)}\right]\left[s + \frac{a_Z^2}{\varepsilon_{ZZ}(\omega)}\right] \tag{11}$$

where a_i are the semiaxes of the ellipsoidal cavity. Another equation for $R_\gamma(\omega)$ may be obtained in the context of the ellipsoidal cavity model if we make the assumption that the permittivity tensor $\varepsilon_{ij}(\omega)$ of the material surrounding the cavity is frequency independent and equal to the static tensor $\varepsilon_{ij}(0)$ [13].

We shall then have

$$R_\gamma(\omega) = \frac{\varepsilon_\gamma(0) - \sigma_\gamma(0)[\varepsilon_\gamma(0) - \varepsilon_\gamma(\omega)]}{\varepsilon_\gamma(\omega) - \sigma_\gamma(\omega)[\varepsilon_\gamma(\omega) - \varepsilon_{\gamma\infty}]} \tag{12}$$

According to these equations, even if the macroscopic autocorrelation functions are known, it is still not possible to calculate the complex permittivity because the depolarization tensor depends on the cavity dimensions, which are arbitrary. Edwards and Madden [14] criticized the ellipsoidal cavity model from the point of view that if one introduces the cavity concept in a theory, it is difficult to see how dipole–dipole correlations can be used systematically to relate the frequency dependence of the permittivity to reorientations of single molecules. They generalized Sullivan and Deutch's [15] theory of dielectric relaxation in isotropic polar media to the anisotropic case. Here, neither a cavity nor an internal field factor is used explicitly. As was shown by Kalmykov [16], Edwards and Madden's theory actually predicts the same results as given by Eq. (4), with the internal field factor defined by Eq. (12) and $\varepsilon_{\gamma\infty} = 1$, Edwards and Madden [14] considered a system of nonpolarizable molecules.

This dependence on the geometry of the sample is an unsatisfactory feature of the analysis that also occurs for isotropic media [13]. As pointed out by Luckhurst and Zannoni [13], however, it does not present a major problem if the cavity contains many molecules, because the cavity shape is then irrelevant. It is, therefore, convenient to select a spherical sample, because the depolarization tensor is then $\sigma_\gamma(\omega) = 1/3$ and independent of the sample size. Hence the relationship between the permittivity tensor and autocorrelation function tensor for the net dipole moment of the spherical cavity is completely defined. Another essential feature of the problem consists in relating the macroscopic autocorrelation function tensor to that of a single particle [13]:

$$C_\parallel(t) = \langle u_Z(0)u_Z(t)\rangle_0 \qquad (13)$$

$$C_\perp(t) = \langle u_X(0)u_X(t)\rangle_0 \qquad (14)$$

where u_X, u_Z are the projections of the unit vector $\mathbf{u}$ along $\boldsymbol{\mu}$ onto the axes X and Z. The same problem is encountered for isotropic media [17], and a number of solutions have been proposed [13,14,16]. The simplest one, by far, is to ignore the correlations between dipole moments of different molecules. The macroscopic dipole autocorrelation function then becomes that of a single molecule [13]:

$$\Phi_\gamma(t) \approx N\mu^2 C_\gamma(t) \qquad (15)$$

where $\boldsymbol{\mu}$ is the dipole moment of the molecule. This assumption, however, has an obvious drawback in that Eq. (15) does not tend to the correct low frequency limit as $\omega \to 0$, which is the Kirkwood-Fröhlich equation for an anisotropic medium [1]:

$$\frac{[\varepsilon_\gamma(0) - \varepsilon_{\gamma\infty}]\{\varepsilon_\gamma(0) - \sigma_\gamma(0)[\varepsilon_\gamma(0) - \varepsilon_{\gamma\infty}]\}}{\varepsilon_\gamma(0)} = \frac{4\pi N g_\gamma \langle (\mu_\gamma^i(0))^2 \rangle_0}{kT} \qquad (16)$$

Here

$$g_\gamma = \left\langle \sum_{i,j=1}^{N} \mu_\gamma^i(0)\mu_\gamma^j(0) \right\rangle_0 \Big/ N\left\langle (\mu_\gamma^i(0))^2 \right\rangle_0 \qquad (17)$$

is the static equilibrium orientational correlation factor (an analogue of the Kirkwood g-factor in isotropic liquids),

$$\langle(\mu_\parallel^i)^2\rangle_0 = \frac{\mu_0^2[\bar{\varepsilon}_\infty + 2]^2[1 - (1 - \cos^2\beta)S]}{27} \tag{18}$$

$$\langle(\mu_\perp^i)^2\rangle_0 = \frac{\mu_0^2[\bar{\varepsilon}_\infty + 2]^2[1 + (1 - \cos^2\beta)S/2]}{27} \tag{19}$$

$$S = \frac{1}{2}(3\langle\cos^2\vartheta\rangle_0 - 1)$$

is the order parameter, β and ϑ are the angles between the molecular axis and vectors μ and n respectively, μ_0 is the gas phase dipole moment, and $\bar{\varepsilon}_\infty = (\varepsilon_{\infty\parallel} + 2\varepsilon_{\infty\perp})/3$.

Yet another approximate equation relating the spectra $\tilde{\Phi}_\gamma(\omega)$ and $\tilde{C}_\gamma(\omega)$ was obtained [16]:

$$\tilde{\Phi}_\gamma(\omega) = \frac{\Phi_\gamma(0)g_\gamma\tilde{C}_\gamma(\omega)}{1 + i\omega(g_\gamma - 1)\tilde{C}_\gamma(\omega)} \tag{21}$$

Eq. (21) was derived by assuming that the dynamic angular velocity correlations of different molecules may be neglected. In the low frequency limit $\omega \to 0$, Eq. (21) coincides precisely with Eq. (16). Moreover, in the high frequency limit, $(\omega \to \infty)$, Eq. (21) has the correct asymptotic behavior

$$\varepsilon_\gamma(\omega) = \varepsilon_{\infty\gamma} + \frac{4\pi N\langle\mu_\gamma^i(0)\ddot{\mu}_\gamma^i(0)\rangle_0}{kT\omega^2} + O(\omega^{-4}) \tag{22}$$

Eq. (21) is in agreement with the so-called micro-macro correlation theorem of Kivelson and Madden [17,18], which states that if the single particle correlation function $C_\gamma(t)$ may be represented as a set of exponentials, then the behavior of the macroscopic correlation function $\Phi_\gamma(t)$ is described by the same set of the exponentials with scaled parameters. For example, if the relaxation of $C_\gamma(t)$ is given by a single exponential, viz.

$$C_\gamma(t) = \langle u_\gamma^2(0)\rangle_0 e^{-t/\tau_\gamma} \tag{23}$$

then according to Eq. (21) [16],

$$\Phi_\gamma(t) = \frac{g_\gamma N}{kT}\langle(\mu_\gamma(0))^2\rangle_0 e^{-t/g_\gamma\tau_\gamma} \tag{24}$$

That is, the behavior of $\Phi_\gamma(t)$ is also given by a single exponential with the relaxation time

$$\tau_\gamma^N = g_\gamma \tau_\gamma \tag{25}$$

In particular, for two exponential relaxations, when

$$C_\gamma(t) = \langle u_\gamma^2(0) \rangle_0 [\delta e^{-t/\tau_{\gamma 1}} + (1 - \delta) e^{-t/\tau_{\gamma 2}}], (\delta \leq 1) \tag{26}$$

we have

$$\Phi_\gamma(t) = \frac{g_\gamma N}{kT} \langle (\mu_\gamma(0))^2 \rangle_0 [\Delta e^{-t/\tau_{\gamma 1}^N} + (1 - \Delta) e^{-t/\tau_{\gamma 2}^N}] \tag{27}$$

with

$$\tau_{\gamma 1}^N = \frac{A_\gamma + B_\gamma}{2}, \tau_{\gamma 2}^N = \frac{A_\gamma - B_\gamma}{2} \tag{28}$$

$$\Delta = \frac{2 g_\gamma [\delta \tau_{\gamma 1} + (1 - \delta) \tau_{\gamma 2}] - A_\gamma + B_\gamma}{2 B_\gamma} \tag{29}$$

where

$$A_\gamma = (1 - \delta) \tau_{\gamma 1} + \delta \tau_{\gamma 2} + g_\gamma [\delta \tau_{\gamma 1} + (1 - \delta) \tau_{\gamma 2}] \tag{30}$$

$$B_\gamma = \sqrt{A_\gamma^2 - 4 g_\gamma \tau_{\gamma 1} \tau_{\gamma 2}} \tag{31}$$

Unfortunately, if more than two relaxation modes are involved, the relationships between single particle and macroscopic relaxation times become much more complicated [16].

Eq. (21) can be used to evaluate the dielectric permittivity spectra $\varepsilon_\gamma(\omega)$, provided the molecular correlation function spectra $\tilde{C}_\gamma(\omega)$ and correlation factors g_γ are known. To calculate spectra $\varepsilon_\gamma(\omega)$ from Eq. (21), one can estimate the values of g_γ from the Kirkwood-Fröhlich (Eq. (16)) and make use of experimental spectra $\tilde{C}_\gamma(\omega)$ obtained from IR rotation/vibration absorption spectra. Unfortunately, direct theoretical evaluation of both $\tilde{C}_\gamma(\omega)$ and g_γ are difficult tasks and can be accomplished only for model systems. In view of these difficulties, we shall suppose that models of molecular reorientation that give simple spectra $\tilde{C}_\gamma(\omega)$ may be applied to semiquantitative evaluations of spectra $\varepsilon_\gamma(\omega)$. As an example we consider the rotational diffusion model in a mean field potential.

III. THREE-DIMENSIONAL ROTATIONAL BROWNIAN MOTION IN AN EXTERNAL POTENTIAL

A. The Langevin Equation Approach to the Maier-Saupe Model

The noninertial three-dimensional rotational Brownian motion of a particle in the presence of an external potential arises in a variety of problems. Examples are dielectric and Kerr-effect relaxation of polar fluids subjected to a constant electric field [19], dielectric relaxation of nematic liquid crystals [3,6], and magnetization relaxation of single domain ferromagnetic particles and ferrofluids [12]. Theories of all these phenomena were mainly based on the Fokker-Planck equation [20]. There is a method of bypassing the problem of constructing and solving the Fokker-Planck equation entirely [5]. It is the purpose of this chapter to show how hierarchies of differential-recurrence relations in the noninertial limit arise naturally from the vector Langevin equations defined as Stratonovich stochastic equations [5,21].

We study the three-dimensional rotational Brownian motion of a rodlike particle in an external potential. The particle contains a rigid electric dipole μ. In this section we confine ourselves to an ensemble of rigid dipolar molecules, where the vector μ is oriented along the symmetry axis of the particle only. (The general case, where the vector μ is oriented at an arbitrary angle β to the direction of the long axis of the particle will be considered later. We take a unit vector $\mathbf{u}(t)$ through the center of the mass of the particle in the direction of μ. Then the rate of change of $\mathbf{u}(t)$ is [5]

$$\frac{d\mathbf{u}(t)}{dt} = \omega(t) \times \mathbf{u}(t) \tag{32}$$

where $\omega(t)$ is the angular velocity of the particle. It should be noted that Eq. (32) is a purely kinematic relation with no particular reference either to the Brownian movement or to the shape of the particle. For simplicity, we specialize it to the rotational Brownian motion of a linear or spherical top molecule by supposing that the angular velocity $\omega(t)$ obeys the Euler-Langevin equation [22,23].

$$I\frac{d}{dt}\omega(t) + \zeta\omega(t) = \mu(t) \times \mathbf{E}(t) + \lambda(t) \tag{33}$$

where I is the moment of inertia of the molecule, μ is the total dipole moment of the molecule, $\zeta\omega$ is the damping torque owing to Brownian movement, and $\lambda(t)$ is the white noise driving torque, again owing to

Brownian movement so that $\lambda(t)$ has the following properties:

$$\overline{\lambda_i(t)} = 0,$$

$$\overline{\lambda_i(t_1)\lambda_j(t_2)} = 2kT\zeta\delta_{ij}\delta(t_1 - t_2) \tag{34}$$

where the overbar means a statistical average over an ensemble of molecules that start at time t with the same angular velocity $\boldsymbol{\omega}$ and orientation $\mathbf{u}$; δ_{ij} is Kronecker's delta; $i, j = 1, 2, 3$, which corresponds to the Cartesian axes x, y, z, of the laboratory coordinate system; and $\delta(t)$ is the Dirac delta function. Here we have also assumed that the damping coefficient ζ is a scalar.

The electric field $\mathbf{E}(t)$ may consist of externally applied electric fields and the mean field. The term $\boldsymbol{\mu}(t) \times \mathbf{E}(t)$ in Eq. (33), is the torque owing to the total electric field acting on the molecule. This torque can be expressed in terms of the potential function $U(\{\mathbf{u}, t\})$ as a function of the components of the vector $\mathbf{u}$

$$\boldsymbol{\mu} \times \mathbf{E} = -\mathbf{u} \times \frac{\partial}{\partial \mathbf{u}} U(\{\mathbf{u}\}) \tag{35}$$

where

$$\frac{\partial}{\partial \mathbf{u}} = i\frac{\partial}{\partial u_x} + j\frac{\partial}{\partial u_y} + k\frac{\partial}{\partial u_z}$$

i, j and k are the unit vectors along the Cartesian axes x, y, and z respectively and u_x, u_y, and u_z are the Cartesian components of the unit vector $\mathbf{u}(t)$. These components are expressed in terms of the polar and azimuthal angles as follows:

$$u_x = \sin\vartheta \cos\varphi, \quad u_y = \sin\vartheta \sin\varphi, \quad u_z = \cos\vartheta$$

Eq. (33) includes the inertia of the molecule. The noninertial (or low frequency) response is that when the moment of inertia I tends to zero or when the friction coefficient ζ becomes very large. In this limit the angular velocity vector may be immediately obtained from Eq. (33) as

$$\zeta\boldsymbol{\omega}(t) = \boldsymbol{\mu}(t) \times \boldsymbol{E}(t) + \boldsymbol{\lambda}(t) \tag{36}$$

On combining Eq. (36) with the kinematic relation Eq. (32) one can obtain

$$\frac{d\mathbf{u}(t)}{dt} = \mu\zeta^{-1}[\mathbf{u}(t) \times \boldsymbol{E}(t)] \times \mathbf{u}(t) + \zeta^{-1}\boldsymbol{\lambda}(t) \times \mathbf{u}(t)$$

which, with the properties of the triple vector product, becomes

$$\frac{d}{dt}\mathbf{u}(t) = -\zeta^{-1}\left[\frac{\partial}{\partial \mathbf{u}}U - \mathbf{u}(t)\left(\mathbf{u}(t)\cdot\frac{\partial}{\partial \mathbf{u}}U\right)\right] + \zeta^{-1}\lambda(t)\times\mathbf{u}(t) \qquad (37)$$

This is the vector Langevin equation for the motion of the vector $\mathbf{u}$ in the noninertial limit. Eq. (37) is equivalent to the three equations for the Cartesian components of $\mathbf{u}$

$$\frac{d}{dt}u_x(t) = \zeta^{-1}\left[\lambda_y(t)u_z(t) - \lambda_z(t)u_y(t) - \frac{\partial}{\partial u_x}U + u_x\left(\mathbf{u}(t)\cdot\frac{\partial}{\partial \mathbf{u}}U\right)\right] \qquad (38)$$

$$\frac{d}{dt}u_y(t) = \zeta^{-1}\left[\lambda_z(t)u_x(t) - \lambda_x(t)u_z(t) - \frac{\partial}{\partial u_y}U + u_y\left(\mathbf{u}(t)\cdot\frac{\partial}{\partial \mathbf{u}}U\right)\right] \qquad (39)$$

$$\frac{d}{dt}u_z(t) = \zeta^{-1}\left[\lambda_x(t)u_y(t) - \lambda_y(t)u_x(t) - \frac{\partial}{\partial u_z}U + u_z\left(\mathbf{u}(t)\cdot\frac{\partial}{\partial \mathbf{u}}U\right)\right] \qquad (40)$$

The stochastic differential Eqs. (38) to (40) contain multiplicative noise terms $\lambda_i(t)u_j(t)$. This poses an interpretation problem for these equations, as discussed by Risken [20]. We recall, taking the Langevin equation for N stochastic variables $\{\xi(t)\} = \{\xi_1(t), \xi_2(t), \dots, \xi_N(t)\}$:

$$\frac{d}{dt}\xi_i(t) = h_i(\{\xi(t)\}, t) + g_{ij}(\{\xi(t)\}, t)\Gamma_j(t) \qquad (41)$$

with

$$\overline{\Gamma_i(t)} = 0,$$
$$\overline{\Gamma_i, (t_1)\Gamma_j(t_2)} = 2D\delta_{ij}\delta(t_1 - t_2) \qquad (42)$$

and interpreting it as a Stratonovich [24] equation, that the averaged equation for the sharp values $\xi_i(t) = x_i$ at time t is [20]:

$$\frac{d}{dt}x_i = \lim_{\tau\to 0}\frac{\overline{\xi_i(t+\tau) - x_i}}{\tau} = h_i(\{\mathbf{x}\}, t) + Dg_{kj}(\{\mathbf{x}\}, t)\frac{\partial}{\partial x_k}g_{ij}(\{\mathbf{x}\}, t) \qquad (43)$$

where $\xi_i(t+\tau)(\tau > 0)$ is the solution of Eq. (41) with the initial conditions $\xi_i(t) = x_i$. In Eqs. (41) and (43) the summation over j and k is understood (Einstein's notation). The last term in Eq. (43) is called the noise-induced or spurious drift [20]. The proof of Eq. (43) can be found elsewhere [20].

In like manner one can prove that the averaged equation for an *arbitrary* differentiable function $f(\{\xi\})$ has the following form (Appendix A):

$$\frac{\mathrm{d}}{\mathrm{d}t}f(\{\mathbf{x}\}) = h_i(\{\mathbf{x}\},t)\frac{\partial}{\partial x_i}f(\{\mathbf{x}\}) + Dg_{kj}(\{\mathbf{x}\},t)\frac{\partial}{\partial x_k}\left[g_{ij}(\{\mathbf{x}\},t)\frac{\partial}{\partial x_i}f(\{\mathbf{x}\})\right]$$

(44)

where summation over i, j and k is also understood.

This theorem and the preceding one in Eq.(43) are direct consequences of the basic *ansatz* underlying the theory of the Brownian movement. For simplicity, we refer to the translational motion of a particle. The *ansatz* is that we may introduce a time interval τ which is so small that the momentum and external deterministic force acting on the particle do not alter during τ. On the other hand, τ is so large that the integrals of the white noise force with respect to time evaluated at t and $t + \tau$ are statistically independent of each other. The integral of the white noise force evaluated between t and $t + \tau$ is called the Wiener process [5]. The statistical independence of the integrals means that the Wiener process has *independent* increments. In the problems under consideration, which involve multiplicative noise, the stochastic differential equations must be supplemented by an interpretation rule. If we imagine that we may model the noise by a random sequence of delta functions, then each delta function jump in the noise causes a jump in the random variable $\xi(t)$. Hence the value of $\xi(t)$ at the time the delta function arrives is indeterminate and consequently so is the multiplicative term at this time also. A problem arises, as the stochastic differential equation does not indicate which value of ξ one should substitute in the multiplicative noise term. Itô [20] interpreted a Langevin equation with multiplicative noise by supposing that in the multiplicative noise term the value of the random variable ξ *before the jump* in the noise is taken. This is in contrast to Stratonovich [20] who takes the value of ξ as the *mean of the values before and after the jump*. In physical applications, the white noise is always the limiting case of a physical noise with finite noise power (corresponding in the time domain to very small but finite noise correlation time), and here the Stratonovich interpretation is the correct one to use. This conclusion is reinforced by the fact [5] that the Stratonovich interpretation of the underlying stochastic differential equation yields on averaging the differential recurrence relations which result from taking the noninertial limit of the inertial response [5]. The Itô-Stratonovich "interpretation problem" does not arise in this response [5] as the position is mean square differentiable.

Note that here we always use the Stratonovich [24] definition of the average of the multiplicative noise term, because that definition constitutes the mathematical idealization of the physical stochastic process of orientational relaxation in the noninertial limit. Thus it is unnecessary to transform the Langevin Eqs. 38 to 40 to Itô [21] equations. Moreover, we can apply the methods of ordinary analysis [21].

In the study of the orientational relaxation, the quantities of interest are the spherical harmonics X_{nm} defined as

$$X_{nm} = e^{im\varphi} P_n^m(\cos \vartheta) = e^{im\varphi}(1 - \cos^2 \vartheta)^{m/2} \frac{d^m P_n(\cos \vartheta)}{d\cos \vartheta^m} \tag{45}$$

where $P_n^m(x)$ are the associated Legendre functions [25]. The X_{nm} are expressed in terms of u_x, u_y, and u_z as follows:

$$X_{nm} = (u_x + iu_y)^m \frac{d^m P_n(u_z)}{du_z^m} \tag{46}$$

Note that according to the Stratonovich definition the conventional rules of transformation of the variables can be used and that

$$\frac{d}{dt} X_{nm} = m(u_x + iu_y)^{m-1} \frac{d^m P_n(u_z)}{du_z^m} \frac{d}{dt} u_x$$

$$+ im(u_x + iu_y)^{m-1} \frac{d^m P_n(u_z)}{du_z^m} \frac{d}{dt} u_y + (u_x + iu_y)^m \frac{d^{m+1} P_n(u_z)}{du_z^{m+1}} \frac{d}{dt} u_z \tag{47}$$

We can thus obtain the equation of motion of the spherical harmonics X_{nm} by cross multiplying Eqs. (38) to (40) by

$$m(u_x + iu_y)^{m-1} \frac{d^m P_n(u_z)}{du_z^m}, im(u_x + iu_y)^{m-1} \frac{d^m P_n(u_z)}{du_z^m}, \quad \text{and}$$

$$(u_x + iu_y)^m \frac{d^{m+1} P_n(u_z)}{du_z^{m+1}}$$

respectively, and then summing them. Thus we have

$$
\begin{aligned}
\frac{d}{dt}&X_{nm}(\{\mathbf{u}(t)\})\\
&= \frac{1}{\zeta}\Bigg\{[u_x(t)+iu_y(t)]^m\frac{d^{m+1}P_n(u_z(t))}{du_z^{m+1}(t)}\left[-\frac{\partial}{\partial u_z}U+u_z(t)\left(\mathbf{u}(t)\cdot\frac{\partial}{\partial\mathbf{u}}U\right)\right]\\
&\quad+ m[u_x(t)+iu_y(t)]^{m-1}\frac{d^mP_n(u_z(t))}{du_z^m(t)}\\
&\qquad\times\left[-\frac{\partial}{\partial u_x}U-i\frac{\partial}{\partial u_y}U+[u_x(t)+iu_y(t)]\left(\mathbf{u}(t)\cdot\frac{\partial}{\partial\mathbf{u}}U\right)\right]\Bigg\}\\
&\quad+\Bigg\{m[u_x(t)+iu_y(t)]^{m-1}\frac{d^mP_n(u_z(t))}{du_z^m(t)}[g_{xj}(\{\mathbf{u}(t)\})+ig_{yj}(\{\mathbf{u}(t)\})]\\
&\quad+[u_x(t)+iu_y(t)]^m\frac{d^{m+1}P_n(u_z(t))}{du_z^{m+1}(t)}g_{zj}(\{\mathbf{u}(t)\})\Bigg\}\lambda_j(t)
\end{aligned}
\tag{48}
$$

where the components of the tensor $\mathbf{g}$ are

$$
\begin{aligned}
g_{xx}&=0 & g_{xy}&=u_z/\zeta & g_{xz}&=-u_y/\zeta\\
g_{yx}&=-u_z/\zeta & g_{yy}&=0, & g_{yz}&=u_x/\zeta\\
g_{zx}&=u_y/\zeta & g_{zy}&=-u_x/\zeta & g_{zz}&=0
\end{aligned}
\tag{49}
$$

On averaging the stochastic Eq. (48) and noting Eq. (44), we have

$$
\begin{aligned}
2\tau_D\frac{d}{dt}X_{nm}&=\frac{1}{kT}m(u_x+iu_y)^{m-1}\\
&\quad\times\frac{d^mP_n(u_z)}{du_z^m}\left[-\frac{\partial}{\partial u_x}U-i\frac{\partial}{\partial u_y}U+(u_x+iu_y)\left(\mathbf{u}\cdot\frac{\partial}{\partial\mathbf{u}}U\right)\right]\\
&\quad+(u_x+iu_y)^m\frac{d^{m+1}P_n(u_z)}{du_z^{m+1}}\left[-\frac{\partial}{\partial u_z}U+u_z\left(\mathbf{u}\cdot\frac{\partial}{\partial\mathbf{u}}U\right)\right]\\
&\quad+\zeta^2g_{kj}\frac{\partial}{\partial u_k}\left[m(u_x+iu_y)^{m-1}\frac{d^mP_n(u_z)}{du_z^m}(g_{xj}+ig_{yj})\right.\\
&\quad\left.+(u_x+iu_y)^m\frac{d^{m+1}P_n(u_z)}{du_z^{m+1}}g_{zj}\right]
\end{aligned}
\tag{50}
$$

where

$$
\tau_D=\frac{\zeta}{2kT}
\tag{51}
$$

is the Debye relaxation time. Note that u_x, x_y, and u_z in Eq. (50) and $u_x(t)$, $u_y(t)$, and $u_z(t)$ in Eqs. (38) to (40) have different meanings, namely, $u_x(t)$, $u_y(t)$, and $u_z(t)$ in Eqs. (38) to (40) are stochastic variables (processes), whereas u_x, u_y, and u_z in Eq. (50) are the sharp (definite) values $u_k(t) = u_k$ at time t. Instead of using different symbols for the two quantities we have distinguished sharp values at time t from stochastic variables by deleting the time argument as in [20].

The right-hand side of Eq. (50) consists of two terms: the deterministic drift and noise-induced (or spurious) drift. Let us first simplify the noise-induced drift (the last term on the right-hand side). We have [26]:

$$\zeta^2 g_{kj} \frac{\partial}{\partial u_k} \left[m(u_x + iu_y)^{m-1} \frac{d^m P_n(u_z)}{du_z^m} (g_{xj} + ig_{yj}) + (u_x + iu_y)^m \frac{d^{m+1} P_n(u_z)}{du_z^{m+1}} g_{zj} \right]$$

$$= u_z \frac{\partial}{\partial u_x} \left[m(u_x + iu_y)^{m-1} \frac{d^m P_n(u_z)}{du_z^m} u_z - (u_x + iu_y)^m \frac{d^{m+1} P_n(u_z)}{du_z^{m+1}} u_x \right]$$

$$- u_y \frac{\partial}{\partial u_x} \left[m(u_x + iu_y)^m \frac{d^m P_n(u_z)}{du_z^m} \right] + u_x \frac{\partial}{\partial u_y} \left[im(u_x + iu_y)^m \frac{d^m P_n(u_z)}{du_z^m} \right]$$

$$+ u_z \frac{\partial}{\partial u_y} \left[im(u_x + iu_y)^{m-1} \frac{d^m P_n(u_z)}{du_z^m} u_z - (u_x + iu_y)^m \frac{d^{m+1} P_n(u_z)}{du_z^{m+1}} u_y \right]$$

$$- iu_y \frac{\partial}{\partial u_z} \left[m(u_x + iu_y)^{m-1} \frac{d^m P_n(u_z)}{du_z^m} u_z + (u_x + iu_y)^m \frac{d^{m+1} P_n(u_z)}{du_z^{m+1}} (u_z + iu_y) \right]$$

$$- u_x \frac{\partial}{\partial u_z} \left[m(u_x + iu_y)^{m-1} \frac{d^m P_n(u_z)}{du_z^m} u_z - (u_x + iu_y)^m \frac{d^{m+1} P_n(u_z)}{du_z^{m+1}} u_x \right]$$

$$= (u_x + iu_y)^m \left[(1 - u_z^2) \frac{d^{m+2} P_n(u_z)}{du_z^{m+2}} - 2(m+1)u_z \frac{d^{m+1} P_n(u_z)}{du_z^{m+1}} \right.$$

$$\left. -m(m+1) \frac{d^m P_n(u_z)}{du_z^m} \right] = -n(n+1)(u_x + iu_y)^m \frac{d^m P_n(u_z)}{du_z^m}$$

$$= -n(n+1)X_{nm} \tag{52}$$

Here we have used [27]

$$(1 - x^2) \frac{d^{m+2} P_n(x)}{dx^{m+2}} - 2(m+1)x \frac{d^{m+1} P_n(x)}{dx^{m+1}}$$

$$+ (n - m)(n + m + 1) \frac{d^m P_n(x)}{dx^m} = 0 \tag{53}$$

Let us now consider the deterministic drift term. We have [26]

$$
\frac{\mu}{kT} \left\{ m(u_x + iu_y)^{m-1} \frac{d^m P_n(u_z)}{du_z^m} [(E_x + iE_y) - (u_x + iu_y)(\mathbf{u} \cdot \mathbf{E})] \right.
$$
$$
\left. + (u_x + iu_y)^m \frac{d^{m+1} P_n(u_z)}{du_z^{m+1}} [E_z - u_z(\mathbf{u} \cdot \mathbf{E})] \right\}
$$
$$
= \frac{\mu(E_x + iE_y)}{2kT(2n+1)} [n(n-m+1)(n-m+2)X_{n+1m-1}
$$
$$
+ (n+1)(n+m-1)(n+m)X_{n-1m-1}]
$$
$$
- \frac{\mu(E_x - iE_y)}{2kT(2n+1)} [nX_{n+1m+1} + (n+1)X_{n-1m+1}]
$$
$$
- \frac{\mu E_z}{kT(2n+1)} [n(n-m+1)X_{n+1m} - (n+1)(n+m)X_{n-1m}] \quad (54)
$$

Here we have used Eq. (53) and the recurrence relations for the associated Legendre functions [25]:

$$
\sqrt{1-x^2} P_n^{m+1}(x) - 2mx P_n^m(x)
$$
$$
+ (n+m)(n-m+1)\sqrt{1-x^2} P_n^{m-1}(x) = 0 \quad (55)
$$
$$
\sqrt{1-x^2} P_n^{m+1}(x) - (n+m+1)x P_n^m(x) + (n-m+1) P_{n+1}^m(x) = 0 \quad (56)
$$
$$
(n-m+1)\sqrt{1-x^2} P_n^{m-1}(x) - x P_n^m(x) + P_{n-1}^m(x) = 0 \quad (57)
$$

and

$$
(2n+1)x P_n^m(x) = (n-m+1) P_{n+1}^m(x) + (n+m) P_{n-1}^m(x) \quad (58)
$$

Thus we obtain the averaged equation of motion of a spherical harmonic as

$$
2\tau_D \frac{d}{dt} X_{nm} + n(n+1)X_{nm}
$$
$$
= \frac{\mu(E_x + iE_y)}{2kT(2n+1)} [n(n-m+1)(n-m+2)X_{n+1m-1}
$$
$$
+ (n+1)(n+m-1)(n+m)X_{n-1m-1}]
$$
$$
- \frac{\mu}{2kT(2n+1)} (E_x - iE_y)[nX_{n+1m+1} + (n+1)X_{n-1m+1}]
$$
$$
- \frac{\mu}{kT(2n+1)} E_z[n(n-m+1)X_{n+1m} - (n+1)(n+m)X_{n-1m}]
$$
$$
(59)
$$

As mentioned, all the averaged quantities X_{nm} and the E_i in Eq. (59) are in general functions of the sharp values u_k, which are themselves random variables with the probability density function W such that $W du_k$ is the probability of finding u_k in the interval $(u_k, u_k + du_k)$. (We remark that finding the average equations of motion for sharp values of u_k effectively corresponds to finding the Fourier coefficients in the fundamental solution (Green's function or transition probability) of the Fokker-Planck equation.) Therefore to obtain equations for the moments that govern the relaxation dynamics of the system we must also average Eq. (59) over W [5]. If, however, the system under consideration is in equilibrium all averages are either constant or zero. Thus we need first to construct from Eq. (59) a set of differential-recurrence relations for equilibrium correlation functions. We shall illustrate this by considering dielectric relaxation of a nematic liquid crystal with uniaxial crystalline anisotropy [28–30].

Let us consider the Maier-Saupe uniaxial crystalline anisotropy potential Eq. (3):

$$U = -K u_z^2 = -K \cos^2 \vartheta \tag{60}$$

According to Eq. (35) the field $\mathbf{E}$ has only the z component given by

$$\mathbf{E} = \frac{2K}{\mu} u_z \mathbf{k} \tag{61}$$

Substituting Eq. (61) into (59) and using Eq. (58), we obtain

$$2\tau_D \frac{d}{dt} X_{nm} + \left[n(n+1) - 2\sigma \frac{n(n+1) - 3m^2}{(2n-1)(2n+3)} \right] X_{nm}$$
$$= 2\sigma \left[\frac{(n+1)(n+m)(n+m-1)}{(2n-1)(2n+1)} X_{n-2m} \right.$$
$$\left. - \frac{n(n-m+1)(n-m+2)}{(2n+1)(2n+3)} X_{n+2m} \right] \tag{62}$$

where

$$\sigma = \frac{K}{kT} \tag{63}$$

Multiplying Eq. (62) by $u_z(0)$ and averaging the resulting equation over the equilibrium distribution function

$$W_0 = C \exp(\sigma \cos^2 \vartheta) \tag{64}$$

at the instant $t = 0$, we obtain differential-recurrence relations for the longitudinal correlation functions $f_n(t)$

$$\frac{2\tau_D}{n(n+1)}\frac{d}{dt}f_n(t) + \left[1 - \frac{2\sigma}{(2n-1)(2n+3)}\right]f_n(t)$$
$$= 2\sigma\left[\frac{(n-1)}{(2n-1)(2n+1)}f_{n-2}(t) - \frac{(n+2)}{(2n+1)(2n+3)}f_{n+2}(t)\right] \quad (65)$$

where $f_n(t)$ are the equilibrium correlation functions defined as

$$f_n(t) = \langle \cos\vartheta(0)P_n(\cos\vartheta(t))\rangle_0 \quad (66)$$

Eq. (65) governs the longitudinal relaxation of the system under consideration. See Figure 1.

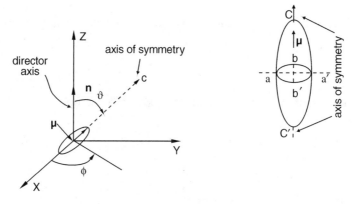

Figure 1. Diagram showing rotational motion of a dipolar molecule with the dipole $\mu\|$ to the long axis cc′ of the molecule. The direction of the dipole is specified by the polar angles ϑ and ϕ. The longitudinal relaxation is a slow hindered rotation about the axis aa′ impeded by the crystalline anisotropy field $\frac{2K}{\mu}\mu_z\mathbf{k}$ and dominated by the slow overbarrier mode (with a small contribution due to fast reorientation of the molecules in the wells of the bistable potential created by this field). In the frequency domain this motion is represented by a pronounced low frequency peak with an exponential relaxation time and a weak high-frequency peak resulting from the motion in the wells. The motion about the Z axis is the transverse motion, that is, in the direction normal to $\mathbf{n}$. It consists of a relatively fast rotation which is dominated by two near degenerate modes, so that the motion appears as a single mode. The Debye relaxation time for the isotropic diffusion is $\tau_D = (2D_\perp)^{-1}$ with $D_\perp$, the rotational diffusion coefficient with respect to the axis (aa′ = bb′) perpendicular to the axis of symmetry (that is, the long axis cc′ of the molecule). The transverse motion appears in the frequency domain as a relatively high frequency Debye-like relaxation.

Similarly, we can obtain dynamic equations for the transverse correlation functions $g_n(t)$ defined as:

$$g_n(t) = \langle \text{Re}\{X_{11}(0)\}\text{Re}\{X_{n1}(t)\}\rangle_0 = \langle \cos \varphi(0)\sin \vartheta(0)\cos \varphi(t)P_n^1(\cos \vartheta(t))\rangle_0$$
(67)

This equation is

$$2\tau_D \frac{d}{dt}g_n(t) + \left[n(n+1) - 2\sigma \frac{n(n+1)-3}{(2n-1)(2n+3)}\right]g_n(t)$$

$$= 2\sigma n(n+1)\left[\frac{(n+1)}{(2n-1)(2n+1)}g_{n-2}(t) - \frac{n}{(2n+1)(2n+3)}g_{n+2}(t)\right] \quad (68)$$

Solving Eqs. (65) and (68) for the Laplace transforms of f_1 and g_1 and then using linear response theory we obtain below exact analytic solutions for the longitudinal and transverse complex susceptibilities and correlation times. This can be accomplished, because the Laplace transform of these systems may be reduced to infinite systems of inhomogeneous linear algebraic equations, the exact solution of which is determined using a method developed by Coffey et al. [5]. This method constitutes a further development of Risken's [20] continued fraction method and allows us to solve the inhomogeneous recurrent relations, which are necessary for the calculation of dielectric permittivity spectra and relaxation times.

B.   The Longitudinal Relaxation

By inspection of Eq. (65) it is obvious that it decomposes into sets for even and odd $f_i(t)$. The odd f_i satisfy

$$\frac{d}{dt}f_{2n+1}(t) = \frac{(2n+1)(n+1)}{\tau_D}\left[\frac{2\sigma}{(4n+1)(4n+5)} - 1\right]f_{2n+1}(t)$$

$$+ \frac{4\sigma(n+1)(2n+1)}{\tau_D(4n+3)}\left[\frac{n}{(4n+1)}f_{2n-1}(t) - \frac{(2n+3)}{2(4n+5)}f_{2n+3}(t)\right]$$
(69)

In our situation, only the odd $f_{2n+1}(t)$ are of interest because we seek the relaxation behavior of $f_1(t)$.

Eq. (69) may be solved numerically by transforming it into the matrix equation

$$\frac{d}{dt}\mathbf{X}(t) = \mathbf{A}\mathbf{X}(t)$$
(70)

where

$$
\mathbf{A} = -\frac{1}{\tau_D}
\begin{pmatrix}
\left(1 - \frac{2}{5}\sigma\right) & \frac{2}{2}\sigma & 0 & 0 & 0 & \cdots \\
-\frac{24}{35}\sigma & \left(6 - \frac{4}{15}\sigma\right) & \frac{20}{21}\sigma & 0 & 0 & \cdots \\
0 & -\frac{40}{33}\sigma & \left(15 - \frac{10}{39}\sigma\right) & \frac{210}{143}\sigma & 0 & \cdots \\
\cdots & \cdots & \cdots & \cdots & \cdots & \cdots
\end{pmatrix},
$$

$$
\mathbf{X}(t) =
\begin{pmatrix}
f_1(t) \\
f_3(t) \\
\cdots \\
f_{2n+1}(t) \\
\cdots
\end{pmatrix}
\tag{71}
$$

In Eq. (70), n is taken large enough (equal to P say) to ensure convergence of the set of equations.

The lowest eigenvalue, which corresponds to the reciprocal of the longest relaxation time, is then the smallest root of the characteristic equation

$$
\det(\lambda \mathbf{l} - \mathbf{A}) = 0
\tag{72}
$$

The relaxation modes of $f_1(t)$ may be found from Eq. (70) by assuming that $\mathbf{A}$ has a linearly independent set of P eigenvectors

$$
(\mathbf{R}_1, \ldots, \mathbf{R}_P)
$$

so that

$$
\mathbf{R}^{-1}\mathbf{A}\mathbf{R} = \mathbf{B}
$$

and

$$
\mathbf{X}(t) = \mathbf{R}e^{\mathbf{B}t}\mathbf{R}^{-1}\mathbf{X}_0
$$

where $\mathbf{R}$ is the matrix consisting of the n eigenvectors of $\mathbf{A}$ and

$$
\mathbf{B} = \frac{1}{\tau_D}
\begin{pmatrix}
\lambda_1 & 0 & 0 & \cdots \\
0 & \lambda_3 & 0 & \cdots \\
0 & 0 & \lambda_5 & \cdots \\
\cdots & \cdots & \cdots & \cdots
\end{pmatrix} \quad \text{and}
$$

$$
e^{\mathbf{B}t} = -\frac{1}{\tau_D}
\begin{pmatrix}
e^{-\lambda_1 t/\tau_D} & 0 & 0 & \cdots \\
0 & e^{-\lambda_3 t/\tau_D} & 0 & \cdots \\
0 & 0 & e^{-\lambda_5 t/\tau_D} & \cdots \\
\cdots & \cdots & \cdots & \cdots
\end{pmatrix}
\tag{73}
$$

and X_0 is the matrix of the initial values of the $f_{2n+1}(t)$. The solution of Eq. (70) may then be exhibited in the form

$$X(t) = b_1 e^{-\lambda_1 t/\tau_D} R_1 + b_2 e^{-\lambda_3 t/\tau_D} R_2 + \ldots + b_P e^{-\lambda_{2P+1} t/\tau_D} R_P \qquad (74)$$

where the b_i are to be determined from the initial conditions. The initial value vector X_0 is determined by the initial value of $f_n(t)$

$$f_n(0) = \frac{\int_0^\pi \cos \vartheta P_n(\cos \vartheta) e^{\sigma \cos^2 \vartheta} \sin \vartheta \, d\vartheta}{\int_0^\pi e^{\sigma \cos^2 \vartheta} \sin \vartheta \, d\vartheta} \qquad (75)$$

The functions $f_n(0)$ may be expressed in terms of the confluent hypergeometric functions $M(a, b, z)$ [31], as shown below.

Eq. (70) may be solved to any desired degree of accuracy to yield $f_1(t)$ in the form

$$\frac{f_1(t)}{f_1(0)} = \frac{\sum_{k=0}^\infty A_{2k+1} e^{-\lambda_{2k+1} t/\tau_D}}{\sum_{k=0}^\infty A_{2k+1}} \qquad (76)$$

The quantity of most interest is the correlation time $\tau_\parallel$, which is the area under the curve of the longitudinal autocorrelation function. The longitudinal autocorrelation function is

$$C_\parallel(t) = \frac{f_1(t)}{f_1(0)} \qquad (77)$$

so that the correlation time $\tau_\parallel$ is

$$\tau_\parallel = \int_0^\infty C_\parallel(t) dt = \tau_D \frac{\sum_k A_{2k+1} \lambda_{2k+1}^{-1}}{\sum_k A_{2k+1}} \qquad (78)$$

where A_k are the weight coefficients (amplitudes). Furthermore, for the one-dimensional Langevin or Fokker-Planck equation the correlation time can always be expressed in closed (integral) form (Appendix B).

Another required quantity is the effective relaxation time, defined as

$$\tau_\parallel^{ef} = \frac{\tau_D \sum_k A_{2k+1}}{\sum_k A_{2k+1} \lambda_{2k+1}} \qquad (79)$$

A review of the effective eigenvalue method was provided by Coffey et al. [32]. The effective relaxation time also includes contributions from all the eigenvalues. Moreover, it gives precise information on the initial relaxation of the polarization in the time domain and occasionally can accurately predict the entire time response of the system. Indeed, if a single eigenvalue dominates the relaxation, then $\tau \cong \tau^{ef}$. If widely different time scales are involved, as happens in activation processes, however, the behaviors of the correlation time and effective relaxation time are very different.

It is usually impossible to evaluate analytically both the correlation time and effective relaxation time from these formulas, because a knowledge of the law of formation of the eigenvalues and their corresponding weights (amplitudes) is required. The approach we use below does not attempt to calculate those times by explicitly calculating the eigenvalue spectrum, as dictated by Eqs. (78) and (79), rather it proceeds in terms of an exact integral representation for the correlation time and in terms of the order parameter S for the effective relaxation time.

We have on taking the Laplace transform of Eq. (69)

$$
\left[\frac{2\tau_D s}{n(n+1)} + 1 - \frac{2\sigma}{(2n-1)(2n+3)}\right]\tilde{f}_n(s) + \frac{2\sigma(n+2)}{(2n+1)(2n+3)}\tilde{f}_{n+2}(s)
$$
$$
= \frac{2\tau_D}{n(n+1)}f_n(0) + \frac{2\sigma(n-1)}{4n^2-1}\tilde{f}_{n-2}(s) \tag{80}
$$

The solution of Eq. (80) allows one to determine $\tilde{f}_1(s)$. The homogeneous Eq. (80) (i.e., with $f_n(0) = 0$) may be readily solved in terms of continued fractions. Here the relevant continued fraction is

$$
\tilde{S}_n(s) = \cfrac{\dfrac{2\sigma(n-1)}{4n^2-1}}{\dfrac{2\tau_D s}{n(n+1)} + 1 - \dfrac{2\sigma}{(2n-1)(2n+3)} + \dfrac{2\sigma(n+2)}{(2n+1)(2n+3)}\tilde{S}_{n+2}(s)} \tag{81}
$$

We seek a complete solution of the inhomogeneous Eq. (80). We can regard Eq. (80) as having a particular solution and a complementary solution. The particular solution of Eq. (81) satisfies the recurrence relation in Eq. (80) with $f_n(0) = 0$. Thus we write [20]:

$$
\tilde{f}_n(s) = \tilde{S}_n(s)\tilde{f}_{n-2}(s) + q_n(s) \tag{82}
$$

Substituting Eq. (82) into (80) and using Eq. (81), we obtain

$$
\left[\frac{2\tau_D s}{n(n+1)} + 1 - \frac{2\sigma}{(2n-1)(2n+3)}\right] q_n
$$
$$
+ \frac{2\sigma(n+2)}{(2n+1)(2n+3)} \left[q_{n+2} + q_n \tilde{S}_{n+2}\right] = \frac{2\tau_D f_n(0)}{n(n+1)} \tag{83}
$$

Eq. (83) may be solved for q_n to get

$$
q_n = a_n \left[\frac{\tau_D}{\sigma} f_n(0) - b_n q_{n+2}\right] \tilde{S}_n(s) \tag{84}
$$

where

$$
a_n = \frac{4n^2 - 1}{n(n^2 - 1)}, \quad b_n = \frac{n(n+1)(n+2)}{(2n+1)(2n+3)} \tag{85}
$$

Recalling Eq. (83), we obtain

$$
\tilde{f}_n(s) = \left\{\tilde{f}_{n-2}(s) + a_n \left[\frac{\tau_D}{\sigma} f_n(0) - b_n q_{n+2}\right]\right\} \tilde{S}_n(s) \tag{86}
$$

According to Eq. (80) we have for $n = 1$

$$
\tilde{f}_1(s) = \frac{1}{G(\sigma, s)} \left[\tau_D f_1(0) - \frac{2}{5}\sigma q_3\right] \tag{87}
$$

where

$$
G(\sigma, s) = s\tau_D + 1 - \frac{2}{5}\sigma + \frac{2}{5}\sigma \tilde{S}_3(s) \tag{88}
$$

Substituting for q_3 from Eq. (84) into Eq. (87) we have

$$
\tilde{f}_1(s) = \frac{1}{G(\sigma, s)} \left\{\tau_D f_1(0) - a_3 b_1 [\tau_D f_3(0) - \sigma b_3 q_5] \tilde{S}_3(s)\right\} \tag{89}
$$

Similarly substituting for q_5 we have

$$
\tilde{f}_1(s) = \frac{1}{G(\sigma, s)} \left\{\tau_D f_1(0) - a_3 b_1 [\tau_D f_3(0) \tilde{S}_3(s)\right.
$$
$$
\left. - b_3 a_5 (\tau_D f_5(0) - \sigma b_5 q_7) \tilde{S}_3(s) \tilde{S}_5(s)]\right\} \tag{90}
$$

Continuing the process we obtain

$$\tilde{f}_1(s) = \frac{\tau_D}{G(\sigma, s)} \left[f_1(0) + \sum_{n=1}^{\infty} (-1)^n f_{2n+1}(0) \prod_{k=1}^{n} a_{2k+1} b_{2k-1} \tilde{S}_{2k+1}(s) \right] \quad (91)$$

Eq. (91) may further be simplified if we write out the product $a_{2k+3} b_{2k+1}$ explicitly. We have

$$\prod_{k=1}^{n} a_{2k+1} b_{2k-1} = \frac{(4n+3)\Gamma\left(n+\frac{1}{2}\right)}{3\Gamma(n+2)\Gamma\left(\frac{1}{2}\right)} \quad (92)$$

where $\Gamma(z)$ is the γ function [31]. So Eq. (91) reduces to

$$\tilde{f}_1(s) = \frac{\tau_D}{G(\sigma, s)} \left[f_1(0) + \frac{4}{3} \sum_{n=1}^{\infty} (-1)^n f_{2n+1}(0) \frac{\left(n+\frac{3}{4}\right)\Gamma\left(n+\frac{1}{2}\right)}{\Gamma(n+2)\Gamma\left(\frac{1}{2}\right)} \prod_{k=1}^{n} \tilde{S}_{2k+1}(s) \right]$$

$$(93)$$

The initial value vector $f_{2n+1}(0)$ is given by Eq. (75), namely:

$$f_{2n+1}(0) = \langle \cos\vartheta P_{2n+1}(\cos\vartheta) \rangle_0 = \frac{\int_0^1 x P_{2n+1}(x) e^{\sigma x^2} dx}{\int_0^1 e^{\sigma x^2} dx} \quad (94)$$

In Eq. (94), the subscript 0 denotes the equilibrium ensemble average. Now, using Eq. (58) for $m = 0$ we obtain

$$f_{2n+1}(0) = \langle x P_{2n+1} \rangle_0 = \frac{1}{4n+3} [2(n+1)\langle P_{2n+2} \rangle_0 + (2n+1)\langle P_{2n} \rangle_0] \quad (95)$$

Now, according to Eq. (62) the equilibrium averages $\langle P_n \rangle_0$ satisfy

$$\left[1 - \frac{2\sigma}{(2n-1)(2n+3)} \right] \langle P_n \rangle_0$$

$$= 2\sigma \left[\frac{n+1}{(2n-1)(2n+1)} \langle P_{n-2} \rangle_0 - \frac{n}{(2n+1)(2n+3)} \langle P_{n+2} \rangle_0 \right] \quad (96)$$

Hence, we have

$$\frac{\langle P_n \rangle_0}{\langle P_{n-2} \rangle_0} = \tilde{S}_n(0) \quad (97)$$

where $\tilde{S}_n(0)$ is given by Eq. (81) at $s = 0$, namely:

$$\tilde{S}_n(0) = \frac{\dfrac{2\sigma(n-1)}{4n^2 - 1}}{1 - \dfrac{2\sigma}{(2n-1)(2n+3)} + \dfrac{2\sigma(n+2)}{(2n+1)(2n+3)}\tilde{S}_{n+2}(0)} \tag{98}$$

Thus, according to Eq. (95), the initial conditions $f_{2n+1}(0)$ may be expressed as

$$f_{2n+1}(0) = \frac{1}{4n+3}\tilde{S}_{2n}(0)\tilde{S}_{2n-2}(0)\ldots\tilde{S}_2(0)[(2n+2)\tilde{S}_{2n+2}(0) + 2n + 1] \tag{99}$$

so yielding the initial conditions entirely in terms of the continued fractions $\tilde{S}_n(0)$.

We require $\tilde{S}_n(0)$ as a ratio of confluent hypergeometric (Kummer) functions $M(a, b, z)$ [31] defined as

$$M(a,b,z) = 1 + \frac{a}{b}z + \frac{a(a+1)}{b(b+1)}\frac{z^2}{2!} + \frac{a(a+1)(a+2)}{b(b+1)(b+2)}\frac{z^3}{3!} + \ldots \tag{100}$$

This is accomplished by noting that Eq. (98) can be rearranged to yield after simple algebra

$$1 - \tilde{S}_n(0) = \cfrac{1}{1 + \cfrac{\dfrac{2\sigma(n-1)}{(2n+1)(2n-1)}}{1 - \dfrac{2\sigma(n+2)}{(2n+1)(2n+3)}(1 - \tilde{S}_{n+2})}} \tag{101}$$

Comparing Eq. (101) with the continued fraction [33]

$$\frac{M(a+1,b+1,z)}{M(a,b,z)} = \cfrac{1}{1 - \cfrac{\dfrac{z(b-a)}{b(b+1)}}{1 + \dfrac{z(a+1)}{(b+1)(b+2)}\dfrac{M(a+2,b+3,z)}{M(a+1,b+2,z)}}} \tag{102}$$

we can see that the fraction (101) is identical to Eq. (102) if

$$z = -\sigma, a = n/2 \text{ and } b = n - 1/2$$

Thus we can write

$$\tilde{S}_n(0) = 1 - \frac{M\left(1 + \frac{n}{2}, n + \frac{1}{2}, -\sigma\right)}{M\left(\frac{n}{2}, n - \frac{1}{2}, -\sigma\right)} = 1 - \frac{M\left(\frac{n-1}{2}, n + \frac{1}{2}, \sigma\right)}{M\left(\frac{n-1}{2}, n - \frac{1}{2}, \sigma\right)} \tag{103}$$

Here we have used the Kummer transformation [31]

$$M(a, b, z) = e^z M(b - a, b, -z) \tag{104}$$

Using the recurrence relation [31]

$$M(a, b - 1, z) - M(a, b, z) = \frac{az}{b(b-1)} M(a + 1, b + 1, z) \tag{105}$$

we can rearrange Eq. (103) as follows:

$$\tilde{S}_n(0) = \frac{2(n-1)\sigma}{(4n^2 - 1)} \frac{M\left(\frac{n+1}{2}, n + \frac{3}{2}, \sigma\right)}{M\left(\frac{n-1}{2}, n - \frac{1}{2}, \sigma\right)} \tag{106}$$

Thus using Eq. (106) in Eq. (99) we obtain

$$f_{2n+1}(0) = \frac{\sigma^n \Gamma\left(n + \frac{3}{2}\right) M\left(n + \frac{3}{2}, 2n + \frac{5}{2}, \sigma\right)}{2\Gamma\left(2n + \frac{5}{2}\right) M\left(\frac{1}{2}, \frac{3}{2}, \sigma\right)} \tag{107}$$

Eq. (93) allows one to calculate the frequency dependence of the longitudinal polarizability [5]

$$\alpha_\parallel(\omega) = \alpha'_\parallel(\omega) - i\alpha''_\parallel(\omega)$$

because, according to linear response theory,

$$\frac{\alpha_\parallel(\omega)}{\alpha'_\parallel(0)} = 1 - i\omega \frac{\tilde{f}_1(i\omega)}{f_1(0)} \tag{108}$$

where

$$\alpha'_{\parallel}(0) = \frac{\mu^2 N}{3kT}(1 + 2S) \tag{109}$$

Thus we have

$$\frac{\alpha_{\parallel}(\omega)}{\alpha'_{\parallel}(0)} = \frac{1}{i\omega\tau_D + 1 - \frac{2}{5}\sigma + \frac{2}{5}\sigma\tilde{S}_3(i\omega)}\left[1 - \frac{2}{5}\sigma + \frac{2}{5}\sigma\tilde{S}_3(i\omega)\right.$$
$$\left. - i\omega\tau_D\frac{4}{3}\sum_{n=1}^{\infty}(-1)^n f_{2n+1}(0)\frac{(n+3/4)\Gamma(n+1/2)}{\Gamma(n+2)\Gamma(1/2)}\prod_{k=1}^{n}\tilde{S}_{2k+1}(i\omega)\right] \tag{110}$$

The most significant feature of Eq. (93), however, is that it is capable of yielding an exact expression for the correlation time. We have, setting $s = 0$ in Eq. (93)

$$\tau_{\parallel} = \frac{\tilde{f}_1(0)}{f_1(0)} = \frac{\tau_D}{1 - \frac{2\sigma}{5} + \frac{2\sigma}{5}\tilde{S}_{2k+1}(0)}$$
$$\times\left[1 + \frac{4}{3}\sum_{n=1}^{\infty}(-1)^n\frac{f_{2n+1}(0)}{f_1(0)}\frac{\left(n+\frac{3}{4}\right)\Gamma\left(n+\frac{1}{2}\right)}{\Gamma(n+2)\Gamma\left(\frac{1}{2}\right)}\prod_{k=1}^{n}\tilde{S}_{2k+1}(0)\right] \tag{111}$$

where $\tilde{S}_{2k+1}(0)$ is given in terms of Kummer's functions by Eq. (106). Using Eqs. (103) and (104) and properties of Kummer's functions [31],

$$M(0, b, z) = 1, \quad bM(a, b, z) - bM(a - 1, b, z) - zM(a, b + 1, z) = 0 \tag{112}$$

we can express the leading term of Eq. (111) as follows:

$$\frac{1}{1 - \frac{2\sigma}{5}[1 - \tilde{S}_3(0)]} = \frac{1}{1 - \frac{2\sigma}{5}\frac{M\left(1, \frac{7}{2}, \sigma\right)}{M\left(1, \frac{5}{2}, \sigma\right)}} = M\left(1, \frac{5}{2}, \sigma\right) \tag{113}$$

Using Eqs. (106) and (113), Eq. (111) may further be simplified to

$$\frac{\tau_{\parallel}}{\tau_D} = \frac{3}{2} \sum_{n=0}^{\infty}$$
$$\times \frac{(-\sigma^2)^n \left(n+\frac{3}{4}\right)\Gamma\left(n+\frac{3}{2}\right)\Gamma\left(n+\frac{1}{2}\right)M\left(n+\frac{3}{2},2n+\frac{5}{2},\sigma\right)M\left(n+1,2n+\frac{5}{2},\sigma\right)}{(n+1)\Gamma^2\left(2n+\frac{5}{2}\right)M\left(\frac{3}{2},\frac{5}{2},\sigma\right)}$$

$$(114)$$

which is the exact solution in terms of known functions. For large σ this has asymptotic behavior [9]

$$\tau_{\parallel} \sim \tau_D \frac{\sqrt{\pi}e^{\sigma}}{2\sigma^{3/2}}\left[1+\frac{1}{\sigma}+O(\sigma^{-2})\right]$$

which is in agreement with Storonkin's [4] asymptotic solution.

Eq. (114) may be used to calculate the relaxation time for all values of σ [34]. Furthermore, the series in Eq. (114) can be summed to yield [10]:

$$\frac{\tau_{\parallel}}{\tau_D} = \frac{3e^{\sigma}}{\sigma^2 M\left(\frac{3}{2},\frac{5}{2},\sigma\right)} \int_0^1 \frac{\cosh[\sigma(z^2-1)]-1}{1-z^2}dz \qquad (115)$$

Note that all the Kummer functions appearing in Eq. (114) may be expressed in terms of the more familiar error functions of real and imaginary arguments [31].

$$\mathrm{erf}(z) = \frac{2}{\sqrt{\pi}}\int_0^z \exp(-t^2)dt, \quad \mathrm{erfi}(z) = \frac{2}{\sqrt{\pi}}\int_0^z \exp(t^2)dt$$

In particular [35]

$$M\left(1,\frac{5}{2},z\right) = \frac{3}{2z}\left[\sqrt{\frac{\pi}{4z}}e^z\mathrm{erf}(\sqrt{z})-1\right]$$
$$M\left(\frac{1}{2},\frac{3}{2},z\right) = \sqrt{\frac{\pi}{4z}}\mathrm{erfi}(\sqrt{z})$$

Equations for the other Kummer functions occurring in Eq. (114) may be obtained from Prudnikov et al. [35], and the recurrence relations for the M functions.

Equations (114) and (115) have the disadvantage, however, of not being easy to manipulate, especially when one desires an explicit expression for $\tau_\parallel$ in terms of σ. This difficulty has been overcome recently [9]; using a combination of a variational principle and curve fitting, the following simple formula was derived that renders a close approximation to the exact solution for all σ [36]:

$$\frac{\tau_\parallel^{ap}}{\tau_D} \approx \frac{e^\sigma - 1}{\sigma} \left(\frac{2\sigma\sqrt{\sigma/\pi}}{1 + \sigma} + 2^{-\sigma} \right)^{-1} \tag{116}$$

The behavior of this formula is compared in Figure 2 with the exact solution and with the earlier formula of Meier and Saupe [8].

$$\frac{\tau_\parallel^{ap}}{\tau_D} \approx \frac{e^\sigma - 1}{\sigma} \tag{117}$$

which is simply the prefactor of Eq. (116). It is apparent that the correction to Eq. (117) provided by Eq. (116) is significant, because the formula considerably overestimates $\tau_\parallel$ for $\sigma \gg 1$.

As mentioned, yet another approximation that has been used to estimate the effective longitudinal relaxation time is the inverse of the effective eigenvalue [5,32,34]. The effective relaxation time is found by evaluating

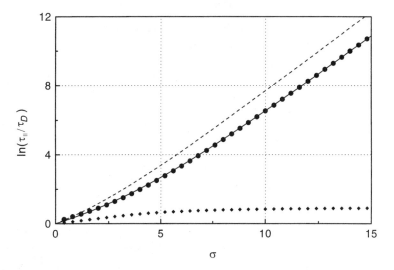

Figure 2. The longitudinal correlation times as a function of σ computed from the exact solution Eq. (114) (*solid line*), approximate formula Eq. (116) (●), Meier and Saupe's [8] formula Eq. (117) (*dashed line*), and the effective eigenvalue formula Eq. (119) (◆).

Eq. (75) for $n = 1$ at $t = 0$. We have

$$\tau_D \frac{d}{dt} f_1(0) + \left[1 - \frac{2\sigma}{5}\right] f_1(0) = -\frac{2\sigma}{5} f_3(0) \tag{118}$$

The effective relaxation time is then

$$\frac{\tau_{\parallel}^{ef}}{\tau_D} = \frac{1}{1 - \frac{2\sigma}{5} + \frac{2\sigma f_3(0)}{5 f_1(0)}} = \frac{1}{1 - \frac{2\sigma}{5} + \frac{12\sigma^2 M\left(\frac{5}{2},\frac{9}{2},\sigma\right)}{175 M\left(\frac{3}{2},\frac{5}{2},\sigma\right)}} = \frac{2S + 1}{1 - S} \tag{119}$$

where

$$S = \tilde{S}_2(0) = \langle P_2 \rangle_0 = \frac{2\sigma M\left(\frac{3}{2},\frac{7}{2},\sigma\right)}{15 M\left(\frac{1}{2},\frac{3}{2},\sigma\right)} \tag{120}$$

is the order parameter. It is apparent from Figure 2 that the effective eigenvalue approach is inadequate when applied to the longitudinal relaxation [34], because the effective relaxation time cannot reproduce the behavior of the correlation time in the large σ limit; the times diverge exponentially from each other because an activation process is involved.

C. The Transverse Relaxation

The set of equations governing the transverse relaxation is given by Eq. (68). The Laplace transform of this equation is

$$\left[2\tau_D s + n(n+1) - 2\sigma \frac{n(n+1) - 3}{(2n-1)(2n+3)}\right] \tilde{g}_n(s) + \frac{2\sigma n^2(n+1)}{(2n+1)(2n+3)} \tilde{g}_{n+2}(s)$$

$$= 2\tau_D g_n(0) + \frac{2\sigma n(n+1)^2}{(2n+1)(2n-1)} \tilde{g}_{n-2}(s) \tag{121}$$

The solution of Eq. (121) will allow us to determine $\tilde{g}_1(s)$.

The homogeneous equation in Eq. (121) (i.e., with $g_n(0) = 0$) may be readily solved in terms of the continued fraction

$$\tilde{S}_n^{\perp}(s) = \cfrac{\cfrac{2\sigma n(n+1)^2}{4n^2 - 1}}{2\tau_D s + n(n+1) - 2\sigma \cfrac{n(n+1) - 3}{(2n-1)(2n+3)} + \cfrac{2\sigma n^2(n+1)}{(2n+1)(2n+3)} \tilde{S}_{n+2}^{\perp}(s)} \tag{122}$$

Just as in the longitudinal case, we seek a solution of Eq. (121) in the form

$$\tilde{g}_n(s) = \tilde{S}_n^\perp(s)\tilde{g}_{n-2}(s) + q_n^\perp(s) \tag{123}$$

where the $\tilde{S}_n(s)$ are given by Eq. (122). Equation (121) then becomes

$$\left[2\tau_D s + n(n+1) - 2\sigma\frac{n(n+1) - 3}{(2n-1)(2n+3)}\right]q_n^\perp$$

$$+ \frac{2\sigma n^2(n+1)}{(2n+1)(2n+3)}[q_{n+2}^\perp + q_n^\perp \tilde{S}_{n+2}^\perp] = 2\tau_D g_n(0) \tag{124}$$

Equation (124) may be solved for $q_n^\perp(s)$ to get

$$q_n^\perp(s) = \frac{2\tau_D g_n(0) - \dfrac{2\sigma n^2(n+1)}{(2n+1)(2n+3)}q_{n+2}^\perp(s)}{2\tau_D s + n(n+1) - 2\sigma\dfrac{n(n+1) - 3}{(2n-1)(2n+3)} + \dfrac{2\sigma n^2(n+1)}{(2n+1)(2n+3)}\tilde{S}_{n+2}^\perp(s)}$$

$$= a_n\left[\frac{\tau_D}{\sigma}g_n(0) - b_n q_{n+2}^\perp(s)\right]\tilde{S}_n^\perp(s) \tag{125}$$

where

$$a_n = \frac{4n^2 - 1}{n(n+1)^2}, \qquad b_n = \frac{n^2(n+1)}{(2n+1)(2n+3)} \tag{126}$$

Noting that $\tilde{g}_{-1}(s) = 0$, we have from Eqs. (123) and (125) for $n = 1$

$$\tilde{g}_1(s) = \frac{g_1(0)\tau_D - 2\sigma q_3^\perp/15}{s\tau_D + 1 + \sigma/5 + 2\sigma\tilde{S}_3(s)/15} \tag{127}$$

Substituting for $q_3^\perp$ in Eq. (127) from Eq. (125) for $n = 3$, we have

$$\tilde{g}_1(s) = \frac{1}{s\tau_D + 1 + \sigma/5 + 2\sigma\tilde{S}_3^\perp(s)/15}\{\tau_D g_1(0) - a_3 b_1[\tau_D g_3(0) - \sigma b_3 q_5^\perp]\tilde{S}_3^\perp(s)\} \tag{128}$$

so that continuing this iterative procedure we obtain

$$\tilde{g}_1(s) = \frac{3\tau_D \tilde{S}_1^\perp(s)}{4\sigma}\left[g_1(0) + \sum_{n=1}^\infty (-1)^n g_{2n+1}(0)\prod_{k=1}^n a_{2k+1}b_{2k-1}\tilde{S}_{2k+1}^\perp(s)\right] \tag{129}$$

Eq. (129) may be further simplified if we write out the product $\prod_{k=1}^{n} a_{2k+1}b_{2k-1}$ explicitly. Using

$$\prod_{k=1}^{n} a_{2k+1}b_{2k-1} = \frac{(4n+3)\Gamma\left(n+\frac{1}{2}\right)}{3(2n+1)(n+1)\Gamma(n+2)\Gamma\left(\frac{1}{2}\right)} \qquad (130)$$

we may arrange Eq. (129) as

$$\tilde{g}_1(s) = \frac{\tau_D}{4\sigma\sqrt{\pi}} \sum_{n=0}^{\infty} \frac{(-1)^n g_{2n+1}(0)(4n+3)\Gamma\left(n+\frac{1}{2}\right)}{(n+1)(2n+1)\Gamma(n+2)} \prod_{k=0}^{n} \tilde{S}_{2k+1}^{\perp}(s) \qquad (131)$$

The initial conditions may be calculated just as in the longitudinal case. We have

$$g_{2n+1}(0) = \langle \sin \vartheta P_{2n+1}^1(\cos \vartheta)\cos^2\varphi\rangle_0 = \frac{1}{2}\langle P_1^1(\cos \vartheta)P_{2n+1}^1(\cos \vartheta)\rangle_0 \qquad (132)$$

We now make use of the recurrence relation [25]:

$$(2n+1)\sqrt{1-x^2}P_n^m(x) = (n-m+1)(n-m+2)P_{n+1}^{m-1}(x)$$
$$- (n+m)(n+m-1)P_{n-1}^{m-1}(x) \qquad (133)$$

so that Eq. (132) reduces to

$$g_{2n+1}(0) = \frac{(n+1)(2n+1)}{4n+3}[\langle P_{2n}(\cos \vartheta)\rangle_0 - \langle P_{2n+2}(\cos \vartheta)\rangle_0] \qquad (134)$$

Eq. (134) may also be computed in terms of Kummer functions. For example for $n=0$ we have

$$g_1(0) = \frac{1}{3}[1 - \langle P_2(\cos \vartheta)\rangle_0] = \frac{1}{3}[1 - S_2(0)] = \frac{M\left(\frac{1}{2},\frac{5}{2},\sigma\right)}{3M\left(\frac{1}{2},\frac{3}{2},\sigma\right)} \qquad (135)$$

Continuing this procedure, we have from Eqs. (104), (106), and (135)

$$g_{2n+1}(0) = \frac{(2n+1)(n+1)}{4n+3}[1 - S_{2n+2}(0)]S_{2n}(0)S_{2n-2}(0)\ldots S_2(0)$$
$$= \frac{(n+1)\left(\sigma^n\Gamma\left(n+\frac{3}{2}\right)M\left(n+\frac{1}{2},2n+\frac{5}{2},\sigma\right)\right)}{2\Gamma\left(2n+\frac{5}{2}\right)M\left(\frac{1}{2},\frac{3}{2},\sigma\right)} \qquad (136)$$

The exact solution for the Laplace transform of the transverse after effect function allows one to calculate the transverse complex polarizability $\alpha_\perp(\omega) = \alpha'_\perp(\omega) - i\alpha''_\perp(\omega)$, because according to linear response theory

$$\frac{\alpha_\perp(\omega)}{\alpha'_\perp(0)} = 1 - i\omega \frac{\tilde{g}_1(i\omega)}{g_1(0)} \tag{137}$$

where

$$\alpha'_\perp(0) = \frac{\mu^2 N}{3kT}(1 - S)$$

Thus we have

$$\frac{\alpha_\perp(\omega)}{\alpha'_\perp(0)} = 1 - \frac{i\omega\tau_D}{\sigma\sqrt{\pi}}\sum_{n=0}^{\infty}(-1)^n\frac{\left(n+\frac{3}{4}\right)\Gamma\left(n+\frac{1}{2}\right)g_{2n+1}(0)}{(n+1)(2n+1)\Gamma(n+2)g_1(0)}\prod_{k=0}^{n}\tilde{S}^\perp_{2k+1}(i\omega) \tag{138}$$

which indicates that the transverse complex susceptibility is made up of an infinite number of Debye-type relaxation mechanisms.

The exact solution for the transverse relaxation time $\tau_\perp$ is obtained by setting $s = 0$ in Eq. (131). We have

$$\tau_\perp = \frac{\tilde{g}_1(0)}{g_1(0)} = \frac{\tau_D}{4\sigma\sqrt{\pi}}\sum_{n=0}^{\infty}\frac{(-1)^n(4n+3)\Gamma\left(n+\frac{1}{2}\right)g_{2n+1}(0)}{(n+1)(2n+1)\Gamma(n+2)g_1(0)}\prod_{k=0}^{n}\tilde{S}^\perp_{2k+1}(0) \tag{139}$$

At first glance it would seem apparent that this equation could be easily represented as a series of products of Kummer functions just as in the longitudinal case (Table I). Unfortunately, it is not at all easy to do this and we cannot find an appropriate mathematical function that generates the continued fraction $\tilde{S}^\perp_n(0)$. A possible reason for this is that the transverse relaxation involves both polar angles ϑ and ϕ. In view of this difficulty, it is best to seek a simple analytic formula for $\tau_\perp$ by some other method. Fortunately, in contrast to the longitudinal case, such a formula is provided by the effective eigenvalue method, which yields $\tau_\perp$ according to Eq. (68) as

$$\frac{\tau_\perp^{ef}}{\tau_D} = -\frac{g_1(0)}{\tau_D\dot{g}_1(0)} = \frac{1}{1 + \dfrac{\sigma}{5} + \dfrac{2\sigma}{15}\dfrac{g_3(0)}{g_1(0)}}$$

$$= \frac{1}{1 - \dfrac{2\sigma}{5} + \dfrac{8\sigma^2}{175}\dfrac{M\left(\frac{3}{2},\frac{9}{2},\sigma\right)}{M\left(\frac{1}{2},\frac{5}{2},\sigma\right)}} = \frac{2M\left(\frac{1}{2},\frac{5}{2},\sigma\right)}{3M\left(\frac{1}{2},\frac{3}{2},\sigma\right) - M\left(\frac{1}{2},\frac{5}{2},\sigma\right)} \tag{140}$$

TABLE I
Numerical Values for the Transverse Correlation Time $\tau_\perp/\tau_D$
as a Function of σ^a

σ	$\tau_\perp/\tau_D$	$\tau_\perp^{ef}/\tau_D$	$1/\lambda_1^\perp$
0	1	1	1
1	0.805	0.799	0.807
2	0.631	0.612	0.638
3	0.486	0.460	0.500
4	0.372	0.347	0.393
5	0.288	0.267	0.312
6	0.227	0.213	0.252
7	0.184	0.175	0.206
8	0.154	0.148	0.172
9	0.132	0.128	0.146
10	0.116	0.113	0.126

^{a}Computed from the effective eigenvalue formula Eq. (143)
and the exact solution Eq. (139).

For large σ this has asymptotic behavior [34]

$$\tau_\perp^{ef} \sim \tau_D/\sigma \tag{141}$$

On the other hand for small σ it becomes [9,10]

$$\tau_\perp^{ef} \approx \tau_D(1 - \sigma/5) \tag{142}$$

Using Eqs. (103) and (120) the effective relaxation time may in turn be expressed in terms of the order parameter S to give

$$\tau_\perp^{ef} = 2\tau_D \frac{1-S}{2+S} \tag{143}$$

It is apparent from Figure 3 that the effective eigenvalue formula yields a close approximation to the exact solution for $\tau_\perp$ for all σ, unlike $\tau_\parallel$ rendered by that method. Thus the problem of expressing the exact solution in terms of hypergeometric functions appears, in this case, to be purely of mathematical interest.

More insight into the decay modes governing the relaxation process may be gained by calculating the dielectric response using the matrix approach. This requires a knowledge of the set of eigenvalues λ_k^γ and the corresponding amplitudes A_k^γ of the relaxation process. We calculated the first three eigenvalues λ_k^γ and amplitudes A_k^γ using the method described earlier (a 12 × 12

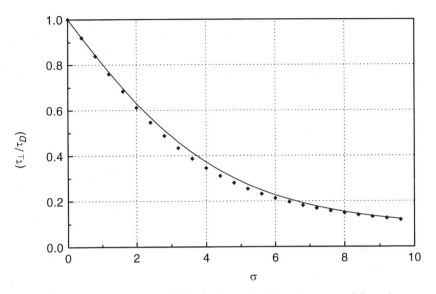

Figure 3. The transverse correlation times as a function of σ computed from the exact solution Eq. (139) (*solid line*) and the effective eigenvalue formula Eq. (143) (◆).

matrix was used for the amplitudes A_k^γ, and a 20×20 matrix was used to ensure convergence of the eigenvalues λ_k^γ). Note that in our notation λ_k differs by a factor of two from that used by Martin et al. [6]. The eigenvalues and amplitudes shown in Tables II and III again reinforce our conclusion that both relaxation processes are effectively described by a single relaxation

TABLE II

Amplitudes $A_{2k+1}^{\parallel}$ and corresponding eigenvalues $\lambda_{2k+1}^{\parallel}$ of the first three decay modes of the longitudinal relaxation as a function of the barrier height parameter σ

σ	$A_1^{\parallel}$	$A_3^{\parallel}$	$A_5^{\parallel}$	$\lambda_1^{\parallel}$	$\lambda_3^{\parallel}$	$\lambda_5^{\parallel}$
0	0.333	0	0	1	6	15
1	0.428	0.000975	9.17×10^{-7}	0.653	5.81	14.8
2	0.528	0.00365	0.0000146	0.404	5.77	14.8
3	0.619	0.00676	0.0000662	0.236	5.91	14.8
4	0.696	0.0088	0.000171	0.13	6.23	15.0
5	0.755	0.00916	0.000319	0.0677	6.74	15.4
6	0.799	0.00824	0.000481	0.0336	7.45	15.9
7	0.832	0.00674	0.000624	0.016	8.37	16.5
8	0.856	0.0052	0.00073	0.00736	9.49	17.3
9	0.875	0.0039	0.000792	0.00329	10.8	18.1
10	0.889	0.00289	0.000812	0.00144	12.3	19.2

TABLE III

Amplitudes $A^{\perp}_{2k+1}$ and corresponding eigenvalues $\lambda^{\perp}_{2k+1}$ of the first three decay modes of the transverse relaxation as a function of the barrier height parameter σ

σ	$A^{\perp}_1$	$A^{\perp}_3$	$A^{\perp}_5$	$\lambda^{\perp}_1$	$\lambda^{\perp}_3$	$\lambda^{\perp}_5$
0	0.333	0	0	1	6	15
1	0.285	0.000773	5.92×10^{-7}	1.24	5.88	14.8
2	0.231	0.00349	0.0000101	1.57	5.89	14.8
3	0.179	0.00793	0.000049	2	6.04	14.9
4	0.135	0.013	0.000134	2.54	6.3	15.2
5	0.1	0.0174	0.000263	3.2	6.66	15.6
6	0.0751	0.0206	0.000408	3.97	7.12	16.1
7	0.0575	0.0224	0.000535	4.85	7.66	16.8
8	0.0473	0.0231	0.000616	5.81	8.27	17.7
9	0.0367	0.0229	0.000636	6.85	8.96	18.8
10	0.0308	0.0222	0.0006	7.95	9.7	20.0
14	0.0191	0.0179	0.000248	12.5	13.1	27.3
16	0.0163	0.0159	0.000131	14.6	15.0	32.3
18	0.0144	0.0142	0.000069	16.8	16.9	38.0
20	0.0129	0.0128	0.000039	18.8	18.9	44.4

mechanism. For the longitudinal relaxation, it is obvious from Table II that the contributions of the modes characterized by the eigenvalues $\lambda^{\|}_3$ and $\lambda^{\|}_5$ are negligible. This conclusion is in agreement with Martin et al. [6]. A different situation is found in the transverse relaxation, in which the modes characterized by $\lambda^{\perp}_1$ and $\lambda^{\perp}_3$ are near degenerate for large σ, and the mode characterized by $\lambda^{\perp}_5$ has almost zero contribution (Table III). Thus the relaxation may again be described by a single mechanism.

IV. FURTHER GENERALIZATION OF THE ROTATIONAL DIFFUSION MODEL

A. Rotational Diffusion Where the Dipole Moment Vector μ Is Not Parallel to the Long Molecular Axis

Now we shall suppose that the dipole moment vector **μ** is oriented at an angle β to the direction of the long axis of the molecule (see Figure 4). We shall also take into account that the rotational diffusion coefficients about short and long molecular axes are different. This case can also be considered in the context of the Langevin equation by means of an appropriate transformation of variables [37,38]. However for convenience and simplicity we will use here the noninertial Fokker-Planck equation, because many important formulas have already been obtained in the context of that approach [3].

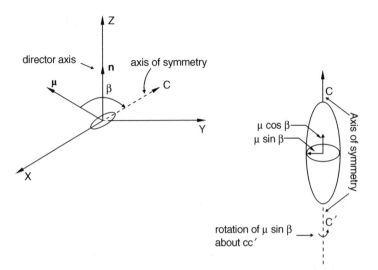

Figure 4. Diagram showing rotational motion of a dipole μ making an angle β with the axis of symmetry (long axis, cc'); the component of μ along the axis of symmetry ($\mu \cos \beta$) gives rise to longitudinal and transverse modes as in $\beta = 0$. The rotation of the perpendicular component of the dipole moment, namely $\mu \sin \beta$, about the long axis cc' yields an additional high frequency Debye-like relaxation characterized by the relaxation time τ_{01}^{ef} (see text) in the longitudinal complex susceptibility $\chi_{\parallel}(\omega)$. Furthermore the rotation of $\mu \sin \beta$ about cc' also gives rise to an additional Debye-like relaxation in the transverse susceptibility $\chi_{\perp}(\omega)$, which is in general not well separated from the transverse motion of $\mu \cos \beta$. The rotation of $\mu \sin \beta$ about the axis of symmetry cc' is characterized by $D_{\parallel}$, the rotational diffusion coefficient with respect to the axis parallel to the axis of symmetry.

The orientation of a moving coordinate system xyz, fixed in the molecule, with respect to the laboratory system XYZ is defined by the Euler angles $\Omega = \{\varphi, \vartheta, \Psi\}$, where φ, ϑ, and Ψ are the azimuthal and polar angles and the angle describing rotation of a molecule around its axis of symmetry, respectively. The components of the dipole vector μ in the Z and X directions of the laboratory coordinate system $OXYZ$ are given by [3]:

$$\mu_z = \sum_{p=-1}^{1} (-1)^p D_{0p}^1(\Omega) \mu^{(1,p)} \tag{144}$$

$$\mu_X = \frac{1}{\sqrt{2}} \sum_{p=-1}^{1} (-1)^p [D_{-1-p}^1(\Omega) - D_{1-p}^1(\Omega)] \mu^{(1,p)} \tag{145}$$

where the irreducible spherical tensor components $\mu^{(1,p)}$ are, in terms of the molecular components μ_x, μ_y, and $\mu_z (\mu_z = \mu\cos\beta, \sqrt{\mu_x^2 + \mu_y^2} = \mu\sin\beta)$:

$$\mu^{(1,0)} = \mu_z \tag{146}$$

$$\mu^{(1,\pm 1)} = \mp\frac{1}{\sqrt{2}}(\mu_x \pm \mu_y) \tag{147}$$

and

$$D_{MM'}^J(\Omega) = e^{-iM\varphi}d_{MM'}^J(\vartheta)e^{-iM'\Psi} \tag{148}$$

are Wigner's D functions. The $d_{MM'}^J(\vartheta)$ in Eq. (148) are given, for example, in Varshalovich et al. [39], and can be expressed in terms of the Legendre polynomials $P_n(\cos\vartheta)$,

$$d_{00}^j(\vartheta) = P_j(\cos\vartheta),$$

$$d_{\pm 11}^j(\vartheta) = \frac{1\pm\cos\vartheta}{j(j+1)}\left\{\frac{dP_j(\cos\vartheta)}{d\cos\vartheta} \pm [1\pm\cos\vartheta]\frac{d^2P_j(\cos\vartheta)}{d\cos\vartheta^2}\right\}, \tag{149}$$

$$d_{\pm 10}^j(\vartheta) = \pm\frac{\sin\vartheta}{\sqrt{j(j+1)}}\frac{dP_j(\cos\vartheta)}{d\cos\vartheta}, \quad d_{0\pm 1}^j(\vartheta) = \pm\frac{\sin\vartheta}{\sqrt{j(j+1)}}\frac{dP_j(\cos\vartheta)}{d\cos\vartheta}$$

$$\tag{150}$$

and so on.

The components of the tensor dipole autocorrelation function from Eqs. (144–149) may be represented as [3]

$$C_\parallel(t) = \cos^2\beta f_{00}^1(t) + \sin^2\beta f_{01}^1(t) \tag{151}$$

$$C_\perp(t) = \cos^2\beta f_{10}^1(t) + \sin^2\beta f_{11}^1(t) \tag{152}$$

where

$$f_{nm}^1(t) = \langle D_{nm}^1(0)D_{nm}^{1*}(t)\rangle_0 = \int D_{nm}^1(\Omega_0)W_0(\Omega_0)d\Omega_0 \int D_{nm}^{1*}(\Omega)W(\Omega,t|\Omega_0)d\Omega \tag{153}$$

are the equilibrium autocorrelation functions, $W_0(\Omega_0) = C\exp(-U/kT)$ and $W(\Omega,t|\Omega_0)$ are, respectively, the equilibrium distribution function and conditional probability density to have a molecule with orientation Ω at an instant t provided that at $t = 0$ that molecule had orientation Ω_0, the asterisk denotes the complex conjugate, $d\Omega \equiv \sin\vartheta d\vartheta d\varphi d\Psi$. The Euler angles are shown in Figure 5.

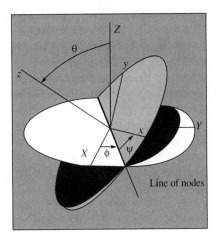

Figure 5. Orientation of a moving coordinate system $x(t)$, $y(t)$, $z(t)$ fixed in the molecule with respect to the laboratory system X, Y, Z, defined by the set of Euler angles $\Omega = \{\phi(t), \vartheta(t), \Psi(t)\}$. Notation follows that of Goldstein: H. Goldstein, *Classical Mechanics*, Addison-Wesley, Reading, MA., 1950.

We seek a solution of the diffusion Eq. (1) in the form

$$W(\Omega, t|\Omega_0) = \sum_{j=0}^{\infty} \sum_{n=-j}^{j} \sum_{m=-j}^{j} (2j+1)c^j_{nm}(t)D^j_{nm}(\Omega) \qquad (154)$$

with the initial condition

$$W(\Omega, 0|\Omega_0) = \delta(\Omega - \Omega_0) \qquad (155)$$

Thus the conditional probability density $W(\Omega, t|\Omega_0)$ is the Green's function of the Fokker-Planck Eq. (1). Noting the representation of the Dirac δ-function [37]:

$$\delta(\Omega - \Omega_0) = \sum_{j=0}^{\infty} \sum_{n=-j}^{j} \sum_{m=-j}^{j} (2j+1)D^j_{nm}(\Omega_0)D^j_{nm}(\Omega) \qquad (156)$$

we have the initial values for $c^j_{nm}(0) = D^j_{nm}(\Omega_0)$. Substituting Eqs. (3) and (154) into Eq. (1) we have

$$\frac{1}{D_\perp} \sum_{jnm} (2j+1)D^j_{nm}(\Omega)\frac{\partial}{\partial t}c^j_{n,m}(t)$$

$$= \sum_{jnm} (2j+1)c^j_{n,m}(t)\{\Delta^2_\Omega D^j_{nm}(\Omega) + 2\sigma[\cos\vartheta \sin\vartheta \frac{\partial}{\partial\vartheta}D^j_{nm}(\Omega)$$

$$+ (3\cos^2\vartheta - 1)D^j_{nm}(\Omega)]\} \qquad (157)$$

Using [39]

$$\cos\vartheta D_{nm}^j(\Omega) = \frac{\sqrt{(j^2-n^2)(j^2-m^2)}}{j(2j+1)}D_{nm}^{j-1}(\Omega) + \frac{nm}{j(j+1)}D_{nm}^j(\Omega)$$
$$+ \frac{\sqrt{[(j+1)^2-n^2][(j+1)^2-m^2]}}{(j+1)(2j+1)}D_{nm}^{j+1}(\Omega) \qquad (158)$$

$$\sin\vartheta \frac{\partial}{\partial\vartheta}D_{nm}^j(\Omega) = -\frac{(j+1)\sqrt{(j^2-n^2)(j^2-m^2)}}{j(2j+1)}D_{nm}^{j-1}(\Omega) - \frac{nm}{j(j+1)}D_{nm}^j(\Omega)$$
$$+ \frac{j\sqrt{[(j+1)^2-n^2][(j+1)^2-m^2]}}{(j+1)(2j+1)}D_{nm}^{j+1}(\Omega) \qquad (159)$$

and [3]

$$\Delta_\Omega^2 D_{nm}^j(\Omega) = -\{j(j+1) + m^2[(D_\parallel/D_\perp) - 1]\}D_{nm}^j(\Omega)$$

where $D_\parallel$ is the rotational diffusion coefficient about the axis of symmetry of the molecule, we have the set of equations for $c_{nm}^j(t)$:

$$\tau_D \frac{d}{dt}c_{nm}^j(t) + \left[\frac{j(j+1)}{2} + m^2\Delta - \sigma\frac{[j(j+1)-3m^2][j(j+1)-3n^2]}{j(j+1)(2j-1)(2j+3)}\right]c_{nm}^j(t)$$

$$= \frac{\sigma}{(2j+1)}\left\{\frac{(j+1)\sqrt{[j^2-n^2][(j-1)^2-n^2][j^2-m^2][(j-1)^2-m^2]}}{j(j-1)(2j-1)}c_{nm}^{j-2}(t)\right.$$

$$+ \frac{mn(j+3)\sqrt{[j^2-n^2][j^2-m^2]}}{j(j-1)(j+1)}c_{nm}^{j-1}(t)$$

$$- \frac{mn(j-2)\sqrt{[(j+1)^2-n^2][(j+1)^2-m^2]}}{j(j+1)(j+2)}c_{nm}^{j+1}(t)$$

$$\left. - \frac{j\sqrt{(j+2)^2-n^2][(j+1)^2-n^2][(j+2)^2-m^2][(j+1)^2-m^2]}}{(j+2)(j+1)(2j+3)}c_{nm}^{j+2}(t)\right\}$$

$$(160)$$

where

$$\Delta = \frac{1}{2}\left[\frac{D_\parallel}{D_\perp} - 1\right]$$

Eq. (160) is equivalent to that derived by Nordio and Busolin [40].

It is now possible to derive from Eqs. (154), (156) and (160), systems of differential-recurrence relations for the equilibrium correlation functions

$$f^j_{00}(t) = \langle D^1_{00}(0)D^j_{00}(t)\rangle_0$$
$$f^j_{01}(t) = \sqrt{j(j+1)/2}\langle D^1_{01}(0)D^{j*}_{01}(t)\rangle_0$$
$$f^j_{10}(t) = \sqrt{j(j+1)/2}\langle D^1_{10}(0)D^{j*}_{10}(t)\rangle_0$$
$$f^j_{11}(t) = \langle D^1_{11}(0)D^{j*}_{11}(t)\rangle_0 \tag{161}$$

These equations are the three 3-term recurrence relations:

$$\frac{2\tau_D}{j(j+1)}\frac{d}{dt}f^j_{00}(t) + \left[1 - \frac{2\sigma}{(2j-1)(2j+3)}\right]f^j_{00}(t)$$
$$= \frac{2\sigma}{2j+1}\left[\frac{j-1}{2j-1}f^{j-2}_{00}(t) - \frac{j+2}{2j+3}f^{j+2}_{00}(t)\right] \tag{162}$$

$$\frac{2\tau_D}{j(j+1)}\frac{d}{dt}f^j_{10}(t) + \left[1 - \frac{2\sigma[1 - 3/j(j+1)]}{(2j-1)(2j+3)}\right]f^j_{10}(t)$$
$$= \frac{2\sigma}{2j+1}\left[\frac{j+1}{2j-1}f^{j-2}_{10}(t) - \frac{j}{2j+3}f^{j+2}_{10}(t)\right] \tag{163}$$

$$\tau_D\frac{d}{dt}f^j_{01}(t) + \left[\frac{j(j+1)}{2} + \Delta - \frac{\sigma[j(j+1) - 3]}{(2j-1)(2j+3)}\right]f^j_{01}(t)$$
$$= \frac{\sigma j(j+1)}{2j+1}\left[\frac{j+1}{2j-1}f^{j-2}_{01}(t) - \frac{j}{2j+3}f^{j+2}_{01}(t)\right] \tag{164}$$

and the five-term recurrence relation

$$\tau_D\frac{d}{dt}f^j_{11}(t) + \left[\frac{j(j+1)}{2} + \Delta - \frac{\sigma[j(j+1) - 3]^2}{j(j+1)(2j-1)(2j+3)}\right]f^j_{11}(t)$$
$$= \frac{\sigma}{2j+1}\left[\frac{(j-2)(j+1)^2}{2j-1}f^{j-2}_{11}(t) + \frac{j+3}{j}f^{j-1}_{11}(t) - \frac{j-2}{j+1}f^{j+1}_{11}(t)\right.$$
$$\left. - \frac{j^2(j+3)}{2j+3}f^{j+2}_{11}(t)\right] \tag{165}$$

with $j = 1, 2, \ldots$ and $f^{-1}_{nm}(t)$, $f^0_{nm}(t) = 0$.

Note that Eqs. (162) and (163) are mathematically identical to those [65,68] describing dielectric relaxation when $\beta = 0$. Such equations were comprehensively described earlier, so we merely quote the main results here. We have the exact solution of Eq. (162) for the Laplace transform

$$\tilde{f}_{00}^1(s) = \int_0^\infty e^{-st} f_{00}^1(t)\,dt \tag{166}$$

in terms of scalar continued fractions, namely:

$$\tilde{f}_{00}^1(s) = \frac{\tau_D}{s\tau_D + 1 - 2\sigma[1 - S_{00}^3(s)]/5}$$
$$\times \left[f_{00}^1(0) + \sum_{n=1}^\infty (-1)^n \frac{(4n+3)\Gamma\left(n+\frac{1}{2}\right) f_{00}^{2n+1}(0)}{3\sqrt{\pi}\,\Gamma(n+2)} \prod_{k=1}^n S_{00}^{2k+1}(s) \right] \tag{167}$$

where $S_{00}^n(s)$ is the continued fraction defined as

$$S_{00}^n(s) = \frac{2\sigma(n-1)}{4n^2 - 1}\left[\frac{2\tau_D s}{n(n+1)} + 1 - \frac{2\sigma}{(2n-1)(2n+3)} \right.$$
$$\left. + \frac{2\sigma(n+2)}{(2n+1)(2n+3)} S_{00}^{n+2}(s) \right]^{-1} \tag{168}$$

The initial conditions $f_{00}^{2n+1}(0)$ are given by

$$f_{00}^{2n+1}(0) \frac{\sigma^n \Gamma\left(n+\frac{3}{2}\right) M\left(n+\frac{3}{2}, 2n+\frac{5}{2}, \sigma\right)}{2\Gamma\left(2n+\frac{5}{2}\right) M\left(\frac{1}{2}, \frac{3}{2}, \sigma\right)} \tag{169}$$

Similarly we have the exact solution for $\tilde{f}_{10}^1(s)$ and $\tilde{f}_{01}^1(s)$ in terms of scalar continued fractions:

$$\tilde{f}_{\{{10 \atop 01}\}}^1(s) = \frac{\tau_D}{4\sigma\sqrt{\pi}} \sum_{n=0}^\infty \frac{(-1)^n(4n+3)\Gamma\left(n+\frac{1}{2}\right)}{(2n+1)(n+1)\Gamma(n+2)} f_{\{{10 \atop 01}\}}^{2n+1}(0) \prod_{k=0}^n S_{\{{10 \atop 01}\}}^{2k+1}(s) \tag{170}$$

where

$$S_{10}^n(s) = \frac{\sigma n(n+1)^2}{4n^2 - 1}\left[\tau_D s + \frac{n(n+1)}{2} - \frac{\sigma[n(n+1) - 3]}{(2n-1)(2n+3)} \right.$$
$$\left. + \frac{\sigma n^2(n+1)}{(2n+1)(2n+3)} S_{10}^{n+2}(s) \right]^{-1} \tag{171}$$

$$S_{01}^n(s) = S_{10}^n\left(s + \frac{\Delta}{\tau_D}\right) \tag{172}$$

$$f_{10}^{2n+1}(0) = f_{01}^{2n+1}(0) = \frac{(n+1)(2n+1)}{4n+3}[\langle P_{2n}(\cos\vartheta)\rangle_0 - \langle P_{2n+2}(\cos\vartheta)\rangle_0]$$

$$= \frac{\sigma^n(n+1)\Gamma\left(n+\frac{3}{2}\right)M\left(n+\frac{1}{2}, 2n+\frac{5}{2}, \sigma\right)}{2\Gamma\left(2n+\frac{5}{2}\right)M\left(\frac{1}{2}, \frac{3}{2}, \sigma\right)} \qquad (173)$$

To obtain $\tilde{f}_{11}^1(s)$ from Eq. (165), we can also apply the general matrix method of solving multi-term differential-recurrence relations developed by Risken [20] and Coffey et al. [10] (Appendix C). Eq. (165) may be transformed to the matrix equation

$$\tau_D\frac{d}{dt}\mathbf{C}_n(t) = \mathbf{Q}_n^-\mathbf{C}_{n-1}(t) + \mathbf{Q}_n\mathbf{C}_n(t) + \mathbf{Q}_n^+\mathbf{C}_{n+1}(t) \qquad (174)$$

with $\mathbf{C}_0(t) = \mathbf{0}$, if we represent that equation as follows

$$\tau_D\frac{d}{dt}\begin{pmatrix} f_{11}^{2j-1}(t) \\ f_{11}^{2j}(t) \end{pmatrix} = \begin{pmatrix} \dfrac{4\sigma j^2(2j-3)}{(4j-1)(4j-3)} & \dfrac{2\sigma(j+1)}{(2j-1)(4j-1)} \\ 0 & \dfrac{2\sigma(j-1)(2j+1)^2}{(4j-1)(4j+1)} \end{pmatrix}\begin{pmatrix} f_{11}^{2j-3}(t) \\ f_{11}^{2j-2}(t) \end{pmatrix}$$

$$+ \begin{pmatrix} \dfrac{\sigma[2j(2j-1)-3]^2}{2j(2j-1)(4j+1)(4j-3)} & -\dfrac{\sigma(2j-3)}{2j(4j-1)} \\ -j(2j-1)-\Delta & \\ \dfrac{\sigma(2j+3)}{2j(4j+1)} & \dfrac{\sigma[2j(2j+1)-3]^2}{2j(2j+1)(4j-1)(4j+3)} \\ & -j(2j+1)-\Delta \end{pmatrix}\begin{pmatrix} f_{11}^{2j-1}(t) \\ f_{11}^{2j}(t) \end{pmatrix}$$

$$+ \begin{pmatrix} -\dfrac{2\sigma(j+1)(2j-1)^2}{(4j-1)(4j+1)} & 0 \\ -\dfrac{2\sigma(j-1)}{(2j+1)(4j+1)} & -\dfrac{4\sigma j^2(2j+3)}{(4j+1)(4j+3)} \end{pmatrix}\begin{pmatrix} f_{11}^{2j+1}(t) \\ f_{11}^{2j+2}(t) \end{pmatrix}$$

$$(175)$$

The exact solution of Eq. (175) in terms of matrix continued fractions is (Appendix C)

$$\begin{pmatrix} \tilde{f}_{11}^1(s) \\ \tilde{f}_{11}^2(s) \end{pmatrix} = \tau_D[\tau_D s\mathbf{I} - \mathbf{Q}_1 - \mathbf{Q}_1^+\mathbf{S}_{11}^2(s)]^{-1}$$

$$\times \left\{\mathbf{C}_1(0) + \sum_{n=2}^{\infty}\left[\prod_{k=2}^{n}\mathbf{Q}_{k-1}^+\mathbf{S}_{11}^k(s)(\mathbf{Q}_k^-)^{-1}\right]\mathbf{C}_n(0)\right\} \qquad (176)$$

where $\mathbf{I}$ is the unit matrix, $\mathbf{Q}_n, \mathbf{Q}_n^\pm$ are the matrices defined by Eqs. (174) and (175), the matrix continued fraction $\mathbf{S}_{11}^n(s)$ is given by

$$\mathbf{S}_{11}^n(s) = [\tau_D s\mathbf{I} - \mathbf{Q}_n - \mathbf{Q}_n^+ \mathbf{S}_{11}^{n+1}(s)]^{-1}\mathbf{Q}_n^- \qquad (177)$$

All the matrices in Eq. (176) are of size 2×2. The initial value vectors $\mathbf{C}_n(0)$ are given by

$$\mathbf{C}_n(0) = \begin{pmatrix} f_{11}^{2n-1}(0) \\ f_{11}^{2n}(0) \end{pmatrix}$$

$$= \frac{1}{2}\begin{pmatrix} \dfrac{1}{4n-1}[2n\langle P_{2n-2}(\cos\vartheta)\rangle_0 + (2n-1)\langle P_{2n}(\cos\vartheta)\rangle_0] \\ \langle P_{2n}(\cos\vartheta)\rangle_0 \end{pmatrix} \qquad (178)$$

Here we have used the equality [39]

$$D_{11}^1(\Omega)D_{-1-1}^j(\Omega) = \frac{1}{2}\left[\frac{j+1}{2j+1}P_{j-1}(\cos\vartheta) + \frac{j}{2j+1}P_{j+1}(\cos\vartheta) + P_j(\cos\vartheta)\right]$$
$$(179)$$

Using Eqs. (167), (170), and (176), we may calculate from Eqs. (151) and (152) the spectra of the dipolar correlation functions:

$$\tilde{C}_\parallel(i\omega) = \cos^2\beta \tilde{f}_{00}^1(i\omega) + \sin^2\beta \tilde{f}_{01}^1(i\omega) \qquad (180)$$

$$\tilde{C}_\perp(i\omega) = \cos^2\beta \tilde{f}_{10}^1(i\omega) + \sin^2\beta \tilde{f}_{11}^1(i\omega) \qquad (181)$$

and so the components of the complex dielectric permittivity tensor from Eq. (4).

The exact solutions (167), (170), and (176) in terms of scalar and matrix continued fractions are convenient for computations. Moreover, they are much simpler than those previously available [3]. They are applicable for any values of the parameters σ, β and $D_\parallel/D_\perp$ (at $\beta = 0$, the results obtained agree with those obtained earlier). Furthermore, they allow us to determine the accuracy of the various approximate solutions [4,41]. Earlier, we obtained simple approximate expressions for the correlation functions at $\beta = 0$ using the concepts of the correlation time and effective eigenvalue. It is also possible to deduce similar approximate formulas for $\beta \neq 0$ and to test their accuracy by comparing them with the exact solutions given above.

B. Comparison of Exact and Approximate Solutions

The behavior of the correlation function $f^1_{00}(t) = \langle D^1_{00}(0)D^1_{00}(t)\rangle_0$ can be approximated by a single exponential as was shown earlier,

$$f^1_{00}(t) \approx f^1_{00}(0)e^{-t/\tau_{00}} = \frac{1+2S}{3}e^{-t/\tau_{00}} \qquad (182)$$

where τ_{00} is the longitudinal relaxation time given by exact Eq. (114) or (115). The approximate formulae for the relaxation time τ_{00} valid for all values of σ is given by Eq. (116):

$$\frac{\tau^{ap}_{00}}{\tau_D} \approx \frac{e^\sigma - 1}{\sigma}\left(\frac{2\sigma\sqrt{\sigma/\pi}}{1+\sigma} + 2^{-\sigma}\right)^{-1} \qquad (183)$$

The main contribution to the relaxation of $f^1_{00}(t)$ is the result of the overbarrier relaxation mode, which has the smallest nonvanishing eigenvalue. Eq. (182), however, ignores high-frequency relaxation inside the wells, which is detected as a very weak peak in the dielectric loss spectrum $\varepsilon''(\omega)$ when $\beta = 0$ and $\sigma \gg 1$. Here a better approximation for $f^1_{00}(t)$ is [42]

$$f^1_{00}(t) \approx e^{-t/\tau_{00}}\{\langle\cos\vartheta\rangle^2_{\text{well}} + [\langle\cos^2\vartheta\rangle_0 - \langle\cos\vartheta\rangle^2_{\text{well}}]e^{-t/\tau_W}\} \qquad (184)$$

where

$$\tau_W \approx \tau_D/2\sigma \qquad (185)$$

is the time characterizing relaxation inside the wells [43]

$$\langle\cos\vartheta\rangle_{\text{well}} = \frac{e^\sigma - 1}{2\sigma M(\frac{1}{2},\frac{3}{2},\sigma)} \qquad (186)$$

$\langle(\cdot)\rangle_{\text{well}}$ means an average in a single potential well (e.g., in the domain $0 \le \vartheta \le \pi/2$).

As shown earlier, the behavior of the (transverse) correlation function $f^1_{10}(t) = \langle D^1_{10}(0)D^{1*}_{10}(t)\rangle_0$ may be accurately described by a single exponential by means of the effective eigenvalue method:

$$f^1_{10}(t) \approx f^1_{10}(0)e^{-t/\tau_{10}} = \frac{(1-S)}{3}e^{-t/\tau_{10}} \qquad (187)$$

where

$$\tau_{10} \approx \tau_{10}^{ef} = -\frac{f_{10}^1(0)}{\dot{f}_{10}^1(0)}$$

$$= \tau_D \left[1 + \frac{\sigma}{5} + \frac{2\sigma f_{01}^3(0)}{15 f_{01}^1(0)} \right]^{-1}$$

$$= 2\tau_D \frac{1-S}{2+S} \tag{188}$$

τ_{10}^{ef} is the effective relaxation time, which is defined from Eq. (163) at $t = 0$.

In like manner, relaxation of both of the correlation functions $f_{01}^1(t)$ and $f_{11}^1(t)$, which mainly characterize the rotation of the molecule about the long molecular axis, can be described by the single exponentials

$$f_{01}^1(t) \approx \frac{(1-S)}{3} e^{-t/\tau_{01}} \tag{189}$$

$$f_{11}^1(t) \approx \frac{(2+S)}{6} e^{-t/\tau_{11}} \tag{190}$$

where

$$\tau_{01} \approx \tau_{01}^{ef} = -\frac{f_{01}^1(0)}{\dot{f}_{01}^1(0)}$$

$$= \tau_D \left[1 + \frac{\sigma}{5} + \Delta + \frac{2\sigma f_{01}^3(0)}{15 f_{01}^1(0)} \right]^{-1}$$

$$= \tau_D \frac{1-S}{1 + (1-S)\Delta + S/2} \tag{191}$$

$$\tau_{11} \approx \tau_{11}^{ef} = -\frac{f_{11}^1(0)}{\dot{f}_{11}^1(0)}$$

$$= \tau_D \left[1 - \frac{\sigma}{10} + \frac{1}{2}\left(\frac{D_\parallel}{D_\perp} - 1\right) + \frac{4\sigma f_{11}^3(0)}{15 f_{11}^1(0)} - \frac{\sigma f_{11}^2(0)}{6 f_{11}^1(0)} \right]^{-1}$$

$$= \tau_D \frac{2+S}{2 + (2+S)\Delta - S/2} \tag{192}$$

τ_{01}^{ef} and τ_{11}^{ef} are the effective relaxation times yielded by Eqs. (164) and (165) respectively at $t = 0$.

The results of the calculation of the relaxation times

$$\tau_{nm} = \frac{1}{f'_{nm}(0)} \int_0^\infty f'_{nm}(t)dt = \frac{\tilde{f}'_{nm}(0)}{f'_{nm}(0)}$$

and τ_{nm}^{ef} are shown in Figure 6, where the exact and approximate relaxation times τ_{nm} and τ_{nm}^{ef} as functions of σ clearly are in complete agreement. The effective eigenvalue method is successful in the evaluation of the decay of $f_{10}^1(t)$, $f_{01}^1(t)$, and $f_{11}^1(t)$ here, because the overbarrier relaxation (activation) mode is not involved in these relaxation processes so that the behavior of the correlation times τ_{01}, τ_{01}, and τ_{11} and τ_{01}^{ef}, τ_{01}^{ef}, and τ_{11}^{ef} is similar. This is not so for $f_{00}^1(t)$ where the effective relaxation time $\tau_{00}^{ef} = -f_{00}^1(0)/\dot{f}_{00}^1(0)$ diverges exponentially from τ_{00} because of the activation process (Fig. 2).

Using Eqs. (4), (15), (151), (152), (182), (187), (189), and (190), we can now derive simple approximate expressions for the normalized complex susceptibility spectra in the approximation of Eq. (15),

$$\chi_\gamma(\omega) = \frac{kT[\varepsilon_\gamma(\omega) - \varepsilon_{\gamma\infty}]}{4\pi\mu^2 N_0 R_\gamma(\omega)} = C_\gamma(0) - i\omega\tilde{C}_\gamma(i\omega), \quad (\gamma = \parallel, \perp) \qquad (193)$$

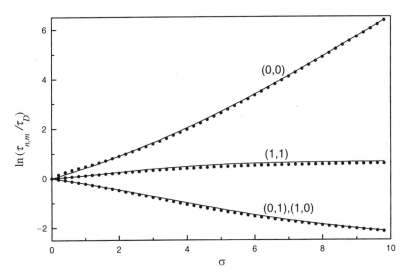

Figure 6. Log (τ_{nm}/τ_D) as a function of σ, calculated from the exact Eqs. (167), (170), (176), and (179) for $D_\parallel/D_\perp = 1$ (*solid lines*). ($\bullet$), Calculation from the approximate Eqs. (183), (188), (191), and (192).

which may be written as

$$\chi_\parallel(\omega) \approx \frac{1}{3} \left[\frac{(1+2S)\cos^2\beta}{1+i\omega\tau_{00}^{ap}} + \frac{(1-S)\sin^2\beta}{1+i\omega\tau_{01}^{ef}} \right] \qquad (194)$$

$$\chi_\perp(\omega) \approx \frac{1}{3} \left[\frac{(1-S)\cos^2\beta}{1+i\omega\tau_{10}^{ef}} + \frac{(1+S/2)\sin^2\beta}{1+i\omega\tau_{11}^{ef}} \right] \qquad (195)$$

with τ_{00}^{ap}, τ_{01}^{ef}, τ_{10}^{ef}, and τ_{11}^{ef} given by Eqs. (152), (153), (191), and (192), respectively. The results of the calculation of the dielectric loss spectra $\chi_\gamma''(\omega)$ from the exact and approximate formulas are shown in Figures 7 to 10. For $\sigma \gg 1$, the longitudinal dielectric loss spectrum $\chi_\parallel''(\omega)$ has two loss peaks. The low frequency peak of $\chi_\parallel''(\omega)$, which usually occurs in the megahertz region, is owing to the overbarrier relaxation mode of the parallel (to the long molecular axis) component of the dipole moment. On the other hand, both high frequency relaxation modes inside the wells and the rotation of the perpendicular component of the dipole moment around the long molecular axis manifest themselves in the high-frequency (GHz) band. In contrast, in the transverse dielectric loss spectrum $\chi_\perp''(\omega)$, the two dispersion

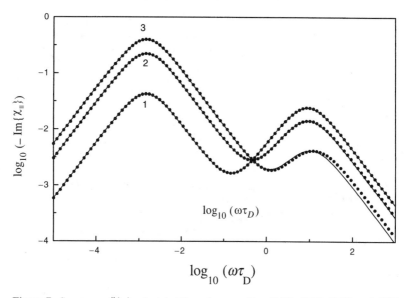

Figure 7. Spectrum $\chi_\parallel''(\omega)$ calculated from the exact Eqs. (167), (170), (180), and (193) at $\sigma = 10$ and $D_\parallel/D_\perp = 1$ (*solid lines*). Curves 1, 2, and 3 correspond to $\beta = \pi/10, \pi/4$ and $2\pi/5$. (●), Calculation from Eq. (194), at $\beta = \pi/10$, Eq. (184) was used.

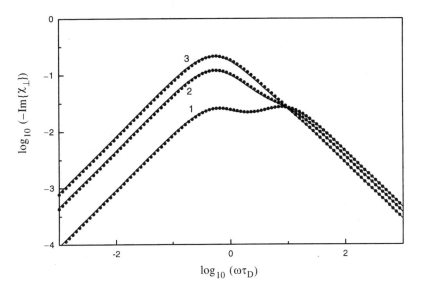

Figure 8. Spectrum $\chi_{\perp}''(\omega)$, calculated from the exact Eqs. (170), (176), (181), and (193) at $\sigma = 10$ and $D_{\parallel}/D_{\perp} = 1$ (*solid lines*). Curves 1, 2, and 3 correspond to $\beta = \pi/10, \pi/4$ and $2\pi/5$. (●), calculation from Eq. (195).

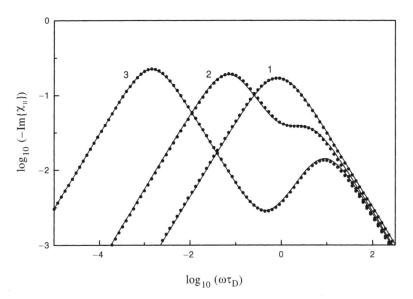

Figure 9. Spectrum $\chi_{\parallel}''(\omega)$, calculated from the exact Eqs. (167), (170), (180), and (193) at $\beta = \pi/4$ and $D_{\parallel}/D_{\perp} = 1$ (solid lines). Curves 1, 2, and 3 correspond to $\sigma = 1, 5$, and 10. (●), calculation from Eq. (194).

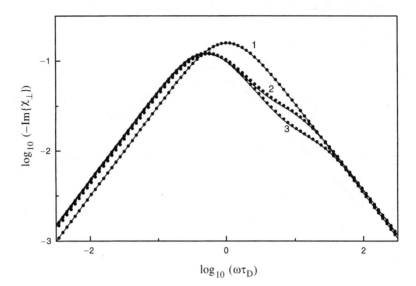

Figure 10. Spectrum $\chi_\perp''(\omega)$, calculated from the exact Eqs. (170), (176), (181), and (193) at $\beta = \pi/4$ and $D_\parallel/D_\perp = 1$ (*solid lines*). Curves 1, 2, and 3 correspond to $\sigma = 1$, 5, and 10. (●), calculation from Eq. (195).

regions are not widely separated because they are both located in the high-frequency region. Nevertheless, if they can be distinguished (e.g., curve 1 in Fig. 9), one frequency dispersion band is associated with the rotation of the perpendicular component of the dipole moment about the long molecular axis, whereas the high-frequency relaxation modes inside the wells contribute to another high-frequency peak (or shoulder) of the spectrum $\chi_\perp''(\omega)$. Despite the large number of high frequency modes involved in both high-frequency mechanisms, essentially two near degenerate modes only (so that they have approximately the same characteristic frequencies) determine both $\chi_\parallel''(\omega)$ and $\chi_\perp''(\omega)$ in the high-frequency relaxation process. Thus both low- and high-frequency processes are still effectively governed by a single relaxation mode. As is apparent from the figures, the results predicted by the approximate Eqs. (194) and (195) agree to good accuracy with the spectra calculated from the exact Eqs. (167), (170), (176), (180), (181), and (193). This means that for $\beta \neq 0$ (that is with the dipole moment vector oriented at an angle to the long axis) both longitudinal and transverse dielectric relaxation can be approximately described by two exponential decays.

Just as for $\beta = 0$, we may formally introduce retardation factors for each effective mode (the retardation factor g_{nm} is defined as the ratio of relaxation times of the corresponding mode in the nematic and isotropic liquid phases).

According to Eqs. (183) and (191) the longitudinal relaxation retardation factors for the low- and high-frequency modes are given by

$$g_{00} = \frac{e^\sigma - 1}{\sigma} \left(\frac{2\sigma\sqrt{\sigma/\pi}}{1 + \sigma} + 2^{-\sigma} \right)^{-1} \tag{196}$$

and

$$g_{01} = \frac{(1 - S)(1 + \Delta)}{1 + (1 - S)\Delta + S/2} \tag{197}$$

respectively. The retardation factors for the low- and high-frequency modes for the transverse relaxation follow from Eqs. (188) and (192) and are

$$g_{11} = \frac{(2 + S)(1 + \Delta)}{2 + (2 + S)\Delta - S/2} \tag{198}$$

and

$$g_{10} = \frac{1 - S}{1 + S/2} \tag{199}$$

In Eq. (196) the retardation factor g_{00} is given as a function of the barrier height parameter σ. One can also express g_{00} as a function of the order parameter S by using the inverse function of Eq. (183) or by using the extrapolating equation [44]

$$\sigma \approx \frac{3S(5 - \pi S)}{2(1 - S^2)} \tag{200}$$

which provides a close approximation to $\sigma(S)$. The results of the calculation of σ as a function of the order parameter S from Eqs. (120) and (200) are shown in Figure 11.

C.   Generalization for an Arbitrary Axially Symmetric Mean Field Potential

The approach we have given for the evaluation of the dielectric parameters of nematics can be generalized to an axially symmetric mean field potential of the form

$$\frac{U}{kT} = -\sum_k \Lambda_k P_{2k}(\cos \vartheta) \tag{201}$$

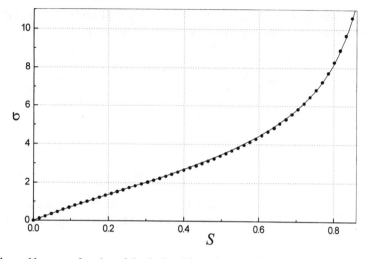

Figure 11. σ as a function of S calculated from the exact Eq. (120) (*solid lines*) and the approximate Eq. (200) ($\bullet$).

Non axially symmetric potentials have also been considered recently on the context of magnetic relaxation [37] [38]. Here instead of Eq. (160) we have [40]

$$\frac{1}{D_\perp}\frac{d}{dt}c_{nm}^j(t) + \left[j(j+1) + m^2\left(\frac{D_\parallel}{D_\perp} - 1\right)\right]c_{nm}^j(t) = \sum_r (R_{nm})_{jr}c_{nm}^r(t)$$

(202)

where

$$(R_{nm})_{jr} = \frac{2r+1}{2(2j+1)}\sum_k \Delta_k[j(j+1) - r(r+1)]$$
$$+ 2k(2k+1)]C(2k,r,j;0,n)C(2k,r,j;0,m)$$

$C(j_1,j_2,j_3;m_1,m_2)$ are the Clebsch-Gordan coefficients [25]. In a like manner, it is possible to derive sets of equations for the equilibrium correlation functions $f_{00}^j(t)$, $f_{01}^j(t)$, $f_{10}^j(t)$, and $f_{11}^j(t)$ from Eqs. (153) and (202), which can be solved exactly by using matrix continued fractions.

Furthermore, the approximate Eqs. (194) and (195) for the components of the normalized susceptibilities can also be applied, with some modification, to a potential of the form of Eq. (201). Indeed, for the calculation of the longitudinal relaxation time τ_{00} it is possible to use the exact integral representation (Appendix B) [10,45].

$$\tau_{00} = \frac{6\tau_D}{(2S+1)Z} \int_{-1}^{1} \frac{e^{U(x)/kT}}{1-x^2} \left[\int_{-1}^{x} xe^{-U(x)/kT}\right]^2 dx \qquad (203)$$

where

$$S = \frac{1}{Z} \int_{-1}^{1} P_2(x)e^{-U(x)/kT} dx$$

$$Z = \int_{-1}^{1} e^{-U(x)/kT} dx \qquad (204)$$

For the Maier-Saupe potential, Eq. (203) reduces to Eq. (115).

Equations (188), (190), and (191) for the effective relaxation times τ_{01}^{ef}, τ_{10}^{ef}, and τ_{11}^{ef} as functions of the order parameter S also remain valid for a potential of the form of Eq. (201) if we write them as follows

$$\tau_{10} \approx \tau_{10}^{ef} = 2\tau_D \frac{1-S}{2+S} \qquad (205)$$

$$\tau_{01} \approx \tau_{01}^{ef} = \tau_D \frac{1-S}{1+(1-S)\Delta + S/2} \qquad (206)$$

$$\tau_{11} \approx \tau_{11}^{ef} = \tau_D \frac{2+S}{2+(2+S)\Delta - S/2} \qquad (207)$$

where the order parameter S is given by Eq. (204). Eq. (205) is valid for any axially symmetric potential, as has been proved by Coffey et al. [32]. This proof may be similarly extended to τ_{01}^{ef} and τ_{11}^{ef} given by Eqs. (206) and (207). Thus, using Eqs. (194), (195), and (203) to (207), we may calculate $\chi_{\parallel}(\omega)$ and $\chi_{\perp}(\omega)$ for any potential of the form of Eq. (201) with S as a parameter. In particular Eqs. (203) and (205) to (207) provide simple analytical solutions for the potential [46]

$$\frac{U}{kT} = -\Lambda_2 P_2(\cos\vartheta) - \Lambda_4 P_4(\cos\vartheta) \qquad (208)$$

and agree in all respects with the numerical calculations of Nordio et al. [3] for this potential. As has also been shown by Nordio et al. [3], by varying the value of Λ_4 it is possible to improve the agreement of the theory with experimental data.

The exact and approximate solutions for the retardation factors g_{00} and g_{10} for the Maier-Saupe potential given by Eqs. (196) and (199) have been recently compared with experiments by Urban and co-workers [47–49,54–56]. They verified that the predictions of the theory are in qualitative agreement with the experiment; however, there are deviations from the experimental data in the temperature dependence of g_{00} and g_{10} predicted by Eqs. (193) and (199). One would expect deviations of this kind because the mean field approximation provides only a qualitative description of relaxation processes in liquid crystals [2].

Note, in the context of dielectric relaxation, that the area of applicability of the rotational diffusion model is restricted to the low-frequency range, because the model under consideration does not include the effects of molecular inertia. A consistent treatment of inertial effects must be carried out using the kinetic equation for the probability density function in configuration-angular velocity space, or using an equivalent approach, for example, the Mori-Zwanzig memory function formalism [22]. The extended rotational diffusion in the Maier-Saupe potential, which takes into account inertial effects, was used by Kalmykov [16,50] for the evaluation of the complex dielectric susceptibility tensor of polar liquid crystals in the whole frequency range of orientational polarization (up to 5 THz). In particular, this model allows one to explain features of the far-infrared spectra of nematics [51–53].

Thus we have shown how one may obtain exact solutions for the longitudinal and transverse components of the complex susceptibility tensor of a nematic liquid crystal having simple uniaxial anisotropy and with the dipole moment vector μ directed at an arbitrary angle β to the molecular axis of symmetry. Both exact and simple approximate analytical formulas for the spectra of the longitudinal and transverse components of the complex dielectric permittivity tensor and dielectric relaxation times are obtained by solving these sets of equations. Furthermore, we have demonstrated that both $\chi_\parallel(\omega)$ and $\chi_\perp(\omega)$ are each effectively governed by two relaxation mechanisms with their respective relaxation times, which may be described by the rather simple formulas in Eqs. (194) and (195).

APPENDIX A

DERIVATION OF EQ. (44)

Noting that the rule for changing of variables in Stratonovich differential equations is the same as in ordinary analysis [21], the equation of motion for an arbitrary differentiable function $f(\{\xi\})$ may be obtained by cross-multiplying the ith Eq. (41) by $\partial f(\{\xi(t)\})/\partial \xi_i$, respectively and then summing them (Einstein's notation). Thus we obtain a stochastic equation for $f(\{\xi(t)\})$:

$$\frac{\mathrm{d}}{\mathrm{d}t} f(\{\xi(t)\}) = h_i(\{\xi(t)\}, t) \frac{\partial}{\partial \xi_i} f(\{\xi(t)\}) + g_{ij}(\{\xi(t)\}, t) \frac{\partial}{\partial \xi_i} f(\{\xi(t)\}) \Gamma_j(t)$$

(A1)

From a mathematical point of view the stochastic differential Eq. (A1) (just as Eq. (41)) with the δ-correlated Langevin forces $\Gamma_j(t)$ is not completely defined [5,21]. The most satisfactory interpretation of Eqs. (41) and (A1) is as the stochastic integral equations [20,21]

$$\xi_i(t + \tau) = x_i + \int_t^{t+\tau} h_i(\{\xi(t')\}, t')\mathrm{d}t'$$
$$+ \int_t^{t+\tau} g_{ij}(\{\xi(t')\}, t')\Gamma_j(t')\mathrm{d}t' \qquad (A2)$$

$$f(\{\xi(t + \tau)\}) = f(\{\mathbf{x}\}) + \int_t^{t+\tau} h_i(\{\xi(t')\}, t') \frac{\partial f(\{\xi(t')\})}{\partial \xi_i} \mathrm{d}t'$$
$$+ \int_t^{t+\tau} g_{ij}(\{\xi(t')\}, t') \frac{\partial f(\{\xi(t')\})}{\partial \xi_i} \Gamma_j(t')\mathrm{d}t' \qquad (A3)$$

Supposing that the integrands in Eqs. (A2) and (A3) can be expanded in Taylor series, we obtain

$$\xi_i(t + \tau) = x_i + \int_t^{t+\tau} h_i(\{\mathbf{x}\}, t')\mathrm{d}t' + \int_t^{t+\tau} [\xi_k(t') - x_k] \frac{\partial}{\partial x_k} h_i(\{\mathbf{x}\}, t')\mathrm{d}t'$$
$$+ \int_t^{t+\tau} g_{ij}(\{\mathbf{x}\}, t')\Gamma_j(t')\mathrm{d}t'$$
$$+ \int_t^{t+\tau} [\xi_k(t') - x_k] \frac{\partial}{\partial x_k} g_{ij}(\{\mathbf{x}\}, t')\Gamma_j(t')\mathrm{d}t' + \ldots \qquad (A4)$$

and

$$f(\{\xi(t + \tau)\}) = f(\{\mathbf{x}\}) + \int_t^{t+\tau} h_i(\{\mathbf{x}\}, t') \frac{\partial f(\{\mathbf{x}\})}{\partial x_i} \mathrm{d}t'$$
$$+ \int_t^{t+\tau} [\xi_k(t') - x_k] \frac{\partial}{\partial x_k} \left[h_i(\{\mathbf{x}\}, t') \frac{\partial f(\{\mathbf{x}\})}{\partial x_i} \right] \mathrm{d}t'$$
$$+ \int_t^{t+\tau} g_{ij}(\{\mathbf{x}\}, t') \frac{\partial f(\{\mathbf{x}\})}{\partial x_i} \Gamma_j(t')\mathrm{d}t'$$

$$+ \int_{t}^{t+\tau} [\xi_k(t') - x_k] \frac{\partial}{\partial x_k} \left[g_{ij}(\{\mathbf{x}\}, t') \frac{\partial f(\{\mathbf{x}\})}{\partial x_i} \right] \Gamma_j(t') dt' + \dots$$

(A5)

Substituting $\xi_k(t') - x_k$ from Eq. (A4) into Eq. (A5) we iterate

$$f(\{\xi(t+\tau)\}) = f(\{\mathbf{x}\}) + \int_{t}^{t+\tau} h_i(\{\mathbf{x}\}, t') \frac{\partial f(\{\mathbf{x}\})}{\partial x_i} dt'$$

$$+ \int_{t}^{t+\tau} \frac{\partial}{\partial x_k} \left[h_i(\{\mathbf{x}\}, t') \frac{\partial f(\{\mathbf{x}\})}{\partial x_i} \right] \int_{t}^{t'} h_k(\{\mathbf{x}\}, t'') dt'' dt'$$

$$+ \int_{t}^{t+\tau} \frac{\partial}{\partial x_k} \left[h_i(\{\mathbf{x}\}, t') \frac{\partial f(\{\mathbf{x}\})}{\partial x_i} \right] \int_{t}^{t'} g_{kn}(\{\mathbf{x}\}, t'') \Gamma_n(t'') dt'' dt'$$

$$+ \int_{t}^{t+\tau} g_{ij}(\{\mathbf{x}\}, t') \frac{\partial f(\{\mathbf{x}\})}{\partial x_i} \Gamma_j(t') dt'$$

$$+ \int_{t}^{t+\tau} \frac{\partial}{\partial x_k} \left[g_{ij}(\{\mathbf{x}\}, t') \frac{\partial f(\{\mathbf{x}\})}{\partial x_i} \right] \Gamma_j(t') \int_{t}^{t'} h_k(\{\mathbf{x}\}, t'') dt'' dt'$$

$$+ \int_{t}^{t+\tau} \frac{\partial}{\partial x_k} \left[g_{ij}(\{\mathbf{x}\}, t') \frac{\partial f(\{\mathbf{x}\})}{\partial x_i} \right] \Gamma_j(t')$$

$$\times \int_{t}^{t'} g_{kn}(\{\mathbf{x}\}, t'') \Gamma_n(t'') dt'' dt' + \dots$$

(A6)

Then averaging Eq. (A6) taking into account the properties of Eq. (42) and retaining only the terms of the order of τ, we have

$$\overline{f(\{\xi(t+\tau)\})} = f(\{\mathbf{x}\}) + \int_{t}^{t+\tau} h_i(\{\mathbf{x}\}, t') \frac{\partial}{\partial x_i} f(\{\mathbf{x}\}) dt'$$

$$+ 2D\delta_{jn} \int_{t}^{t+\tau} \frac{\partial}{\partial x_k} \left[g_{ij}(\{\mathbf{x}\}, t') \frac{\partial}{\partial x_i} f(\{\mathbf{x}\}) \right]$$

$$\times \int_{t}^{t'} g_{kn}(\{\mathbf{x}\}, t'') \delta(t' - t'') dt'' dt' + 0(\tau)$$

(A7)

After obvious transformations in Eq. (A7), we obtain

$$\frac{\overline{f(\{\xi(t+\tau)\})} - f(\{\mathbf{x}\})}{\tau} = h_i(\{\mathbf{x}\}, t + \tau\Theta_{iii}^{(1)}) \frac{\partial}{\partial x_i} f(\{\mathbf{x}\})$$

$$+ Dg_{kj}(\{\mathbf{x}\}, t + \tau\Theta_{ijk}^{(2)}) \frac{\partial}{\partial x_k} \left[g_{ij}(\{\mathbf{x}\}, t + \tau\Theta_{ijk}^{(2)}) \frac{\partial}{\partial x_i} f(\{\mathbf{x}\}) \right] + 0(\tau)$$

(A8)

Where $\Theta_{ijk}^{(n)}$ are constants $(0 \leq \Theta_{ijk}^{(n)} \leq 1)$. Here we have also used the property of the δ-function

$$y(b) = 2 \int_a^b \delta(b - x)y(x)\mathrm{d}x$$

Taking the limit $\tau \to 0$ in Eq. (A8), we have Eq. (44).

APPENDIX B

INTEGRAL REPRESENTATION FOR THE CORRELATION TIME

We recall that for a stochastic system with dynamics obeying the one-variable Fokker-Planck equation [20]

$$\frac{\partial}{\partial t} W(x, t) = L_{FP} W(x, t) \tag{A9}$$

with

$$L_{FP}(x) = \frac{\partial}{\partial x} \left(D^{(2)}(x) e^{-U(x)} \frac{\partial}{\partial x} e^{U(x)} \right)$$

where $D^{(2)}(x)$ is the diffusion coefficient and U is called a generalized potential, the correlation time τ_{AA} of the equilibrium (stationary) autocorrelation function $C_{AA}(t) = \langle A(x(0))A(x(t))\rangle_0 - \langle A \rangle_0^2$ of a dynamic variable $A(x)$ is [20]

$$\tau_{AA} = \frac{1}{C_{AA}(0)} \int_0^\infty C_{AA}(t)\mathrm{d}t = \frac{1}{C_{AA}(0)} \int_{x_1}^{x_2} \frac{f^2(x)\mathrm{d}x}{D^{(2)}(x)W_0(x)}$$

where W_0 is the equilibrium (stationary) distribution function,

$$f(x) = \int_{x_1}^x [A(x') - \langle A \rangle_0] W_0(x')\mathrm{d}x'$$

Risken [20] considered only the autocorrelation function for the model of one-dimensional translational Brownian motion in a potential. For orientational relaxation problems the case $A = B$ was considered by others [5,45,52,57]. The generalization of this approach to the calculation of the correlation time τ_{AB} of the correlation function $C_{AB}(t)$ of two dynamic variables $A(x)$ and $B(x)$ is straightforward and is given below.

The equilibrium (stationary) correlation function $C_{AB}(t)$ of two dynamic variables $A(x)$ and $B(x)$ is given by

$$C_{AB}(t) = \langle A(x(0))B(x(t))\rangle_0 - \langle A\rangle_0\langle B\rangle_0$$
$$= \int_{x_2}^{x_1} [A(x) - \langle A\rangle_0]e^{L^0_{FP}t}[B(x) - \langle B\rangle_0]W_0(x)dx$$

where $L^0_{FP}(x)$ is the Fokker-Planck operator in the equilibrium (stationary in general) state (i.e., in the absence of perturbations). The correlation time τ_{AB} is defined as the area under the curve of the normalized correlation function

$$\tau_{AB} = \frac{1}{C_{AB}(0)} \int_0^\infty C_{AB}(t)dt \tag{A10}$$

We shall now show how to derive an analytic expression for τ_{AB}.

The definition of $C_{AB}(t)$ implies that we can seek a solution of the Fokker-Planck equation in the form [20]

$$W(x,t) = W_0(x) + w(x,t) \tag{A11}$$

with the initial conditions

$$w(x,0) = [B(x) - \langle B\rangle_0]W_0(x)$$

and with $w(x, \infty) = 0$.

Noting that $L^0_{FP}W_0 = 0$ we have

$$w(x,t) = e^{L^0_{FP}t}[B(x) - \langle B\rangle_0]W_0(x) \tag{A12}$$

so that

$$\tau_{AB} = -\frac{\int_{x_1}^{x_2}[A(x) - \langle A\rangle_0](L^0_{FP})^{-1}[B(x) - \langle B\rangle_0]W_0(x)dx}{\langle AB\rangle_0 - \langle A\rangle_0\langle B\rangle_0} \tag{A13}$$

The evaluation of

$$(L^0_{FP})^{-1}[B(x) - \langle B\rangle_0]W_0(x) \tag{A14}$$

can be accomplished by taking the Laplace transform of the Fokker-Planck equation (A9). Thus we obtain

$$s\tilde{w}(x, s) - [B(x) - \langle B \rangle_0] W_0(x) = L^0_{FP}\tilde{w}(x, s) \tag{A15}$$

where

$$\tilde{w}(x, s) = \int_0^\infty w(x, t)e^{-st}dt$$

In the limit $s \to 0$ using the final value theorem of Laplace transformation,

$$\lim_{s \to 0} s\tilde{w}(x, s) = \lim_{t \to \infty} w(x, t) = 0$$

we have a formal solution of Eq. (A15)

$$\tilde{w}(x, 0) = -(L^0_{FP})^{-1}[B(x) - \langle B \rangle_0] W_0(x) \tag{A16}$$

and Eq. (A13) becomes

$$\tau_{AB} = \frac{\int_{x_1}^{x_2} [A(x) - \langle A \rangle_0]\tilde{w}(x, 0)dx}{\langle AB \rangle_0 - \langle A \rangle_0 \langle B \rangle_0} \tag{A17}$$

The quantity $\tilde{w}(x, 0)$ in Eq. (A17) can be evaluated as follows. Using Eq. (A15) and taking account of Eq. (A9), one obtains

$$-[B(x) - \langle B \rangle_0] W_0(x) = \frac{d}{dx}\left(D^{(2)}(x)\left[\frac{d}{dx}\tilde{w}(x, 0) + \tilde{w}(x, 0)\frac{d}{dx}U(x)\right]\right) \tag{A18}$$

Eq. (A18) can be integrated to yield

$$\tilde{w}(x, 0) = -W_0(x)\int_{x_1}^x \frac{f_B(x')dx'}{D^{(2)}(x')W_0(x')} \tag{A19}$$

where

$$f_B(x) = \int_{x_1}^x [B(y) - \langle B \rangle_0] W_0(y)dy \tag{A20}$$

Thus we obtain from Eqs. (A17) and (A19)

$$\tau_{AB} = -\frac{1}{\langle AB \rangle_0 - \langle A \rangle_0 \langle B \rangle_0} \int_{x_1}^{x_2} [A(x) - \langle A \rangle_0] W_0(x) \int_{x_1}^{x} \frac{f_B(x')}{D^{(2)}(x') W_0(x')} dx' dx$$

or an integration by parts

$$\tau_{AB} = \frac{1}{\langle AB \rangle_0 - \langle A \rangle_0 \langle B \rangle_0} \int_{x_1}^{x_2} \frac{f_A(x) f_B(x) dx}{D^{(2)}(x) W_0(x)} \tag{A21}$$

where

$$f_A(x) = \int_{x_1}^{x} [A(y) - \langle A \rangle_0] W_0(y) dy \tag{A22}$$

For $A = B$ and $x_1 = -\infty$, $x_2 = \infty$ Eq. (A21) reduces to the equation presented by Risken [20].

Insofar as we are concerned with the problem of the relaxation of the longitudinal component of the dipole moment, the relevant Fokker-Planck (Smoluchowski) equation for the distribution function W of the orientations of the particles is given by [5,57]

$$2\tau_D \frac{\partial}{\partial t} W = \frac{\beta}{\sin \vartheta} \frac{\partial}{\partial \vartheta} \left(\sin \vartheta W \frac{\partial}{\partial \vartheta} U \right) + \frac{1}{\sin \vartheta} \frac{\partial}{\partial \vartheta} \left(\sin \vartheta \frac{\partial}{\partial \vartheta} W \right), \beta = (kT)^{-1}$$

or introducing a new variable $x = \cos \vartheta$

$$2\tau_D \frac{\partial}{\partial t} W = \frac{\partial}{\partial x} \left[(1 - x^2) \left(\frac{\partial}{\partial x} W + \beta W \frac{\partial}{\partial x} U \right) \right]$$

In this case $W_0(x)$ is the Maxwell-Boltzmann distribution function

$$W_0(x) = e^{-\beta U(x)} / Z, \quad Z = \int_{-1}^{1} e^{-\beta U(x)} dx \tag{A23}$$

$x_1 = -1$, $x_2 = 1$, and the function $D^{(2)}(x)$ is given by

$$D^{(2)}(x) = (1 - x^2)/2\tau_D \tag{A24}$$

Thus with Eqs. (A23) and (A24), Eq. (A21) yields

$$\tau_{AB} = \frac{2\tau_D}{Z[\langle AB \rangle_0 - \langle A \rangle_0 \langle B \rangle_0]} \int_{-1}^{1} \frac{e^{\beta U(z)} f_A(z) f_B(z) dz}{1 - z^2} \tag{A25}$$

where

$$f_A(z) = \int_{-1}^{z} [A(z') - \langle A \rangle_0] e^{-\beta U(z')} dz'. \tag{A26}$$

$$f_B(z) = \int_{-1}^{z} [B(z'') - \langle B \rangle_0] e^{-\beta U(z'')} dz'' \tag{A27}$$

For the problem under consideration $A(x) = B(x) = x$ and $\langle A(x) \rangle_0 = \langle B(x) \rangle_0 = 0$ and we arrive at Eq. (203).

APPENDIX C

MATRIX CONTINUED FRACTION SOLUTION OF THREE TERM MATRIX RECURRENCE RELATIONS

We have mentioned that in the majority of problems the Langevin or Fokker-Planck equation may not be reduced to a scalar three-term recurrence relation. Hence the method based on conversion of the recurrence relation to an ordinary continued fraction no longer applies. Examples of this are problems that involve diffusion in phase space and diffusion in configuration space in which the form of the potential gives rise to a five- or higher-order term recurrence relation. These difficulties may, however, be circumvented, because we have a method of converting a multiterm scalar recurrence relation to a three-term matrix one [20]. We shall confine our discussion to matrix continued fractions, because the scalar continued fraction is simply a special case of the matrix one.

Such a matrix recurrence relation, in the notation of Risken [20], may be written in a general form as

$$\frac{d}{dt} \mathbf{C}_p(t) = \mathbf{Q}_p^- \mathbf{C}_{p-1}(t) + \mathbf{Q}_p \mathbf{C}_p(t) + \mathbf{Q}_p^+ \mathbf{C}_{p+1}(t) \tag{A28}$$

where the $\mathbf{C}_p(t)$ are column vectors that, in general, have P components; and $\mathbf{Q}_p^-, \mathbf{Q}_p^+$, and are $\mathbf{Q}_p$ time-independent noncommutative matrices. A general method of solution of a tridiagonal vector recurrence relation in terms of matrix continued fractions is described by Risken [20]. We now present another method for the solution of Eq. (A28), which has the merit of being considerably simpler than the previously available algorithm [5].

The solution of such a recurrence relation may be accomplished as follows. Taking the Laplace transform of Eq. (A28) we have

$$(s\mathbf{I} - \mathbf{Q}_p)\tilde{\mathbf{C}}_p(s) = \mathbf{C}_p(0) + \mathbf{Q}_p^- \tilde{\mathbf{C}}_{p-1}(s) + \mathbf{Q}_p^+ \tilde{\mathbf{C}}_{p+1}(s) \tag{A29}$$

Let us further, following Risken (20), seek the solution of Eq. (A29) in the form

$$\tilde{\mathbf{C}}_p(s) = \tilde{\mathbf{S}}_p(s)\tilde{\mathbf{C}}_{p-1}(s) + \mathbf{q}_p(s) \qquad (A30)$$

where the matrix continued fraction $\tilde{\mathbf{S}}_p(s)$ satisfies the homogeneous Eq. (A29) with $\mathbf{C}_p(0) = \mathbf{0}$,

$$(s\mathbf{I} - \mathbf{Q}_p - \mathbf{Q}_p^+\tilde{\mathbf{S}}_{p+1}(s))\tilde{\mathbf{S}}_p(s) = \mathbf{Q}_p^- \qquad (A31)$$

or

$$\tilde{\mathbf{S}}_p(s) = (s\mathbf{I} - \mathbf{Q}_p - \mathbf{Q}_p^+\tilde{\mathbf{S}}_{p+1}(s))^{-1}\mathbf{Q}_p^- \qquad (A32)$$

The matrix continued fraction of $\tilde{\mathbf{S}}_p(s)$ represents the complementary solution of the recurrence relation. The particular solution may be found as follows. We have from Eqs. (A29) to (A31)

$$(s\mathbf{I} - \mathbf{Q}_p)\mathbf{q}_p(s) - \mathbf{Q}_p^+(\tilde{\mathbf{S}}_{p+1}(s)\mathbf{q}_p(s) + \mathbf{q}_{p+1}(s)) = \mathbf{C}_p(0) \qquad (A33)$$

and thus we have an expression for $\mathbf{q}_p(s)$,

$$\begin{aligned} \mathbf{q}_p(s) &= (s\mathbf{I} - \mathbf{Q}_p - \mathbf{Q}_p^+\tilde{\mathbf{S}}_{p+1}(s))^{-1}[\mathbf{C}_p(0) + \mathbf{Q}_p^+\mathbf{q}_{p+1}(s)] \\ &= \tilde{\mathbf{S}}_p(s)(\mathbf{Q}_p^-)^{-1}[\mathbf{C}_p(0) + \mathbf{Q}_p^+\mathbf{q}_{p+1}(s)] \end{aligned} \qquad (A34)$$

Using Eqs. (A37) and (A30), we have

$$\tilde{\mathbf{C}}_p(s) = (s\mathbf{I} - \mathbf{Q}_p - \mathbf{Q}_p^+\tilde{\mathbf{S}}_{p+1}(s))^{-1}[\mathbf{Q}_p^-\tilde{\mathbf{C}}_{p-1}(s) + \mathbf{C}_p(0) + \mathbf{Q}_p^+\mathbf{q}_{p+1}(s)] \qquad (A35)$$

which is the general solution in the frequency domain of the recurrence relation Eq. (A28). This may be written out explicitly by expanding the $\mathbf{q}_{p+1}(s)$,

$$\begin{aligned} \tilde{\mathbf{C}}_p(s) = {}&\tilde{\mathbf{S}}_p(s)\tilde{\mathbf{C}}_{p-1}(s) + \tilde{\mathbf{S}}_p(s)(\mathbf{Q}_p^-)^{-1}\mathbf{C}_p(0) + \tilde{\mathbf{S}}_p(s)(\mathbf{Q}_p^-)^{-1} \\ &\times \mathbf{Q}_p^+\tilde{\mathbf{S}}_{p+1}(s)(\mathbf{Q}_{p+1}^-)^{-1}\mathbf{C}_{p+1}(0) + \tilde{\mathbf{S}}_p(s)(\mathbf{Q}_p^-)^{-1}\mathbf{Q}_p^+\tilde{\mathbf{S}}_{p+1}(s)(\mathbf{Q}_{p+1}^-)^{-1} \\ &\times \mathbf{Q}_{p+1}^+\mathbf{q}_{p+2}(s) \end{aligned} \qquad (A36)$$

If we let the subscript of $\mathbf{q}$ tend to infinity we obtain

$$\tilde{\mathbf{C}}_p(s) = \tilde{\mathbf{S}}_p(s)\tilde{\mathbf{C}}_{p-1}(s) + \tilde{\mathbf{S}}_p(s)(\mathbf{Q}_p^-)^{-1}$$

$$\times \left[\mathbf{C}_p(0) + \sum_{k=1}^{\infty} \left(\prod_{n=1}^{k} \mathbf{Q}_{p+n-1}^+ \tilde{\mathbf{S}}_{p+n}(s)(\mathbf{Q}_{p+n}^-)^{-1} \right) \mathbf{C}_{p+k}(0) \right] \qquad (A37)$$

Equation (A37) constitutes the solution of Eq. (A28) rendered as a sum of products of matrix continued fractions in the s domain.

In many physical applications, the initial conditions $\mathbf{C}_p(0)$ can be expressed in terms of equilibrium (stationary in the general case) averages, which are in fact the equilibrium (stationary) solution of the corresponding Fokker-Planck equation or equivalently the vector recurrence relation

$$\mathbf{Q}_{p-1}^- \mathbf{C}_{p-1}^0 + \mathbf{Q}_p \mathbf{C}_p^0 + \mathbf{Q}_p^+ \mathbf{C}_{p+1}^0 = \mathbf{0} \qquad (A38)$$

This has a tridiagonal form so it is possible to express the equilibrium averages $\mathbf{C}_p^0$ in terms of the matrix continued fraction $\tilde{\mathbf{S}}_p(0)$ from Eq. (A32).

Acknowledgments

The support of this work by the Ministry of Science and Technological Policy of the Russian Federation and INTAS (grant 96-0663) is gratefully acknowledged. One of us (Y.P.K.) thanks the Queen's University of Belfast for a Visiting Professorship and the Royal Irish Academy for a Senior Visiting Fellowship (1998). WTC thanks Enterprise Ireland for funding under Basic Research grant SC/97/701 and the Research collaboration fund 1998–1999.

References

1. W. H. de Jeu, *Physical Properties of Liquid Crystalline Materials*, Gordon & Breach, New York, 1980.

2. L. M. Blinov, *Electro-Optical and Magneto-Optical Properties of Liquid Crystals*. John Wiley & Sons, Chichester, UK, 1983.

3. P. L. Nordio, G. Rigatti, and U. Segre, *Mol. Phys.*, **25**, 129 (1973).

4. B. A. Storonkin, *Kristallografiya*, **30**, 841 (1985).

5. W. T. Coffey, Y. P. Kalmykov, and J. T. Waldron, *The Langevin Equation*, World Scientific, Singapore, 1996.

6. A. J. Martin, G. Meier, and A. Saupe, *Symp. Faraday Soc.*, **5**, 119 (1971).

7. P. L. Nordio, G. Rigatti, and U. Segre, *J. Chem. Phys.*, **56**, 2117 (1971).

8. G. Meier, and A. Saupe, *Mol. Cryst.* **1**, 515 (1966).

9. P. J. Cregg, D. S. F. Crothers, and A. W. Wickstead, *J. Appl. Phys.*, **76**, 4900 (1994).

10. W. T. Coffey and D. S. F. Crothers, *Phys. Rev. E.*, **54**, 4768 (1996).

11. W. F. Brown Jr., *Phys. Rev.*, **130**, 1677 (1963).

12. Y. L. Raikher and M. I. Shliomis, *Adv. Chem. Phys.*, **87**, 595. (1994).

13. G. Luckhurst and C. Zannoni, *Proc. Roy. Soc. Lond. A.* **343**, 389 (1975).

14. D. M. F. Edwards and P. Madden, *Mol. Phys.*, **48**, 471 (1983).

15. D. E. Sullivan and J. M. Deutch *J. Chem. Phys.*, **62**, 2130 (1975).

16. Y. P. Kalmykov, *Liq. Cryst.*, **10**, 519 (1991).

17. P. Madden and D. Kivelson, in I. Prigogine and S. A. Rice, eds., *Advances in Chemical Physics*, Vol. 56, 467, John Wiley & Sons, New York, 1984.

18. D. Kivelson and P. Madden, *Mol. Phys.* **30**, 1749 (1975).

19. H. Watanabe and A. Morita, in I. Prigogine and S. A. Rice, eds., *Advances in Chemical Physics*, Vol. 56, John Wiley & Sons, New York, 1984.

20. H. Risken, *The Fokker-Planck Equation*, 2nd ed., Springer, Berlin, 1989.

21. C. W. Gardiner, *Handbook of Stochastic Methods*, Springer, Berlin, 1985.

22. M. W. Evans, G. J. Evans, W. T. Coffey, and P. Grigolini, *Molecular Dynamics*, John Wiley & Sons, New York, 1982.

23. J. R. McConnell, *Rotational Brownian Motion and Dielectric Theory*, Academic Press, London, 1980.

24. R. L. Stratonovich, *Conditional Markov Processes and Their Application to the Theory of Optimal Control*, Elsevier, New York, 1968.

25. E. M. Rose, *Elementary Theory of Angular Momentum*, John Wiley & Sons, New York, 1967.

26. Y. P. Kalmykov, *J. Mol. Liq.* **69**, 117 (1996).

27. E. W. Hobson, *The Theory of Spherical and Ellipsoidal Harmonics*, Cambridge University Press, Cambridge, UK, 1931.

28. W. Maier and A. Saupe, *Z. Naturforsch.*, **13a**, 564 (1958).

29. W. Maier and A. Saupe, *Z. Naturforsch.*, **14a**, 882 (1959).

30. W. Maier and A. Saupe, *Z. Naturforsch.*, **15a**, 287 (1960).

31. M. Abramowitz and I. Stegun, eds., *Handbook of Mathematical Functions*, Dover, New York, 1965.

32. W. T. Coffey, Y. P. Kalmykov, and E. S. Massawe, in I. Prigogine and S. A. Rice, eds., *Advances in Chemical Physics*, Vol. 85, John Wiley & Sons, New York, 1993.

33. H. S. Wall, *Continued Fractions*, Van Nostrand, Princeton, N. J., 1969.

34. W. T. Coffey, D. S. F. Crothers, Y. P. Kalmykov, and J. T. Waldron, *Physica A*, **213**, 551 (1994).

35. A. P. Prudnikov, Y. A. Brychkov, and O. I. Marichev, *Integrals and Series. Additional Chapters*, Science, Moscow, 1986.

36. W. T. Coffey, Y. P. Kalmykov, and J. T. Waldron, *Liq. Cryst.*, **18**, 677 (1995).

37. Yu. P. Kalmykov and S. V. Titov, *Phys. Rev. Letts.* 82, 2967 (1999).

38. J. L. Déjardin, P. M. Déjardin, Y. P. Kalmykov, and S. V. Titov, *Phys. Rev. E* 60, 1475 (1999).

39. D. A. Varshalovich, A. N. Moskalev, and V. K. Khersonskii, *Quantum Theory of Angular Momentum*, Science, Leningrad, 1975.

40. P. L. Nordio and P. Busolin, *J. Chem. Phys.*, **55**, 5485 (1971).

41. A. Jakli and A. Buka, *Rotational Brownian Motion and Dielectric Permittivity in Nematic Liquid Crystals*. Preprint Central Research Institute Physics, Budapest, 1984.

42. A. Perico, R. Pratolongo, K. F. Freed, R. W. Pastor, and A. Szabo, *J. Chem. Phys.*, **98**, 564 (1993).

43. D. A. Garanin, *Phys. Rev. E.*, **54**, 3250 (1996).

44. Y. P. Kalmykov, unpublished data.

45. G. Moro and P. L. Nordio, *Mol. Phys.*, **56**, 255 (1985).

46. R. L. Humphries, P. G. James, and G. R. Luckhurst, *J. C. S. Faraday Trans. II*, **68**, 1031 (1972).

47. S. Urban, B. Gestblom, T. Brückert, and A. Würflinger, *Z. Naturforsch.*, **50a**, 984 (1995).

48. B. Gestblom and S. Urban, *Z. Naturforsch.* **50a**, 595 (1995).

49. S. Urban and A. Würflinger, in I. Prigogine and S. A. Rice, eds., *Advances in Chemical Physics*, Vol. 98, John Wiley & Sons, New York, 1997.

50. Y. P. Kalmykov, *Khim. Fizika.*, **6**, 592 (1987).

51. S. Venugonalan and S. N. Prasad, *J. Chem. Phys.*, **71**, 5293 (1979).

52. U. M. S. Murthy and J. K. Vij, *Liq. Cryst.* **4**, 529 (1989); J. K. Vij and U. M. S. Murthy, *J. Mol. Liq.* **43**, 109 (1989).

53. T. S. Perova, in I. Prigogine and S. A. Rice, eds., *Advances in Chemical Physics*, Vol. 87, John Wiley & Sons, New York, 1994.

54. S. Urban, D. Busing, A. Würflinger, and B. Gestblom, *Liquid Crystals*, **25**, 253, 1998.

55. A. Würflinger and S. Urban, *Z. Naturforsch.*, **53a**, 883, 1998.

56. S. Urban, B. Gestblom, R. Dabrowski, H. Kresse, *Z. Naturforsch.*, **53a**, 134, 1998.

57. A. Szabo, *J. Chem. Phys.*, **72**, 4620 (1980).

AUTHOR INDEX

Numbers in parentheses are reference numbers and indicate that the author's work is referred to although his name is not mentioned in the text. Numbers in *italic* show the pages on which the complete references are listed.

SUBJECT INDEX

571